Prentice Hall Advanced Reference Series

Physical and Life Sciences

AGNEW AND MCCREERY, EDS. *Neural Prostheses: Fundamental Studies*

BALLENTINE *Quantum Mechanics*

BINKLEY *The Pineal: Endocrine and Nonendocrine Function*

BLACKWELL *Poisonous and Medicinal Plants*

CAROZZI *Carbonate Rock Depositional Models: A Microfacies Approach*

CHAFETZ *Nutrition and Neurotransmitters: The Nutrient Bases of Behavior*

EISEN *Mathematical Methods and Models in the Biological Sciences: Linear and One-Dimensional Theory*

EISEN *Mathematical Methods and Models in the Biological Sciences: Nonlinear and Multidimensional Theory*

FRASER *Clastic Depositional Sequences: Process of Evolution and Principles of Interpretation*

HAZELWOOD *The Endocrine Pancreas*

JEGER, ED. *Spatial Components of Plant Disease Epidemics*

MCLENNAN *Introduction to Nonequilibrium Statistical Mechanics*

MEFFE AND SNELSON *Ecology and Evolution of Livebearing Fishes (Poeciliidae)*

PLISCHKE AND BERGERSEN *Equilibrium Statistical Physics*

SORBJAN *Structure of the Atmospheric Boundary Layer*

VALENZUELA AND MYERS *Adsorption Equilibrium Data Handbook*

WALDEN *Genetic Transformation in Plants*

WARREN *Evaporite Sedimentology: Importance in Hydrocarbon Accumulation*

WARD *Fermentation Biotechnology*

WONG *Introductory Nuclear Physics*

INTRODUCTORY NUCLEAR PHYSICS

Samuel S. M. Wong

University of Toronto

Prentice Hall, Englewood Cliffs, New Jersey 07632

Library of Congress Cataloging-in-Publication Data

Wong, S. S. M. (Samuel Shaw Ming),
Introductory nuclear physics / S.S.M. Wong.
p. cm.
Bibliography: p.
Includes index.
ISBN 0-13-491168-7
1. Nuclear physics. I. Title.
QC776.W66 1990
539.7--dc20 89-8755
CIP

Editorial/production supervision: **Brendan M. Stewart**
Cover design: **Wanda Lubelska**
Manufacturing buyer: **Mary Ann Gloriande**

Printed in the United States of America
10 9 8 7 6 5 4 3 2 1

ISBN 0-13-491168-7

Prentice-Hall International (UK) Limited, *London*
Prentice-Hall of Australia Pty. Limited, *Sydney*
Prentice-Hall Canada Inc., *Toronto*
Prentice-Hall Hispanoamericana, S.A., *Mexico*
Prentice-Hall of India Private Limited, *New Delhi*
Prentice-Hall of Japan, Inc., *Tokyo*
Simon & Schuster Asia Pte. Ltd., *Singapore*
Editora Prentice-Hall do Brasil, Ltda., *Rio de Janeiro*

Contents

Preface

Nuclear physics is a subject basic to the curriculum of modern physics. There are several good reasons for this to be so. First and foremost is the intrinsic interest of the subject itself: the study of atomic nuclei has historically given us many of the first insights into modern physics. Furthermore, the potential of future discoveries remains very promising. In the second place, nuclear physics is closely associated with several other active branches of research: particle physics, in terms of the large overlap of interests in fundamental interactions and symmetries, and condensed matter physics, through the many-body nature of the problems involved. Third, nuclear physics may be usefully applied to other fields: chronology in geophysics and archaeology, tracer element techniques and nuclear medicine, just to name a few.

The diversity of interest in nuclear physics also makes it very difficult to cover the entire subject in any satisfactory manner; some philosophy and guiding principles had to be adopted in selecting the material to be presented. The basic principle used for this book was to include what I believe every serious student of physics should know about the atomic nucleus. It was not always possible to live up to this principle. In the first place, an appreciation of nuclear physics today will require not only a good knowledge of quantum mechanics and many-body theory, but also quantum field theory. This, in general, is too much to expect for the average reader and some sacrifice must be made. In the second place there are many interesting techniques, both experimental and theoretical, that form a part of the subject itself. Any reasonable coverage of these technical aspects will greatly expand the size of the book and make it useless in practice.

On the other hand, it is not possible to give a true flavor of nuclear physics without some background in quantum mechanics. In preparing this volume I have assumed that the student has the equivalent of a one-year undergraduate course in quantum mechanics or is taking concurrently an advanced quantum mechanics course at the level of one of the textbooks listed as general reference at the end. A basic knowledge of electromagnetic theory is also assumed; it is, however, unlikely that the background required here will be a problem to most students.

Some effort has been devoted to make the book as self-contained as possible. For this purpose, references to the literature are kept to a minimum. A specific paper published in scientific journals is mentioned only if a direct quotation is taken from it or if there is some historical interest associated with it. If references are needed, the first preference has been given to books that are readily available. However, this is not always possible. As a second choice, review articles are cited because a student starting out in the field may better comprehend this type of article than the original paper. Conference proceedings are used only as a last resort since it is difficult to expect standard libraries to be stocked with the multitude of proceedings published every year. One result of adhering to this philosophy is that very few of the excellent papers of my colleagues have been cited. I have also had some difficulty in selecting standard textbooks for reference in subjects such as quantum mechanics, classical mechanics, electromagnetism, and statistical mechanics. Here, I have relied purely on my own biases without guidance from a general philosophy, as I have done with papers.

One decision that had to be made concerns the system of units used for equations involving electromagnetism. The SI or MKS system would have been the more correct choice since essentially all students have been exposed to it and are more likely to be familiar with it. However, many of the advanced treatments on the subject, and nearly all the standard references on the topic in subatomic physics, are written using cgs units. It is therefore more practical to use the latter system here so that it is easier for a reader to make use of other references. For the convenience of those who are more comfortable with SI units, most of the equations (except those in §V.2) have the necessary additional factor enclosed in large square brackets to convert the expressions to SI units. In most cases, it is possible to write the equations involving electromagnetism in a form independent of the system of units by making use of the fine structure constant α and by measuring charge in units of e, the absolute value of charge carried by an electron, and magnetic dipole moment in units of μ_N, the nuclear magneton.

The book is aimed at physics students in their final year of undergraduate or first year of graduate studies in nuclear physics. There is enough material for a one-year course though it could be used for a one-semester course by leaving out some of the detail and peripheral topics. The selection of material is guided in part by current interests in the field; no attempt has been made to give a complete account of everything that is known in nuclear physics. However, sufficient knowledge is provided here so that a student may then go to the library and obtain information on a particular nucleus or a special aspect of a topic.

S. S. M. Wong

USEFUL CONSTANTS

Quantity	Symbol	Value			
Universal constants:					
Speed of light	c	299792458	m s^{-1}		
Unit of charge	e	$1.60217733(49) \times 10^{-19}$	C	$4.8032068 \times 10^{-10}$	esu
Planck's constant	h	$6.6260755(40) \times 10^{-34}$	J s	4.135669×10^{-21}	MeV s
(reduced)	$\hbar = \frac{h}{2\pi}$	$1.05457266(63) \times 10^{-34}$	J s	$6.5821220 \times 10^{-22}$	MeV s
fine structure constant	$\alpha = \frac{e^2}{[4\pi\epsilon_0]\hbar c}$	1/137.0359895(61)			
	$\hbar c$			197.327053(59)	MeV fm
Conversion of units:					
length	fm	10^{-15}	m		
area	barn	10^{-28}	m^2		
energy	eV	$1.60217733(49) \times 10^{-19}$	J		
mass	eV c^{-2}	$1.78266270(54) \times 10^{-36}$	kg		
	u	$1.6605402(10) \times 10^{-27}$	kg	931.49432(28)	MeVc^{-2}
charge	C	2.99792458×10^{9}	esu		
Masses:					
electron	m_e	$9.10938974 \times 10^{-31}$	kg	0.511099906(15)	MeV c^{-2}
muon	m_μ	$1.8835327 \times 10^{-28}$	kg	105.65839(6)	MeV c^{-2}
pions	$\pi^\pm$	$2.4880187 \times 10^{-28}$	kg	139.56755(33)	MeV c^{-2}
	π^0	2.406120×10^{-28}	kg	134.9734(25)	MeV c^{-2}
proton	M_p	$1.6726231 \times 10^{-27}$	kg	938.27231(28)	MeV c^{-2}
				1.007276470(12)	u
neutron	M_n	$1.6492860 \times 10^{-27}$	kg	939.56563(28)	MeV c^{-2}
				1.008664904(14)	u
Lengths:					
classical electron radius	$r_e = \frac{\alpha\hbar}{m_e c}$	$2.81794092(35) \times 10^{-15}$	m		
Bohr radius	$a_0 = \frac{r_e}{\alpha^2}$	$5.29177249(24) \times 10^{-11}$	m		
Compton wave length					
electron	$\lambda_{\mathrm{Ce}} = \frac{h}{m_e c}$	$2.426310585(22) \times 10^{-12}$	m		
proton	$\lambda_{\mathrm{Cp}} = \frac{h}{M_p c}$	1.32141×10^{-17}	m		
Others:					
Rydberg energy	$Ry = \frac{1}{2} m_e c^2 \alpha^2$			13.6056981(40)	eV
Bohr magneton	$\mu_B = \frac{e\hbar[c]}{2m_e c}$			$5.78838263(52) \times 10^{-11}$	MeV T^{-1}
Nuclear magneton	$\mu_N = \frac{e\hbar[c]}{2M_p c}$			$3.15245166(28) \times 10^{-14}$	MeV T^{-1}
magnetic dipole moment:					
electron	μ_e			1.001159652193(10)	μ_B
proton	μ_p			2.792847386(63)	μ_N
neutron	μ_n			−1.91304275(45)	μ_N
Fermi coupling constant	$\frac{G_F}{(\hbar c)^3}$	$1.43584(3) \times 10^{-62}$	J m^3	$1.16637(2) \times 10^{-5}$	GeV^{-2}
Gamow-Teller to Fermi coupling constants	$\frac{G_A}{G_V}$	−1.259(4)			
Avogadro number	N_A			$6.0221367(36) \times 10^{26}$	mol^{-1}
Boltzmann constant	k	1.380658×10^{-23}	J K^{-1}	$8.617385(73) \times 10^{-11}$	MeV K^{-1}
permittivity, free space	ϵ_0	$8.854187817 \times 10^{-12}$	$C^2 N^{-1} m^{-2}$		
permeability, free space	μ_0	$4\pi \times 10^{-7}$	N A^{-2}	$\epsilon_0 \mu_0 = c^{-2}$	

Chapter 1

INTRODUCTION

Nuclear physics is the study of the structure of nuclei and the interaction between nucleons. The basic building blocks of all nuclei are protons and neutrons, two different aspects of the same particle, the nucleon. The fact that a large variety of nuclei are constructed out of nucleons makes the subject an interesting one. The diversity of observed phenomena is the result of both the fundamental interactions operating between subatomic particles and the basic symmetries governing their behavior. It is therefore appropriate to start a study of nuclear physics with a review of the fundamental interactions in nature. This will also serve to relate the study of the atomic nucleus to the overall pursuit of physics.

§1-1 FUNDAMENTAL INTERACTIONS

The dominant force acting between nucleons is an aspect of the strong interaction. In addition, both electromagnetic and weak interactions also play an important role in determining the properties of nuclei. These three interactions: strong, electromagnetic, and weak together with gravitation interaction, form the four fundamental interactions in nature.

The modern view of force between particles is based on field theoretical ideas. A particle feels the presence of another one through the exchange of one or more field quanta, little "bundles" of energy. For example, two charged particles feel the presence of each other by exchanging photons between them. The field quanta are necessarily *bosons*, particles governed by Bose-Einstein statistics in quantum mechanics, so that they may be absorbed and emitted by the interacting particles without being constrained by the Pauli exclusion principle. The energy E associated with a field quantum is related to the range of the interaction it carries. This can be seen from the Heisenberg uncertainty principle. When a quantum of energy E is emitted, the state of the particle that emits the quantum is changed by the process. If the field quantum exists only for a time $t \leq \hbar/E$, where $\hbar$ is the Planck's constant divided by 2π, we need not be concerned with energy conservation. Furthermore, since the particle exists only for a short time, it cannot be observed directly. For this reason the field quantum is called a *virtual* particle. In contrast, a real particle has a definite energy and the amount can be measured in the laboratory.

For the purpose of estimating the range of a force, we may consider the field quanta to be travelling essentially at the speed of light c. The distance a quantum can travel, and hence its range, is therefore $r_0 \approx ct$, where t is the amount of time the field quantum existed. From the uncertainty principle $Et \leq \hbar$, we have the relation

$$r_0 \approx ct \approx \frac{\hbar c}{E}$$

If the field quantum has a non-zero rest mass m, it must have an amount of energy no less than its rest mass energy mc^2 and the range of the interaction it carries is limited to a distance

$$r_0 = \frac{\hbar}{mc} \tag{1-1}$$

Since $\hbar c \simeq 197$ MeV-fm (1 MeV $= 10^6$ eV and 1 fm $= 10^{-15}$ m), a range of 2 fm is obtained for a particle with rest mass energy mc^2 of the order of 100 MeV. On the other hand, if the field quantum is massless, the range is infinite as in the case of gravitational and electromagnetic interactions.

Mesons are the field quanta exchanged between nucleons when they interact with each other. However, nucleons and mesons are not fundamental particles: they are members of the hadron family, consisting of all the strongly interacting particles that are made of quarks. The quarks feel the presence of each other mainly through the strong interaction mediated by the exchange of gluons as the field quanta. The ultimate building blocks of nuclei are therefore quarks. Gaining an understanding of nuclear physics using quarks and gluons as the starting point is, however, impractical, and would be analogous to studying chemistry starting from electromagnetic interactions. While we shall attempt, where relevant, to make connection with the underlying quark structure, we shall think of nuclei primarily in terms of nucleons and mesons.

Even though the strength of strong interactions is much larger, electromagnetic and weak interactions do play an important role in nuclei. The reason for this comes in part from the very different nature of these three fundamental interactions. For our purposes here, the most important point is the difference in their range, as can be seen from Table 1-1. Except for its part in initiating the process of nucleosynthesis, the creation of essentially all the nuclei beyond the proton, the gravitational interaction can be ignored in nuclear physics discussions. Among the remaining three interactions, we find the field quanta involved in each case are quite different from each other. In the case of electromagnetic interactions, the photon is massless and, hence, the range of the force is infinite. On the other hand, the field quanta of weak interactions, $W^{\pm}$ and Z^0 particles, are extremely massive, with rest mass energy of the order of 100 GeV (10^{11} eV). This should

Table 1-1 Fundamental interactions.

Interaction	Field quantum	Range (m)	Relative strength	Typical cross section (m^2)	Typical time scale (s)
Strong	Gluon	10^{-15}	1	10^{-30}	10^{-23}
Weak	$W^{\pm}, Z^0$	10^{-18}	10^{-5}	10^{-44}	10^{-8}
Electro-magnetic	Photon	∞	$\alpha = \frac{1}{137}$	10^{-33}	10^{-20}
Gravity	Graviton	∞	10^{-38}	–	–

be contrasted with the rest mass energy of the nucleons which is only of the order of 1 GeV. As a result, weak interactions behave as if they are "contact," or zero-range, interactions.

In the absence of electromagnetic interactions, nuclei having the same number of nucleons A but different in proton number Z and neutron number N are very similar to each other. Strictly speaking, this is true only if nuclear force itself cannot distinguish between a proton and a neutron. There is considerable evidence in support of such a symmetry in nuclear force. For example, certain states in different members of an isobar, *i.e.,* nuclei having the same A, are found to have very similar properties and they may be regarded as the "isobaric analogue" of each other. The invariance of nuclear properties under an exchange between a neutron and a proton is an aspect of *isospin* symmetry. In general, the concept of symmetry is an important one in subatomic physics and we shall often return to it later.

Even though electromagnetic interactions are much weaker than the nuclear force, the role they play in nuclear physics is, nevertheless, quite prominent. Since each proton carries a unit of positive charge e, electrostatic repulsion between protons reduces the binding energy between a pair of protons compared with that between a pair of neutrons. Because of its long range, electrostatic repulsion in a nucleus increases with the number of proton pairs and, thus, quadratically with the number of protons. In contrast, nuclear force has a short range and, as we shall see later in §4-10, nuclear binding energy, on the whole, increases only linearly with the number of nucleons. As a result, electromagnetic interactions become an important consideration in the stability of heavy nuclei and in understanding the small binding energy difference between members of an isobar. Furthermore, the motion of protons constitutes an electric current flowing inside a nucleus. This current, in turn, produces a magnetic field which, together with the intrinsic magnetic dipole moments of nucleons, provides most nuclei with a finite magnetic moment. As a result, a nucleus can interact with an external electromagnetic field, causing both emission and absorption of γ-rays.

Weak interaction in nuclear physics is associated with β-decay, nuclear transmutation accompanied by the emission of an electron or positron together with a neutrino. The most obvious case of nuclear weak interaction is the decay of a free neutron into a proton plus an electron and an (electron anti-)neutrino. The inverse reaction of converting a proton into a neutron cannot take place for a free proton, since the rest mass of a proton is smaller than that of a neutron. However, for a proton bound in a nucleus, it can decay into a neutron as, for example, in the β-decay of ^{11}C to ^{11}B

$$^{11}\mathrm{C} \rightarrow {}^{11}\mathrm{B} + e^{+} + \nu_e \tag{1-2}$$

Both nuclei are made of eleven nucleons ($A = 11$), indicated by the superscripts preceding the chemical symbol for the element. In the decay, a positron (e^+), the antiparticle of an electron, and an electron neutrino (ν_e) are emitted when one of the six protons in the nucleus ^{11}C is converted to a neutron, resulting in a ^{11}B nucleus consisting of five protons and six neutrons. Such a reaction is possible as long as there is sufficient energy difference between the parent and the daughter nuclei for the transition to take place.

§1-2 A BRIEF HISTORY OF NUCLEAR PHYSICS

The beginnings of nuclear physics may be traced to the discovery of radioactivity in 1896 by Becquerel, who noticed that well-wrapped photographic plates were blackened when placed near certain minerals. Two years later Pierre and Marie Curie succeeded in separating a naturally occurring radioactive element, radium ($Z = 88$), from the ore. Soon afterwards, it was realized that the activity changed the chemical properties of the element. Furthermore, three different types of activity were established when radioactive sources were placed in a magnetic field: the trajectories of some of the particles emitted were found to be deflected by the magnetic field to one direction, some to the opposite direction, and some not affected at all. These were named α-, β- and γ-rays since nothing more was known about these particles until much later. It is now common knowledge that α-particles are ^{4}He nuclei, β-rays are made of electrons or positrons, and γ-rays are nothing but electromagnetic radiations; the names assigned by the early workers, however, stayed with us and are still in common use today.

The nuclear model for atoms was first proposed by Rutherford in 1911. In this model, the nucleus occupies only a very small part of the volume taken up by the atom. The radii of a few heavy nuclei were measured by Chadwick in 1920 and they were found to be of the order of 10^{-14} m, much smaller than the order of 10^{-10} m for atomic radii. The experiments involved the scattering of positively charged α-particles from such heavy elements as copper, silver, and gold, and the nuclear radii were inferred from the departure of the measured cross sections from values given by the Rutherford formula for Coulomb scattering of point charges.

Our knowledge of the basic ingredients of nuclei was more or less complete fifteen years or so later with the discovery of the neutron in 1932 by Chadwick, Curie and Joliot, and with Yukawa making the postulate of meson exchange for nuclear interaction in 1935.

The great discoveries of nuclear physics in the first part of this century have often also served as milestones in the progress of modern physics, a world that is remote from our direct senses and, for the most part, involves phenomena created in the laboratory. What is perhaps even more puzzling to the uninitiated is that these phenomena can no longer be described by classical mechanics that have been so successful in understanding all the macroscopic phenomena surrounding our daily lives. Instead, quantum mechanics was required, together with all its seemingly bizarre probabilistic interpretations. The microscopic world of subatomic physics is, however, a very rich one where many fundamental laws of nature are revealed to us for the first time.

The study of the atomic nucleus has taught us many new things about nature. In addition to radioactivity itself, there was another mystery in the early days of modern physics concerning the long lifetimes of radioactive decay involving the emission of α-particles. The observed lifetimes varied over a wide range of values covering some thirty orders of magnitude. For example, the half-life of ^{238}U is 4.47×10^9 years, whereas for a neighboring nucleus, ^{238}Pu, with the same number of nucleons except with two neutrons replaced by two protons, the half-life is only 87.8 years. The kinetic energies of the α-particles emitted are usually in the range of a few MeV, 4.20 MeV for ^{238}U and 5.50 MeV for ^{238}Pu, much smaller than the Coulomb barrier, given by the energy required to overcome the electrostatic repulsion encountered in bringing an α-particle, having two units of positive charge, from infinity to the surface of a nucleus. For an estimate, we may take the nucleus to be a uniformly charged sphere of radius $R = 1.2A^{1/3}$ fm, where A is the nucleon number (see §4-3). For example, for ^{208}Pb with $A = 208$ and proton number $Z = 82$, a value of 35 MeV is obtained. Since the α-particle was originally bound to the nucleus before the decay, one can view it as being contained in an attractive potential well, with height given by the Coulomb energy.

In terms of classical mechanics concepts, the α-particle must acquire an amount of energy, for example, through random collisions with the other constituents of the nucleus, to bring it to the top of the potential barrier before it can be free from the nucleus. If this were the case, the kinetic energy of α-particles emitted would be equal to or greater than the barrier height. Furthermore, in terms of present day language, α-decay belongs to the domain of strong interaction where the typical time scale is of the order of 10^{-23} s, much shorter than the typical lifetime of α-decay observed. Both the energy and lifetime of heavy nucleus α-decay were puzzling and eventually led to the discovery of quantum mechanical tunnelling, which becomes a strong support for the wave nature of particles. In the quantum picture, an α-particle does not have to jump over the Coulomb barrier; instead, its wave function leaks through the barrier with a probability that decreases exponentially with the height and width of the barrier (see also §5-5); hence, the long lifetime and low kinetic energy of the emitted particle.

Before the discovery of the neutron in 1932, it was assumed that a nucleus was made of protons and electrons, with the positive charge of some of the protons neutralized by the electrons. This was necessary to account for the fact that, except for hydrogen, the number of positive charges in a nucleus is smaller than the number of nucleons. In terms of what we know now, this picture of having only protons and electrons inside a nucleus was an impossible one. Nuclei with odd number of nucleons are known to have half-integer value spins, the total angular momentum of all the nucleons, and nuclei with even numbers of nucleons have integer value spins. Since particles with half-integer spins are fermions, particles that obey Fermi-Dirac statistics, an odd-A nucleus must be a fermion. Both electrons and protons are also fermions by virtue of the fact that the values of their intrinsic spins are half. An electron and a proton may be combined to form an electrically neutral object, but their total spin must be an integer and the combined object, as a result, cannot be a fermion. If there were no neutrons, the question of whether the spin of a nucleus takes on integer or half-integer values would be determined entirely by whether the charge of a nucleus in units of e is even or odd. This is not true in practice and a model of the nucleus made of protons and electrons cannot be correct, since it is in violation of the fundamental relationship between spin and quantum statistics. Indeed, the description of a nucleus was very awkward, as can be recognized, for example, by reading a nuclear physics book written before 1932.

Atomic nuclei have been, and remain, an important "laboratory" for weak interaction studies. The neutrino was proposed by Pauli in 1931 and used by Fermi in 1933 to explain the puzzle in a nuclear β-decay that electrons are emitted with a continuous spectrum of energy up to a maximum known as the end-point energy. This seemed to violate energy conservation, since there is a definite energy difference ΔE between the parent and daughter nuclei. If the final state of the decay involves only two particles, an electron and the daughter nucleus, the kinetic energy of the electron is fixed and completely specified by conservation of energy and momentum in the reaction. A continuous energy spectrum of the electrons emitted from a β-decay nucleus is in violation of this simple argument. It was a bold move on the part of Pauli to propose that the final state of the decay involves a third particle, the neutrino, unobserved since it carries no charge and very little mass. This "unobserved" particle, as we know now, is even more elusive than the neutron: it hardly interacts with any other particles and is so light that even today we are still uncertain whether it is massless or not.

The concept of parity violation, the first one of a series of "broken" symmetries found in physics, was confirmed through nuclear β-decay. Both strong and electromagnetic interactions are known to conserve parity, *i.e.,* experiments give the same results whether they are viewed in right-handed coordinate systems or left-handed coordinate systems. In the early 1950s, there was no reason to doubt that weak interactions should be any different. However, experimental data involving particles which seemed to be identical except for their parity were baffling. The concept of parity violation, proposed by Lee and Yang in 1957, was confirmed by a β-decay experiment on ^{60}Co in which it was observed that more electrons were emitted with momentum components opposite to the orientation of the nuclear spin than along it (for more details see §4-4). This is a clear violation of the invariance of operations under space inversion, *i.e.,* a reflection through the origin of the coordinate system used, and has led to a better understanding of the weak interaction itself. At the same time, the discovery led us to reexamine some of the symmetries of physical laws which we have often taken for granted.

§1-3 POSSIBLE FUTURE DIRECTIONS

While it is relatively easy to make a long list of past achievements of nuclear physics, it is much harder to predict what will be of interest in the future. However, based on the scientific activities of today, it is perhaps safe to make the following forecast.

The measurement of neutrino mass is one of the fundamental problems in physics today. Besides the interest in the nature of neutrino itself, the answer also has a bearing on cosmology and our views of the fundamental laws of physics. For example, if neutrinos have finite masses, no matter how small, the aggregate of their huge numbers, possibly several orders of magnitude larger than the number of nucleons, will make up a significant fraction of the total mass in the universe. Other questions such as the relation between different types of neutrino (neutrino oscillation) and between neutrinos and antineutrinos (Majorana or Dirac particles) are also related to the question of whether the neutrino has a finite mass. Another related development is that neutrinos are emitted from every star, whether it is an ordinary one like our sun or a short-lived, exotic one like a supernova. If our knowledge of and our detection methods for neutrinos can be improved, it is not hard to imagine that neutrino astronomy will become an important tool in astrophysics, alongside with optical and radio telescope observations.

Advances in physics are strongly coupled with progress in technology. In recent years, the possibility of building accelerators for heavy ions, a term used to refer to projectiles made of nuclei from lithium to the actinide series, and detection instruments that can identify the multitude of particles produced, has made it possible to study the collision between heavy nuclei. Several interesting directions can be pursued with these new instruments. The most obvious one is to make new elements in the laboratory that are even further away from naturally occurring, stable isotopes. The additional nuclear species are of interest by themselves and these unstable nuclei may reveal to us physical laws that have so far been hidden from us. Moreover, the collision of heavy ions also involves interactions between intense Coulomb fields; as a result, the laws of quantum electrodynamics, believed to be the best known laws in physics, will be tested under conditions never encountered before. There are already indications, as we shall see briefly in §7-7, that new insights on electrodynamics may be gained from such interactions.

The present goal of heavy ion accelerators is to reach energies far in excess of the rest mass energy of the particles themselves, of the order of several hundred GeV per nucleon. When nuclei collide with such highly relativistic energies, individual nucleons inside these nuclei may lose their identity and the concentration of hadronic matter may be regarded as a plasma of quarks and gluons. For the short duration that the two heavy nuclei are fused together we have a region of extremely high energies approaching conditions which existed in the primordial fireball just before the big bang that created the universe. There does not seem to be any other convenient way in the laboratory to reach this important state of matter, the study of which is essential for an understanding of cosmology as well as basic physical laws.

Double β-decay is another problem in nuclear physics that has a bearing on a wide range of topics including the unification of all fundamental interactions, the central theme of grand unification theories. For many nuclei the binding energies are such that decay from (A, Z) to $(A, Z+2)$ by emitting two electrons or two positrons (with or without two neutrinos as we shall see later) is possible. However, such decays are very slow, since second-order weak interaction processes are involved. For example, the lifetime of ^{76}Ge decay to ^{76}Se is estimated to be around 10^{21} years. Measurements of such long lifetimes are difficult; the problem is, however, interesting for several reasons.

One of the arguments for the longevity of double β-decay nuclei is the five-body final state involved, for example,

$$^{76}\text{Ge} \rightarrow {}^{76}\text{Se} + 2e^- + 2\bar{\nu}_e \tag{1-3}$$

For simplicity we may think of the process as the β^--decay of one neutron after another, with each step accompanied by the emission of an electron and an electron antineutrino. If a neutrino is its own antiparticle, it is possible to have the neutrino from the decay of one of the two neutrons annihilating the neutrino from the decay of the other neutron. Such neutrinos are called Majorana neutrinos and if they exist, lifetimes of double β-decay may be shortened greatly, since no neutrinos are actually emitted and the final state of the decay involves only three particles, two electrons and the daughter nucleus. On the other hand, if neutrinos are strictly Dirac particles, with antiparticles different from their corresponding particles, neutrinoless double β-decay is not possible.

One of the most intriguing questions concerning nuclei is the effect of quarks or, rather, the lack of any direct evidence for the presence of quarks among the observed properties. If for the moment we do not question the origin of nuclear force, we find that there is almost no need to invoke the quark degrees of freedom for understanding essentially all nuclear properties. Everything seems to be adequately explained in terms of nucleons and mesons. It is likely that we are not asking the right questions. Since quarks are mostly confined in nucleons, their influence will be evident only at very short distances and very high energies. Furthermore, since quark effects have not yet been noticed in conventional measurements, they must be small at best and must be studied under circumstances where competing phenomena from other processes are known to precision better than the anticipated contributions from quarks. Lepton scattering turns out to be an ideal probe here and this is the aim of a large number of investigations centring around the EMC effect, first noted in 1983 by the European Muon Collaboration (see §6-5). High energy protons have also been used for similar studies, since well-defined, intense beams are readily available. The interpretation of the proton results is, however, complicated by the multitude of hadrons produced through the strong interaction mechanism and leptons through the Drell-Yan process. This is part of the interest of the new field of high energy nuclear physics which may lead to many new and interesting investigations.

What does it mean *if* all nuclear physics can be understood without having to invoke quarks? One possibility is that the only effect of quarks in nuclei is a change of the scale associated with nucleons. In other words, when a nucleon is imbedded in the nuclear medium, the quarks inside readjust themselves in such a way that all the nucleon properties "scale" in some straightforward manner such as, for example, a change of the nucleon size. If everything scales by a simple factor, it is not easy for us to notice anything except that free and bound nucleons are somewhat different. There is some evidence that bound nucleons inside a nucleus behave slightly differently from free or "bare" nucleons and this may well be the only quark effect we can observe apart from the nuclear force itself. If this is the case, it will be of interest to find out this *scaling* law and the physical reasons behind it.

At the level of nucleons, questions concerning the presence of mesons in nuclei are still waiting further clarification. It is well established that virtual mesons are exchanged to provide the interaction between nucleons. Consequently there must a constant flow of mesons between nucleons inside a nucleus. Such a *mesonic current* should also contribute to observed nuclear properties. Unlike the presence of quarks, there is very little doubt that mesonic currents exist in nuclei: the difficulty is that the observed contributions seem to be too small. In other words, most of the nuclear properties can be understood by considering bound nucleons to be essentially identical to free nucleons. One possible explanation for the apparent absence of a stronger effect from mesonic currents is that their effects are, to a large extent, cancelled by other factors operating at the same time. If this is the case, such a cancellation cannot be simply attributed to an accident of nature; some fundamental symmetry must be operating here and waiting for us to discover.

Given sufficient energy, a nucleon may be excited into a Δ-particle, a strong resonance in the pion-nucleon channel (see §2-6). A Δ-particle is "distinguishable" from a nucleon and, as a result, it does not suffer from the Pauli exclusion principle that excludes a nucleon from taking up single-particle states already occupied by other nucleons in the nucleus. In particular, the effect of Δ-excitation should be noticeable in intermediate energy reactions where the energies involved are sufficiently high for the excitation to take place. Before one can identify the presence of Δ-particles, one must first be able to account for the purely nuclear effects sufficiently well that the anticipated contributions from Δ-particles can be isolated in an unambiguous way. This, in turn, requires a good knowledge of nuclear wave functions, especially at smaller internucleon distances. Short interaction distances imply high momenta and, consequently, high energies. Under such circumstances, relativistic effects become important. As a result, today there is intense interest in the relativistic treatment of the nuclear many-body problem.

Elementary particle physics shares some of the interests of nuclear physics. In fact, for certain topics, such as neutrino mass measurements and relativistic heavy ion collisions, the distinction between nuclear and particle physics is quite meaningless. Historically, these two fields have not been separated and even today they are often referred to together as the field of *subatomic* physics. The division between particle and nuclear physics became most apparent in the 1960s when particle physicists concentrated primarily on the study of the structure of nucleons and other hadrons while nuclear physicists were more interested in the structure of and reactions between nuclei.

Table 1-2 De Broglie wave lengths of γ-ray, electron and proton at different kinetic energies.

Energy (MeV)	Wave length (fm)		
	Photon	Electron	Proton
0.1	1.2×10^4	3701	90
0.5	2.5×10^3	1421	40
1	1.2×10^3	872	29
10	1.2×10^2	118	9
100	1.2×10	12	2.8
1000	1.2	1	0.7

These days, distinctions between different branches of subatomic physics are sometimes made on the basis of the energies involved. Elementary particle and high energy nuclear physics require energies of the order of 10^{12} eV (TeV); intermediate energy nuclear physics uses energies in the range of 10^8 eV (100 MeV) to 10^9 eV (1 GeV); and low energy nuclear physics is mainly concerned with tens of MeV and below. The separation by energy is meaningful in one sense. From de Broglie wave length ($\lambda = h/p$) considerations, the scale with which we can examine subatomic particles depends on the energy used: examples of the wave length for different kinetic energies are given in Table 1-2. If our interest is in the structure of nuclei, excitation energies of the order of 1 MeV and wave length of the order of 1 fm are quite adequate. To examine phenomena involving scales much smaller than this, such as in the internal structure of nucleons, higher energies are required. In elementary particle physics, high energies are essential in order to probe the behavior of quarks and to create new and massive particles, such as the $W^{\pm}$ and Z bosons. It is here that particle physics interests become quite distinct from those of nuclear physics. The classification of different interests by energy is, however, not completely satisfactory either. For example, if we are interested in the quark effects in nuclei, both high energy and a good knowledge of nuclear physics are necessary. For our purposes then, we shall restrict ourselves to topics that are of interest to nuclear physics and define nuclear physics vaguely as physics research carried out by nuclear physicists.

§1-4 NUCLEOSYNTHESIS

Where do nuclei come from? As far as we can tell, only a minimal amount of primordial nucleosynthesis took place in the first million years or so after the big bang, before the end of the radiation era. The bulk of matter was in the form of protons and electrons with minute amounts of d (deuteron), ^{3}He, ^{4}He, and ^{7}Li. Most of the elements heavier than deuteron were formed in the next 10^{10} years as stars evolved from one stage of their existence to another. The process of nucleosynthesis, the creation of all nuclei, and hence all chemical elements, is an operation that started soon (on a cosmological time scale) after the big bang and continues today in all stars.

Hydrogen burning. In a typical young star like our own sun, the bulk of its energy comes from "burning" protons, *i.e.*, converting protons into ^{4}He nuclei (α-particles). The reaction is accompanied by the emission of two positrons (e^+) to conserve charge and two electron neutrinos (ν_e) to conserve lepton number. This is the fusion process that converts approximately 0.7% of mass into energy, the most efficient energy production mechanism known.

Although the process is an exothermic one, the probability for four protons being fused together directly into a ^{4}He nucleus is too small to be of any significance in solar energy production. Since there is a shortage of heavier elements in a young star, the more likely process of turning protons into ^{4}He in the sun is a chain of reactions starting with the weak interaction process,

$$p + p \rightarrow d + e^+ + \nu_e \tag{1-4}$$

which converts two protons into a deuteron, a nucleus made of a proton and a neutron bound to each other by 2.2 MeV (see §3-1). A positron is also emitted to balance the charge and a neutrino is emitted to balance the lepton number in the reaction.

Once there is enough deuteron present in a star, conversion of a deuteron into a ^{3}He by capturing a proton becomes probable,

$$p + d \rightarrow {}^3\text{He} + \gamma \tag{1-5}$$

This is known as a (p, γ) reaction through which a proton is captured by a nucleus, followed by the emission of a γ-ray to discard any excess energy in the final nucleus formed and to maintain energy and momentum conservation for the reaction. When enough ^{3}He accumulate, some of them may collide with each other to form the more stable ^{4}He through the reaction,

$$^3\text{He} + {}^3\text{He} \rightarrow {}^4\text{He} + p + p \tag{1-6}$$

To make these two ^{3}He, six protons were used as well as two positrons and two electron neutrinos were emitted. Since two protons are released at the end, the net effect of this chain of reactions, (1-4) to (1-6), is to change four protons into a ^{4}He nucleus (as well as two positrons and two neutrinos) accompanied by the release of 28 MeV of energy. This is known in astrophysics as the PPI chain, one of several possible chains of reaction to convert protons into α-particles.

When a sufficient amount of ^{4}He is accumulated in a star, the reaction

$$^3\text{He} + {}^4\text{He} \rightarrow {}^7\text{Be} + \gamma \tag{1-7}$$

through which a ^{3}He is captured by a ^{4}He nucleus, becomes probable. This is the start of an alternate route to convert four protons into a ^{4}He nucleus, known as the PPII chain. The ^{7}Be produced, in turn, captures an electron and converts itself into a ^{7}Li nucleus,

$$e^- + {}^7\text{Be} \rightarrow {}^7\text{Li} + \nu_e \tag{1-8}$$

A neutrino is emitted here so as to satisfy lepton number conservation. The ^{7}Li then captures a proton to form two ^{4}He nuclei,

$$p + {}^7\text{Li} \rightarrow {}^4\text{He} + {}^4\text{He} \tag{1-9}$$

The entire PPII chain consumes an electron, a proton, and a ^{3}He nucleus, formed originally by three protons (with the emission of a positron and an electron neutrino), and produces a ^{4}He and an electron neutrino. The net result is essentially the same as that of the PPI chain except that ^{4}He is used as a catalyst here.

It is worthwhile mentioning another route for proton burning that is going on in the sun, the PPIII chain. The ^{7}Be produced in the PPII chain in (1-7) may also capture a proton to form ^{8}B which, in turn, β^+-decays to ^{8}Be. Since ^{8}Be is unstable toward decay into two ^{4}He (see §4-10), we have again a conversion of four protons into a ^{4}He nucleus together with two positrons and two electron neutrinos,

$$\begin{aligned} p + {}^7\text{Be} &\rightarrow {}^8\text{B} + \gamma \\ {}^8\text{B} &\rightarrow {}^8\text{Be} + e^+ + \nu_e \\ {}^8\text{Be} &\rightarrow {}^4\text{He} + {}^4\text{He} \end{aligned} \tag{1-10}$$

The interest of this reaction is in the high energy (7 MeV on the average) neutrino released in the β^+-decay of ^{8}B. The energy is sufficient to change ^{37}Cl to ^{37}Ar through the reaction,

$$\nu_e + {}^{37}\text{Cl} \to e^- + {}^{37}\text{Ar} \tag{1-11}$$

This process is used in the study of solar neutrinos and solar energy production. The rate of neutrino production from ^{8}B-decay in the sun may be deduced from the amount of argon collected in a large tank of ^{37}Cl in the chemical form of cleaning fluid (C_2Cl_4). This is the principle used in many solar neutrino detectors (see also Problem 1-7).

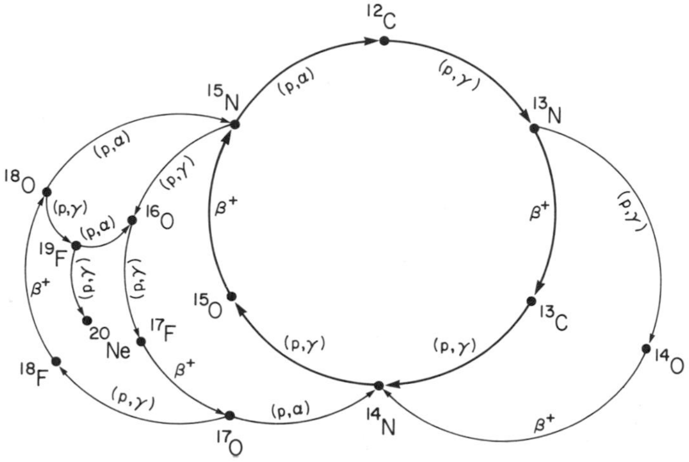

Fig. 1-1 The carbon-nitrogen-oxygen (CNO) cycle of nucleosynthesis showing the different reactions involved in converting protons into ^{4}He.

In the evolution of a young star, some nuclei heavier than α-particles are produced through processes we shall see later. If sufficient amounts of heavier elements, such as ^{12}C, are present, a more efficient proton burning process may take place. This is known as the CNO (carbon-nitrogen-oxygen) cycle shown schematically in Fig. 1-1. Let us concentrate first on the main cycle represented by the circle in the middle. If we start at the top from the ^{12}C, the cycle may be viewed as a chain of (p,γ) reactions to capture four protons one after another, and convert two of them into neutrons through two β^+-decays,

$$^{12}\text{C}(p,\gamma)^{13}\text{N}(\beta^+)^{13}\text{C}(p,\gamma)^{14}\text{N}(p,\gamma)^{15}\text{O}(\beta^+)^{15}\text{N}(p,\alpha)^{12}\text{C}$$

In the last reaction of the chain, the ^{15}N captures a proton and releases a ^{4}He through a (p,α) process. The net result is again the conversion of four protons into a ^{4}He plus two positrons and two electron neutrinos, the same final result as the PP chains except that now the process is more efficient, with ^{12}C used as the catalyst.

There are several side chains to the main CNO cycle that are of interest. The ^{13}N produced by the (p,γ) reaction on ^{12}C may be converted through another (p,γ) reaction into ^{14}O which then β^+-decays to ^{14}N,

$$^{13}\mathrm{N}(p,\gamma)^{14}\mathrm{O}(\beta^+)^{14}\mathrm{N}$$

The final product returns the process to the main CNO cycle in the form of ^{14}N. Similarly, some of the ^{15}N near the end of the main cycle may be converted back to ^{14}N by the following chain of reactions:

$$^{15}\mathrm{N}(p,\gamma)^{16}\mathrm{O}(p,\gamma)^{17}\mathrm{F}(\beta^+)^{17}\mathrm{O}(p,\alpha)^{14}\mathrm{N}$$

Again a ^{4}He nucleus is made from four protons by this procedure. The ^{17}O in the intermediate step above may return some ^{15}N to the main cycle and produce a ^{4}He nucleus through the chain of reaction,

$$^{17}\mathrm{O}(p,\gamma)^{18}\mathrm{F}(\beta^+)^{18}\mathrm{O}(p,\alpha)^{15}\mathrm{N}$$

The ^{18}O in the intermediate stage here may also undergo further proton capture and produce ^{19}F through the process,

$$^{18}\mathrm{O}(p,\gamma)^{19}\mathrm{F}(p,\alpha)^{16}\mathrm{O}$$

Some of the ^{19}F, in turn, may be converted into ^{20}Ne by a further (p,γ) reaction

$$^{19}\mathrm{F}(p,\gamma)^{20}\mathrm{Ne}$$

In this way we see that heavier and heavier elements are produced by capturing protons one at a time.

In order for PP chains and other nucleosynthesis reactions to take place, both high density and high temperature are required. High density ensures that nuclei will collide with each other frequently through their random motion in the star and high temperature ensures that the collisions will be energetic. In this way the two colliding nuclei will have a reasonable probability of penetrating the Coulomb barrier, bringing them in close proximity for the short-range nuclear force to be effective.

Consider two nuclei, having proton numbers z and Z, respectively. When these two nuclei are separated by a distance R between their centers greater than the sum of their radii, the height of the repulsive Coulomb barrier between is given by the energy required to bring them together to a distance R:

$$V_c = \left[\frac{1}{4\pi\epsilon_0}\right]\frac{(ze)(Ze)}{R} \tag{1-12}$$

The formula is in cgs units if the factor in square brackets is ignored and in SI units when the factor is included. To avoid any dependence on the system of electromagnetic units adopted, we can make use of the fine structure constant

$$\alpha = \left[\frac{1}{4\pi\epsilon_0}\right]\frac{e^2}{\hbar c} \approx \frac{1}{137} \tag{1-13}$$

to replace factors in (1-12) that depend on the system adopted,

$$V_c = \alpha\hbar c\frac{zZ}{R} \approx 1.44\frac{zZ}{R(\text{fm})}\ \text{MeV} \tag{1-14}$$

since $\alpha\hbar c \approx 1.44$ MeV-fm. Between two protons, we have $z = Z = 1$, and R may be taken to be of the order 1 fm. The height of the Coulomb barrier for reactions in the PPI chain is estimated in this way to be of the order of 1 MeV. This is a large amount of energy. In order for an average proton to have this amount of kinetic energy, the temperature of a star must be of the order of $T = E/k \approx 10^{10}$ K, where $k = 8.6 \times 10^{-11}$ MeV/K is the Boltzmann constant. The interior of our sun is much cooler than this value by roughly three orders of magnitude. Most of the nuclear reactions in the sun must therefore take place through barrier penetration due to quantum mechanical tunnelling rather than by overcoming the Coulomb barrier.

For reactions between heavier nuclei, such as those in processes we shall see below, the Coulomb barrier is even higher, increasing roughly in relation to the product between proton numbers z and Z of the two nuclei involved. As a result, even higher temperatures are required so that the barrier penetration probability can become sufficiently high for nuclear "burning" to take place. The source of the kinetic energy comes from the gravitational contraction of a star. As the raw material for forming a star gradually coalesces into a smaller volume, the decrease in gravitational potential energy is converted into heat. Only a part of the thermal energy is radiated into space and the remainder serves to heat up the star. Once the temperature of the star reaches a certain value, nuclear reactions start and the thermonuclear energy produced supplies a pressure in the form of random motion of the particles to counteract the gravitational force, preventing further contraction.

Helium burning and formation of nuclei up to $A = 56$. When most of the protons in a young star are converted into ^{4}He, the production of fusion energy drops significantly and the star starts to cool down due to radiation loss of energy to the surrounding space. As the temperature drops, thermal pressure is reduced; gravitational contraction resumes once again until the star is heated up to a temperature higher than the proton burning stage. At 10^8 K, corresponding to kinetic energy of the order of 10 keV, collisions between ^{4}He becomes sufficiently energetic that nuclear reactions involving ^{4}He begin to be significant. This starts the ^{4}He burning process, the next stage of nucleosynthesis and thermonuclear energy production.

The basic reaction in ^{4}He burning is the conversion of three ^{4}He nuclei into a ^{12}C nucleus, analogous to the conversion of four protons into a ^{4}He nucleus in proton burning. Again the probability of a three-body reaction to fuse three ^{4}He nuclei directly into a ^{12}C is too small to be of any significance in our discussion. The majority of the reaction takes place in two stages. The first is the formation of a ^{8}Be nucleus from two ^{4}He nuclei through the process,

$$^4\text{He} + {}^4\text{He} \rightarrow {}^8\text{Be}$$

Although ^{8}Be is unstable, the lifetime of 2.6×10^{-16} s in the ground state is sufficiently long that the reaction,

$$^4\text{He} + {}^8\text{Be} \rightarrow {}^{12}\text{C} + \gamma \tag{1-15}$$

in the second stage can take place with enough frequency. This is due in part to the large cross section for the $^8\text{Be}(\alpha,\gamma)^{12}\text{C}$ reaction, in particular, to the 0^+ excited state in ^{12}C at 7.644 MeV. From the point of view of nuclear structure, the 7.644 MeV state is essentially made of three α-particles coupled together in a straight line (in contrast, the ground state of ^{12}C is more akin to three α's located at the vertices of a triangle) and its wave function has a large overlap with the loosely bound ^{8}Be plus a ^{4}He.

When the amount of ^{12}C accumulated in a star is large enough, ^{4}He capture by ^{12}C becomes significant. The (α,γ) reaction,

$$^4\text{He} + {}^{12}\text{C} \rightarrow {}^{16}\text{O} + \gamma \tag{1-16}$$

also has a large cross section, since ^{16}O is a nucleus whose ground state wave function has an α-cluster structure as well. The production rates for ^{12}C in (1-15) and ^{16}O in (1-16) are so high that the abundance of ^{12}C and ^{16}O in the solar system is second only to those of proton and α-particle. Other light nuclei with α-cluster structure, such as ^{20}Ne and ^{24}Mg, are also formed in significant quantities by successive capture of α-particles. (Most of the observed elements heavier than ^{4}He in the solar system are not produced by the sun; they are primarily the remnants of stars which have gone through their life cycles before the formation of our solar system.)

When all the ^{4}He is used up, the star goes through another stage of gravitational contraction and heating. As the temperature rises above 10^9 K, corresponding to 100 keV in average kinetic energy per particle, reactions involving the conversion of ^{12}C into heavier nuclei become possible. For example,

$$\begin{aligned} {}^{12}\text{C} + {}^{12}\text{C} &\rightarrow {}^{24}\text{Mg} + \gamma \\ &\rightarrow {}^{23}\text{Na} + p \\ &\rightarrow {}^{23}\text{Mg} + n \\ &\rightarrow {}^{20}\text{Ne} + \alpha \\ &\rightarrow {}^{16}\text{O} + \alpha + \alpha \end{aligned} \tag{1-17}$$

At even higher temperatures, conversion of ^{16}O to heavier elements becomes important, such as the reactions

$$\begin{aligned} {}^{16}\mathrm{O} + {}^{16}\mathrm{O} &\rightarrow {}^{32}\mathrm{S} + \gamma \\ &\rightarrow {}^{31}\mathrm{P} + p \\ &\rightarrow {}^{31}\mathrm{S} + n \\ &\rightarrow {}^{28}\mathrm{Si} + \alpha \\ &\rightarrow {}^{24}\mathrm{Mg} + \alpha + \alpha \end{aligned} \tag{1-18}$$

On the other hand, the reaction $^{12}\mathrm{C} + {}^{16}\mathrm{O}$ is not considered to be crucial in nucleosynthesis since nearly all the ^{12}C will be exhausted before the temperature is high enough for the reaction to become significant. (See, *e.g.,* D.D. Clayton, *Principles of Stellar Evolution and Nucleosynthesis*, McGraw-Hill, New York, 1968).

The process of forming successively heavier elements through exothermic thermonuclear reactions stops around mass $A = 56$ (^{56}Fe and ^{56}Ni). As we shall see in §4-10, nuclear binding energy reaches a maximum here. The net binding energy of a nucleus is the result of the balance between attractive nuclear forces and the repulsive Coulomb force. Although nuclear forces are stronger, they have only a short range. As a result, nuclear binding energies increase only linearly with the number of nucleons, whereas contributions from the repulsive Coulomb force increase quadratically with the number of protons. Since nuclear forces are, on the average, more attractive between a neutron and a proton than between a pair of identical nucleons, the number of protons and neutrons in a stable nucleus tends to be roughly equal. For light nuclei ($A < 56$), we find that the Coulomb force is relatively unimportant and, as a result, a nucleus gains energy by acquiring more nucleons. However, when A increases beyond 56, electrostatic repulsion becomes a significant factor. In fact, for heavy nuclei, Coulomb energy may become comparable to nuclear binding energy. We can make an estimate of the Coulomb energy in a nucleus using a uniformly charged sphere of radius R as a model. The amount of energy required to overcome the electrostatic repulsion in assembling Z protons into a sphere of radius R is

$$E_c = \left[\frac{1}{4\pi\epsilon_0}\right]\frac{(Ze)^2}{R} = \alpha\hbar c\frac{Z^2}{R} \rightarrow \alpha\hbar c\frac{Z(Z-1)}{R} \tag{1-19}$$

where, in the final form, we have excluded from E_c the Coulomb energy associated with individual protons. The nuclear radius may be taken to be $R = 1.2A^{1/3}$ fm, as we shall see later in §4-3. The result is, for example, $E_c = 1.3$ GeV for ^{208}Pb or 6.5 MeV per nucleon. This value should be compared with the maximum average binding energy per nucleon of the order of 9 MeV found around $A = 56$. Because of the fast increase in Coulomb repulsion, it is energetically not profitable to synthesize heavy elements. In fact, heavy nuclei, such as uranium, release energy by fission into two or more lighter fragments, the principle behind nuclear fission.

Sources of heavy nuclei. Nuclei with $A > 56$ are generally referred to as "heavy nuclei" in discussions of nucleosynthesis. Since these nuclei cannot be formed by exothermic processes, no further thermonuclear reaction can take place once all the nuclei are converted to, for instance, ^{56}Ni. When gravitational contraction resumes again there is no longer the thermal pressure generated by nuclear reactions to counteract gravitational attraction. If the mass of a star is small, it will stop shining when all the gravitational energy is released and the star comes to a gentle end. On the other hand, the gravitational force in a massive star is so overwhelming that the collapse may take place very quickly, in the order of seconds, and the process may compress matter in the star for a fraction of a second to be denser than nuclear matter density. As a result, a part of the gravitational energy in the collapse is transformed into potential energy in the form of super-compressed nuclear matter. Since nuclear matter is rather stiff against compression, the excess potential energy will cause the star to explode in the form of a supernova and, as we shall soon see, such an explosion is an important source of heavy element formation.

Some heavy elements must be formed in the normal course of stellar evolution prior to the end of life. A star is in thermal equilibrium for most of its lifetime and the probability of finding a particle with kinetic energy E at a given stellar temperature T is given by the Boltzmann factor,

$$P(E) \sim e^{-E/kT}$$

At any given temperature there is, then, some finite probability of having charged particles with enough energy to tunnel through the Coulomb barrier of a heavy nucleus. This allows the formation of even heavier nuclei by capture. The reaction is an endothermic one and the necessary energy input may be taken from the thermal energy. It is therefore possible, in principle, for heavy elements to be made, for instance, by successive α-particle capture starting from the most stable nuclear species around $A = 56$. The probability for forming heavy elements through a large number of α-particle captures is, however, too small to account even for the relatively small amounts of heavy elements observed.

As an example, let us estimate the probability for α-particle capture of ^{204}Hg to form ^{208}Pb by calculating the height of Coulomb barrier involved. Using (1-14) with $z = 2$, $Z = 80$, and taking $R = 1.2A^{1/3} \approx 5.9$ fm, we obtain a height of ~ 32 MeV. In contrast, the barrier height for capturing an α-particle by ^{52}Fe to form ^{56}Ni is ~ 15 MeV. Since the barrier height enters into the rate of α-particle capture through quantum mechanical tunnelling roughly as an exponential factor, the probability of α-capture is reduced by several orders of magnitude in heavy nuclei. When this factor is folded in with the number of α-particle captures required to form heavy nuclei, we find that the amount of heavy elements produced by this mechanism is minuscule. Other processes must be responsible for the production of the bulk of the observed heavy nuclei.

One such process is neutron capture which does not suffer from the Coulomb repulsion between charged particles. However, under normal stellar conditions, the neutron flux is too small for this slow process – hence the name *s*-process – to be effective enough to make all the heavy elements. The *r*-process, or rapid process, which takes place in a high neutron flux environment, is the more likely candidate for forming the bulk of heavy elements observed in the universe today. During a supernova explosion, the neutron flux is extremely high and neutron capture becomes an important process for making new elements — in particular, those with a large neutron excess. We have strong supporting evidence that this is the case from the differences in the relative abundance of heavy elements observed in different parts of our galaxy, as measured from meteorites and spectroscopic data of stars. Since supernova explosions are relatively rare events, we cannot expect a uniform distribution of such events to have taken place in the past throughout different parts of the galaxy. As a result, the source of heavy elements, and hence their relative abundance, is inhomogeneous even in our own galaxy.

The main processes of nucleosynthesis described so far do not explain the relative amounts of all the elements observed. In particular, some of the lighter elements, such as lithium, beryllium and boron, are more abundant than would be expected from stellar sources. The production of heavy elements through *r*- and *s*-processes alone is not quite adequate either. More complicated procedures must be invoked (see, *e.g.,* J. Audouze and S. Vauclair, *An Introduction to Nuclear Astrophysics*, Reidel Publishing Co., Dordrecht, Holland, 1980). However, there is no doubt that the basic processes for the formation of all the nuclei are known even though there are still many unanswered questions remaining.

§1-5 THE STUDY OF NUCLEAR PHYSICS

As we have seen earlier, nuclear physics was inspired by natural occurring radioactivity. Indeed, radioactive decay, including that from elements created in the laboratory, remains an important source of information on nuclei. For example, most of the neutrino mass measurements these days involve the decay of tritium (half-life 12 years). As a group, radioactive decay tells us not only about the lifetime of the excited state, but also about the relation between the initial and final states involved.

The detection of solar neutrinos mentioned earlier is another good example of experimental nuclear physics without using an accelerator. In fact, advances in detection instruments, including those located high above the earth's atmosphere in satellites and deep in underground mines, make it possible to carry out observations that were impossible a decade ago. Experiments in nuclear physics without the use of accelerators are often referred to as non-accelerator physics.

Passive observations, however, constitute only a part of the experimental work carried out in nuclear physics. A large fraction of the data are accumulated by probing and exciting various nuclei. In this way, new states are formed and new elements are created. Such work usually requires an accelerator. In fact, much of the development in nuclear physics is influenced by the improvements in accelerator technology. All charged particles can in principle be accelerated to high energies; however, for nuclear physics investigations, high intensity and good energy resolution are also essential. In this respect, electron accelerators have an advantage. Because of their small rest mass, electrons at energies of interest to nuclear physics are already relativistic particles and, as a result, their rest mass does not enter to complicate the design of an accelerator. Partly for this reason, electron scattering has been a good source of high precision information.

Hadronic probes of nuclei are also extensively used, and they may be divided into four categories. Historically light ions, nucleon, deuteron, triton, ^{3}He, and α-particle have been used primarily because it is much easier to accelerate them. Each one of these particles has some unique property that is useful for a certain type of study. For example, both the spin and isospin of an α-particle are zero and, as a result, α-particle scattering is ideal for exciting a nucleus without changing the isospin. On the other hand, deuterons are loosely bound systems of protons and neutrons and are useful for transferring a nucleon to the target nucleus. These light ions form the first category of hadronic probes.

The second category consists of heavy ions. We have seen earlier that heavy ions can create new and exotic nuclei, intense electromagnetic fields and even quark-gluon plasma. In addition, heavy ion reactions are also capable of transferring large amount of mass and angular momentum, useful as we shall see in §6-3 in creating states of extremely high angular momentum.

A third category of hadronic probes consists of mesons. Since nuclear force is known to be mediated by mesons, meson scattering and meson absorption by nuclei become integral parts of nuclear physics investigations. Intense beams of pions, produced by striking high energy protons on a target made of heavy nuclei, have been available for some years. There are also plans to build accelerators that can produce intense beams of kaons, and this will bring us to the realm of probing the nucleus with "strange" particles as well.

Antiprotons form the fourth category of hadronic probes. Even though the precision that can be achieved today with antiproton scattering off nucleons and nuclei is far less than what can be obtained with protons, the information collected provides us with the opportunity to check our understanding of nucleon-nucleon interaction. Progress with antiproton beams and detectors will certainly make this line of investigation even more exciting in the future.

Weak interaction, or lepton, probes have been used for certain types of nuclear studies, such as the EMC effect mentioned earlier. Although electrons are also leptons, their interaction with nuclei is dominated by its charge, and weak interaction effects are overshadowed by those coming from electromagnetic interactions. Neutrinos, being neutral leptons, are more useful as weak interaction probes. The usefulness of neutrino scattering is limited by the small reaction cross sections, more than 15 orders of magnitude smaller than strong interactions at comparable low energies.

In general, nuclear physics, like its cousin elementary particle physics, is becoming more and more involved with large scale and expensive setup, even for nonaccelerator experiments. In fact, the instruments used in nuclear physics are so sophisticated that they can no longer be dealt within a textbook on nuclear physics — they belong, more naturally, in a special treatise by themselves.

The synthesis of such large quantities of diverse experimental information is the role of theoretical nuclear physics. Partly because nuclei are many-body systems, model construction has become an essential step toward understanding the observed data. Although nuclear models are very successful in correlating large amounts of experimental data, we are not yet at the stage of having an unified, all-encompassing theory of nuclear structure and nuclear reaction. Perhaps the nature of strong interaction is complicating the many-body problem. In either case, there is much work still left to be done in theoretical nuclear physics in spite of the advances already made.

Interest in nuclear physics is not confined to intellectual curiosity alone. The most visible aspect of applied nuclear physics is nuclear power. Many other applications, such as nuclear medicine and nuclear tracer element techniques, have potential far beyond their present scope. Many nuclear phenomena can also be used to further our knowledge in other fields. For example, measurements of the relative abundance of isotopes, such as the ratio of ^{14}C to ^{12}C, can give us a good indication of the age of an archaeological object. Other ratios, such as the abundance of iridium, tell us that some deposits of material on earth may have origins outside the solar system (see, *e.g.*, L.W. Alvarz, *Physics Today* **40** [July 1987] 24). Ratios of long-lived elements provide us with the yard stick of geochronology, the study of the history of the earth, and give us some of the best evidence of the age of the solar system. It is fair to say that applications of nuclear physics are still in their infancy and many more new and sophisticated ones will be developed in the future.

In this book, our focus will be confined to the fundamental aspects of nuclear physics. We shall take the view that the large variety of nuclei are laboratories to be used for investigating basic principles of physics. For this purpose, the atomic nucleus is an excellent choice. In addition to several hundred naturally-occurring stable nuclei, we also have over a thousand and an ever-increasing number of man-made, unstable isotopes to work with. Since the primary interaction between nucleons that make up the nucleus derives from the strong interaction between quarks, the nucleus is one of the places for examining the strong interaction. The nucleus is also a many-body system, involving anywhere between two to over 250 nucleons. As a result, techniques of the many-body problem can be tested in a variety of situations.

§1-6 COMMONLY USED UNITS

In studying atomic nuclei, we are dealing with length scales that are extremely small and time scales that are extremely short compared with standard measures of daily life. Instead of the meter, a more suitable unit of length is femtometer, abbreviated as fm (1 fm= 10^{-15} m). For example, the typical range of nuclear force is of the order 1 fm. A typical area, such as the cross section of strong interaction processes, is of the order of 1 fm^2. Another commonly used unit for area is a barn, equal to 10^{-28} m^2. Often cross sections for nuclear reactions are given in units of millibarn (10^{-1} fm^2) and microbarn (10^{-4} fm^2). Where possible we shall keep fm^2 as the unit for cross sections even though this is not yet the common practice in the literature.

The mass of a nucleon is 1.67×10^{-27} kg, with a neutron more massive than a proton by about 0.14%. The convenient unit for mass is the atomic mass unit, commonly abbreviated as u (or amu), and 1 u is $1.6605402 \times 10^{-27}$ kg. Perhaps a more useful way to express mass is in terms of the equivalent rest mass energy. The convenient unit for energy in nuclear physics, as we have already seen, is MeV, or million electron volt, and 1 MeV is $1.60217733 \times 10^{-13}$ J. For example, the rest mass energy of a neutron is 939.56563 MeV.

A large variety of time scales enter into nuclear physics. In Table 1-1 we have seen that the typical reaction time for strong interactions is 10^{-23} second or 10^{-23} s using the standard abbreviation for seconds. On the other hand, we also find naturally occurring radioactive elements that were made prior to the formation of the solar system. The lifetimes of these radioactive nuclei must be of the order of 10^9 years or longer, as anything with much shorter lives would have almost completely decayed away. However, such a long time scale is not typical of nuclear interactions. For lifetimes of the order of 10^{-15} to 10^{-23} s, it is more convenient to represent them in terms of their inverse in energy units. Because of the uncertainty principle, $\Delta E \Delta t = \hbar$, a state that lives only for a time Δt can have its energy measured only up to an uncertainty no better than $\Delta E \sim \hbar/\Delta t$. This gives a width $\varGamma = \hbar/\overline{T}$ in the probability distribution of the energy of the state. Here, $\overline{T}$ is the lifetime, or mean life, of the state (for more details, see §5-1). Since $\hbar = 6.58 \times 10^{22}$ MeV-s, lifetimes of the order of 10^{-23} s correspond to $\varGamma$ of the order of 100 MeV, and time scale of the order of 10^{-15} s corresponds to widths of the order of 1 eV.

Because of the small size of an atomic nucleus, it is often more convenient to adopt other special units, such as nuclear magneton for the magnetic dipole moment. We shall define each one of these units as they appear in the discussion.

After this brief introduction to the subject of nuclear physics, we shall begin in Chapter 2 with a study of the structure of nucleons in terms of quarks so that we can make the connection between nuclear interactions and the fundamental strong interaction. Chapters 3 and 4 are concerned with the experimental information available on nuclei. Because of the large amount of data collected, a very stringent selection of relevant results has been made. The guiding principle used here is to discuss only those measured quantities that have direct bearing on understanding the basic structure of nuclei. This information is used in Chapter 6 to construct models, the stepping stones to a future theory. Nuclear reactions are discussed in Chapter 7. The subject is a vast one and, again, a choice has to be made on the topics to be presented. Here an overview of the subject is emphasized rather than the detail of each type of reaction mechanism. Many other topics, such as angular momentum theory and scattering theory, and some background preparation, such as time-dependent perturbation theory, are useful for a better understanding of nuclear physics. These are presented in the form of appendices.

Problems

1-1. Given that the radius of a nucleon inside a nucleus is $R = 1.2$ fm, calculate the density of nuclear matter. From this, evaluate the radius of the sun (mass $= 2 \times 10^{30}$ kg) if it collapses into a neutron star without losing any of its present mass.

1-2. From the uncertainty relation, find the minimum kinetic energy of a nucleon in ^{208}Pb. Use $R = 1.2A^{1/3}$ fm for the value of nuclear radius.

1-3. The barrier penetration factor for an α-particle through the Coulomb field of the residual nucleus may be estimated by a triangular-shaped, one-dimensional barrier as shown in the figure on the right. For a heavy nucleus with $Z = 72$, we can use $R = 9$ fm. The height of barrier V_0 may be taken as the Coulomb energy at distance R and the distance R_1 may be estimated by the point where the Coulomb energy is equal to the kinetic energy E of the α-particle. Find the probability for an α-particle of $E =$ 5 MeV to escape from the nucleus.

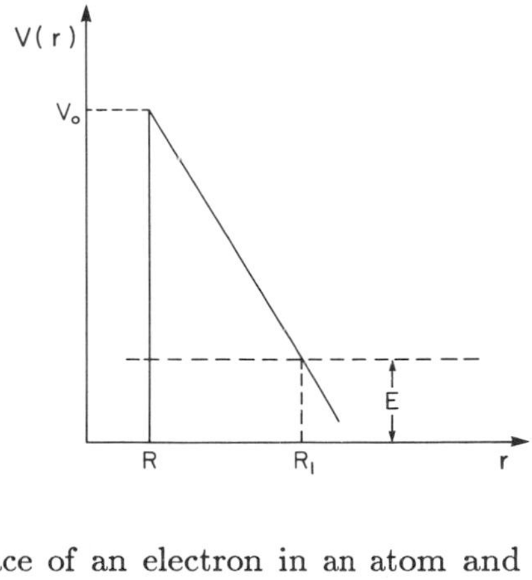

1-4. A negative muon (μ^-) can take the place of an electron in an atom and form a muonic atom. Since the muon is 207 times more massive than an electron, the radii of its orbits in an atom are much smaller. For a heavy atom, the low-lying muonic orbits may have substantial overlap with the nuclear wave function as a result. Calculate the fraction of time a muon is inside a spherical nucleus of radius $R = 1.2A^{1/3}$ fm, when the muonic atom is in the lowest energy state. Take $A = 200$ and $Z = 70$.

1-5. If the cross section for neutrino interaction with iron is $\sigma \approx 10^{-48}$ m^2, find the mean free path of a neutrino through solid iron.

1-6. Use conservation of energy and momentum to calculate the maximum kinetic energies of electrons released in the decay of a free neutron,

$$n \to p + e^- + \overline{\nu}_e$$

and in the decay of a free muon

$$\mu^- \to e^- + \nu_\mu + \overline{\nu}_e$$

at rest in the laboratory.

1-7. The inverse β-decay reaction,

$$\nu_e + {}^{37}\text{Cl} \to e^- + {}^{37}\text{Ar}$$

is used to detect neutrinos from the sun. The number of solar neutrinos produced may be estimated from the solar constant (1350 W/m^2). Assume that 10% of the thermonuclear energy is carried by neutrinos of mean energy 1 MeV each. Only about 1% of the neutrino is energetic enough to convert ^{37}Cl to ^{37}Ar. If a detector containing 400 m^3 of tetrachloroethylene (C_2Cl_4) is used, estimate the average number of ^{37}Ar produced in a day if the density of C_2Cl_4 is 1.5 g/cm^3 and about a quarter of the chlorine is ^{37}Cl. The cross section for the reaction may be taken to be 10^{-49} m^2.

Chapter 2

NUCLEON STRUCTURE

All nuclei are made of neutrons and protons, the two lightest members of the baryon family. Since nucleons are not elementary particles, a large fraction of the present day interest of nuclear physics is related in one way or another to the underlying quark's degree of freedom in nucleons. Such a study is, in turn, a part of the larger subject of quantum chromodynamics (QCD), the study of quarks and the interaction between them. It is still too early at this stage of the development of QCD to demand a complete description of nuclear physics starting from first principles; nevertheless, an understanding of the nucleus cannot be achieved without an awareness of quarks and their interactions. We shall attempt here only an introduction to certain aspects of QCD essential to nuclear physics.

There are also good practical reasons to examine the relationship between quarks before those between nucleons. One of the dominating considerations in subatomic physics is the role of symmetries. In this respect, there are many similarities between quarks and nucleons. Often it is easier to study these symmetry principles using quarks rather than nucleons, since the number of quarks inside a hadron is much more restricted than the possible number of nucleons inside a nucleus. For this reason, we shall devote a large part of this chapter to the study of symmetry relations between strongly interacting fermions using quarks as the example.

§2-1 QUARKS AND LEPTONS

The search for the fundamental building blocks of all matter in the universe has always been a central issue in physics. As our understanding of physical laws improves, our view changes on what constitutes the elementary particles, particles that cannot be made as composites of other particles. These days, the accepted view is that all matter is made of quarks and leptons. The only exceptions are photons, $W^{\pm}$ and Z^0 bosons, gluons and gravitons, particles mediating electromagnetic, weak, strong, and gravitational interactions, respectively.

Quarks. Quarks are the basic building blocks of hadrons, particles that interact with each other through strong interactions. In nuclear physics, we are mostly concerned with the lightest members of the hadron family: nucleons that make up all the nuclei and pions that are the main carriers of nuclear force. There are six different kinds or *flavors* of quarks, u (up), d (down), c (charm), s (strange), t (top) and b (beauty, or bottom). These six particles may be arranged according to their masses into three pairs, with one member of each pair having a charge of $\frac{2}{3}e$ and the other member a charge of $-\frac{1}{3}e$, as shown in Table 2-1. Since quarks have not been observed in isolation – they appear either as bound pairs of a quark and an antiquark in the form of mesons, or bound groups of three quarks in the form of baryons – the names assigned to them, up, down, strange, etc. are only mnemonic symbols to identify the different quarks. The word "flavor" is used to distinguish between the quarks for convenience, not because it has anything to do with taste. Besides flavor, quarks also come in three different *colors*; for example, red, green, and blue. Color and flavor are quantum-mechanical labels or quantum numbers, very similar to spin and parity, that are needed to differentiate between the different states in which a quark finds itself. Since there are no classical analogues to the flavor and color degrees of freedom, there are no observables that can be directly associated with them. In this respect, they are similar to the parity label of a state which must be "observed" through indirect evidence such as the decay modes of certain particles. For quarks, direct observation of any of their properties is made even harder by the fact that they appear only in groups of two or more. However, there is, by now, a large amount of evidence for the presence of flavor, color, and other degrees of freedom associated with quarks, and we shall examine some of these properties in this chapter.

Table 2-1 Quarks and leptons.

Quarks			
$Q/e = \frac{2}{3}$	u	c	t
$Q/e = -\frac{1}{3}$	d	s	b
Leptons			
$Q/e = -1$	e	μ	τ
$Q/e = 0$	ν_e	ν_μ	ν_τ

Leptons. Although quarks make up the bulk of observed mass in the universe, they are not the only elementary building block of particles with finite rest mass. *Leptons*, or light particles, are not made of quarks. Furthermore, they participate in electromagnetic and weak interactions but not in strong interactions. The number of different types of known leptons is also six and can be arranged into three pairs, as shown in Table 2-1. The electron (e), the muon (μ), and the tau lepton (τ) carry a charge $-e$ each, but the electron neutrino (ν_e), the muon neutrino (ν_μ), and the tau neutrino (ν_τ) are neutral. The masses of the leptons are much less than those of quarks, with $m_e c^2 = 0.511$ MeV, $m_\mu c^2 = 106$ MeV, and $m_\tau c^2 = 1784$ MeV. The neutrinos are known to be very much lighter and their rest masses may even be zero. A large amount of effort has been devoted in recent years to measuring the mass of ν_e, and the best estimate at the moment is that $m_{\nu_e} c^2 \lesssim 30$ eV, although much larger values as well as lower values for the upper limit have also been reported. For the other two types of neutrino, only the upper limits of their masses are known: $m_{\nu_\mu} < 0.25$ MeV and $m_{\nu_\tau} < 70$ MeV.

In nuclear physics, leptons make their presence felt through nuclear β-decay and other weak transitions. In general, only electrons and electron neutrinos are involved; occasionally muons may enter, such as in the case of muonic atom where a muon replaces one of the electrons in the atom. Because of its larger mass and its more recent discovery, the τ lepton has yet to enter nuclear physics studies.

Lepton number conservation. The number of leptons is conserved in a reaction. For example, a free neutron decays with a mean life of 896 ± 10 s through the reaction

$$n \to p + e^- + \overline{\nu}_e \tag{2-1}$$

The bar over ν_e indicates that it is an electron antineutrino, the antiparticle of an electron neutrino. On the left hand side of the equation, only a neutron is present. Since there is no lepton, we can assign it a *lepton number* $L = 0$. On the right hand side of the equation, we have one electron, which carries a lepton number $L = 1$. An antiparticle is given a particle number of the same magnitude as the particle with which it is associated, but with the opposite sign. This is necessary since an antiparticle combines with a particle to form a state without particle. Hence the lepton number of $\overline{\nu}_e$ is -1. The total lepton number on the right hand side of (2-1) is $L = 1 + (-1) = 0$. With these assignments, the lepton number is conserved in the reaction.

Conservation of lepton numbers in (2-1) depends on the recognition of the fact that the neutral lepton ($\overline{\nu}_e$) produced in the reaction is an antineutrino rather than a neutrino. This is not merely a gimmick to balance the lepton number of the two sides of (2-1). The two types of neutrinos, ν_e and $\overline{\nu}_e$, are two different particles, related by a transformation between a particle and its antiparticle, or *charge conjugation* for short. Electron neutrinos, ν_e, can be obtained, for example, from the reaction

$$p_{\text{bound}} \to n + e^+ + \nu_e \tag{2-2}$$

Such a process is not energetically possible for a free proton, the nucleus of a hydrogen atom, since a neutron is more massive ($M_n c^2 = 939.566$ MeV) than a proton ($M_p c^2 = 938.272$ MeV). However a bound proton, p_{bound}, inside a nucleus can undergo the reaction described by (2-1). The necessary energy conservation is now between the parent nucleus containing the bound proton (as well as other nucleons) and the daughter nucleus containing the neutron. As long as there is enough of an energy difference between the parent and daughter nuclei to create the two leptons, a positron e^+ and an electron neutrino ν_e, the reaction is possible (see §5-4 for detail). Since a positron is the antiparticle of an electron, its lepton number is $L = -1$. In order to conserve the charge, the charged lepton on the right hand side of (2-1) must be a positron and, in order to conserve the lepton number, the reaction must be accompanied by an electron neutrino in the final state.

If ν_e and $\overline{\nu}_e$ were the same particle, we could make use of the electron neutrino obtained from reaction (2-1) to induce the inverse of reaction (2-1),

$$\nu_e + p \rightarrow e^+ + n \tag{2-3}$$

To conserve charge, the charged lepton on the right hand side must be a positron with $L = -1$. The failure to observe such a reaction is a testimony of the importance of lepton number conservation. In contrast, the reaction,

$$\overline{\nu}_e + p \rightarrow e^+ + n \tag{2-4}$$

is observed. This establishes that ν_e and $\overline{\nu}_e$ are two different particles as well as confirming that the lepton number must be conserved (see §5-5 for more detail). In fact, the conservation of leptons (*i.e.,* that leptons cannot be created or annihilated, except in pairs of a lepton and an antilepton) is a fundamental conservation law not too different from the conservation of energy and momentum. Our convention of assigning lepton numbers starts by giving $L = +1$ to an electron. Once this is fixed, all the other lepton numbers are determined by conservation requirements in reactions.

Particles that are distinct from their antiparticles are called Dirac particles. This is to distinguish them from Majorana particles which are identical to their antiparticles. As we have seen in Chapter 1, and will in §5-5, one of the interests in double β-decay, nuclear decay through the emission of two electrons or positrons, is to see whether neutrinos can be Majorana particles. So far all the evidence seems to suggest that they are strictly Dirac particles.

The conservation of lepton numbers applies separately to each one of the three groups of leptons, e and ν_e, μ and ν_μ, and τ and ν_τ. That is, the number of leptons in the electron family L_e, the number of leptons in the muon family L_μ, and the number of leptons in the tau family L_τ are conserved separately in a reaction. For example, muons decay with a mean life of 2.2 μs through the reaction,

$$\mu^- \rightarrow e^- + \overline{\nu}_e + \nu_\mu \tag{2-5}$$

Since only a muon appears on the left hand side of this reaction, we have $L_e = 0$ (as well as $L_\tau = 0$) and $L_\mu = 1$. On the right hand side, the muon number is conserved by the appearance of ν_μ; the electron number must be zero to conserve L_e, and this requires the presence of both e^- and $\overline{\nu}_e$. The fact that the reaction produces two neutrinos, a muon neutrino and an electron antineutrino, rather than, for example, ν_e and $\overline{\nu}_e$ or two γ-rays, is good evidence for the conservation of L_e and L_μ separately.

For most interests in nuclear physics we are concerned primarily with leptons in the electron family, with only occasional reference to the other leptons. The context will usually make it clear with which of the three lepton families we are dealing, without having to state this explicitly.

Baryon number conservation. The numbers of different types of quarks, u, d, s, etc. are also conserved in strong interaction processes; one type of quark cannot be changed into another except through weak interactions. This is to say that flavor is a good quantum number if the weak force can be ignored. Unless we are dealing with the quark contents of hadrons, it is more convenient to examine instead the baryon number, which is known to be conserved under the influence of weak interactions as well. The only exception is the possible decay of protons through reactions such as

$$p \rightarrow e^+ + \pi^0 \tag{2-6}$$

allowed under theories for the grand unification of all forces. At present, the observed limit on the lifetime of a proton is longer than 10^{30} years: we shall therefore not be concerned with this possibility and we shall take the baryon number to be conserved in all the reactions in which we are interested.

There is no conservation law for the number of mesons: they can decay into other mesons, baryon and antibaryon pairs, or lepton and antilepton pairs. The lightest members of the meson family are the pions with rest mass around 140 MeV/c^2. It is stable on the time scale of strong interactions, since it cannot decay into another hadron. However, through weak interactions, charged pions decay predominantly to muons,

$$\pi^+ \rightarrow \mu^+ + \nu_u \qquad\qquad \pi^- \rightarrow \mu^- + \overline{\nu}_u \tag{2-7}$$

with a mean life of 2.6×10^{-8} s, and a neutral pion decays 99% of the time to two γ-rays,

$$\pi^0 \rightarrow \gamma + \gamma \tag{2-8}$$

with a mean life of 8.4×10^{-17} s. Both lifetimes are much longer, by something around six to fourteen orders of magnitude, than the typical time scale of strong interactions. Note also that in all three modes of decay the lepton numbers are conserved and, as we shall see later, the total number of quarks is also conserved.

§2-2 QUARKS, THE BASIC BUILDING BLOCK OF HADRONS

Quark masses. Among the six quarks listed in Table 2-2 the least massive members are the u- and d-quarks. These two quarks are believed to have essentially the same mass, around 0.39 GeV/c^2. The lightest baryons, nucleons and Δ-particles, and the lightest mesons, pions, must therefore be made exclusively of these two quarks. The s-quark is more massive, around 0.51 GeV/c^2. The unique feature of the s-quark is that it carries a quantum number called *strangeness* and is therefore a necessary constituent of particles with nonzero strangeness such the K-mesons, or kaons, and the baryon Λ. The c-quark is even more massive. It was first found by the discovery of J/ψ-meson in 1974 as a narrow resonance in the annihilation of a positron e^+ with an electron e^- at 3.1 GeV center-of-mass energy. Since a meson is made of a quark and an antiquark, a new quark, heavier than the three known at the time, must be postulated in order to understand this new meson. This is the c- or charm quark, having a rest mass around 1.65 GeV/c^2, far greater than those of u, d, and s quarks. The existence of the c-quark was subsequently confirmed by other experiments, including the discovery of excited states of J/ψ. The birth of the c-quark prompted the search for even heavier quarks. In this way, the presence of b-quark was found in the Υ-meson at 10 GeV. This puts the rest mass of a b-quark at around 5 GeV/c^2. The t-quark, postulated to be even heavier, has so far escaped experimental confirmation. Perhaps a new generation of higher energy accelerators are needed to create one. At the same time one may also wonder whether a fourth generation of quarks may exist beyond the three groups of quarks outlined so far.

Table 2-2 Quantum numbers of quarks.

Flavor	$\mathcal{A}$	t	t_0	$\mathcal{S}$	$\mathcal{C}$	$\mathcal{B}$	$\mathcal{T}$	$Q(e)$	M(GeV)
u (up)	$\frac{1}{3}$	$\frac{1}{2}$	$\frac{1}{2}$	0	0	0	0	$+\frac{2}{3}$	$\approx$ 0.39
d (down)	$\frac{1}{3}$	$\frac{1}{2}$	$-\frac{1}{2}$	0	0	0	0	$-\frac{1}{3}$	$\approx$ 0.39
s (strange)	$\frac{1}{3}$	0	0	-1	0	0	0	$-\frac{1}{3}$	$\approx$ 0.51
c (charm)	$\frac{1}{3}$	0	0	0	1	0	0	$+\frac{2}{3}$	$\approx$ 1.6
b (beauty)	$\frac{1}{3}$	0	0	0	0	-1	0	$-\frac{1}{3}$	$\approx$ 5.4
t (top)	$\frac{1}{3}$	0	0	0	0	0	1	$+\frac{2}{3}$	$\approx$ 20

$\mathcal{A}$: baryon number t: isospin $\mathcal{S}$: strangeness
$\mathcal{C}$: charm $\mathcal{B}$: beauty $\mathcal{T}$: top

Associated with each quark is an antiquark. All hadrons are made of these six quarks and their antiquarks. The properties of quarks are deduced from measurements made on mesons and baryons, since observations on individual quarks cannot be carried out. The assignment of masses, magnetic moments, and other properties to the quarks is therefore inferred from what we know of the properties of mesons and baryons. Our ability to make such deductions depends on our understanding of QCD, which is still incomplete at the moment, and especially inadequate at low energies where the majority of the experimental observations are made. As a result, one must resort to models in order to make these inferences. For example, in order to obtain the masses of quarks from the known hadron masses, we also need to know the strength of the interaction between quarks in the hadron. Since this is poorly known, the quark masses listed in Table 2-2 are only the *constituent* masses, *i.e.,* the masses as they appear in the hadrons which may or may not be closely related to their true masses.

Fermions and bosons. Hadrons are subdivided into two classes, baryons and mesons. Besides nucleons, we have Δ-, Λ-, and a large number of heavier particles in the baryon family. Among mesons, we have already encountered pions, kaons, J/ψ and Υ, and there are many others.

Baryons are distinguished by the fact that they are fermions, particles that obey Fermi-Dirac statistics. Because of this property, two identical baryons cannot occupy the same quantum mechanical state. The fact that baryons are fermions implies that quarks must also be fermions, since one cannot construct fermions except from odd numbers of fermions. Furthermore, if we accept that a quark cannot exist as a free particle, the lightest fermion in the hadron family must be made of three quarks.

As fermions, baryons must have half-integer intrinsic spins. For example, the intrinsic spin of a nucleon is $\frac{1}{2}$ and the intrinsic spin of a Δ-particle is $\frac{3}{2}$. This means that quarks must also have half-integer intrinsic spins. Just as electrons in atoms tend to have the lowest energy if their orbital angular momenta ℓ are at their minimum values allowed by selection rules, quarks in the lightest hadrons are also in their lowest possible orbital angular momentum states. The exact energy level spectrum of three quarks in a baryon depends on the interaction between the quarks, a subject dealt with by the quark model to be introduced later.

Among the baryons, we are mostly concerned in nuclear physics with the lightest pair, the neutron and the proton. From charge considerations alone, we can deduce that a proton, which carries one unit of positive charge, must be made of two u-quarks, each of which has a charge $\frac{2}{3}e$, and one d-quark, which has a charge $-\frac{1}{3}e$. The quark wave function of a proton may be represented in the form

$$|p\rangle = |uud\rangle \tag{2-9}$$

Similarly, the quark wave function of a neutron must be

$$|n\rangle = |udd\rangle \tag{2-10}$$

so that the total charge of a neutron is $(\frac{2}{3} - \frac{1}{3} - \frac{1}{3}) = 0$ in units of e. Nuclear physics is usually not concerned with any of the heavier baryons, except perhaps for Δ- and Λ-particles. This comes from the fact that we are normally dealing with very low energy phenomena, a few GeV per nucleon or less: there is seldom the energy to excite nucleons to become heavier baryons.

Bosons, particles obeying Bose-Einstein statistics, may be made from any even number of fermions. This means that mesons may be constructed of an even number of quarks. Since on the one hand bosons can be created or annihilated under suitable conditions, and on the other hand the number of quarks must be conserved under strong interactions, a meson must be made of an equal number of quarks (q) and antiquarks ($\overline{q}$). The simplest mesons is then made of a quark-antiquark pair ($q\overline{q}$). For example, pions, or π-mesons, the lightest members among the mesons, are made of a quark, either u or d, and an antiquark, either $\overline{u}$ and $\overline{d}$.

Quark charge. Since many hadrons are charged particles, quarks must also carry electric charge. In nature all observed charges are in multiples of the electron charge $e = 1.6021773 \times 10^{-19}$ C, with the charge on an electron being $-e$, and on a proton $+e$. The most convenient assignment of charge to the quarks is for u, c, and t quarks to have $+\frac{2}{3}e$ and d, s, and b quarks to have $-\frac{1}{3}e$. This assignment of multiples of $\frac{1}{3}e$ to quarks seems, on the surface, to violate the notion that e is a fundamental or indivisible unit of charge. However, there is no reason to assume that $\frac{1}{3}e$ cannot be the more fundamental unit of charge instead of e. Furthermore, this is not a problem if free quarks do not exist and all the observed charges are then integer multiples of e.

§2-3 ISOSPIN

The nucleon. A proton and a neutron may be considered as two different aspects of a single particle, the nucleon. Both of them have spin $\frac{1}{2}$ and their masses, 939.566 MeV/c^2 for a neutron and 938.272 MeV/c^2 for a proton, differ only by about 0.1%. The main distinction between these two particles is in their electromagnetic properties: namely, charge and magnetic dipole moment (see §2-8). If we are dealing only with strong interactions, such differences do not appear; that is, in the absence of an electromagnetic interaction, a proton cannot be distinguished from a neutron. This is similar to the case of particles with different values of m_s, projections of the intrinsic spin $\boldsymbol{s}$ on the quantization axis. In order to illustrate a similarity with the neutron-proton system, let us consider a spin-$\frac{1}{2}$ particle. In the absence of a magnetic field $\boldsymbol{B}$, particles with the two possible values of m_s, $\pm\frac{1}{2}$, are degenerate in energy and, consequently, are indistinguishable from each other. On the other hand, once a finite magnetic field is introduced, the degeneracy is removed and particles are observed to have different energies, depending on whether their intrinsic spins $\boldsymbol{s}$ are aligned parallel or antiparallel to $\boldsymbol{B}$. The difference between a proton and a neutron is analogous to the difference between particles with $m_s = \pm\frac{1}{2}$ if we substitute the Coulomb field with a magnetic field.

If protons and neutrons are considered as identical particles, we need a new label to distinguish between them. For this purpose, the concept of ***isospin*** is introduced. Since there are only two possible states for a nucleon, the proton state and the neutron state, we can assign isospin $\boldsymbol{t} = \frac{1}{2}$ to a nucleon, using the analogy that a spin-$\frac{1}{2}$ system can have two different substates. The two nucleons are distinguished by $t_0 = \pm\frac{1}{2}$, the expectation value of the third component of isospin operator $\boldsymbol{t}$. It is a matter of convention whether we consider the $|t{=}\frac{1}{2}, t_0{=}+\frac{1}{2}\rangle$ state to be a proton state and the $|t{=}\frac{1}{2}, t_0{=}-\frac{1}{2}\rangle$ state to be a neutron state, or the other way around. Both conventions are in use and we shall adopt the more popular one with

$$|p\rangle \equiv \left|t{=}\tfrac{1}{2}, t_0{=}+\tfrac{1}{2}\right\rangle \qquad |n\rangle \equiv \left|t{=}\tfrac{1}{2}, t_0{=}-\tfrac{1}{2}\right\rangle \tag{2-11}$$

where $|p\rangle$ and $|n\rangle$ represent the wave functions of a proton and a neutron, respectively. For a nucleus consisting of several nucleons, the total isospin is given by the vector sum of the isospin of each individual nucleon,

$$\boldsymbol{T} = \sum_{i=1}^{A} \boldsymbol{t}(i) \tag{2-12}$$

where A is the number of nucleons. This is identical to the rule for the addition of angular momentum.

In the absence of electromagnetic interactions, we expect isospin to be a constant of motion; that is, the eigenstates of the Hamiltonian are also the eigenstates of the square of the isospin operator $\boldsymbol{t}^2$ and its third component $\boldsymbol{t}_0$. As a result, each eigenstate may also be labelled by t (or T) and t_0 (T_0), where $t(t+1)$ (or $T(T+1)$) is the expectation value of $\boldsymbol{t}^2$ (or $\boldsymbol{T}^2$) and t_0 (T_0) is the expectation value of $\boldsymbol{t}_0$ ($\boldsymbol{T}_0$) for the eigenstate. In dealing with nuclei, the main source of isospin symmetry breaking comes from Coulomb interactions between protons. A less severe but nevertheless noticeable source is the difference between the masses of the neutral and charged mesons exchanged between two nucleons (see §3-6). The possibility of more fundamental isospin breaking terms in the nuclear force, for instance, due to a possible small difference between the masses of u- and d-quarks, is not yet well established but has not been completely ruled out either.

From a purely mathematical point of view, spin and isospin have very similar structures. Let us concentrate on isospin-$\frac{1}{2}$ systems for the moment and study them by analogy with spin-$\frac{1}{2}$ systems. A particle with $\boldsymbol{s} = \frac{1}{2}$ and projection along the quantization axis $m = +\frac{1}{2}$ may be represented by a two-component column matrix in the following form

$$\left|s{=}\tfrac{1}{2}, m{=}+\tfrac{1}{2}\right\rangle = \begin{pmatrix} 1 \\ 0 \end{pmatrix} \tag{2-13}$$

Similarly, a spin-$\frac{1}{2}$ particle with $m = -\frac{1}{2}$ may be represented in the form

$$\left|s{=}\tfrac{1}{2}, m{=}-\tfrac{1}{2}\right\rangle = \begin{pmatrix} 0 \\ 1 \end{pmatrix} \tag{2-14}$$

The isospin wave functions of nucleons may be written in an analogous way

$$|p\rangle = |t{=}\tfrac{1}{2}, t_0{=}+\tfrac{1}{2}\rangle = \begin{pmatrix} 1 \\ 0 \end{pmatrix}_t \tag{2-15}$$

$$|n\rangle = |t{=}\tfrac{1}{2}, t_0{=}-\tfrac{1}{2}\rangle = \begin{pmatrix} 0 \\ 1 \end{pmatrix}_t \tag{2-16}$$

where the subscript t, which we shall omit in the future unless required for reasons of clarity, identifies that the column matrices are for isospin. Using the convention that a proton has $t_0 = +\frac{1}{2}$ and a neutron has $t_0 = -\frac{1}{2}$, we can relate the *charge number*, *i.e.*, Q in units of e, on a nucleon to t_0,

$$Q = t_0 + \tfrac{1}{2}$$

When we extend the concept of isospin to antiparticles and to systems of several nucleons, the relation between Q and t_0 depends also on A, the number of baryons in the system,

$$Q = t_0 + \tfrac{1}{2}A \tag{2-17}$$

A more general relation involving strangeness and other quantum numbers is given later in (2-32).

Isospin operators for $t = \frac{1}{2}$ systems can be constructed from Pauli spin matrices $\boldsymbol{\sigma}$ in the same way as angular momentum operators for a spin-$\frac{1}{2}$ system. For example, we can write

$$\boldsymbol{\tau}_1 = \begin{pmatrix} 0 & 1 \\ 1 & 0 \end{pmatrix} \qquad \boldsymbol{\tau}_2 = \begin{pmatrix} 0 & -i \\ i & 0 \end{pmatrix} \qquad \boldsymbol{\tau}_3 = \begin{pmatrix} 1 & 0 \\ 0 & -1 \end{pmatrix} \tag{2-18}$$

for the x-, y-, and z-components of the isospin operator $\boldsymbol{\tau}$. The matrices obey the relation

$$\boldsymbol{\tau}_i \boldsymbol{\tau}_j = \delta_{ij} \boldsymbol{I} + i\epsilon_{ijk} \boldsymbol{\tau}_k \tag{2-19}$$

where $\boldsymbol{I}$ is the 2×2 unit matrix and ϵ_{ijk} is the three-dimensional Levi-Civita symbol with $\epsilon_{ijk} = 1$ if the order of i, j, and k is an even permutation of 1, 2, and 3; -1 if the order is an odd permutation; and zero if any two or more of the three indices are the same.

For a nucleon it is easily seen that the wave functions given by (2-15,16) are the eigenfunctions of the $\boldsymbol{\tau}_3$ operator, or $\boldsymbol{\tau}_0$ operator in spherical coordinates,

$$\boldsymbol{\tau}_0 \begin{pmatrix} 1 \\ 0 \end{pmatrix} = \boldsymbol{\tau}_3 \begin{pmatrix} 1 \\ 0 \end{pmatrix} = \begin{pmatrix} 1 & 0 \\ 0 & -1 \end{pmatrix} \begin{pmatrix} 1 \\ 0 \end{pmatrix} = + \begin{pmatrix} 1 \\ 0 \end{pmatrix}$$

$$\boldsymbol{\tau}_0 \begin{pmatrix} 0 \\ 1 \end{pmatrix} = \boldsymbol{\tau}_3 \begin{pmatrix} 0 \\ 1 \end{pmatrix} = \begin{pmatrix} 1 & 0 \\ 0 & -1 \end{pmatrix} \begin{pmatrix} 0 \\ 1 \end{pmatrix} = - \begin{pmatrix} 0 \\ 1 \end{pmatrix}$$

The value of the third component of isospin, t_0, is equal to half of the expectation value of $\boldsymbol{\tau}_0$, the same relation as that between m_s, the projection of the intrinsic spin of a nucleon (and other spin-$\frac{1}{2}$ particles) on the quantization axis, and the expectation value of the third component of the Pauli spin operator $\boldsymbol{\sigma}_0$. By the same token, the expectation values of $\boldsymbol{\tau}^2$ is 3, four times the value of $t(t+1)$ for a nucleon.

From the form of $\boldsymbol{\tau}$ given in (2-18), we can construct the isospin raising (τ_+) and lowering (τ_-) operators which transform, respectively, a neutron to a proton and a proton to a neutron,

$$\tau_+ = \frac{1}{2}(\tau_1 + i\tau_2) = \begin{pmatrix} 0 & 1 \\ 0 & 0 \end{pmatrix} \qquad \tau_- = \frac{1}{2}(\tau_1 - i\tau_2) = \begin{pmatrix} 0 & 0 \\ 1 & 0 \end{pmatrix} \tag{2-20}$$

In the same way as for angular momentum raising and lowering operators, $\tau_\pm$ changes t_0 without changing the isospin t itself

$$\tau_\pm \big| t, t_0 \big\rangle = \sqrt{t(t+1) - t_0(t_0 \pm 1)}\ \big| t, t_0 \pm 1 \big\rangle \tag{2-21}$$

The definition used here for τ_+ and τ_- is the more general form used in textbooks and differs somewhat from the convention for spherical tensor operators given in §B-2.

For nuclei made of several nucleons, the isospin operators may be constructed out of the single-nucleon operators $\boldsymbol{I}$, τ_1, τ_2, and τ_0 $(= \tau_3)$. For example,

$$\boldsymbol{T}_0 = \frac{1}{2}\sum_{i=1}^{A} \tau_0(i) \tag{2-22}$$

where $\tau_0(i)$ acts only on the isospin wave function of the i-th nucleon.

The usefulness of isospin is not restricted to the economy gained in treating formally a proton and a neutron as two different states of the same particle. Since isospin is a constant of motion under strong interactions, it is a fundamental symmetry, essentially on the same footing as flavor, parity, etc. Isospin is useful in classifying hadrons in general. For example, as we shall see in §2-5, pions come in three different charge states, π^+ with rest mass energy 139 MeV, π^0 with rest mass energy 135 MeV, and π^- with the same mass as π^+. They may be treated as the three projections $t_0 = +1$, 0, and -1 of an isospin $t = 1$ system. Since pions are not baryons, the baryon number is $A = 0$; we see that the relation between charge number Q and the third component of the isospin given in (2-17) holds here as well. In §2-7 we shall see the case of a quartet of baryons, the Δ-particles which appear in four different charge states, Δ^{++}, Δ^+, Δ^0, and Δ^-, with charge number $Q = 2$, 1, 0, and -1. It is therefore a $t = \frac{3}{2}$ system of baryons. We shall return later for a discussion of the isospin wave function of hadrons and nuclei.

§2-4 ISOSPIN OF ANTIPARTICLES

Particles and antiparticles. An antiparticle may characterized by the fact that it can annihilate the particle with which it is associated. Energy and momentum conservation are maintained in the process; for example, by the emission of two γ-rays or the creation of a different particle-antiparticle pair. Since the final state of an annihilation process is an electrically neutral state, a particle and its antiparticle must have opposite charges in order to conserve electric charge. For example, an electron has charge $-e$, whereas its antiparticle, the positron, has charge $+e$. Similarly, the conservation of other scalar quantum numbers, such as lepton numbers, or baryon numbers, requires that these labels for particles and antiparticles are equal in magnitude but opposite in sign, as we have seen in earlier examples. For vector quantities, such as intrinsic spin and isospin, the rules of addition require that magnitudes be the same for a particle and its antiparticle so that they can be coupled together to form scalars.

Let us take the case of proton-antiproton annihilation at rest with the emission of two photons as an example:

$$p + \overline{p} \rightarrow \gamma + \gamma$$

Since a photon is an isospin zero, or *isoscalar* particle, the total isospin on the right hand side of the reaction is zero. Conservation of isospin requires that the proton and the antiproton are coupled to a $T = 0$ state on the left hand side. Since $\boldsymbol{t} = \frac{1}{2}$ for a proton, the antiproton $\overline{p}$ must also have $\boldsymbol{t} = \frac{1}{2}$. On the other hand, the third component of the isospin for a proton is $t_0 = +\frac{1}{2}$ by the convention we have adopted. For an isoscalar system, the sum of the third components of the isospin of all the quantities involved must also be zero. From this, we conclude that $t_0 = -\frac{1}{2}$ for an antiproton, the opposite in sign from that for its particle.

We see that the relation between charge number Q and the third component of the isospin given by (2-17) applies to antiparticles as well. Since a proton has charge $+e$, the charge carried by an antiproton must be $-e$. For an antiproton, the baryon number is $A = -1$ and we obtain $Q = -1$ from (2-17) using the result of $t_0 = -\frac{1}{2}$ deduced above.

The transformation of wave functions under charge conjugation (particle-antiparticle transformation) is slightly more complicated than simply changing the sign of the third component of isospin. We see from §A-2 that it is also necessary to change the overall sign of the antineutron wave function (but not that of the antiproton) in order to maintain a proper tensorial relation between particle and antiparticle wave functions. It is also well known that the intrinsic parity of an antiparticle is opposite to that of its particle (see §A-1).

§2-5 ISOSPIN OF QUARKS

One of the consequences of treating a proton and a neutron as two different isospin states of a nucleon is that we can change a proton into a neutron, and vice versa, by the application of isospin lowering and raising operators given in (2-21) to the appropriate wave functions,

$$\tau_+|n\rangle = \begin{pmatrix} 0 & 1 \\ 0 & 0 \end{pmatrix}\begin{pmatrix} 0 \\ 1 \end{pmatrix} = \begin{pmatrix} 1 \\ 0 \end{pmatrix} = |p\rangle$$

$$\tau_-|p\rangle = \begin{pmatrix} 0 & 0 \\ 1 & 0 \end{pmatrix}\begin{pmatrix} 1 \\ 0 \end{pmatrix} = \begin{pmatrix} 0 \\ 1 \end{pmatrix} = |n\rangle \tag{2-23}$$

In terms of quarks, we have already seen in (2-9,10) that

$$|p\rangle = |uud\rangle \qquad\qquad |n\rangle = |udd\rangle$$

Hence, (2-23) imply that

$$\tau_+|udd\rangle = |uud\rangle \qquad\qquad \tau_-|udd\rangle = |uud\rangle \tag{2-24}$$

Since a proton and a neutron are considered here to be identical to each other except for the third component of their isospin, the other parts of the wave functions are not changed by the isospin raising and lowering operations. In terms of quarks, the only difference between a proton and a neutron is the replacement of one of the two u-quarks by a d-quark. The equalities expressed by (2-24) imply that the isospin raising operator acts on the quarks, transforming a d-quark to a u-quark and the other way around for the isospin lowering operator. Since no other quarks are involved here, we conclude that d- and u-quarks also form an isospin doublet, analogous to the proton-neutron pair. Furthermore, the third component of isospin is a scalar quantity; the sum of t_0 of two u-quarks and one d-quark in a proton must be $+\frac{1}{2}$ and that of one u-quark and two d-quarks in a neutron must be $-\frac{1}{2}$. To satisfy both requirements we must assign $t_0 = +\frac{1}{2}$ to a u-quark and $t_0 = -\frac{1}{2}$ to a d-quark. The relation between charge number Q and t_0 is still given by (2-17). Note that a quark has baryon number $A = \frac{1}{3}$ from the fact that it takes three quarks to make a baryon.

More formally, we can write the $\tau_\pm$ operator for nucleon wave functions as a sum of isospin raising or lowering operators acting on each one of the three quarks in the nucleon,

$$\tau_\pm(\text{nucleon}) \longrightarrow \sum_{i=1}^{3} \tau_\pm(q_i)$$

where $\tau_\pm(q_i)$ acts on the isospin of the i-th quark only. Ignoring for the moment any antisymmetrization requirement between the three quarks in a nucleon, we can write the first equation of (2-24) in the following form:

$$\tau_+|n\rangle = \{\tau_+(q_1) + \tau_+(q_2) + \tau_+(q_3)\}\ |u(1)d(2)d(3)\rangle$$

where we have assumed that the first quark in the neutron is a u-quark, and the remaining two d-quarks. Since $\boldsymbol{\tau}_+|\,u\,\rangle = 0$ (a u-quark has $t_0 = +\frac{1}{2}$), the first term vanishes. The second and third terms give the results

$$\boldsymbol{\tau}_+(q_2)\big|u(1)d(2)d(3)\big\rangle = \big|u(1)u(2)d(3)\big\rangle$$

$$\boldsymbol{\tau}_+(q_3)\big|u(1)d(2)d(3)\big\rangle = \big|u(1)d(2)u(3)\big\rangle$$

Upon antisymmetrization these two terms produce identical results which we shall represent generically as $|\,uud\,\rangle$.

Quark wave functions of pions. We can check the isospin assignment to the u- and d-quarks by examining the structure of mesons formed of these two quarks and their antiquarks. It is simplest to start from π^- which, as mentioned earlier, has $t = 1$ and $t_0 = -1$. Since we cannot use any quarks other than u, d, $\overline{u}$ and $\overline{d}$, the only way to form a $t_0 = -1$ system is to adopt the $\overline{u}d$ combination. We can easily deduce that this pair of quarks must be a $\boldsymbol{t} = 1$ system by elimination. Two isospin-$\frac{1}{2}$ particles can only be coupled to total isospin 0 and 1. The $\overline{u}d$ system cannot be $\boldsymbol{t} = 0$ as it has $t_0 = -1$. As a result, we can make the identification

$$\big|\pi^-\big\rangle = \big|\overline{u}d\big\rangle \tag{2-25}$$

since there is no other way to form a $t = 1$, $t_0 = -1$ state with u, d, $\overline{u}$ and $\overline{d}$.

In general, it is possible to find several different linearly independent components corresponding to the same t and t_0. In such cases, linear combinations of all the components must be made. The appropriate linear combination for a given situation is guided by angular momentum coupling rules for spin, isospin, and the requirement that the wave function is antisymmetric among the quarks and is an eigenstate of the Hamiltonian. For our interest in this section, we shall only be concerned with isospin coupling.

From the wave function of π^-, one can construct the wave function of π^0 using the isospin raising operator,

$$\frac{1}{N}\big|\pi^0\big\rangle = \boldsymbol{\tau}_+\big|\pi^-\big\rangle = \sum_{i=1,2} \boldsymbol{\tau}_+(q_i)\big|\overline{u}d\big\rangle \tag{2-26}$$

where N is a normalization factor to be determined later. The operator $\boldsymbol{\tau}_+$ acts on the wave function of each quark. We have already seen that

$$\boldsymbol{\tau}_+\big|d\big\rangle = |\,u\,\rangle \tag{2-27}$$

However, for the antiquarks,

$$\boldsymbol{\tau}_+\big|\overline{u}\big\rangle = -\big|\overline{d}\,\big\rangle \tag{2-28}$$

where the additional negative sign comes from the symmetry requirement under charge conjugation as explained in §A-2.

The normalization factor N in (2-26) may be determined using (2-21). Since $|\pi^-\rangle$ is a $t = 1$, $t_0 = -1$ system,

$$\tau_+|1,-1\rangle = \sqrt{2}|1,0\rangle$$

we obtain a value $N = \sqrt{2}$. The final result for the wave function of π^0 is then

$$|\pi^0\rangle = \frac{1}{\sqrt{2}}\Big\{|u\overline{u}\rangle - |d\overline{d}\rangle\Big\} \tag{2-29}$$

The same result can also be obtained by coupling the two isospin-$\frac{1}{2}$ particles using Clebsch-Gordan coefficients described in §B-3. Since the values of both Clebsch-Gordan coefficients for coupling two isospin-$\frac{1}{2}$ particles to total isospin 1 are

$$\langle\tfrac{1}{2},\pm\tfrac{1}{2};\tfrac{1}{2},\mp\tfrac{1}{2}|10\rangle = \frac{1}{\sqrt{2}}$$

we obtain the same result as given in (2-29) after inserting an "extra" minus sign arising from the transformation from $|d\rangle$ to $|\overline{d}\rangle$ under charge conjugation.

The wave function for π^+ in terms of quarks is

$$|\pi^+\rangle = -|u\overline{d}\rangle \tag{2-30}$$

This result may be arrived at either by applying an isospin raising operator on the quark wave function of π^0 obtained above or by constructing a $(t, t_0) = (1, +1)$ system from the quarks and antiquarks given, in the same manner as we have just done for the π^- system. Again, the overall minus sign is due to the charge conjugation between d and $\overline{d}$.

One question still remains concerning the $t_0 = 0$ wave function for a quark-antiquark pair. There are two different ways, $u\overline{u}$ and $d\overline{d}$, to form a $t_0 = 0$ state from the two quarks and two antiquarks given. Besides the linear combination of these two $t_0 = 0$ components given in (2-29), we can also take the linear combination

$$|\eta_0\rangle = \frac{1}{\sqrt{2}}\Big\{|u\overline{u}\rangle + |d\overline{d}\rangle\Big\} \tag{2-31}$$

It is orthogonal to $|\pi^0\rangle$ given in (2-29) and must therefore describe a meson other than π^0. Each component, $u\overline{u}$ and $d\overline{d}$, is a system with $t_0 = 0$ but a mixture of $(t, t_0) = (1, 0)$ and $(t, t_0) = (0, 0)$. A particular linear combination was taken in (2-29) so as to have the correct isospin of $t = 1$ for the π^0-meson. The linear combination given in (2-31) is a different one and must correspond to an isospin zero system as a result, a fact that can also be seen from the explicit values of the Clebsch-Gordan coefficients required to construct a $t = 0$ system. Such an isospin-singlet meson may be identified with the η-meson which has a rest mass ~ 550 MeV/c^2.

The four particles, π^+, π^0, π^-, and η, exhaust all the observed mesons in the form of a quark-antiquark pair that can be constructed out of u, d, $\overline{u}$, and $\overline{d}$ in their lowest possible energy states. To obtain other mesons, we must either introduce excitations in the quark-antiquark system or invoke the s- and other more massive quarks. We shall return to this point later.

Other quarks. Let us examine briefly the isospin of the other quarks, c, s, t, b and their antiparticles. It is perhaps tempting to assume that each one of the remaining two pairs of quarks forms also an isospin doublet. This, however, is not the case. As can be seen from Table 2-2, these four quarks are isoscalar particles. Two questions are raised here: How are the assignments of $t = 0$ made to these quarks? What is the relation between their values of Q and t_0? Since quarks are not particles observed in isolation, assignment of isospin (as well as other quantum numbers) must be done through the hadrons they make up. This is what we have done for the u- and d-quarks and we shall see how it is carried out for the s-quarks as an example.

After u- and d-quarks, the next particle in order of increasing mass is the s-quark. They are found in hadrons with nonzero strangeness $\mathcal{S}$. For our purposes here we can regard $\mathcal{S}$ as a label to identify the number of strange antiquarks in the hadron. We shall return to this and other quantum numbers associated with the heavy quarks in the next section.

The lightest strange mesons are the kaons or K-meson. They come as two isospin doublets ($t = \frac{1}{2}$ systems), one doublet consisting of $K^+(u\overline{s})$ and $K^0(d\overline{s})$, and the other $K^-(\overline{u}s)$ and $\overline{K}^0(\overline{d}s)$. Since u-and d-quarks have isospin $t = \frac{1}{2}$, the s-quark must have integer isospin 0 or 1 in order to form kaons with $t = \frac{1}{2}$. The assignment of $t = 1$ may be ruled out on the grounds that, if this were the case, we should be able to form $t = \frac{3}{2}$-mesons; for example, made of a s-quark and an antiquark, either $\overline{u}$ or $\overline{d}$. The fact that such mesons have not been found implies that the isospin of the s-quark is zero. The assignment of isospin to the other quarks may be carried out in a similar way and we shall not go into them here.

With the assignment of $t = 0$ to the heavy quarks, we need now to modify the relation between the charge number Q of a particle and the third component of the isospin t_0. The relation given by (2-17) was derived for u- and d-quarks and must be changed now as the other quarks have different relations between Q and t_0. The more general form of (2-17) is given by

$$Q = t_0 + \tfrac{1}{2}(A + \mathcal{S} + \mathcal{C} + \mathcal{B} + \mathcal{T}) \qquad \text{(2-32)}$$

where assignments of baryon number A, strangeness $\mathcal{S}$, charm quantum number $\mathcal{C}$, beauty quantum number $\mathcal{B}$, top quantum number $\mathcal{T}$ for each of the six flavors of quarks are given in Table 2-2.

§2-6 STRANGENESS AND OTHER QUANTUM NUMBERS

In strong interaction processes, the total number of each type of quark, u, d, s, c, b, and t, is conserved. However, through weak interactions, quarks can be transformed from one flavor to another, such as the example shown in Fig. 2-1. In terms of observed particles, the flavor degree of freedom in quarks shows its presence by separating hadrons into different groups with transitions between groups allowed only through weak interactions. As a result, transition rates between members of the same group, characterized by the fact that they have the same quark content, are fast and typical of strong interaction processes: transitions between members of different groups involve the transformation of one type of quark to another and are much slower, with lifetimes more typical of weak interaction processes.

Fig. 2-1 Transformation of a d-quark into a u-quark through weak interaction. The reaction emits a virtual W^--boson which subsequently decays into a pair of leptons, an electron e^- and an electron antineutrino $\overline{\nu}_e$.

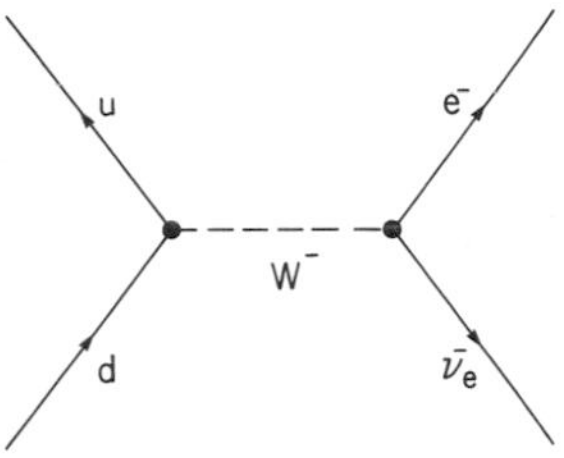

Each group of hadrons is characterized by a definite number of quarks of a particular flavor. For example, the K^+-meson, with mass 494 MeV/c^2, is made of the quark-antiquark pair $u\overline{s}$. The dominant mode of decay, 63.5% of the time, is into leptons, $\mu^+ + \overline{\nu}_\mu$. A less prominent mode, 21.2% of the time, is into a pair of pions $\pi^+ + \pi^0$. In either decay mode, the total number of quarks is conserved. However, there is no strange quark among the end products of the decay. One way to eliminate the $\overline{s}$ without changing the net number of quarks involved is to let it decay to a $\overline{u}$, which then annihilates with the u-quark in K^+. Alternatively, the $\overline{s}$ may β-decay to a $\overline{d}$ instead and form a part of the pions in the end product. The mean life of the K^+-meson, 1.2×10^{-8} s, is typical for weak decays.

Strangeness, charm and beauty. Among baryons, the Λ is a particle with quark structure (uds) and mass 1116 MeV/c^2 . It is produced in reactions such as

$$\pi^- + p \rightarrow \Lambda + K^0$$

where the meson K^0, with quark structure $d\overline{s}$, is the isospin partner of K^+. On the left hand side of the reaction, there is no strange quark since both π^- and p are made exclusively of u's and d's. On the right hand side, we see that the production of an s-quark in the Λ-particle is accompanied by an $\overline{s}$ in the K^0 meson. In terms of the observed hadrons, we find that the production of a Λ and other hadrons containing an s-quark is always accompanied by a hadron containing an $\overline{s}$. This type of association may be accounted for by assigning a *strangeness* quantum number, $\mathcal{S}$, to count the number of s-quarks. In strong interaction processes, we can say that strangeness is a conserved quantity to indicate the fact that the numbers of s and $\overline{s}$ produced must be the same. For historical reasons, an s-quark is assigned $\mathcal{S} = -1$ and $\overline{s}$ is assigned $\mathcal{S} = +1$ (and $\mathcal{S} = 0$ for the other quarks).

As with the strangeness quantum number, we can assign a *charm* quantum number $\mathcal{C}$ to account for the number of c-quarks with $\mathcal{C} = 1$ for a c-quark and $\mathcal{C} = -1$ for a $\bar{c}$-quark (and $\mathcal{C} = 0$ for all other quarks). In order to account for the number of b-quarks, a *beauty* quantum number $\mathcal{B}$ is used with $\mathcal{B} = -1$ for a b-quark and $\mathcal{B} = +1$ for a $\bar{b}$-quark (and $\mathcal{B} = 0$ for all other quarks). Thus, for example, a hadron with n c-quarks has $\mathcal{C} = n$ and hadrons with m b-quarks has $\mathcal{B} = -m$. Whether a quark or an antiquark of a given flavor should take on the positive sign for the quantum number representing that flavor is somewhat arbitrary. The convention described here is the commonly used one and satisfies the relation between Q and t_0 given in (2-32).

Table 2-3 Lifetimes of ϕ, J/ψ and Υ mesons.

Meson	Rest mass energy MeV	Width Γ (MeV)	Mean life $\overline{T}$ (s)	quark content
ϕ	1019.41 ± 0.01	4.41 ± 0.05	1.49×10^{-22}	$s\bar{s}$
J/ψ	3096.93 ± 0.09	0.068 ± 0.010	0.97×10^{-20}	$c\bar{c}$
Υ	9460.32 ± 0.22	0.052 ± 0.003	1.27×10^{-20}	$b\bar{b}$

Earlier, we have seen the long lifetime of kaons as an example of strangeness conservation. Similarly long lifetimes of the order 10^{-13} s are observed for the analogous situation of charm and beauty conservation in the decay of D- and B-mesons, the lightest mesons containing a c- or b-quark (or antiquark). The rest masses of D (1869 MeV/c^2 for $D^{\pm}$) and B (5278 MeV/c^2 for $B^{\pm}$) are, however, much larger than those for K-mesons, reflecting the larger masses of c- and b-quarks. Relatively long lifetimes are also observed for mesons made of heavy quark-antiquark pairs as, for example, those shown in Table 2-3. Since these particles are not stable, they are observed as resonances when their production cross sections are plotted as functions of the bombarding energy. For this reason, it is more common to characterize the stability of such "particles" by the widths Γ of their resonance curves, related to their mean life $\overline{T}$ through the uncertainty relation,

$$\Gamma = \frac{\hbar}{\overline{T}}$$

where $\hbar = 6.58 \times 10^{-22}$ MeV-s is the Planck constant (see §5-1 for more detail).

It is worthwhile noting that lifetimes for J/ψ and Υ mesons are very long, or the widths Γ very narrow, in view of the high energies where these events take place. This is caused by the special circumstance that there is not enough energy available for a $J/\psi(c\bar{c})$-particle (rest mass 3097 MeV/c^2) to decay to a $D^+(c\bar{d})$ and a $D^-(\bar{c}d)$ particle which have a combined rest mass energy of 2×1869 MeV. Similarly, an Υ-particle (rest mass 9460 MeV/c^2) cannot decay to a $B^+(\bar{b}u)$ and a

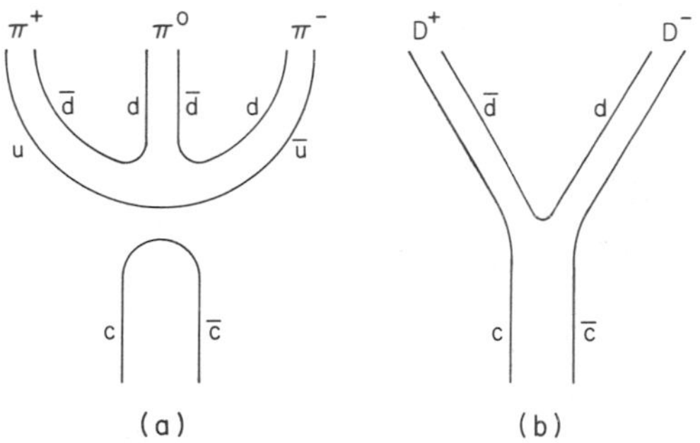

Fig. 2-2 Decay of J/ψ-particle. Transition to D^+D^-, shown in (*b*) is forbidden since the total mass of the final product (2×1869 MeV/c^2) is greater than the rest mass of J/ψ (3097 MeV/c^2). The meson, made of $c\bar{c}$, must decay through processes involving several hadrons (branching ratio 86%) such as the three-pion process shown in (*a*) and leptons (e^+e^- and $\mu^+\mu^-$, branching ratio 14%).

$B^-(b\bar{u})$ particle having a combined rest mass energy of 2×5278 MeV. As a result, J/ψ and Υ must decay through much slower processes involving three or more lighter hadrons, as shown for example in Fig. 2-2(*a*), and into lepton pairs. In contrast, the analogous ϕ-meson (mass 1019 MeV/c^2), made of $s\bar{s}$, can decay to a K^+ and a K^- with a combined rest mass energy of 2×493.6 MeV. The narrow widths of J/ψ and Υ particles are quite astonishing in view of the high energies involved. As a result, they are also useful as energy calibrations and as signatures of special events in high-energy nuclear physics as, for example, those discussed in Chapter 7.

Color. Besides flavor, each quark has another important degree of freedom, generally known as *color*. The need of this additional quantum mechanical label can most readily be seen by examining the quark wave function of a Δ-particle. As we have seen earlier, Δ is an isospin $t = \frac{3}{2}$ particle with four different charge states, Δ^{++}, Δ^+, Δ^0, and Δ^-. Since it is a non-strange baryon ($\mathcal{S} = 0$), it must be made of u- and d-quarks alone. For Δ^{++}, the member with the highest charge state, there is only one possible combination of quarks, (uuu), to make a baryon with $Q = 2$. The intrinsic parity of the Δ is positive. This, together with other evidence, requires the spatial part of the wave function for the three u-quarks in Δ^{++} to be symmetric. The intrinsic spin of Δ is $\frac{3}{2}$ and hence the intrinsic spin part of the wave function for the three u-quarks is also symmetric. Similarly, the isospin part of the wave must also be symmetric in order to have $t = \frac{3}{2}$. As a result, the product of space, intrinsic spin, and isospin parts of the wave function of Δ^{++} is symmetric under a permutation among the three u-quarks. On the other hand, quarks are fermions and the Pauli exclusion principle requires that the total wave function of the three identical quarks be antisymmetric with respect to a permutation of any two of the three quarks. The wave function we have obtained so far for Δ^{++} is in contradiction to this fundamental principle of quantum mechanics. There are two possible ways to get out of this dilemma: Either the Pauli principle is wrong, which is very unlikely, or else we have missed one of the degrees of freedom for quarks.

Fig. 2-3 Total cross of charged pions scattering off protons showing the strong resonance found in $(J^\pi, T) = (3/2^+, 3/2)$ pion-nucleon channel. Since the resonance occurs in the $\ell = 1$ channel at total energy 1232 MeV, it is also known as the P_{33}-resonance or $\Delta(1232)$-particle. Note also that the $\pi^- + p$ cross section at the resonance is much smaller than the corresponding $\pi^+ + p$ value due to the fact that the $\pi^- + p$ system is a mixture of $T = \frac{3}{2}$ and $T = \frac{1}{2}$ channels.

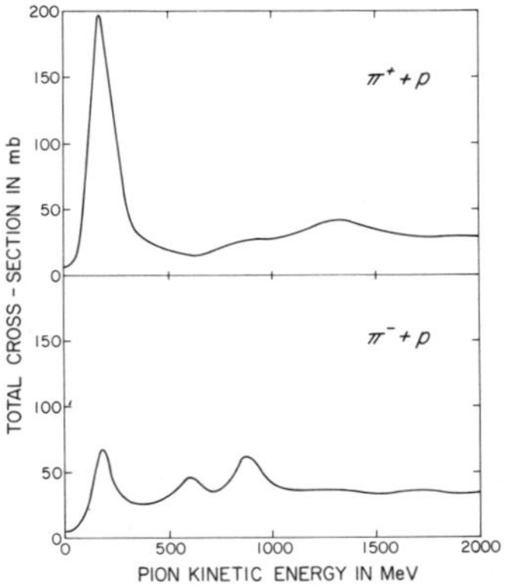

This new degree of freedom for quarks is given the name "color" and hence the name *quantum chromodynamics* for the theory dealing with strong interactions involving "colored" quarks. To account for this new degree of freedom, a color is assigned to each quark, for example, R (red), G (green), and B (blue). From the example of the Δ^{++}-particle we can deduce that the quarks in hadrons must be antisymmetric in the color degree of freedom; that is, the net color in a hadron must vanish. We can also reach this conclusion from another point of view. Since color has not been an observed property, all hadrons must be colorless objects: the color degree of freedom of the constituents inside a hadron must somehow neutralize each other. For mesons, this is easy to achieve since an antiquark has the opposite color of a quark. For baryons made of three identical quarks, such as Δ^{++}, the different colors cancel by being antisymmetric with respect to each other.

In nuclear physics, we shall not be involved explicitly with the color degree of freedom. However the Δ-particle is an important one. It was discovered by Fermi and Anderson in 1949 as a resonance in π^+ scattering off protons at pion kinetic energy $T_\pi = 195$ MeV, as shown in Fig. 2-3, and corresponds to a mass of the pion-proton system of 1232 MeV. Since this takes place in the $\ell = 1$ reaction channel with both spin and isospin $\frac{3}{2}$, it is also known as the P_{33}-resonance. Because it is a very strong resonance at a relatively low energy, nucleons inside a nucleus may be excited relatively easily to become a Δ and, as we shall see later, such excitations may have a strong effect in processes involving energies comparable to those required to change a nucleon into a Δ-particle.

§2-7 STATIC QUARK MODEL OF HADRONS

In principle, a quark model of the hadrons should involve all six different flavors. This can be a rather complicated picture as a large number of hadrons can be constructed from six different quarks and six different antiquarks. Fortunately c- and b-quarks are so much more massive than u-, d-, and s-quarks that they are important primarily in heavy hadrons. In fact, because of the high energy required for their production, baryons with one or more c- or b-quarks are rarely observed experimentally. By the same token, no hadrons involving the t-quark have been reported to date. For most of the known hadrons, in particular those of interest in nuclear physics, only the three light quarks, u, d, and s, and their antiquarks, are involved. For an understanding of the low-lying hadrons, it is therefore quite adequate to consider a model involving only these three light quarks and their antiquarks.

Mesons. Let us start with the simpler case of mesons. Although mesons can be made with any number of quark-antiquark ($q\overline{q}$) pairs, most of the mesons observed may be understood by considering only a single $q\overline{q}$-pair. A simple quark model of the mesons therefore involves a quark and an antiquark moving with respect to each other with orbital angular momentum $\boldsymbol{\ell}$. The total angular momentum of the system is $\boldsymbol{J} = \boldsymbol{\ell} + \boldsymbol{S}$ where $\boldsymbol{S} = \boldsymbol{s}_q + \boldsymbol{s}_{\overline{q}}$ is the sum of the intrinsic spins of the quark and the antiquark. Since $\boldsymbol{s}_q = \boldsymbol{s}_{\overline{q}} = \frac{1}{2}$, the possible values of S for the $q\overline{q}$ system is either 0 (singlet state) or 1 (triplet state). As for the spatial part of the wave, it has been found that mesons with relative orbital angular momentum $\ell = 0$ are lower in energy, similar to atomic energy levels. We shall restrict ourselves to these low-lying mesons as they contain all those of interest to nuclear physics.

Pseudoscalar mesons. We have already seen that pions are the least massive particles among mesons. Since both orbital angular momentum ℓ and total intrinsic spin S are zero, the total angular momentum J of a pion is also zero. They are therefore "scalar" particles, as their wave functions are invariant under a rotation of the spatial coordinate system. However, unlike ordinary scalars, their wave functions change sign under a parity transformation. This may be seen in the following way. The parity of the pion wave function is given by the product of the intrinsic parities of the quark (+1) and the antiquark (−1) and the parity of the spatial wave function of the ($q\overline{q}$)-pair. The property of spatial wave function under a parity transformation is related to the orbital angular momentum ℓ and is given by $(-1)^\ell$, the same as spherical harmonics of order ℓ as discussed in §A-1. Since $\ell = 0$, the parity of the pion wave function is negative. The pion wave function therefore behaves like a pseudoscalar quantity, one that is invariant under a rotation but changes sign under an inversion of the coordinate system. For this

S \ t_0	-1	$-\frac{1}{2}$	0	$+\frac{1}{2}$	$+1$
$+1$		K^0 $(d\overline{s})$		K^+ $(u\overline{s})$	
0	π^- $(d\overline{u})$		π^0 η_8 $(u\overline{u},\ d\overline{d},\ s\overline{s})$		π^+ $(u\overline{d})$
-1		K^- $(s\overline{u})$		$\overline{K}^0$ $(s\overline{d})$	

Table 2-4 Pseudoscalar mesons.

reason pions, and other $J = 0$, negative-parity mesons, are called *pseudoscalar* mesons.

We have already seen that with two quarks, u and d, and two antiquarks, $\overline{u}$ and $\overline{d}$, a total of $2 \times 2 = 4$ pseudoscalar mesons can be constructed, three pions and one η-meson. When the strange quark s and its antiquark, $\overline{s}$, are included in addition, a total of nine pseudoscalar mesons can be formed. These are shown in Table 2-4. The nine mesons may be separated into two groups. Eight of the nine particles form an octet which transform into each other under a rotation in the flavor space. That is, when we make an interchange among u, d, and s, the wave functions of the eight mesons transform into each other as an irreducible representation of an SU_3 group, special unitary group of dimension three. Mathematically the transformation is very similar to, for instance, a rotation of the spatial coordinate axes by some Euler angles. The various components of a spherical tensor of a given rank, *e.g.*, spherical harmonics $Y_{\ell m}(\theta, \phi)$ differing only in the values of m, are changed because of the rotation. However, the relation between the various components of $Y_{\ell m}(\theta, \phi)$, and spherical tensors in general is such that, in the rotated system, the new $Y_{\ell m}(\theta, \phi)$ can always be expressed in terms of the spherical harmonics of the same order ℓ in the old system, as shown in §B-2. In group theoretical language, the $(2\ell + 1)$ components of a spherical tensor of rank ℓ form an irreducible representation. Members of the meson octet also form such a group representation except that the rotation is now in the flavor space of u, d and s; that is, the transformation is from quarks of one flavor to another.

The remaining meson, η_0, is invariant under any such interchanges among the three quarks and therefore forms an irreducible representation by itself. In this way, the nine mesons in the model flavor space of u-, d-, and s-quarks and their antiquarks may be classified into an octet and a singlet according to their SU_3 symmetry in flavor transformation. We shall soon see that, although this symmetry in SU_3(flavor) is not exactly preserved in strong interactions, it is nevertheless useful as a classification scheme for both mesons and baryons.

It is a simple matter to write down the wave functions of the nine mesons in terms of $(q\overline{q})$-pairs. The pion wave functions have already been given in §2-4. There is no ambiguity in writing down the kaon wave functions, since each one must involve either an s or $\overline{s}$. The flavor of the other quark for the $\mathcal{S}=1$ kaons or antiquark for the $\mathcal{S}=-1$ kaons must be either u or d, or $\overline{u}$ or $\overline{d}$, and the choice is completely determined by the charge carried by each kaon. The results are shown in Table 2-4.

The wave functions of the two isoscalar mesons in the table, η_8 and η_0, are slightly more complicated but may be deduced from symmetry arguments. Since η_0 is invariant under a transformation among the three flavors, its wave function must be a linear combination of $(u\overline{u})$, $(d\overline{d})$ and $(s\overline{s})$ with equal weight for each $(q\overline{q})$-pair,

$$|\eta_0\rangle = \frac{1}{\sqrt{3}}\Big\{|u\overline{u}\rangle + |d\overline{d}\,\rangle + |s\overline{s}\rangle\Big\} \tag{2-33}$$

where the factor $1/\sqrt{3}$ comes from normalization. The η_0-meson is an "extension" of the η_0-meson constructed out of u- and d-quarks (and their antiquarks) given in (2-31). Similar to the two-flavor case, the wave function of η_8, the isoscalar meson in the octet, may be obtained by requiring it to be an isoscalar and orthogonal to both $|\,\pi^0\,\rangle$ and $|\,\eta_0\,\rangle$,

$$|\eta_8\rangle = \frac{1}{\sqrt{6}}\Big\{|u\overline{u}\rangle + |d\overline{d}\,\rangle - 2|s\overline{s}\rangle\Big\} \tag{2-34}$$

The derivation is left as an exercise (see Problem 2-5).

Two isospin $t=0$ pseudoscalar mesons are known at low energies, the η-meson with mass 548.8 MeV/c^2 and the η'-meson with mass 957.5 MeV/c^2. Since the SU_3(flavor) symmetry is not an exact one, the observed mesons are mixtures of η_0 and η_8 given above. The mixing coefficient is usually expressed in terms of an angle θ, known as the Cabibbo angle,

$$\begin{aligned}\eta &= \eta_8\cos\theta + \eta_0\sin\theta\\ \eta' &= -\eta_8\sin\theta + \eta_0\cos\theta\end{aligned}$$

For the pseudoscalar mesons, the value is $\theta \simeq 10°$.

Vector mesons. Instead of $S=0$, the total intrinsic spin of the quark-antiquark pair in a meson may be coupled to $S=1$. For $\ell=0$, the total angular momentum of the pions produced is now $J=1$ but the parity remains negative. Similar to the pseudoscalar mesons, we now have a set of nine vector mesons whose wave functions behave like an ordinary vector under transformations of the spatial coordinate system.

The structure of the set of vector mesons is very similar to that of the pseudoscalar mesons, as can be seen by comparing Tables 2-5 and 2-4. Corresponding to the pions, we have an isospin triplet of ρ-mesons, and instead of the strange pseudoscalar mesons, K^0, K^+, K^- and $\overline{K}^0$, we now have the strange vector mesons, K^{*0}, K^{*+}, K^{*-} and $\overline{K}^{*0}$. The two isoscalar vector mesons with definite SU_3 symmetry are ϕ_0 and ϕ_8. The observed isoscalar vector mesons, ϕ and ω, have much larger SU_3(flavor) mixing between ϕ_0 and ϕ_8 with $\theta \sim 40°$, as compared to $\theta \sim 10°$ for the pseudoscalar mesons.

S \ t_0	-1	$-\frac{1}{2}$	0	$+\frac{1}{2}$	$+1$
$+1$		K^{*0} $(d\overline{s})$		K^{*+} $(u\overline{s})$	
0	ρ^- $(d\overline{u})$		ρ^0 ϕ_8 $(u\overline{u},\ d\overline{d},\ s\overline{s})$		ρ^+ $(u\overline{d})$
-1		K^{*-} $(s\overline{u})$		$\overline{K}^{*0}$ $(s\overline{d})$	

Table 2-5 $J^\pi = 1^-$ vector mesons.

The vector mesons are more massive than their pseudoscalar counterparts. For example, the ρ-meson has a rest mass energy of 767 MeV and ω, 782 MeV. In contrast, the pion rest mass energies are 140 MeV for $\pi^\pm$ and 135 MeV for π^0. As far as their wave functions are concerned, the vector and pseudoscalar mesons differ only in their total intrinsic spin, with $S = 1$ for the former and $S = 0$ for the latter. The large difference in their masses must result from the difference in the interaction between a quark and an antiquark in the $S = 0$ and 1 states. We see here an example of the important role of the interaction between quarks which we have ignored for the sake of simplicity in most of our discussions.

Because of their larger masses, the ρ- and ω-mesons can decay via strong interactions to pions with lifetimes at least six orders of magnitude shorter than the lifetimes of pions. The ρ-meson decays to two pions with a mean life of 4×10^{-24} s (or width $\Gamma = 153$ MeV) and the ω-meson decays 90% of the time to three pions with a lifetime of 8×10^{-23} s ($\Gamma = 8.5$ MeV). As we shall see later, both ρ- and ω-mesons play a special role in the interaction between nucleons.

Baryons. With three flavors, we can construct a total of $3\times3\times3 = 27$ baryons for a given set of (ℓ, S)-values. They can be classified according to their SU_3(flavor) symmetries into four groups consisting of 10, 8, 8, and 1 members each. The group of ten baryons (decuplet) is completely symmetric under a transformation in flavor and the group of one baryon (singlet) is completely antisymmetric. The other two groups, consisting of eight members each (octets), have mixed SU_3(flavor) symmetry, neither completely symmetric nor completely antisymmetric. Similar to the case of mesons, we can make use of the SU_3(flavor) symmetry to construct the quark wave functions of baryons.

The baryon wave functions are slightly more complicated to derive than those for mesons for the simple reason that in general it is more involved to find a linear combination of products of three objects having a given symmetry than that of two objects. It is convenient here to treat all the quarks as identical particles distinguished only by their flavor and color labels. Since hadrons are color neutral objects, their quark wave functions must be antisymmetric in color. This means that the rest of the wave functions, formed of the product of flavor, spin, and spatial parts, must be symmetric under a permutation of any two quarks.

Consider first the decuplet. The ten members of the group, together with their quark contents, are listed in Table 2-6. Because of the SU_3(flavor) symmetry, it is relatively simple to construct the quark wave functions. We have already encountered one of the members in this group, Δ^{++}, in introducing the color degree of freedom for quarks. As mentioned previously, both the intrinsic spin and isotopic parts of the Δ-particle wave function must be completely symmetric in order to couple to the $S = \frac{3}{2}$ and $t = \frac{3}{2}$ state. Furthermore, for a symmetric product of spin, isospin, and spatial parts, the spatial part of the wave function of three u-quarks must also be symmetric.

S \ t_0	$-\frac{3}{2}$	-1	$-\frac{1}{2}$	0	$+\frac{1}{2}$	$+1$	$+\frac{3}{2}$
0	Δ^- (ddd)		Δ^0 (udd)		Δ^+ (uud)		Δ^{++} (uuu)
-1		Σ^{*-} (dds)		Σ^{*0} (uds)		Σ^{*+} (uus)	
-2			Ξ^{*-} (dss)		Ξ^{*+} (uss)		
-3				Ω^- (sss)			

Table 2-6 $J^\pi = \frac{3}{2}^+$ baryon decuplet.

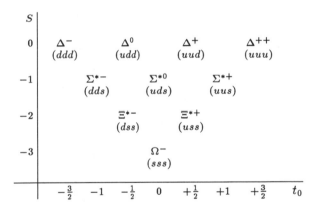

Once the quark wave function of Δ^{++} is known, the wave functions of the other three members of the isospin quartet, Δ^+, Δ^0, and Δ^-, may be obtained using the isospin-lowering operator $\boldsymbol{\tau}_-$ on $|\Delta^{++}\rangle$, in the same way as we have done earlier when obtaining the pion wave functions. With each application of the $\boldsymbol{\tau}_-$-operator, one of the u-quarks is changed into a d-quark. This gives us the correct isospin structure of all four members of the Δ. However, unlike the pion case where two non-identical fermions are involved, a quark and an antiquark, we are dealing here with three identical particles. In addition to isospin coupling, we must also ensure the proper symmetry between the quarks under a permutation between any two of them.

When two identical particles are said to be in a symmetrical state under an exchange of the two particles, we mean that the wave function is unchanged when we permute the particle labels 1 and 2, for example,

$$\Psi_S(1,2) = \frac{1}{\sqrt{2}}\{\xi(1)\zeta(2) + \zeta(1)\xi(2)\} \tag{2-35}$$

where ξ and ζ are single-particle wave functions and $1/\sqrt{2}$ is the normalization factor for the case where the two single-particle wave functions are different. Under a permutation between particle number 1 and particle number 2, the two particles exchange the single-particle states they occupy. Let us denote this operation by the permutation operator P_{12}. It is obvious that, for the wave function $\Psi_S(1,2)$ defined above, we have

$$P_{12}\Psi_S(1,2) = \Psi_S(2,1) = \Psi_S(1,2)$$

Similarly an antisymmetrical two-particle wave function $\Psi_A(1,2)$ may be written in the form

$$\Psi_A(1,2) = \frac{1}{\sqrt{2}}\{\xi(1)\zeta(2) - \zeta(1)\xi(2)\} \tag{2-36}$$

By inspection, we see that

$$P_{12}\Psi_A(1,2) = \Psi_A(2,1) = -\Psi_A(1,2)$$

For $\zeta = \xi$, the symmetric wave function reduces to $\xi(1)\xi(2)$ and the antisymmetric wave function vanishes as required under the Pauli exclusion principle.

For the wave function of Δ^{++} we have the simple situation that all three quarks have the same flavor. The completely symmetric wave function in flavor is the symmetric product of the wave functions of three u-quarks,

$$|\Delta^{++}\rangle = |u(1)\rangle|u(2)\rangle|u(3)\rangle \tag{2-37}$$

We shall now see in detail how to obtain the wave function of Δ^+ using the isospin lowering operator. When $\boldsymbol{\tau}_-$ is applied to $|\,\Delta^{++}\,\rangle$ to form $|\,\Delta^+\,\rangle$, we have a choice of changing any one of the three u-quarks into a d-quark. Since there is no way to make a distinction between these choices, the wave function of Δ^+ must be a linear combination of all three possibilities with equal weight. The normalized and symmetrized (since the color degree of freedom is omitted) wave function for Δ^+ is therefore

$$|\Delta^+\rangle = \frac{1}{\sqrt{3}}\Big\{|d(1)\rangle|u(2)\rangle|u(3)\rangle + |u(1)\rangle|d(2)\rangle|u(3)\rangle + |u(1)\rangle|u(2)\rangle|d(3)\rangle\Big\} \tag{2-38}$$

To simplify the notation, we shall write (2-38) in the shorthand form

$$|\Delta^+\rangle = \frac{1}{\sqrt{3}}\Big\{|duu\rangle + |udu\rangle + |uud\rangle\Big\} \tag{2-39}$$

where it is implied that the first symbol in each term is for quark number 1, the second one for quark number 2, and the third one for quark number 3. In cases where we wish only to indicate the quark content of a hadron without having to display the permutation symmetry explicitly, the notation can be shortened further to (uud), for example, as done in Table 2-6.

Following this rule, the wave functions of Δ^0 and Δ^- can be written in the following manner:

$$|\Delta^0\rangle = \frac{1}{\sqrt{3}}\Big\{|ddu\rangle + |dud\rangle + |udd\rangle\Big\} \tag{2-40}$$

$$|\Delta^-\rangle = |ddd\rangle \tag{2-41}$$

They are obtained by applying the τ_- operator to the wave function of Δ^+, once for $|\Delta^0\rangle$ and twice for $|\Delta^-\rangle$. Alternatively, we can start from the only possibility to construct the wave function for $|\Delta^-\rangle$, the $t_0 = -\frac{3}{2}$ member of the isospin quartet, as we have done earlier for Δ^{++} and then apply the isospin raising operator to produce $|\Delta^0\rangle$.

The wave functions of the three strangeness $\mathcal{S} = -1$ baryons in the decuplet may be obtained by beginning with $|\Sigma^{*+}\rangle$, the $t_0 = 1$ member of the isospin triplet. We can use $|\Delta^{++}\rangle$ given in (2-37) as the starting point and replace one of the three u-quarks with an s-quark. This is similar to the way we obtained $|\Delta^+\rangle$ from $|\Delta^{++}\rangle$ by replacing a u-quark with a d-quark. Here, instead of isospin, we are lowering the strangeness by replacing a d-quark with an s-quark. Again, from symmetry requirements, the normalized wave function is

$$|\Sigma^{*+}\rangle = \frac{1}{\sqrt{3}}\Big\{|suu\rangle + |usu\rangle + |uus\rangle\Big\} \tag{2-42}$$

Next we apply the τ_- operator to $|\Sigma^{*+}\rangle$ and obtain $|\Sigma^{*0}\rangle$ and thence $|\Sigma^{*-}\rangle$, the wave functions for the other two members of the isospin triplet. Now since the s-quark is an isospin zero particle, it vanishes when acted upon by either the τ_+ or the τ_- operator. The only effect of the isospin lowering operator is then to change one of the u-quarks to a d-quark. As a result, we obtain

$$|\Sigma^{*0}\rangle = \frac{1}{\sqrt{6}}\Big\{|dus\rangle + |uds\rangle + |dsu\rangle + |usd\rangle + |sdu\rangle + |sud\rangle\Big\} \tag{2-43}$$

$$|\Sigma^{*-}\rangle = \frac{1}{\sqrt{3}}\Big\{|dds\rangle + |dsd\rangle + |sdd\rangle\Big\} \tag{2-44}$$

This completes the wave functions for the three $\mathcal{S} = -1$ members.

It is trivial to obtain the wave functions for the strangeness $\mathcal{S} = -2$ members of the decuplet since now only one of the three quarks carries a nonzero isospin and, as a result, only an isospin doublet can be constructed. Their wave functions are given by the following expression:

$$|\Xi^{*+}\rangle = \frac{1}{\sqrt{3}}\Big\{|uss\rangle + |sus\rangle + |ssu\rangle\Big\} \tag{2-45}$$

$$|\Xi^{*-}\rangle = \frac{1}{\sqrt{3}}\Big\{|dss\rangle + |sds\rangle + |ssd\rangle\Big\} \tag{2-46}$$

For $\mathcal{S} = -3$ there is only one possibility,

$$|\Omega^-\rangle = |\,sss\,\rangle \tag{2-47}$$

Although it is an isoscalar particle, and consequently $t_0 = 0$, it is a charged particle. This can be seen from (2-32). Since the baryon number is $A = 1$ and strangeness $\mathcal{S} = -3$, the charge number of the particle is -1. The same result can also be obtained from the fact that each s-quark carries a charge $-\frac{1}{3}e$. The particle therefore carries one unit of negative charge and hence the negative sign in the superscript.

Baryon singlet. A state of three quarks completely antisymmetric in flavor is also simple to construct. The quark content in this case must be uds, one of each flavor. However there are $3! = 6$ possible choices, three choices in arranging which one of the three quarks has flavor label u times two choices in arranging which one of the remaining two quarks has flavor label d. (The last one takes the label s.) There is no reason to favor any one of the six possibilities and a linear combination of all six terms is required. The particular linear combination, however, must be antisymmetric with respect to a permutation between any two quarks in order to satisfy the requirement of being a singlet state. We can arrive at the correct linear combination by starting from any one of the six terms; for instance, $u(1)d(2)s(3)$. To this, we add terms generated from it by applying all the possible linearly independent permutations among the indices 1, 2, and 3. For the three odd permutations P_{12}, P_{23}, and P_{31} which produce the arrangements (dus), (usd), and (sdu), we must take them with the negative sign in order to satisfy the requirement of being symmetric. Similarly, the two even permutations, $P_{12}P_{23}$ and $P_{31}P_{23}$, which produce the arrangements (dsu) and (sud), must be taken with the positive sign. The normalized singlet quark wave function is then

$$|\Lambda_1\rangle = \frac{1}{\sqrt{6}}\Big\{|uds\rangle + |dsu\rangle + |sud\rangle - |dus\rangle - |usd\rangle - |sdu\rangle\Big\} \tag{2-48}$$

Except for an overall sign, this is the only unique way to form the required antisymmetric linear combination. Since there is no other way to construct a wave function with the same symmetry, $|\,\Lambda_1\,\rangle$ forms an irreducible SU_3(flavor) representation by itself.

Let us now examine the symmetry of the isospin part of the wave function. Since the s-quark is an isoscalar quantity, the isospin of the wave function is determined by u- and d-quarks. To illustrate this point, we can rewrite the wave function for the singlet in the following form, always putting the s-quark at the end,

$$\begin{aligned}|\Lambda_1\rangle = \frac{1}{\sqrt{6}}\Big\{&\big(|\,u(1)\,\rangle|\,d(2)\,\rangle - |\,d(1)\,\rangle|\,u(2)\,\rangle\big)|\,s(3)\,\rangle\\ &+\big(|\,u(2)\,\rangle|\,d(3)\,\rangle - |\,d(2)\,\rangle|\,u(3)\,\rangle\big)|\,s(1)\,\rangle\\ &+\big(|\,u(3)\,\rangle|\,d(1)\,\rangle - |\,d(3)\,\rangle|\,u(1)\,\rangle\big)|\,s(2)\,\rangle\Big\}\end{aligned}$$

This is a $t = 0$ linear combination since the wave function is antisymmetric in the isospin-carrying parts. The singlet SU_3(flavor) representation therefore describes an isoscalar particle. The isospin is, however, completely determined by the symmetry in flavor of the quark wave function and is not an independent degree of freedom in the wave function.

So far we have not explicitly put in the intrinsic spin part of the wave function. We shall carry out an example involving the intrinsic spin wave function later for the nucleon. Here, we shall simply give the result that $J^\pi = \frac{1}{2}^+$ for the Λ_1 baryon.

As mentioned earlier, the SU_3(flavor) symmetry is not an exact one. The observed strangeness $\mathcal{S} = -1$, isoscalar, $J^\pi = \frac{1}{2}^+$ particle Λ is a mixture of the $|\,\Lambda_1\,\rangle$ and $|\,\Lambda_8\,\rangle$, the latter being a member of the $J^\pi = \frac{1}{2}^+$ baryon octet. This is similar to the admixture in pseudoscalar and in vector meson wave functions we have seen earlier.

Baryon octet. The remaining 16 members of the 27 possible baryons that can be constructed from u-, d-, and s-quarks have mixed symmetry in flavor. They may be classified as two octets distinguished by their symmetries under the simultaneous interchange of both flavor and spin. We shall be interested only in the octet with the lower energy as it contains protons and neutrons as members. The waves function of each member of this group is antisymmetric under the combined exchange of both flavor and intrinsic spin. The members of the octet together with their quark contents are shown in Table 2-7.

$\mathcal{S}$ \ t_0	-1	$-\frac{1}{2}$	0	$+\frac{1}{2}$	$+1$
0		n (udd)		p (uud)	
-1	Σ^- (dds)		Σ^0 Λ_8 (uds)		Σ^+ (uus)
-2		Ξ^- (dss)		Ξ^+ (uss)	

Table 2-7 $J^\pi = \frac{1}{2}^+$ baryon octet.

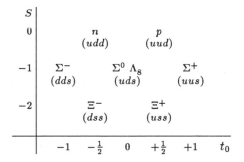

Let us try to construct the proton wave function as an example. Since the intrinsic spin and parity of a proton are $\frac{1}{2}^+$, we can start by coupling the intrinsic spins of the three quarks to the value $\frac{1}{2}$. There are several ways to achieve this, and we shall take the simplest approach by coupling the intrinsic spins of the first two quarks to 0 and then couple the third one with spin up to form a system with $(S, S_0) = (\frac{1}{2}, +\frac{1}{2})$,

$$|\tfrac{1}{2}, +\tfrac{1}{2}\rangle = \frac{1}{\sqrt{2}}\Big(|q(1)\uparrow\rangle|q(2)\downarrow\rangle - |q(1)\downarrow\rangle|q(2)\uparrow\rangle\Big)|q(3)\uparrow\rangle \tag{2-49}$$

where the up-arrow symbol $\uparrow$ represents a quark with intrinsic spin up $(+\frac{1}{2})$ and the down-arrow symbol $\downarrow$ a quark with spin down $(-\frac{1}{2})$. The assignment of flavor to each of the quarks q will be made later. A second possibility to form an $(S, S_0) = (\frac{1}{2}, +\frac{1}{2})$-system is to couple the first two quarks to spin 1 instead of 0 as we have done above. This choice is complicated by the fact that the final system is a mixture of total spin $\frac{3}{2}$ and $\frac{1}{2}$. To project out the desired $S = \frac{1}{2}$ part, a linear combination must be taken of the two possibilities, $\{(q_1q_2)_{1,1}(q_3)_{1/2,-1/2}\}$ and $\{(q_1q_2)_{1,0}(q_3)_{1/2,+1/2}\}$, where the first one of the two subscripts indicates S and the second one the value of its third component.

The combined symmetry of the spin and flavor parts of the wave function may be determined after assigning a flavor to each one of the quarks involved, subject to the condition that, for a proton, each term must consist of two u-quarks and one d-quark. Let us start by assigning the first two quarks with different flavors, and (2-49) becomes

$$|\tfrac{1}{2}, +\tfrac{1}{2}\rangle = \frac{1}{\sqrt{2}}\Big(|u(1)\uparrow\rangle|d(2)\downarrow\rangle - |u(1)\downarrow\rangle|d(2)\uparrow\rangle\Big)|u(3)\uparrow\rangle \tag{2-50}$$

The combination of spin and flavor in (2-50) may be symmetrized in two stages. First we shall carry out the process only for the first two quarks and obtain

$$\begin{aligned}|\tfrac{1}{2}, +\tfrac{1}{2}\rangle = \frac{1}{2}\Big(&|u(1)\uparrow\rangle|d(2)\downarrow\rangle - |u(1)\downarrow\rangle|d(2)\uparrow\rangle \\ &+|d(1)\downarrow\rangle|u(2)\uparrow\rangle - |d(1)\uparrow\rangle|u(2)\downarrow\rangle\Big)|u(3)\uparrow\rangle\end{aligned} \tag{2-51}$$

Next we shall generate all the other terms by making the permutations P_{31} and P_{32} on each of the four terms in (2-51). This gives us a total of twelve terms. On grouping identical terms together, we obtain the quark wave function for a proton with spin orientations of all the quarks indicated explicitly,

$$\begin{aligned}|p\rangle = \frac{1}{\sqrt{18}}\Big\{&2\Big(|u\uparrow u\uparrow d\downarrow\rangle + |u\uparrow d\downarrow u\uparrow\rangle + |d\downarrow u\uparrow u\uparrow\rangle\Big) \\ &-\Big(|u\uparrow u\downarrow d\uparrow\rangle + |u\uparrow d\uparrow u\downarrow\rangle + |d\uparrow u\uparrow u\downarrow\rangle \\ &+|u\downarrow u\uparrow d\uparrow\rangle + |u\downarrow d\uparrow u\uparrow\rangle + |d\uparrow u\downarrow u\uparrow\rangle\Big)\Big\}\end{aligned} \tag{2-52}$$

To simplify the notation, we have dropped the labels for quark number and rely on the order each quark appears instead. The fact this wave function is antisymmetrical under a simultaneous interchange of flavor and spin between two quarks can be established by inspection.

The neutron wave function can be written down from that for the proton by simply substituting all the u-quarks by d-quarks and vice versa. Similarly, the wave functions of the strangeness $\mathcal{S} < 0$ members of the octet can be built from that of the proton in the same way as we have carried out for members of the decuplet by starting from $|\Delta^{++}\rangle$. These are left as exercises.

§2-8 MAGNETIC DIPOLE MOMENT OF THE BARYON OCTET

The hadron wave functions obtained in the previous section are based solely on symmetry considerations. Since they are not the eigenfunctions of a realistic Hamiltonian involving interaction between quarks, we cannot expect them to be able to describe any of the dynamic properties of hadrons with great accuracy. Nevertheless, calculations for some simple quantities can be carried out and the results will show whether they are useful as zeroth-order approximations to the true wave functions. Besides charge number, spin, isospin, and strangeness, of which we have already made use in obtaining the wave functions, the magnetic dipole moment is one such quantity to consider. We shall only deal here with members of the baryon octet given in Table 2-7, since more is known about them.

The magnetic dipole moment of a baryon comes from two sources, the intrinsic dipole moments of the constituent quarks and the orbital motion of the quarks. For the baryon octet of interest here, all the members have $J^\pi = \frac{1}{2}^+$. In the simple model adopted above for our discussion, the three quarks are symmetric in the spatial parts of their wave functions, with relative motion between them in the $\boldsymbol{\ell} = 0$ states. As a result, no contribution to the magnetic dipole moment comes from quark orbital motion.

Quark magnetic dipole moments. Contributions to the hadron magnetic dipole moment from the quark intrinsic spin arise from the fact that, associated with the intrinsic spin of a particle, there is an intrinsic magnetic dipole. The operator for the dipole moment is given by

$$\boldsymbol{\mu} = g\boldsymbol{s}\mu_{\mathrm{D}} \tag{2-53}$$

where $\boldsymbol{s}$ is the operator for intrinsic spin. The dipole moment is measured in units of μ_{D} which, in terms of the quark charge q and mass m_q, may be expressed in the form $q\hbar/2m_qc$ in cgs units or $q\hbar/2m_q$ in SI units. The quantity g is called the gyromagnetic ratio, the ratio between the magnetic dipole moment and the spin of the particle. For a Dirac particle, *i.e.*, a particle without any internal structure, having intrinsic spin $\boldsymbol{s} = \frac{1}{2}$, we have $g = 2$. In practice, no particle is observed to be a purely Dirac particle. For example, electrons and muons emit and absorb virtual photons. The contributions from these virtual processes give rise to an "anomalous" magnetic dipole moment such that the observed value of g is $2 \times 1.001159652193(10)$ for an electron and $2 \times 1.001165923(8)$ for a muon. The small corrections to the simple Dirac particle values for the charged leptons are well understood and can be calculated to very high accuracy in quantum electrodynamics.

Since a quark is an elementary particle, we can take it as a simple Dirac particle to start with. The relation between intrinsic magnetic moment and spin is then given by (2-53) with $g = 2$. However, we do not know the quark masses: it is therefore not possible to deduce the values of $\boldsymbol{\mu}$ in any simple way. (For this reason there is no point in considering here corrections to g due to anomalous magnetic dipole moment.) However, if we assume that the masses of u- and d-quarks are equal, the ratio between their magnetic dipole moments is then given by the ratio of their charges. This gives us the result

$$\mu_u = -2\mu_d \tag{2-54}$$

As we shall soon see, this is useful in getting an idea of the values of the intrinsic magnetic dipole moments of the u- and d-quarks.

Nucleons. The contribution to the magnetic dipole moment of a baryon from quark intrinsic moment depends also on the orientation of the spin of each quark. Since the quark orbital motion does not enter here, the magnetic dipole moment reduces to the simple situation given only by the number of quarks of each flavor in each one of the two possible spin orientations. For a proton, we can count the numbers of $u\uparrow$, $u\downarrow$, $d\uparrow$ and $d\downarrow$ explicitly using the wave function given earlier in (2-52). This may be done by calculating the expectation value of the number operator for each type of quark of a given spin orientation. The results are simply given by the sums of the squares of the coefficients in the wave function for each one of the four possible combinations of flavor and spin orientations, $\frac{5}{3}$ for the number of u-quark with spin up, $\frac{1}{3}$ for the number of u-quarks with spin down, $\frac{1}{3}$ for the number of d-quarks with spin up and $\frac{2}{3}$ for the number of d-quarks with spin down. The net contribution from u-quarks to the proton magnetic dipole moment is then $\frac{5}{3} - \frac{1}{3} = \frac{4}{3}$, and that from d-quark is $\frac{1}{3} - \frac{2}{3} = -\frac{1}{3}$. In this simple model, the final result is

$$\mu_p = \tfrac{4}{3}\mu_u - \tfrac{1}{3}\mu_d \tag{2-55}$$

for the magnetic dipole moment of a proton in terms of those for u- and d-quarks.

For a neutron, we can again interchange the roles of u- and d-quarks in (2-55) above and obtain the result,

$$\mu_n = \tfrac{4}{3}\mu_d - \tfrac{1}{3}\mu_u \tag{2-56}$$

If we now make the assumption that the masses of the u- and d-quarks involved are equal and their ratio of magnetic dipole moment is given by (2-54), we obtain the ratio between the magnetic dipole moments of a neutron and a proton,

$$\frac{\mu_n}{\mu_p} = \frac{\frac{4}{3}\mu_d - \frac{1}{3}\mu_u}{\frac{4}{3}\mu_u - \frac{1}{3}\mu_d} = \frac{-2\mu_d}{3\mu_d} = -\frac{2}{3} \tag{2-57}$$

This is in good agreement with the observed value of $-1.913/2.793 = -0.685$.

Baryons with $\mathcal{S} < 0$. For the other six members of the octet, there is at least one s-quark involved. As a result, we need the intrinsic magnetic dipole moment of the strange quark and its contributions to the magnetic dipole moment of these baryons. Since the s-quark is known to be more massive than the u- and d-quarks, we cannot easily relate the intrinsic magnetic dipole moment of the s-quark to those of the u- or d-quarks in the way we have done in (2-54) between u- and d-quarks. On the other hand, eight magnetic dipole moments are known for members of the octet and they can all be given in terms of the intrinsic magnetic dipole moments of the three quarks in this simple model. A least-square fitting procedure may be used to deduce the three unknown quark values from these eight pieces of known data. To carry out this procedure, we must first express the baryon magnetic dipole moments in terms of those for the three quarks, as we have done above for the nucleons.

Although we do not have the quark wave functions written out in detail for the $\mathcal{S} < 0$ members of the octet as we have done in (2-52) for the proton, we can nevertheless count the number of quarks of each flavor with spin up and the number with spin down, starting with the quark content of each baryon given in Table 2-7. This is particularly simple for those strange baryons involving only two different flavors, *i.e.*, those made of s- and u-quarks only or s- and d-quarks only. For example, the Σ^+ baryon is made of two u-quarks and one s-quark. Compared with the proton, (uud), we see that the only difference between the quark structure of these two baryons is that, in the place of a d-quark in proton, we have an s-quark in Σ^+. Since all members of the octet have the same combined symmetry for spin and flavor, the proton and Σ^+ must have very similar quark wave functions except for the replacement of d with s. Hence, the magnetic dipole moment of the Σ^+-baryon is given by the expression

$$\mu_{\Sigma^+} = \tfrac{4}{3}\mu_u - \tfrac{1}{3}\mu_s \tag{2-58}$$

Similarly, the quark content of Σ^- is (dds). Comparing it with a neutron, we find that the expression for the magnetic dipole moment of a Σ^--baryon is the same except that, in the place of u-quarks in a neutron, we have the contributions from s-quarks. This gives us the result

$$\mu_{\Sigma^-} = \tfrac{4}{3}\mu_d - \tfrac{1}{3}\mu_s \tag{2-59}$$

Using similar methods the expressions for the magnetic dipole moments of the two $\mathcal{S} = -2$ members of the octet, $\Xi^-(dss)$ and $\Xi^+(uss)$, can be obtained and the results are given in Table 2-8.

For the two remaining members of the octet, Σ^0 and Λ_8, the quark contents are (uds) for both. Since three different flavors are involved, an alternative approach is required in order to express their magnetic dipole moments in terms of those for the three quarks. For Λ_8 and Σ^0, we can make use of their isospin

Table 2-8 Magnetic dipole moment of baryon octet.

Octet member	Quark Content u	d	s	Best Fit (μ_N)	Observed (μ_N)
p	$\frac{4}{3}$	$-\frac{1}{3}$	0	2.793	2.792847386(63)
n	$-\frac{1}{3}$	$\frac{4}{3}$	0	−1.913	−1.91304275(45)
Λ	0	0	1	−0.581	−0.613(4)
Σ^+	$\frac{4}{3}$	0	$-\frac{1}{3}$	2.663	2.42(5)
Σ^-	0	$\frac{4}{3}$	$-\frac{1}{3}$	−1.102	−1.157(25)
Ξ^0	$-\frac{1}{3}$	0	$\frac{4}{3}$	−1.392	−1.250(14)
Ξ^-	0	$-\frac{1}{3}$	$\frac{4}{3}$	−0.451	−0.69(4)
$\Sigma^0 \to \Lambda$	$-\sqrt{\frac{1}{3}}$	$\sqrt{\frac{1}{3}}$	0	−1.630	−1.59(9)
Ω^-			3	−1.744	
u	1			1.852	
d		1		−0.972	
s			1	−0.581	

difference to derive their wave functions. For this purpose we can ignore the s-quark for the moment since it is an isoscalar particle not involved in any isospin considerations.

Let us start with Λ_8. Since it is an isospin singlet, *i.e.*, there is no other Λ-particle with a different charge state, the isospin is $t = 0$. In the discussions given earlier for the quark wave functions of π^0- and η_0-mesons, we have seen that a system consisting of a u- and a d-quark has $t_0 = 0$ and is a mixture of isospin 1 and 0. In order to project out the isospin $t = 0$ part, we need an antisymmetric linear combination of the two possible arrangements of u and d,

$$|(t,t_0)=(0,0)\rangle = \frac{1}{\sqrt{2}}\Big\{|u(1)d(2)\rangle - |d(1)u(2)\rangle\Big\} \tag{2-60}$$

The spins of these two quarks cannot be both up since such an arrangement will be antisymmetric under the simultaneous exchange of spin and flavor. The only possibility is therefore

$$\begin{aligned}|(t,t_0)=(0,0);(s,m_s)=(0,0)\rangle = \frac{1}{2}\Big\{&\big(|u(1)\!\uparrow d(2)\!\downarrow\rangle - |d(1)\!\uparrow u(2)\!\downarrow\rangle\big)\\ &+\big(|d(1)\!\downarrow u(2)\!\uparrow\rangle - |u(1)\!\downarrow d(2)\!\uparrow\rangle\big)\Big\}\end{aligned} \tag{2-61}$$

Note that the total spin of the two quarks is also zero as a result of the symmetry requirement. We can now couple the s-quark to the product and form a spin-$\frac{1}{2}$ system of three quarks,

$$\begin{aligned}|(t,t_0)=(0,0);(s,m_s)=(\tfrac{1}{2},+\tfrac{1}{2})\rangle = \frac{1}{2}\Big\{&\big(|u(1)\!\uparrow d(2)\!\downarrow\rangle - |d(1)\!\uparrow u(2)\!\downarrow\rangle\big)\\ &+\big(|d(1)\!\downarrow u(2)\!\uparrow\rangle - |u(1)\!\downarrow d(2)\!\uparrow\rangle\big)\Big\}|s(3)\!\uparrow\rangle\end{aligned} \tag{2-62}$$

The wave function is not properly antisymmetrized with respect to the third quark. However, for the purpose of calculating the magnetic dipole moment, this is not necessary; all that is needed is to count the number of quarks of each flavor with spin up and the number with spin down, and this is independent of the symmetrization among the three quarks beyond those given in (2-62) above. Furthermore, it is also obvious from the structure of the wave function that the net contributions from both u- and d-quarks are zero since there are equal numbers of each with spin pointing up as there are with spin pointing down. As a result, we obtain

$$\mu_\Lambda = \mu_s \tag{2-63}$$

for the magnetic dipole moment of Λ_8. Because of the crudeness of the model used here, there is no point in considering any SU_3(flavor) symmetry-breaking effects and the resulting difference between Λ_8 and the observed Λ-baryon.

The isospin of Σ^0 is unity since it is a member of an isospin triplet, Σ^+, Σ^0 and Σ^-. The quark wave function is somewhat more complicated than what we have derived for Λ_8, since the u- and d-quarks must now be coupled to a spin-1 state. We shall leave the calculation of μ_{Σ^0} in terms of quark magnetic dipole moments as an exercise since the value of μ_{Σ^0} is not known and we cannot make use of it in our calculation. On the other hand, the decay of Σ^0 through the reaction,

$$\Sigma^0 \to \Lambda + \gamma \tag{2-64}$$

is similar to a magnetic dipole transition (see §5-2) and, as a result, the transition probability is proportional to $|\mu_{\Sigma^0\to\Lambda}|^2$ where the matrix element of the magnetic dipole transition operator $\boldsymbol{O}(M1)$ has the value,

$$\mu_{\Sigma^0\to\Lambda} = \langle \Sigma^0 | \boldsymbol{O}(M1) | \Lambda \rangle = -\frac{1}{\sqrt{3}}(\mu_u - \mu_d) \tag{2-65}$$

The absolute value of this matrix element is determined experimentally to be $1.59 \pm 0.09\mu_N$ (P.C. Peterson et al., *Phys. Rev. Lett.* **57** [1986] 949) and this result, together with the negative sign obtained from other sources, may be used as a piece of datum for our calculation.

Table 2-8 summarizes the contribution from each one of the three quarks to the magnetic dipole moments of the members of the baryon octet together with $\mu_{\Sigma^0\to\Lambda}$. The observed values are listed in the last column in units of nuclear magnetons, $\mu_N = e\hbar/(2M_pc)$ in cgs units or $e\hbar/(2M_p)$ in SI units. As mentioned earlier, we can deduce the values of μ_u, μ_d and μ_s by fitting these three unknown values to the eight measured dipole moments. Since the accuracies that can be achieved for the measured values of the various baryons differ by a large margin, the eight pieces of data that went into the calculation as input have been weighed inversely according to their experimental uncertainties, indicated in brackets in the table. The results of the calculation are shown under the column labelled "Best Fit." The calculated values for the baryon magnetic dipole moments agree quite well with observation, especially in view of the crude model used. The discrepancies are in general less than $0.2\mu_N$ except for Ξ^-. This close agreement has two implications: the first is that the model used to deduce the moments in terms of those of the three quarks is a reasonable one, otherwise much larger differences would have resulted; the second is that the values deduced for the quark magnetic dipole moments are physically meaningful.

We expect several corrections to our simple analysis above. One of the assumptions we have made is that the wave functions have only $\ell = 0$ components. This is true if orbital angular momentum is conserved by the interaction between quarks. As we shall see in the next chapter in an analogous discussion on the ground state of deuteron, the orbital angular momentum is not a constant of motion. Consequently, it is unreasonable for us to expect that the ground states of members of the baryon octet to be purely $\ell = 0$. In general, some configuration mixing from $\ell > 0$ terms is present and this may be the dominant correction to our simple model. A more detailed discussion can be found in a recent status report on the topic of "baryon magnetic moment in the quark model" by Brekke and Rosner (*Comm. Nucl. Part. Phys.* **18** [1988] 83).

The values of the magnetic dipole moments of the three lighter quarks obtained from the least-square-fitting procedure are given at the bottom of the table. Although there are no observed values for quarks to compare with here, we can, nevertheless, get a rough idea of whether the results are reasonable. In the first place, the ratio $\mu_u/\mu_d = -1.91$ is fairly close to the value of -2 obtained earlier by assuming that the masses of u- and d-quarks are identical to each other and both of them have the same gyromagnetic ratios. We can take this naive analysis one step further and deduce the masses of the three quarks involved from the values of their magnetic moments. Using the values $\mu_u = 1.852$, $\mu_d = -0.972$ and $\mu_s = -0.581$ in nuclear magnetons given in Table 2-8, we obtain $m_u c^2 = 0.34$ GeV, $m_d c^2 = 0.32$ GeV, and $m_s c^2 = 0.54$ GeV. We do not expect these values to be identical to those given in Table 2-2. The discrepancy is at the level of roughly 30%, consistent with the amount of other possible components in the wave functions we have ignored.

§2-9 HADRON MASS AND QUARK-QUARK INTERACTION

Another striking feature in hadron spectroscopy is the systematics in their masses. In Table 2-9 the measured masses for some of the low-lying members are given, together with their uncertainties in the last digits in brackets. First of all, we notice that the values for the members of the $J^\pi = \frac{1}{2}^+$ baryon octet are well correlated with their values of strangeness quantum numbers. That is, the mass differences between members with the same strangeness are much smaller than those between members of different strangeness. For example, the mass difference between a proton and a neutron is less than 2 MeV/c^2, whereas the difference between the mass of a Λ-baryon ($\mathcal{S} = -1$) and a neutron is around 176 MeV/c^2, and between a Ξ^0-baryon ($\mathcal{S} = -2$) and a Λ-baryon is around 200 MeV/c^2. The obvious conclusion one can draw from such comparisons is that the rest mass energies of the underlying u-and d-quarks are the same within a few MeV and that the rest mass energy of the s-quark is larger than those of the u- and d-quarks by the order of 100 to 200 MeV. Further support for s-quarks being more massive can be found in the mass differences between members of the baryon decuplet, between members of the pseudoscalar mesons, and between members of the vector mesons. As for the charm and beauty quarks, we have already seen evidence in earlier sections that they are even more massive from the masses of J/ψ and Υ mesons. From a comparison of the masses of hadrons made of different quarks, the masses of the various quarks can be deduced (except for the t-quark yet to be identified experimentally). The results were given earlier in Table 2-2.

The small mass differences between hadrons having the same strangeness can come from either electromagnetic effects or from a small difference between the masses of u- and d-quarks. However, our present knowledge of the strong interaction is not able to elucidate on this question. In spite of our ignorance, the small difference between the masses of proton and neutron and between $\pi^\pm$ and π^0 are important in understanding some of the nuclear phenomena, such as isospin symmetry breaking in the nuclear force.

Table 2-9 Low-lying hadron masses in MeV/c^2.

Baryons		Mesons	
$\mathcal{S} = 0$		Pseudoscalar mesons	
p	938.27231(28)	$\pi^\pm$	139.56755(33)
n	939.56563(28)	π^0	134.9734(25)
$\mathcal{S} = -1$		$K^\pm$	493.646(9)
Λ	1115.63(5)	$K^0, \overline{K}^0$	497.671(30)
Σ^+	1189.37(6)	η	548.8(6)
Σ^0	1192.55(9)	η'	957.50(24)
Σ^-	1197.43(6)	Vector mesons	
$\mathcal{S} = -2$		ρ	766.9(12)
Ξ^0	1314.9(6)	K^*	892.09(30)
Ξ^-	1321.32(13)	ω	781.99(13)
		ϕ	1019.414(10)

It is important to emphasize here again that, since quarks have not been observed in isolation outside hadrons, the masses deduced from hadron spectra are not necessarily their true masses. The observed hadron masses depend on the intrinsic masses of the quarks as well as the binding energy between the quarks. If the binding energies are known, it is a trivial matter to deduce the quark masses from hadron masses. As we shall see later in the analogous situation of nuclear masses, binding energy calculations require a knowledge of the interaction between the constituents. Even in the nuclear case it is not easy to obtain great accuracy in binding energy calculations, partly because of our incomplete understanding of the interaction between nucleons and partly because of the difficulties of the many-body problem.

For quarks, the situation is further complicated by several factors. In the first place, the quark-quark interaction is known to be very strong at energies with which we are concerned here. We have seen an example of this from the mass differences between the pseudoscalar mesons (sum of quark intrinsic spins $S = 0$) and the vector mesons ($S = 1$). For example, the quark contents of the π- and ρ-mesons are the same, but their masses are quite different, with $m_\pi c^2 \approx 140$ MeV and $m_\rho c^2 \approx 767$ MeV. The large difference must be mainly attributed to the dependence of the interaction between quarks on the total intrinsic spin of the quark-antiquark pair. This is quite different from the usual situation in quantum systems where the interesting physics often arises from small parts of the complete interaction. For example, in atomic physics the main contribution to the binding energies of electrons comes from the electrostatic attraction between the nucleus and each one of the electrons. The properties of an atom, on the other hand, are sensitive mainly to small *perturbations* caused by the interaction between electrons. As a result, a number of perturbation methods have been developed over the years and they are found to be quite successful in handling such problems.

For quarks, the interaction is very strong at low energies where nuclear physics operates and where most of the experimental observations are made. Because of what is generally known as *asymptotic freedom*, the quark-quark interaction is weak only at high energies. As a result, perturbative techniques in QCD apply only at extremely high energies, far beyond the realm of nuclear physics and low-lying hadron spectroscopy. For the low-energy regions, new methods other than perturbative approaches must be developed before we can properly link QCD calculations to observations. Until good methods are found, we must be satisfied with model calculations which give us a fairly good idea of what is the most likely picture of strong interactions.

A second problem is the question of *confinement*. Again, since quarks are not observed in isolation, their mutual interaction must have a component that grows stronger as the distance of separation between them increases. This is in contradiction to the macroscopic world where interactions, such as gravitational and electromagnetic interactions, grow weaker as the distance of separation between the interacting objects is increased, following the inverse square law. As a result, we must devise new methods of handling the confinement problem. One way is to impose confinement as a boundary condition. In other words, the quarks are considered to be inside a "bag" which prevents them from escaping to the outside. Such a *bag model*, together with its many variants, has had a large degree of success in improving our understanding of the structure of hadrons and in linking the quark-quark interaction with the interaction between nucleons. Other methods of attack, such as lattice gauge calculations and soliton models, are yet to be fully explored but may well yield interesting results in the near future. Until then, these topics are beyond the scope of a textbook on nuclear physics.

Problems

2-1. Show that conservation of energy and momentum requires at least two γ-rays to be emitted in the annihilation of an electron by a positron.

2-2. Show that the nucleon isospin wave functions given in (2-15,16) are the eigenfunctions of the operator

$$\boldsymbol{\tau}^2 = \tau_1^2 + \tau_2^2 + \tau_3^2$$

with eigenvalues 3. Express $\boldsymbol{\tau}^2$ in terms of τ_+, τ_- and τ_0 and calculate the expectation values in this form for the isospin wave function of a proton and a neutron.

2-3. Antiprotons are created when a beam of high-energy protons strikes a hydrogen target. What is the minimum proton kinetic energy in the laboratory system required for the reaction to take place?

2-4. Construct the quark wave function of π^- by applying the isospin lowering operator to the wave function of π^0 given in (2-29). Use the same technique to construct the quark wave function of Σ^{*0} by applying the isospin lowering operator to that of Σ^{*+} given in (2-42).

2-5. The meson η_8 is a neutral, isospin singlet particle made of a linear combination of quark-antiquark pairs taken from u-, d- and s-quarks and their antiquarks. Construct the quark wave function of η_8 by requiring it to be normalized and orthogonal to those of π^0 and η_0 given in (2-29,33).

2-6. Show that the magnetic dipole moment of an electron moving in a circular orbit is given by the relation

$$\boldsymbol{\mu} = -\frac{e\hbar[c]}{2m_e c}\boldsymbol{\ell}$$

where $\boldsymbol{\ell}$ is the angular momentum in units of $\hbar$ and the factor $[c]$ converts the formula from cgs to SI units. Assume that the charge and mass of the electron is distributed uniformly along the orbit.

2-7. The Σ^0-particle is a baryon made of a u-quark, a d-quark, and an s-quark coupled to total intrinsic spin $S = \frac{3}{2}$ and isospin $t = 1$. Assume that the orbital angular momentum is $\ell = 0$, show that the magnetic dipole momentum of Σ^0 is given in the quark model by the following expression:

$$\mu_{\Sigma^0} = \tfrac{2}{3}(\mu_u + \mu_d) - \tfrac{1}{3}\mu_s$$

where μ_u, μ_d and μ_s are the intrinsic magnetic dipole moments of the u-, d- and s-quark respectively.

2-8. Use the quark model to show that the magnetic dipole moments of the vector mesons ρ^+ and ρ^- are equal in magnitude but opposite in sign.

2-9. A neutron star is a compact, dense object made of degenerate neutrons having a density similar to that in the central part of a heavy nucleus.

(a) If the density of nuclear matter is 0.17 nucleons/fm^3 or 2.8×10^{17} kg/m^3, what is the radius of a neutron star having a mass one and half times that of the sun? (One solar mass $= 2.0 \times 10^{30}$ kg.)

(b) A neutron star is one of the possible remanents of a supernova explosion such as SN 1987a, the one which took place in the Large Magellanic Cloud 15,000 light years away and first observed on earth on February 24, 1987. When the core of a large star exhausts its nuclear fuel, there is no longer the thermal pressure to counterbalance the gravitational force, and the core of the star collapses. For simplicity, we can consider that all the material in the core of the collapsing star is in the form of ^{56}Ni made of 28 neutrons and 28 protons. A nickel atom is electrically neutral with 28 electrons outside. Because of the tremendous gravitation force, the protons in ^{56}Ni change into neutrons by capturing the atomic electrons through the reaction

$$p + e^- \to n + \nu_e$$

Calculate the number of neutrinos released in converting 1.5 solar mass of ^{56}Ni atoms into neutrons during the gravitational collapse.

(c) If the total cross section for a neutrino to interact with each nucleon is 10^{-48} m^2, how many reactions due to the neutrinos from such a gravitational collapse can one expect in a detector on earth made of 3000 tons of water? Compare this with the number of events (12) observed with a similar detector at Kamioka due to supernova SN 1987a.

(d) Assuming that the average energy of each neutrino is 10 MeV, calculate the total amount of energy carried away by the neutrinos from the gravitational collapse. Compare this value with the rest mass energy of the sun.

2-10. If the magnetic dipole moment of a u-quark is $1.852\mu_N$, d-quark, $-0.972\mu_N$, and s-quark, $-0.581\mu_N$, what are the values of magnetic dipole moments of their antiquarks? The ρ^+-meson is a vector meson with $J^\pi = 1^-$ and isospin $T = 1$. Calculate the magnetic dipole moment using the values from quarks assuming the orbital angular momentum to be the lowest value possible.

Chapter 3

NUCLEAR FORCE AND TWO-NUCLEON SYSTEMS

The interaction between two nucleons is one of the central questions in nuclear physics. In a 1953 *Scientific American* article (September issue, p. 58), Bethe estimated that "in the past quarter century physicists have devoted a huge amount of experimentation and mental labour to this problem – probably more man-hours than have been given to any other scientific question in the history of mankind." In the generation after Bethe wrote these words, even more effort has been expended on the topic than before and much progress has been made. We now know that nucleons are not elementary particles and their interactions derive from the force acting between quarks that make them up. While quantum chromodynamics gives a fairly good description of the structure of hadrons in terms of quarks, it is far less certain how the interaction between nucleons is quantitatively related to the fundamental quark-quark interaction.

In this chapter, we shall examine the problem from a mostly phenomenological point of view. We shall concentrate on the two-nucleon system and make use of its simplicity to illustrate some of the problems we have to face in nuclear studies. First we shall examine the deuteron, the only bound system formed of two nucleons. Far more information is provided by the scattering of one nucleon off another and we shall see what we can learn about the nucleon-nucleon interaction from such studies.

§3-1 THE DEUTERON

Binding energy. The deuteron is a very unique nucleus in many respects. It is only loosely bound, having a binding energy much less than the average value between a pair of nucleons in all the other stable nuclei. The binding energy E_B of a nucleus is given by the mass difference between the neutral atom and the sum of the masses of neutrons and protons (in the form of a neutral hydrogen atom, see §4-10 for more detail). The mass of a deuteron M_d is 1876.125 MeV/c^2. The binding energy is then the difference between M_d and the sum of the masses of a neutron M_n and a hydrogen atom M_H

$$\begin{array}{rrl} M_n c^2 = & 939.566 & \text{MeV} \\ + M_{\rm H} c^2 = & 938.783 & \text{MeV} \\ \hline & 1878.349 & \text{MeV} \\ - M_d c^2 = & 1876.124 & \text{MeV} \\ \hline E_B = & 2.225 & \text{MeV} \end{array} \tag{3-1}$$

A more precise value, E_B =2.22457312(22) MeV, is obtained using the radiative capture of a neutron by hydrogen. In the reaction $H(n,\gamma)d$, a slow neutron is captured by a hydrogen atom followed by the emission of a γ-ray (see Greene et al., *Phys. Rev. Lett.* **56** [1986] 819 for details). If the energy of the incident neutron in the reaction is negligible, the energy of the γ-ray emitted gives the binding energy of the deuteron. It is usually far easier to determine γ-ray energies accurately than measurements of atomic masses; as a result, binding energies are often better known than absolute masses.

Table 3-1 Ground state properties of deuteron.

Ground State Property		Value	
Binding energy	E_B	2.22457312(22) MeV	
Spin and parity	J^π	1^+	
Isospin	T	0	
Magnetic dipole moment	μ_d	0.857406(1)	μ_N
Electric quadrupole moment	Q_d	0.28590(30)	efm^2
Radius	r_d	1.963(4)	fm

Partly because of the small binding energy, the deuteron has no excited state: all observations on the deuteron are made on the ground state. The results of the more important measured quantities are listed in Table 3-1. In spite of the small number of independent pieces of data available, we stand to learn a great deal about the two-nucleon system from the deuteron. Furthermore, because of their fundamental importance, many careful and sophisticated measurements have been carried out on the deuteron and the available values represent some of the best that can be obtained for the type of measurement. In this section we shall make use only of the spin, parity, and isospin of the deuteron, leaving the study of the magnetic dipole moment and electric quadrupole moment to the next two sections.

Parity, spin and isospin. The parity of a state describes the behavior of its wave function under a reflection of the coordinate system through the origin as shown in §A-1. For the deuteron, it is known that the parity of the ground state wave function is positive. Let us see what we can learn from this piece of experimental information. For this purpose, it is useful to separate the wave function into a product of three parts: the intrinsic wave function of the proton, the intrinsic wave function of the neutron, and the orbital wave function for the relative motion between the proton and the neutron. Since a proton and a neutron are just two different states of a nucleon, their intrinsic wave functions have the same parity. As a result, the product of their intrinsic wave functions has positive parity regardless of the parity of the nucleon. This leaves the parity of the deuteron to be determined solely by the relative motion between the two nucleons.

For states with a given orbital angular momentum $\boldsymbol{L}$, the angular dependence of the wave function is given by $Y_{LM}(\theta\phi)$, spherical harmonics of order L. Under an inversion of the coordinate system, spherical harmonics transform according to the relation given by (B-14),

$$Y_{LM}(\theta,\phi) \xrightarrow[P]{} Y_{LM}(\pi-\theta,\pi+\phi) = (-1)^L Y_{LM}(\theta,\phi)$$

The parity of $Y_{LM}(\theta,\phi)$ is therefore $(-1)^L$. The fact that the parity of the deuteron is known to be positive implies that the orbital angular momentum must be even.

The spin of the ground state of deuteron is $J = 1$ where $\boldsymbol{J} = \boldsymbol{L} + \boldsymbol{S}$. The possible values of $\boldsymbol{S}$, the sum of the intrinsic spins of the two nucleons, are 0 and 1. We can eliminate $S = 0$ since it is not possible to couple $S = 0$ with any of the allowed even values of L to form a $J = 1$ state. Furthermore, we can also rule out any L values greater than 2 by the same argument. From the fact that the spin and parity of a deuteron are $J^\pi = 1^+$, we find that the only possible values of (L,S) are $(0,1)$ and $(2,1)$. We shall see later that the dominant part of the ground state wave function is the $L = 0$ component; however, there is also a small but significant amount of $L = 2$ admixture.

Through symmetry arguments, we can deduce the isospin T for the deuteron from the values of L and S. Since the projection of isospin on the quantization axis is $t_0 = +\frac{1}{2}$ for a proton and $-\frac{1}{2}$ for a neutron, the deuteron is a state with the sum of the isospin projections $T_0 = 0$. The isospin of such a system of two nucleons can be coupled together to either $T = 0$ or 1. For a light nucleus such as the deuteron, the isospin is expected to be a good quantum number and the ground state of the deuteron can take on only one of these two values.

If, again, we regard a proton and a neutron as two different isospin states of a nucleon, the proton and the neutron in a deuteron may be treated as identical particles and the total wave function of the system must be antisymmetric under a permutation of the indices of the two (Fermi-Dirac) particles,

$$P_{12}\Psi(1,2) = \Psi(2,1) = -\Psi(1,2) \tag{3-2}$$

The wave function $\Psi(1,2)$ may be decomposed into a product of spatial, spin, and isospin parts. For the spatial part of the wave function, a permutation of the indices means that

$$\boldsymbol{r} \equiv \boldsymbol{r}_1 - \boldsymbol{r}_2 \xrightarrow[P_{12}]{} -\boldsymbol{r}$$

In spherical polar coordinate system, this corresponds to the transformation,

$$(r, \theta, \phi) \xrightarrow[P_{12}]{} (r, \pi - \theta, \phi + \pi)$$

Since the radial coordinate r is unchanged by the transformation, the symmetry of the spatial wave function is given by the angular dependence of the wave function and, consequently, that of the spherical harmonics. The transformation is, then, mathematically the same as that under a parity change. For $L = 0$ or 2, the spatial wave function is, then, symmetric under permutation.

It is also easy to see that the intrinsic spin part of the deuteron wave function is even in the $S = 1$ state. Consider the state with $M_S = 1$ among the triplet of states of $M_S = 0$ and ± 1 for $S = 1$. The intrinsic spin wave function may be written as the product of the intrinsic spin wave functions of the two nucleons,

$$|S{=}1, M_S{=}1\rangle = |s{=}\tfrac{1}{2}, m_s{=}\tfrac{1}{2}\rangle_1 |s{=}\tfrac{1}{2}, m_s{=}\tfrac{1}{2}\rangle_2 \tag{3-3}$$

The function on the right hand side of the equation is obviously even under a permutation of the indices of the two nucleons indicated by the subscripts. Since there is no other way to construct an $(S, M_S) = (1, 1)$ state, the function given by (3-3) is the intrinsic spin wave function for the state. The wave functions of the other two $S = 1$ states, with $M_S = 0$ and -1, may be generated from the $M_S = 1$ state using an angular momentum lowering operator. Since the operator is symmetric with respect to the two nucleons, the resulting wave functions retain the symmetry of the $(S, M_S) = (1, 1)$ state we have started with; consequently, they are also symmetric under a permutation of the two nucleons. From this, we establish that the intrinsic spin part of the deuteron wave function is even under a permutation of the two nucleons.

With both spatial and spin parts of the deuteron wave function being symmetric, the isospin part of the wave function must be antisymmetric in order to maintain the product of all three parts to be antisymmetric under a permutation of the two nucleons, as required by the Pauli exclusion principle. The algebra of intrinsic spin and isospin are identical to each other. From the discussion above on the intrinsic spin wave function, we can conclude that the $T = 1$ state of two nucleons is also symmetric under permutation. On the other hand, the antisymmetric linear combination,

$$|T{=}0, T_0{=}0\rangle = \frac{1}{\sqrt{2}} \Big\{ |t{=}\tfrac{1}{2}, t_0{=}+\tfrac{1}{2}\rangle_1 |t{=}\tfrac{1}{2}, t_0{=}-\tfrac{1}{2}\rangle_2 - |t{=}\tfrac{1}{2}, t_0{=}+\tfrac{1}{2}\rangle_2 |t{=}\tfrac{1}{2}, t_0{=}-\tfrac{1}{2}\rangle_1 \Big\} \tag{3-4}$$

describes a $T = 0$ state. This can be seen either by examining the explicit values of the Clebsch-Gordan coefficients involved or by the fact that the right hand side of (3-4) vanishes when either isospin raising or lower operators are applied to it. The requirement that the isospin part of the two-nucleon system is antisymmetric then implies that the ground state of the deuteron is in a $T = 0$ state.

We can also arrive at the same conclusion by a different set of arguments. If the ground state of the deuteron were $T = 1$, we would expect similar bound states in the other two $T = 1$ two-nucleon systems, the two-proton system ($T_0 = 1$) and the two-neutron system ($T_0 = -1$). However, no such bound states have been observed. We can perhaps eliminate the two-proton bound state on the grounds that Coulomb repulsion between two protons is of the order of 1 MeV at the distance of deuteron radius. Since this value is a large fraction of the binding energy for the deuteron, it is difficult to expect that a bound state can be formed of two protons. This limitation, however, does not apply to a system of two neutrons. Since no bound state is observed for two neutrons either, we come to the conclusion that it is not possible to have a $T = 1$ bound state for two nucleons. The neutron-proton system can be either an isoscalar ($T = 0$) or an isovector ($T = 1$) state. Since there does not seem to be a bound state for the $T = 1$ system, the deuteron ground state must have isospin $T = 0$. We may also conclude from the same argument that there is an isospin dependence in the force between a pair of nucleons that is attractive only in the $T = 0$ state.

In summary, we have established through consideration of symmetry that the ground state of the deuteron has $S = 1$ and $T = 0$. There remain, however, two possibilities, $L = 0$ and $L = 2$, for the spatial part of the wave function. In spectroscopic notation, the $L = 0$, $S = 1$ state is represented as $^3S_1$ (triplet-S state) and the $L = 2$, $S = 1$ as $^3D_1$ (triplet-D state). If L and S are good quantum numbers, *i.e.*, if the nuclear Hamiltonian H commutes with $\boldsymbol{L}^2$ and $\boldsymbol{S}^2$, the ground state of the deuteron would have to be in either one of these two states. There is, however, no fundamental reason to expect that this has to be true. In fact, we shall soon see that there is clear evidence that both the $^3S_1$- and the $^3D_1$-components must be present in the deuteron ground state. This, in turn, leads to the conclusion that the nuclear force mixes different L components in an eigenstate.

§3-2 DEUTERON MAGNETIC DIPOLE MOMENT

Magnetic dipole operator. The magnetic dipole moment of a nucleus arises from a combination of two different sources. First, each nucleon has a intrinsic magnetic dipole moment coming from the intrinsic spin and the orbital motion of quarks (see §2-8). Second, since each proton carries a net positive charge, its orbital motion constitutes a current loop. If, for simplicity, we assume that the proton charge is distributed evenly along its orbit, we can use classical electromagnetic theory to obtain its contribution to the magnetic dipole moment of a nucleus,

$$\boldsymbol{\mu}_i^{(\text{orbital})} = \frac{e\hbar[c]}{2M_pc}\boldsymbol{\ell}_i \tag{3-5}$$

where $\boldsymbol{\ell}_i$ is the orbital angular momentum of the i-th proton in units of $\hbar$ and M_p is its mass (see Problem 2-6). As usual, (3-5) is in cgs units if the factor inside the square brackets is ignored and in SI units if included.

It is more convenient to express the contributions to nuclear magnetic dipole moment from individual nucleons in terms of gyromagnetic ratios $g(i)$. For orbital motion, we can define $g_\ell(i)$ by the relation

$$\boldsymbol{\mu}_i^{(\text{orbital})} = g_\ell(i)\,\boldsymbol{\ell}_i \tag{3-6}$$

with

$$g_\ell(i) = \begin{cases} 1\ \mu_N & \text{for a proton} \\ 0 & \text{for a neutron} \end{cases}$$

to reflect the fact that only protons carry a net charge and, consequently, can contribute to the nuclear magnetic dipole moment. The use of nuclear magneton μ_N as the unit avoids any explicit dependence of the form of the equation on the system of electromagnetic units used. Similarly, contributions from the intrinsic spin of each nucleon may be expressed in the form

$$\boldsymbol{\mu}_i^{(\text{spin})} = g_s(i)\,\boldsymbol{s}_i \tag{3-7}$$

Since $s = \frac{1}{2}$, the gyromagnetic ratio for a free nucleon is

$$g_s(i) = \begin{cases} g_p = 2\mu_p = 5.585695\ \mu_N & \text{for a proton} \\ g_n = 2\mu_n = -3.826085\ \mu_N & \text{for a neutron.} \end{cases} \tag{3-8}$$

If we assume that the structure of a bound nucleon inside a nucleus is the same as in its free state, we may use g_p and g_n, the *bare* nucleon values, as those for $g_s(i)$ in nuclei as well.

In terms of gyromagnetic ratios, the magnetic dipole operator may be written as a function of the orbital angular momentum operator $\boldsymbol{\ell}_i$ and the intrinsic spin operator $\boldsymbol{s}_i$ of each nucleon. For a deuteron, only two nucleons are involved, and the operator takes on a particular simple form,

$$\boldsymbol{\mu}_d = g_p\boldsymbol{s}_p + g_n\boldsymbol{s}_n + \boldsymbol{\ell}_p$$

where $\boldsymbol{\ell}_p$ is the angular momentum of the proton, and $\boldsymbol{s}_p$ and $\boldsymbol{s}_n$ are, respectively, the intrinsic spin operators acting on the proton and the neutron wave functions. To simplify the expression, we have made use of the fact that $g_\ell = 1$ for a proton and 0 for a neutron. Since the masses of a proton and a neutron are roughly equal to each other, we may assume that each one of the two nucleons carries one half the orbital angular momentum associated with the relative motion between them, *i.e.*, $\boldsymbol{\ell}_p = \frac{1}{2}\boldsymbol{L}$, and we have the result

$$\boldsymbol{\mu}_d = g_p\boldsymbol{s}_p + g_n\boldsymbol{s}_n + \tfrac{1}{2}\boldsymbol{L} \tag{3-9}$$

where $\boldsymbol{L}$ is the orbital angular momentum of the deuteron.

Contribution from the $^3S_1$-state. For the $^3S_1$-state of the deuteron, $L = 0$, and the expectation value of the magnetic dipole operator, μ_d, reduces to a sum of the intrinsic dipole moments of a proton and a neutron,

$$\mu_d(^3S_1) = \mu_p + \mu_n = 0.879805\ \mu_N \tag{3-10}$$

The details of this calculation are given later in (3-14). The final result of (3-10) is almost the same as the observed value of $0.857406\mu_N$. The small difference,

$$\mu_d - \mu_d(^3S_1) = 0.857406 - 0.879805 = -0.022399\ \mu_N$$

is, however, worth more careful consideration.

We can think of at least three possible causes for the small departure of the measured value of the deuteron magnetic dipole moment from the expectation value in the $^3S_1$-state:

(1) The internal structures of the proton and the neutron are modified by the fact the two nucleons are in a bound state. As a result, the gyromagnetic ratios for intrinsic spin may be different from g_p and g_n given in (3-8) for free nucleons.

(2) There are contributions from charged mesons exchanged between the proton and the neutron and these have not been included in (3-10).

(3) There is a small admixture of the $^3D_1$-state in the ground state of the deuteron.

Item 1 is extremely unlikely since the deuteron is only a loosely bound system: the binding energy of 2.22 MeV can hardly be expected to affect the motion of quarks inside a nucleon that are bound by energies of the order of hundreds of MeV. The effect of *mesonic current* suggested in item 2 is possible. In fact, it has been shown that the contributions from the mesonic current are important in understanding magnetic dipole moments in odd-mass nuclei (see §4-8). However, we shall not discuss the subject here, partly for the reason that item 3 is more likely to be the major cause for the discrepancy between μ_d and $\mu_d(^3S_1)$. Furthermore, it is not easy to distinguish between items 2 and 3 from a more fundamental, field-theoretical point of view. For simplicity, we shall consider only item 3 and concentrate on the admixture of the $^3D_1$-state as the source for the small discrepancy between the observed and the calculated $^3S_1$-state value of the deuteron magnetic dipole moment.

Expectation value of the magnetic dipole operator. Let us calculate next the expectation value of the magnetic dipole operator $\boldsymbol{\mu}_d$ in the $^3D_1$ state of the deuteron using the form given in (3-9). The expectation value of $\boldsymbol{\mu}_d$ depends on the M-value, the projection of spin $\boldsymbol{J}$ of the state on the z-axis. By convention, the magnetic moment, similar to other static electromagnetic moments, is defined as the expectation value of the z-component, or the $q = 0$-component in spherical tensor notation of Appendix B, of the operator in the substate of maximum M, *i.e.*, $M = J$. For the magnetic dipole moment of the deuteron, we have

$$\mu_d = \langle J, M\!=\!J|\mu_0|J, M\!=\!J\rangle \tag{3-11}$$

Since both μ_0 and J_0 are similar operators as far as their angular momentum properties are concerned, their expectation values must be proportional to each other, as shown in §B-6.

The constant of proportionality is given by the Landé formula (B-62) in terms of the expectation value of the projection of $\boldsymbol{\mu}$ on $\boldsymbol{J}$,

$$\begin{aligned}\langle J, M|\mu_0|J, M\rangle &= \frac{1}{J(J+1)}\langle J, M|(\boldsymbol{\mu}\cdot\boldsymbol{J})\, J_0|J, M\rangle \\ &= \frac{M}{J(J+1)}\langle J, M|(\boldsymbol{\mu}\cdot\boldsymbol{J})|J, M\rangle \end{aligned} \tag{3-12}$$

In order to calculate the expectation value of the scalar product between $\boldsymbol{\mu}$ and $\boldsymbol{J}$, we shall first rewrite (3-9) in terms of operators $\boldsymbol{S} = \boldsymbol{s}_p + \boldsymbol{s}_n$ and $\boldsymbol{J} = \boldsymbol{L} + \boldsymbol{S}$,

$$\boldsymbol{\mu}_d = \tfrac{1}{2}\{(g_p + g_n)\boldsymbol{S} + (g_p - g_n)(\boldsymbol{s}_p - \boldsymbol{s}_n) + \boldsymbol{L}\} \tag{3-13}$$

Since the operator $(\boldsymbol{s}_p - \boldsymbol{s}_n)$ acts on proton and neutron spins with opposite signs, it can only connect between two states, one with $S = 1$ and the other with $S = 0$ and therefore cannot contribute to the expectation value in which we are interested here. This reduces $\boldsymbol{\mu}_d$ to a function of $\boldsymbol{L}$ and $\boldsymbol{S}$ only.

On substituting the simplified form of $\boldsymbol{\mu}_d$ in (3-13) into (3-12), we obtain

$$\langle J, M|\mu_0|J, M\rangle = \frac{M}{J(J+1)}\langle J, M|\tfrac{1}{2}\{(g_p + g_n)(\boldsymbol{S}\cdot\boldsymbol{J}) + (\boldsymbol{L}\cdot\boldsymbol{J})\}|J, M\rangle$$

The scalar products in the expression may be written in terms of $\boldsymbol{J}^2$, $\boldsymbol{L}^2$ and $\boldsymbol{S}^2$,

$$\boldsymbol{S}\cdot\boldsymbol{J} = \boldsymbol{S}\cdot(\boldsymbol{L}+\boldsymbol{S}) = \boldsymbol{S}^2 + \tfrac{1}{2}(\boldsymbol{J}^2 - \boldsymbol{L}^2 - \boldsymbol{S}^2) = \tfrac{1}{2}(\boldsymbol{J}^2 - \boldsymbol{L}^2 + \boldsymbol{S}^2)$$

$$\boldsymbol{L}\cdot\boldsymbol{J} = \boldsymbol{L}\cdot(\boldsymbol{L}+\boldsymbol{S}) = \boldsymbol{L}^2 + \tfrac{1}{2}(\boldsymbol{J}^2 - \boldsymbol{L}^2 - \boldsymbol{S}^2) = \tfrac{1}{2}(\boldsymbol{J}^2 + \boldsymbol{L}^2 - \boldsymbol{S}^2)$$

and the value of the magnetic dipole moment in a state of given J, L, S and $M = J$ reduces to

$$\begin{aligned}\mu_d = \frac{1}{4(J+1)}\Big\{&(g_p + g_n)\big(J(J+1) - L(L+1) + S(S+1)\big) \\ &+\big(J(J+1) + L(L+1) - S(S+1)\big)\Big\}\end{aligned} \tag{3-14}$$

For the $^3S_1$-state, $L = 0$ and $J = S = 1$, we recover the result $\mu_d(^3S_1) = \mu_p + \mu_n$ given earlier in (3-10). For the $^3D_1$-state, we have $L = 2$ and this gives us the result

$$\mu_d(^3D_1) = \frac{1}{8}\{(g_p + g_n)(-2) + 6\} = 0.310\mu_N \tag{3-15}$$

Since this is smaller than the value of $\mu_d(^3S_1) = 0.880\mu_N$ given earlier in (3-10), any admixture of the $^3D_1$-state in the ground state of the deuteron will reduce the calculated value of μ_d from that given by $\mu_d(^3S_1)$).

Admixture of $^3D_1$-state. We can make a simple estimate of the amount of $^3D_1$-component in the ground state of the deuteron using the measured value of μ_d and the calculated values of $\mu_d(^3S_1)$ and $\mu_d(^3D_1)$ obtained earlier. We can take that the ground state wave function of the deuteron is given by the form

$$|\psi_d\rangle = a\,|^3S_1\rangle + b\,|^3D_1\rangle \tag{3-16}$$

with the normalization condition

$$a^2 + b^2 = 1 \tag{3-17}$$

Since there is no off-diagonal matrix element of $\boldsymbol{\mu}$ between $^3S_1$- and $^3D_1$-states, the deuteron magnetic dipole moment is given by

$$\mu_d = a^2\mu_d(^3S_1) + b^2\mu_d(^3D_1) = 0.857\mu_N \tag{3-18}$$

The values of a and b may be obtained by solving (3-17,18) as a set of two simultaneous equations. The value for b^2 turns out to be around 0.04, suggesting that there is a 4% admixture of the $^3D_1$ in the ground state of deuteron. As we shall see later, this is consistent, though somewhat on the lower side, with the evidence of admixture of the $^3D_1$-component from other measured properties of the deuteron.

§3-3 DEUTERON ELECTRIC QUADRUPOLE MOMENT

In electrostatics, it is well known that the potential due to an arbitrary charge distribution at points far away from the source is characterized by the distance and the moments of a multipole expansion of the source. For a microscopic object like a nucleus, it is not possible to observe the distribution directly and the measured multipole moments may be used to infer something about the source. For a nuclear charge distribution, the lowest non-vanishing multipole moment is the quadrupole moment, since the expectation value of the electric dipole operator and of all the other odd multipole electric operators vanishes due to the fact that the operators change sign under space inversion (see §4-7). For the deuteron, the measured quadrupole moment is $Q_d = 0.28590\ e\mathrm{fm}^2$, as shown in Table 3-1.

Quadrupole operator. For a spherical nucleus, the expectation values of the square of the distance from the center to the surface along x-, y-, and z-directions are equal to each other,

$$\langle x^2 \rangle = \langle y^2 \rangle = \langle z^2 \rangle$$

As a result, the expectation value of $r^2 = x^2 + y^2 + z^2$ is

$$\langle r^2 \rangle = \langle x^2 + y^2 + z^2 \rangle \xrightarrow[\text{spherical}]{} 3\langle z^2 \rangle$$

The charge quadrupole operator, which measures the lowest order departure from a spherical charge distribution in a nucleus, is defined in terms of the difference between $3z^2$ and r^2,

$$\boldsymbol{Q}_0 = e(3z^2 - r^2) \tag{3-19}$$

In the case of a spherical nucleus, we have $\langle \boldsymbol{Q}_0 \rangle = 0$ as a result. If a nucleus bugles out along the equatorial direction and flattens in the polar region, $\langle z^2 \rangle$ will be smaller than the average expectation value of the squares of the distance along the other two axes and the quadrupole moment will be negative. A positive quadrupole moment of $0.29\ e\,\text{fm}^2$ indicates that the deuteron is slightly elongated along the z-axis, like an olive.

The operator $\boldsymbol{Q}_0$ is a spherical tensor (see Appendix B) of rank two and carries two units of angular momentum. This can be seen by its expression in terms of spherical harmonics,

$$\boldsymbol{Q}_0 = e(3z^2 - r^2) = er^2(3\cos^2\theta - 1) = \sqrt{\frac{16\pi}{5}}\, er^2 Y_{20}(\theta, \phi) \tag{3-20}$$

The electric quadrupole moment of a nuclear state is defined as the expectation value of $\boldsymbol{Q}_0$ in the substate of maximum M,

$$Q_A = \langle J\, M{=}J | \boldsymbol{Q}_0 | J\, M{=}J \rangle \tag{3-21}$$

in the same way as the definition of magnetic dipole moment we have seen earlier.

From the point of view of angular momentum, any nuclear state with $J < 1$ cannot have a quadrupole moment different from zero. The expectation value $\langle J, M | \boldsymbol{Q}_0 | J, M \rangle$ vanishes if the three angular momenta involved, J, 2, and J, cannot be coupled together to form a closed triangle. At the same time, since $\boldsymbol{Q}_0$ operates only in the coordinate space, it is independent of the total intrinsic spin S. This means that the orbital angular momentum L of the state must also be greater or equal to 1. For this reason, the expectation value of $\boldsymbol{Q}_0$ vanishes in the ${}^3S_1$-state; the existence of a nonvanishing quadrupole moment is therefore a direct evidence of the presence of ${}^3D_1$-component in the deuteron ground state.

Expectation value of the quadrupole operator. Let us work out the connection between the spatial part of the wave function and the quadrupole moment, assuming for the time being that the deuteron is in a state of definite orbital angular moment L. Such a wave function $|LS;JM\rangle$ may be represented by the product of a spatial part $|LM_L\rangle$ and an intrinsic spin part $|SM_S\rangle$ coupled together to total angular momentum (J, M),

$$|LS;JM\rangle = \sum_{M_L M_S} \langle LM_L\, SM_S|JM\rangle |LM_L\rangle |SM_S\rangle \tag{3-22}$$

where $\langle LM_L\, SM_S|JM\rangle$ is the Clebsch-Gordan coefficient. For the expectation value of $\boldsymbol{Q}_0$, we have

$$\begin{aligned} Q_d(L) &= \langle LS;JM|\boldsymbol{Q}_0|LS;JM\rangle \\ &= \sum_{M_L M_S} \sum_{M'_L M'_S} \langle LM_L\, SM_S|JM\rangle \langle LM'_L\, SM'_S|JM\rangle \langle LM_L SM_S|\boldsymbol{Q}_0|LM'_L SM'_S\rangle \end{aligned}$$

Since the operator does not act on the intrinsic spin, we can remove this part of the wave function from the matrix element by making use of the orthogonal relation between the intrinsic spin wave functions,

$$\langle SM_S|SM'_S\rangle = \delta_{M_S M'_S}$$

As a result, the expectation value of $\boldsymbol{Q}_0$ is reduced to a matrix element involving the spatial part of the wave function only,

$$Q_d(L) = \sum_{M_L} \langle LM_L\, S(M-M_L)|JM\rangle^2 \langle LM_L|\boldsymbol{Q}_0|LM_L\rangle \tag{3-23}$$

where we have made use of the fact that the Clebsch-Gordan coefficients vanish if $M_S \neq (M - M_L)$.

We may simplify the expression for the matrix element of $\boldsymbol{Q}_0$ on the right hand side of (3-23) further by writing the spatial part of the wave function as a product of radial and angular parts,

$$|LM_L\rangle = R_L(r)\, Y_{LM_L}(\theta\phi)$$

where the angular part is given by the spherical harmonics $Y_{LM_L}(\theta\phi)$ and the radial part satisfies the normalization condition,

$$\int_0^\infty R_L^*(r) R_L(r)\, r^2\, dr = 1$$

Using the explicit form of the quadrupole moment operator given in (3-19), we obtain the result

$$\begin{aligned} \langle LM_L|\boldsymbol{Q}_0|LM_L\rangle &= e\sqrt{\frac{16\pi}{5}} \int_0^\infty R_L^*(r) r^2 R_L(r) r^2\, dr \\ &\quad \times \int_0^{2\pi}\int_0^\pi Y_{LM_L}^*(\theta\phi) Y_{20}(\theta\phi) Y_{LM_L}(\theta\phi) \sin\theta\, d\theta\, d\phi \end{aligned} \tag{3-24}$$

The radial integral here is the expectation value of r^2 and therefore must be a positive quantity; however, we cannot evaluate the integral without making some assumptions concerning the radial wave function.

The angular integral over three spherical harmonics may be expressed in terms of Clebsch-Gordan coefficients in the form,

$$\int_0^{2\pi}\int_0^{\pi} Y^*_{LM_L}(\theta\phi)Y_{20}(\theta\phi)\,Y_{LM_L}(\theta\phi)\sin\theta\,d\theta\,d\phi$$
$$= (-1)^{M_L}\begin{pmatrix} L & 2 & L \\ -M_L & 0 & M_L \end{pmatrix}(2L+1)\sqrt{\tfrac{5}{4\pi}}\begin{pmatrix} L & 2 & L \\ 0 & 0 & 0 \end{pmatrix} \tag{3-25}$$

For $L = 2$, the numerical values are $-1/7\sqrt{5/\pi}$, $+1/14\sqrt{5/\pi}$ and $+1/7\sqrt{5/\pi}$, respectively, for $M_L = 2$, 1, and 0. Before we can insert these values into (3-24) and obtain a value for $Q_d(L = 2)$, we also need the square of the three Clebsch-Gordan coefficients $\langle LM_L\, S\,(M-M_L)|JM\rangle^2$ for $S = 1$, $L = 2$, $M = J = 1$, and $M_L = 2$, 1 and 0. These can be found using Table B-2, and the values are 6/10, 3/10, and 1/10, respectively. With these results, we obtain the result

$$\begin{aligned} Q_d(^3D_1) &= \langle ^3D_1 M=1|\boldsymbol{Q}_0|^3D_1 M=1\rangle \\ &= \frac{6}{10}\langle L=2, M_L=2|\boldsymbol{Q}_0|L=2, M_L=2\rangle \\ &\quad + \frac{3}{10}\langle L=2, M_L=1|\boldsymbol{Q}_0|L=2, M_L=1\rangle \\ &\quad + \frac{1}{10}\langle L=2, M_L=0|\boldsymbol{Q}_0|L=2, M_L=0\rangle \\ &= -\frac{1}{5}e\langle r^2\rangle_D \end{aligned}$$

where

$$\langle r^2\rangle_D = \int_0^\infty R^*_D(r)r^2 R_D(r)r^2\,dr$$

If, as an estimate, we take the value of $\langle r^2\rangle_D$ as the square of the deuteron radius, we obtain a value $Q_d(^3D_1) = -0.77\,e\,\text{fm}^2$ if the deuteron were in a $^3D_1$-state. Since the sign even disagrees with the measured value, it is unlikely that the ground state of the deuteron is made up entirely of the $^3D_1$-state.

For a more realistic model, we shall take a linear combination of $^3S_1$- and $^3D_1$-components as we have done earlier in (3-16) for magnetic dipole moment calculations. The deuteron electric quadrupole moment now has the form,

$$\begin{aligned} Q_d = {} & a^2\langle ^3S_1 M=1|\boldsymbol{Q}_0|^3S_1 M=1\rangle + b^2\langle ^3D_1 M=1|\boldsymbol{Q}_0|^3D_1 M=1\rangle \\ & + 2ab\langle ^3S_1 M=1|\boldsymbol{Q}_0|^3D_1 M=1\rangle \end{aligned}$$

The first term vanishes, since $L = 0$. The main contribution is likely to come from the last term, since $^3D_1$-component is only a few percent of the total and $|a| > |b|$

as a result. This term involves an off-diagonal matrix element and depends on the radial integral,

$$\langle {}^3S_1 M=1|Q_0|{}^3D_1 M=1\rangle \propto \int_0^\infty R_S^*(r) r^2 R_D(r)\, r^2\, dr$$

The value of the integral is sensitive to the detailed shapes of the radial wave functions and, as a result, it is difficult to put a firm value on the amount of ${}^3D_1$-component in the ground state of the deuteron from electric quadrupole moment calculations. Most of the estimates put the value of $|b|^2$ as in the range of 4% to 7%.

§3-4 TENSOR FORCE AND THE DEUTERON *D*-STATE

We have seen from the analyses of both magnetic dipole and electric quadrupole moments that the deuteron ground state is a linear combination of ${}^3S_1$- and ${}^3D_1$-states. Since there is no definite L-value that we can associate with an observed state, orbital angular momentum is not a good quantum number. This, in turn, implies that the nucleon-nucleon interaction potential does not commute with the operator $\boldsymbol{L}^2$. The presence of ${}^3D_1$-component in the deuteron ground state therefore gives us one clear piece of information concerning the property of nuclear force.

The Hamiltonian for the deuteron problem may be written in the form

$$H = -\frac{\hbar^2}{2\mu}\nabla^2 + V \tag{3-26}$$

where the first term is the kinetic energy in the center of mass of the two-nucleon system and μ is the reduced mass. The second term expresses the interaction between the two nucleons in terms of a potential V. The ground state wave function, ψ_d, is the eigenfunction of the Schrödinger equation

$$H|\psi_d\rangle = E_d|\psi_d\rangle \tag{3-27}$$

with eigenvalue E_d. We have deduced earlier in §3-1 from symmetry arguments that only $L = 0$ and 2 can contribute to the wave function, and the eigenfunction therefore has the form given in (3-16),

$$|\psi_d\rangle = a\,|{}^3S_1\rangle + b\,|{}^3D_1\rangle$$

where a and b are coefficients to be determined by solving the Schrödinger equation (3-27).

It may be convenient for us here to think in terms of a matrix approach to the eigenvalue problem (see also §6-6). Since we are only interested in finding the amount of mixing between $^3S_1$- and $^3D_1$-states, we may use these two states as the basis to construct the Hamiltonian matrix,

$$\{H\} = \begin{pmatrix} H_{11} & H_{12} \\ H_{21} & H_{22} \end{pmatrix} \tag{3-28}$$

where the matrix elements are given by the following definitions:

$$H_{11} = \langle {}^3S_1|H|{}^3S_1\rangle \qquad H_{22} = \langle {}^3D_1|H|{}^3D_1\rangle \qquad H_{12} = H_{21} = \langle {}^3D_1|H|{}^3S_1\rangle$$

On diagonalizing this real, symmetric matrix we can obtain the energy E_d and the coefficients a and b. However, this is not our interest at the moment; we are more concerned with the type of Hamiltonian which can produce a mixing between $^3S_1$ and $^3D_1$ states.

If the off-diagonal matrix elements H_{12} and H_{21} vanish, the Hamiltonian matrix is diagonal in the basis states; the two eigenstates are then $|\,^3S_1\,\rangle$ and $|\,^3D_1\,\rangle$, respectively, without any admixtures between them. The fact that the ground state of the deuteron is a mixture of these two basis states implies that the off-diagonal matrix elements of the Hamiltonian are not zero:

$$\langle {}^3D_1|H|{}^3S_1\rangle \neq 0 \tag{3-29}$$

Since the kinetic energy term of the Hamiltonian contributes only to the diagonal matrix elements in the two-dimensional Hilbert space we are working in here, (3-29) must be the result of the interaction potential V. This leads to the conclusion

$$\langle {}^3D_1|V|{}^3S_1\rangle \neq 0 \tag{3-30}$$

that is, the nuclear potential is not diagonal in a basis span by states with definite orbital angular momentum and can therefore mix $^3S_1$- and $^3D_1$-states.

In order to have a non-vanishing matrix element, the potential V must have a spatial part that is a spherical tensor of rank 2 so as to be able to connect an S- to a D-state as required by (3-30). Again, let us express the deuteron wave function as a product of spatial, intrinsic spin, and isospin parts. Similarly, the matrix element of V above may also be written as a product of three matrix elements; one in coordinate space, one in intrinsic spin space, and one in isospin space. For simplicity, in the following discussion we shall ignore any dependence on the isospin in the matrix element of V.

Since the nuclear Hamiltonian conserves the total angular momentum of the system, the potential V must be a scalar in spin J. However, if the spherical rank of the spatial part of the operator is two, we must find an operator of the same rank for the intrinsic spin part of the wave function so that a scalar product of these two rank-two operators can be constructed. For this purpose, let us examine first the possible spherical tensor operators that we can construct in the intrinsic spin space.

For a spin-$\frac{1}{2}$ system, an arbitrary operator may be expressed as a linear combination of the Pauli spin matrices,

$$\boldsymbol{\sigma}_x = \begin{pmatrix} 0 & 1 \\ 1 & 0 \end{pmatrix} \qquad \boldsymbol{\sigma}_y = \begin{pmatrix} 0 & -i \\ i & 0 \end{pmatrix} \qquad \boldsymbol{\sigma}_z = \begin{pmatrix} 1 & 0 \\ 0 & -1 \end{pmatrix} \tag{3-31}$$

and the two-dimensional unit matrix. In terms of spherical components, the Pauli matrices may be written in the form

$$\begin{aligned} \boldsymbol{\sigma}_{+1} &= -\frac{1}{\sqrt{2}}(\boldsymbol{\sigma}_x + i\boldsymbol{\sigma}_y) = \sqrt{2}\begin{pmatrix} 0 & -1 \\ 0 & 0 \end{pmatrix} \\ \boldsymbol{\sigma}_{-1} &= +\frac{1}{\sqrt{2}}(\boldsymbol{\sigma}_x - i\boldsymbol{\sigma}_y) = \sqrt{2}\begin{pmatrix} 0 & 0 \\ 1 & 0 \end{pmatrix} \\ \boldsymbol{\sigma}_0 &= \boldsymbol{\sigma}_z = \begin{pmatrix} 1 & 0 \\ 0 & -1 \end{pmatrix} \end{aligned} \tag{3-32}$$

These are the three components of the operator $\boldsymbol{\sigma}$, having a spherical tensor of rank unity, that acts only on the intrinsic spin part of the wave function of a nucleon. It carries one unit of angular momentum, similar to other vector operators such as $\boldsymbol{L}$.

A two-body operator in the nucleon intrinsic spin space may be constructed from a product of $\boldsymbol{\sigma}(1)$ and $\boldsymbol{\sigma}(2)$, the intrinsic spin operator for particles 1 and 2, respectively. Since each one is a vector, the spherical tensor rank of the product is the vector sum of $\boldsymbol{\sigma}(1)$ and $\boldsymbol{\sigma}(2)$ and may carry zero, one, or two units of angular momentum as shown in §B-3. The first two possibilities are, respectively, analogous to the usual scalar and vector products in multiplications of ordinary vectors. The last possibility is a new one and is often loosely referred to as the "tensor product" of two vector operators. The name should not be confused with the more general product of two spherical tensors.

The scalar product of two vectors is a familiar quantity. For example, in a cartesian coordinate system, a scalar two-body operator in the intrinsic spin space may be expressed in the form,

$$\boldsymbol{\sigma}(1) \cdot \boldsymbol{\sigma}(2) = \boldsymbol{\sigma}_x(1)\boldsymbol{\sigma}_x(2) + \boldsymbol{\sigma}_y(1)\boldsymbol{\sigma}_y(2) + \boldsymbol{\sigma}_z(1)\boldsymbol{\sigma}_z(2) \tag{3-33}$$

In terms of the spherical components given in (3-32), the same product may be written in the following way,

$$\boldsymbol{\sigma}(1) \cdot \boldsymbol{\sigma}(2) = \boldsymbol{\sigma}_0(1)\boldsymbol{\sigma}_0(2) - \boldsymbol{\sigma}_{+1}(1)\boldsymbol{\sigma}_{-1}(2) - \boldsymbol{\sigma}_{-1}(1)\boldsymbol{\sigma}_{+1}(2) \tag{3-34}$$

as done in (B-58). We can also make use of the explicit values of the following Clebsch-Gordan coefficients given in §B-3,

$$\langle 1{+}1\,1{-}1|00\rangle = \langle 1{-}1\,1{+}1|00\rangle = +\frac{1}{\sqrt{3}} \qquad \langle 1010|00\rangle = -\frac{1}{\sqrt{3}}$$

and write the scalar product in spherical tensor notation,

$$\begin{aligned}\boldsymbol{\sigma}(1)\cdot\boldsymbol{\sigma}(2) &= -\sqrt{3}\sum_q \langle 1q\,1-q|00\rangle \boldsymbol{\sigma}_q(1)\boldsymbol{\sigma}_{-q}(2) \\ &= -\sqrt{3}\big(\boldsymbol{\sigma}(1)\times\boldsymbol{\sigma}(2)\big)_{00}\end{aligned} \tag{3-35}$$

where the dummy index of summation q is over values -1, 0, and $+1$. The last form expresses the two-body operator in intrinsic spin space as a product of two Pauli spin operators with angular momentum coupled together to zero.

In general a product of two operators with tensorial ranks r and s, coupled to form a tensor of rank t, may be written in the form of an angular momentum coupled product,

$$\big(T_r \times U_s\big)_{tm} \equiv \sum_{pq} \langle rpsq|tm\rangle\, T_{rp}U_{sq} \tag{3-36}$$

as given in (B-32). Thus, the vector product of $\boldsymbol{\sigma}(1)$ and $\boldsymbol{\sigma}(2)$ is given by

$$\big(\boldsymbol{\sigma}(1)\times\boldsymbol{\sigma}(2)\big)_{1m} = \sum_{pq}\langle 1p1q|1m\rangle \boldsymbol{\sigma}_p(1)\boldsymbol{\sigma}_q(2) \tag{3-37}$$

It is left as an exercise to show that this is equivalent to the vector product of $\boldsymbol{\sigma}(1)$ with $\boldsymbol{\sigma}(2)$ in cartesian coordinates (see Problem 3-5).

Table 3-2 Values of Clebsch-Gordan coefficients $\langle 1p1q|2m\rangle$.

m	p	q	$\langle 1p1q\|2m\rangle$	m	p	q	$\langle 1p1q\|2m\rangle$
0	1	−1	$\sqrt{\frac{1}{6}}$	1	1	0	$\sqrt{\frac{1}{2}}$
0	−1	1	$\sqrt{\frac{1}{6}}$	1	0	1	$\sqrt{\frac{1}{2}}$
0	0	0	$\sqrt{\frac{2}{3}}$	−2	−1	−1	1
−1	−1	0	$\sqrt{\frac{1}{2}}$	2	1	1	1
−1	0	−1	$\sqrt{\frac{1}{2}}$				

By the same token, we can also write out the components of the second rank tensor product of $\boldsymbol{\sigma}(1)$ and $\boldsymbol{\sigma}(2)$ using (3-36),

$$\big(\boldsymbol{\sigma}(1)\times\boldsymbol{\sigma}(2)\big)_{2m} = \sum_{pq}\langle 1p1q|2m\rangle \boldsymbol{\sigma}_p(1)\boldsymbol{\sigma}_q(2)$$

Each of the components may be expressed explicitly in the form

$$\begin{aligned}\big(\boldsymbol{\sigma}(1)\times\boldsymbol{\sigma}(2)\big)_{20} &= \frac{1}{\sqrt{6}}\{\boldsymbol{\sigma}_1(1)\boldsymbol{\sigma}_{-1}(2)+\boldsymbol{\sigma}_{-1}(1)\boldsymbol{\sigma}_1(2)+2\boldsymbol{\sigma}_0(1)\boldsymbol{\sigma}_0(2)\} \\ \big(\boldsymbol{\sigma}(1)\times\boldsymbol{\sigma}(2)\big)_{2\pm1} &= \frac{1}{\sqrt{2}}\{\boldsymbol{\sigma}_{\pm1}(1)\boldsymbol{\sigma}_0(2)+\boldsymbol{\sigma}_0(1)\boldsymbol{\sigma}_{\pm1}(2)\} \\ \big(\boldsymbol{\sigma}(1)\times\boldsymbol{\sigma}(2)\big)_{2\pm2} &= \boldsymbol{\sigma}_{\pm1}(1)\boldsymbol{\sigma}_{\pm1}(2)\end{aligned} \tag{3-38}$$

using the explicit values of the Clebsch-Gordan coefficients given in Table 3-2.

We can now return to the form of the operator for the intrinsic spin part of the nuclear potential V. Since the product of $\boldsymbol{\sigma}(1)$ and $\boldsymbol{\sigma}(2)$ can only be coupled together to form a scalar, a vector, or an operator of spherical rank two, the maximum rank of V that we can have in intrinsic spin space is two. Furthermore, since the intrinsic spin part and the orbital angular momentum part must have the same rank in order to form a scalar product in J, the maximum rank of the orbital angular momentum part of V must also be two. As we have seen earlier, this is adequate for our purposes since, from the admixture of ${}^3D_1$-component in the deuteron ground state, we have concluded in (3-30) that there must be a component in V with spherical tensor rank two in L.

Tensor operator. An operator, formed by the scalar product of a second rank operator in intrinsic spin space and a similar one in coordinate space, is called a *tensor operator* and is generally written in the form

$$S_{12} = \frac{3}{r^2}(\boldsymbol{\sigma}_1 \cdot \boldsymbol{r})(\boldsymbol{\sigma}_2 \cdot \boldsymbol{r}) - \boldsymbol{\sigma}_1 \cdot \boldsymbol{\sigma}_2 \tag{3-39}$$

where we have used subscripts to indicate which one of the two nucleons each Pauli spin operator acts on. We shall follow this practice for single-particle operators in general wherever there is no need to indicate the spherical tensor component of the operator. The context will always make it clear whether the subscript on an operator is for spherical tensor rank or particle number. It should also be pointed out that the form of the tensor operator given in (3-39) is only the L- and S-dependent part: the strength of the force and its radial and isospin dependence must be put in separately.

The fact that there is a small admixture of the ${}^3D_1$-component in the predominantly ${}^3S_1$ deuteron ground state implies that there must be a tensor component in the nucleon-nucleon potential. Although we cannot say much more about this component of the nuclear force from the properties of the deuteron alone, the clear evidence for the existence of such a component is an indication of the richness of the deuteron problem. In the next section, we shall also see that besides the tensor force, the nuclear potential also contains terms that have tensorial ranks zero and one in intrinsic spin and spatial coordinates, as well as other operators of rank two.

§3-5 SYMMETRY AND NUCLEAR FORCE

Nucleons interact with each other through two-body interactions; that is, the force between nucleons acts only between a pair of them at a time. Perhaps the best way to understand this point is by comparing it with the force experienced by an atomic electron. In an atom, the electrons are bound to a central electrostatic potential provided by the protons in the nucleus. As a result, the force experienced by an electron may be divided into two parts, a one-body part provided by the nucleus and a two-body part arising from the mutual interaction with another electron. In a nucleus, there is no external source to provide a force on the individual nucleon and, as a result, there is no "fundamental" one-body interaction. The only one-body operator in a nuclear Hamiltonian is the kinetic energy operator connected with the motion of each nucleon. For the convenience of carrying out calculations, one may, on occasion, construct an "effective" one-body term in the nuclear potential by taking an average of the two-body interaction a nucleon experiences in the presence of all the other nucleons in the nucleus. An example of such a term may be found in the Hartree-Fock approximation discussed in §6-7. The source of the effective one-body potential is, however, the two-body interaction between nucleons.

On the other hand, it is not possible to rule out three-body and higher particle rank terms in the nuclear interaction. A three-body force is one which is felt only when there are at least three particles present. For example, in a three-nucleon system such as a triton, the nucleus of tritium formed of one proton and two neutrons, or a ^{3}He made of two protons and one neutron, a two-body force acts between nucleons 1 and 2, between nucleons 2 and 3, and between nucleons 3 and 1. If, after taking away the sum of the interactions between these three pairs of nucleons and any one-body terms that may be present, there is still a residual force left in the system, we can then say that there is a three-body force between nucleons. All the available evidence indicates that such a force, if present, must be very much weaker than the two-body force. With the possible exception of three-nucleon systems, it is unlikely that our present experimental equipment and theoretical knowledge can detect the effect of three-body forces in nuclei. For this reason we shall ignore any possible three-body forces from now on. The same applies to other many-body terms in the nuclear potential.

One way to study nuclear two-body interactions is to make use of two-nucleon systems such as the deuteron. However, as we have already seen, the deuteron is a very limited system having only one bound state. A comprehensive investigation requires a more general system constructed of two nucleons: this can be achieved with the scattering of one nucleon off another one. Before going into the details of nucleon-nucleon scattering, we shall first ascertain the restrictions imposed on the nuclear interaction by the symmetry requirement of a two-nucleon system.

Charge independence. We shall assume that the nuclear force is *charge independent*: that is, the only difference in the interaction between a pair of protons and a pair of neutrons is the Coulomb interaction between the protons. This is, again, an assumption based on experimental evidence. There is no fundamental reason to rule out a charge-symmetry breaking term in the nuclear force itself. As we shall see in the next section, the difference in mass between charged and neutron pions alone implies a small but significant difference between proton-neutron interaction and the interaction between a pair of identical nucleons. However, from a practical point of view, the observed strengths of charge-symmetry breaking terms are, except for a few highly specialized cases, smaller than or comparable to the accuracy we can achieve at the moment in handling much stronger electromagnetic effects. For simplicity, we shall ignore from now on any charge-symmetry breaking term that may be present in nuclear force. Furthermore, for a discussion of nuclear force here, we shall also ignore any charge-symmetry breaking effects due to electromagnetic forces.

Since charge is related to the third component of the isospin operator $\boldsymbol{T}$, charge independence of the nuclear force implies the following commutation relation:

$$[H, \boldsymbol{T}_0] = 0 \tag{3-40}$$

where H is the nuclear Hamiltonian. This means that the eigenvectors $|\psi\rangle$ of a nuclear Hamiltonian can also be the eigenvectors of $\boldsymbol{T}_0$ at the same time,

$$\boldsymbol{T}_0|\psi\rangle = \tfrac{1}{2}(Z - N)|\psi\rangle \tag{3-41}$$

where Z is the proton number and N the neutron number of the nucleus. This is, however, a trivial statement since both Z and N are fixed in a nucleus. The important point to be made here is that nuclear force is invariant under a rotation in the isospin space.

Isospin Invariance. In addition to $\boldsymbol{T}_0$, the nuclear Hamiltonian commutes also with the square of the isospin operator,

$$[H, \boldsymbol{T}^2] = 0 \tag{3-42}$$

In other words, the eigenfunctions of the nuclear Hamiltonian are also eigenfunctions of operator $\boldsymbol{T}^2$,

$$\boldsymbol{T}^2|\psi\rangle = T(T+1)|\psi\rangle \tag{3-43}$$

Physically, it means that wave functions of a given isospin T are unchanged if we replace some of the protons by neutrons or vice versa. Such a transformation between the two states of a nucleon takes us from one member of the isobar, a nucleus of the same nucleon number, to another. Furthermore, since nothing else is changed, these two states must have essentially the same properties except for the replacement of some of the protons by neutrons or vice versa. Mathematically, the wave functions of these two states are related to each other through isospin raising or lowering operations. A group of states related by a rotation in the isospin space is known as *isobaric analogue states* (IAS) of each other. Many groups of such states have been identified among the light nuclei through the fact they have very similar properties even though they are found in nuclei with different numbers of neutrons and protons (but the same total number of nucleons). The energy level spectra for the two examples found in $A = 11$ and $A = 21$ are shown in Fig. 3-1. The members of each pair shown are *mirror nuclei* of each other; that is, they have the same number of nucleons except that the number of protons in one is equal to the number of neutrons in the other and vice versa. In a sense, they are the image of each other in a mirror that turns protons into neutrons and neutrons into protons. It is no surprise, then, that because of the charge independence of nuclear force, the energy level spectra, as well as the properties of various states of mirror nuclei, are very similar to each other; most of the differences may be attributed to the Coulomb interaction, which we have ignored in our discussion above.

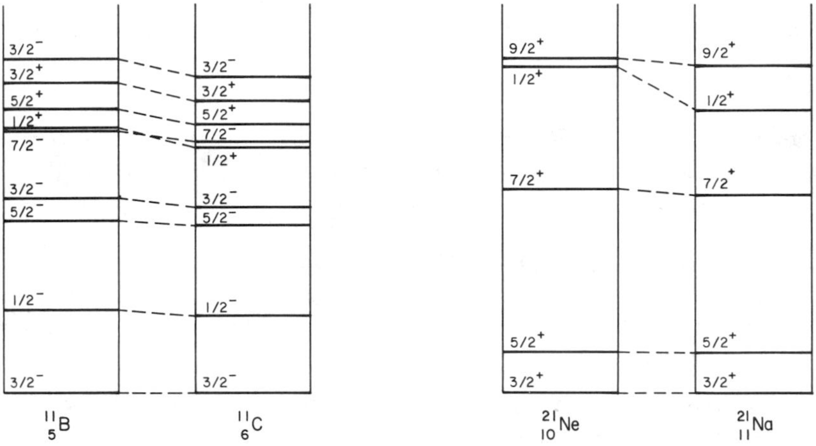

Fig. 3-1 A comparison of the low-lying spectra of members of the $A = 11$ and $A = 21$ isobars, showing the similarity of their level structure. The existence of isobaric analogue states, identical states in nuclei having the same A but different in N and Z, is a good indication of the isospin invariance of nuclear force.

In heavy nuclei, Coulomb repulsion becomes very strong because of the large number of protons present. Since electromagnetic interactions do not conserve isospin, we find that the nuclear states are no longer eigenstates of $\boldsymbol{T}^2$. In contrast with (3-43),

$$\boldsymbol{T}^2|\psi_i\rangle = \sum_j a_j|\psi_j\rangle$$

that is, when $\boldsymbol{T}^2$ acts on an eigenstate of the nuclear Hamiltonian, the result is a linear combination of eigenstates. The situation may described by saying that the strength of the IAS is split among several states. When this happens, it is difficult to find any direct evidence for the existence of IAS and the concept of isospin ceases to be useful. In spite of the difficulty caused by electromagnetic interactions, the evidence for isospin invariance of the nuclear force itself is quite strong. As we shall soon see, we may be able to make use of isospin symmetry to limit the possible forms of the nuclear potential of interest to us here.

Isospin operators. The property of isospin invariance is quite different from isospin dependence of the nuclear force. We have already seen evidence that the nuclear force is different depending on the isospin of the two-nucleon system; for example, a bound state is found for $T = 0$, the deuteron, but not for $T = 1$. This means that the nuclear potential must have an isospin-dependent component. Let us investigate the possible forms of isospin operators that can enter in a nucleon-nucleon potential.

For a single nucleon, the operator $\boldsymbol{\tau}$ may be written in terms of Pauli matrices,

$$\tau_1 = \begin{pmatrix} 0 & 1 \\ 1 & 0 \end{pmatrix} \qquad \tau_2 = \begin{pmatrix} 0 & -i \\ i & 0 \end{pmatrix} \qquad \tau_3 = \begin{pmatrix} 1 & 0 \\ 0 & -1 \end{pmatrix} \tag{3-44}$$

as we have done in (2-18). Alternatively, they can be written in terms of spherical components analogous to (3-31) except here, the operators act on the isospin part of the wave function only. Since there are only two isospin states for a nucleon, the only possible operators for the isospin part of the wave function are $\boldsymbol{\tau}$ and the identity operator

$$\mathbf{1} = \begin{pmatrix} 1 & 0 \\ 0 & 1 \end{pmatrix} \tag{3-45}$$

All other single-nucleon isospin operators can be expressed as a linear combination of $\mathbf{1}$ and $\boldsymbol{\tau}$. For example, the eigenvalue of $\boldsymbol{\tau}^2$ in the space of a nucleon is always 3, as can be seen by explicit calculation using the nucleon isospin wave functions given in (2-15,16). As a result, the operator $\boldsymbol{\tau}^2$ may be replaced by the form 3 times the identity operator $\mathbf{1}$.

For a system consisting of A nucleons, the isospin operator is the sum over those acting on individual nucleons,

$$\boldsymbol{T} = \tfrac{1}{2} \sum_{i=1}^{A} \boldsymbol{\tau}_i \tag{3-46}$$

For the two-nucleon system we are mainly interested in here, we may write

$$\boldsymbol{T} = \tfrac{1}{2}(\boldsymbol{\tau}_1 + \boldsymbol{\tau}_2) \tag{3-47}$$

Since nuclear force is two-body in nature and is a scalar in isospin space, we must now construct such an operator using $\mathbf{1}$ and $\boldsymbol{\tau}$ acting on each one of the two nucleons. The operator $\boldsymbol{T}$ is a one-body operator since it operates on a nucleon at a time. Furthermore, it is a vector in isospin space. One way to construct a two-body, isoscalar operator using $\boldsymbol{T}$ is to take a scalar product with itself. From (3-47), we have

$$\boldsymbol{T}^2 \equiv \boldsymbol{T} \cdot \boldsymbol{T} = \frac{1}{4}(\boldsymbol{\tau}_1{}^2 + \boldsymbol{\tau}_2{}^2 + 2\boldsymbol{\tau}_1 \cdot \boldsymbol{\tau}_2) \tag{3-48}$$

The first two terms on the right hand side are one-body operators, seen by the fact that they do not vanish even when there is only one nucleon present. Only the third term, $\boldsymbol{\tau}_1 \cdot \boldsymbol{\tau}_2$, is a two-body operator since it vanishes unless it is acting on a state with both nucleons 1 and 2 present. The operator $\boldsymbol{T}^2$ therefore has mixed particle rank of one and two. The only purely two-body operators in the isospin space are, then, the unity operator and $\boldsymbol{\tau}_1 \cdot \boldsymbol{\tau}_2$: all other two-body isoscalar operators may be expressed as linear combinations of these two operators.

From (3-48), we have the relation,

$$\boldsymbol{\tau}_1 \cdot \boldsymbol{\tau}_2 = 2\boldsymbol{T}^2 - \tfrac{1}{2}\boldsymbol{\tau}_1{}^2 - \tfrac{1}{2}\boldsymbol{\tau}_2{}^2$$

For a single nucleon (isospin $\frac{1}{2}$), the expectation of $\boldsymbol{\tau}^2$ is 3 as we have seen earlier. The expectation value of $\boldsymbol{\tau}_1 \cdot \boldsymbol{\tau}_2$ in the space of two nucleons with total isospin T is then

$$\langle T|\boldsymbol{\tau}_1 \cdot \boldsymbol{\tau}_2|T\rangle = \begin{cases} -3 & \text{for } T = 0 \\ \ \ 1 & \text{for } T = 1 \end{cases} \tag{3-49}$$

From this result, we can see that the operator $\boldsymbol{\tau}_1 \cdot \boldsymbol{\tau}_2$ is able to distinguish a two-nucleon state with isospin $T = 0$ from one with $T = 1$. In contrast, the identity operator has the same expectation value, unity, in both $T = 0$ and $T = 1$ states. These properties may be used to express the isospin dependence of nuclear force.

Other symmetries and general form of nuclear potential. The force between two nucleons must be invariant under a translation of the two-nucleon system; that is, the interaction can only depend on the relative position of the two nucleons and not on their absolute positions with respect to some arbitrary coordinate system. Consequently, the nuclear potential V is independent of the location of the center of mass of the two-nucleon system and only the relative coordinate

$$\boldsymbol{r} = \boldsymbol{r}_1 - \boldsymbol{r}_2 \tag{3-50}$$

can enter as one of the arguments. The potential may have a dependence on the momenta $\boldsymbol{p}_1$ and $\boldsymbol{p}_2$ of the two particles. On the other hand, since the sum $\boldsymbol{P} = \boldsymbol{p}_1 + \boldsymbol{p}_2$ corresponds to the center-of-mass momentum of the two-nucleon system, it cannot appear as an argument of the interaction between the two particles. The only possible momentum dependence V can have is on the relative momentum between two nucleons,

$$\boldsymbol{p} = \tfrac{1}{2}(\boldsymbol{p}_1 - \boldsymbol{p}_2) \tag{3-51}$$

This is known as the Galilean invariance of the system.

In additional to isospin, translational, and Galilean invariance, a nuclear potential must also be unchanged under a rotation of the coordinate system, time-reversal, space reflection (parity), and a permutation between the two nucleons. In terms of independent variables, the potential can only be a function of $\boldsymbol{\sigma}_1$, $\boldsymbol{\sigma}_2$, $\boldsymbol{\tau}_1$, $\boldsymbol{\tau}_2$ in addition to $\boldsymbol{r}$ and $\boldsymbol{p}$. As we have demonstrated with isospin operators, only a very limited number of linearly independent two-body operators can be constructed with this set of single-nucleon operators that satisfy the symmetry requirements of a nuclear potential. For example, the single-particle orbital angular momentum operator $\boldsymbol{\ell}$ is not an independent variable since it is the vector product of $\boldsymbol{r}$ and $\boldsymbol{p}$. Okubo and Marshak (*Ann. Phys.* **4** [1958] 166) have shown that the most general two-body potential under these conditions must take on the form:

$$\begin{aligned}V(r;\boldsymbol{\sigma}_1,\boldsymbol{\sigma}_2,\boldsymbol{\tau}_1,\boldsymbol{\tau}_2) =& V_0(r) + V_\sigma(r)\boldsymbol{\sigma}_1\cdot\boldsymbol{\sigma}_2 + V_\tau(r)\boldsymbol{\tau}_1\cdot\boldsymbol{\tau}_2 + V_{\sigma\tau}(r)(\boldsymbol{\sigma}_1\cdot\boldsymbol{\sigma}_2)(\boldsymbol{\tau}_1\cdot\boldsymbol{\tau}_2)\\ &+ V_{LS}(r)\boldsymbol{L}\cdot\boldsymbol{S} + V_{LS\tau}(r)(\boldsymbol{L}\cdot\boldsymbol{S})(\boldsymbol{\tau}_1\cdot\boldsymbol{\tau}_2)\\ &+ V_T(r)S_{12} + V_{T\tau}(r)S_{12}\,\boldsymbol{\tau}_1\cdot\boldsymbol{\tau}_2\\ &+ V_Q(r)Q_{12} + V_{Q\tau}(r)Q_{12}\,\boldsymbol{\tau}_1\cdot\boldsymbol{\tau}_2\\ &+ V_{PP}(r)(\boldsymbol{\sigma}_1\cdot\boldsymbol{p})(\boldsymbol{\sigma}_2\cdot\boldsymbol{p}) + V_{PP\tau}(r)(\boldsymbol{\sigma}_1\cdot\boldsymbol{p})(\boldsymbol{\sigma}_2\cdot\boldsymbol{p})(\boldsymbol{\tau}_1\cdot\boldsymbol{\tau}_2)\end{aligned} \tag{3-52}$$

In addition to the tensor operator S_{12} given earlier in (3-39), we have two other operators that are constructed from elementary single-nucleon operators: the two-body spin-orbit operator,

$$\boldsymbol{L}\cdot\boldsymbol{S} = \tfrac{1}{2}(\boldsymbol{\ell}_1 + \boldsymbol{\ell}_2)\cdot(\boldsymbol{\sigma}_1 + \boldsymbol{\sigma}_2) \tag{3-53}$$

and the quadratic spin-orbit operator,

$$Q_{12} = \tfrac{1}{2}\{(\boldsymbol{\sigma}_1\cdot\boldsymbol{L})(\boldsymbol{\sigma}_2\cdot\boldsymbol{L}) + (\boldsymbol{\sigma}_2\cdot\boldsymbol{L})(\boldsymbol{\sigma}_1\cdot\boldsymbol{L})\} \tag{3-54}$$

The radial dependence and strength of each one of the twelve terms are given by the twelve functions $V_0(r)$, $V_\sigma(r)$, $\cdots$. To determine the forms of these functions, we will need information in addition to that generated from symmetry arguments above. For example, we can make use of our knowledge of the basic nature of the nuclear force such as the meson-exchange picture of Yukawa discussed in the next section, or we can use a semi-empirical procedure by fitting some assumed forms of the radial dependence to experimental data. When our understanding of QCD is fully developed in the future, it will be possible to determined these functions from first principles.

The twelve terms in (3-52) may be divided into five groups. The first four terms,

$$V_{\text{central}} = V_0(r) + V_\sigma(r)\boldsymbol{\sigma}_1\cdot\boldsymbol{\sigma}_2 + V_\tau(r)\boldsymbol{\tau}_1\cdot\boldsymbol{\tau}_2 + V_{\sigma\tau}(r)(\boldsymbol{\sigma}_1\cdot\boldsymbol{\sigma}_2)(\boldsymbol{\tau}_1\cdot\boldsymbol{\tau}_2) \tag{3-55}$$

are the *central force* terms, since the tensorial ranks of the spatial parts of all four operators are zero. The first term $V_0(r)$ depends only on the radial distance r and is therefore invariant under a rotation of the coordinate system. The spatial dependence of the second term is also on r only, but it is different from the previous term by its dependence on the intrinsic spin through the operator $\boldsymbol{\sigma}_1\cdot\boldsymbol{\sigma}_2$. Analogous to (3-49), we have

$$\langle S|\boldsymbol{\sigma}_1\cdot\boldsymbol{\sigma}_2|S\rangle = \begin{cases} -3 & \text{for } S=0\\ \;\;1 & \text{for } S=1\end{cases} \tag{3-56}$$

for two nucleons with total intrinsic spin S. The operator $\boldsymbol{\sigma}_1 \cdot \boldsymbol{\sigma}_2$ therefore distinguishes a pair of nucleons in the triplet state ($S = 1$) from a pair in the singlet state ($S = 0$), in the same way as the operator $\boldsymbol{\tau}_1 \cdot \boldsymbol{\tau}_2$ separates an isovector nucleon pair from an isoscalar pair. Similar to the isospin case, there are only two linearly independent scalar two-boy operators, the identity operator and $\boldsymbol{\sigma}_1 \cdot \boldsymbol{\sigma}_2$, acting in the intrinsic spin space of two nucleons. The product of these two operators with two similar linearly independent operators in the isospin space gives the four separate terms in the central force. The first two terms of (3-55) are independent of isospin, but the third and fourth terms have isospin dependence through the operator $\boldsymbol{\tau}_1 \cdot \boldsymbol{\tau}_2$. Similarly, the first and third terms are independent of intrinsic spin. Since all four terms are scalars in intrinsic spin and, hence, in orbital angular momentum as well, a central force commutes with $\boldsymbol{S}^2$, $\boldsymbol{L}^2$, and $\boldsymbol{J}^2$.

The other terms in (3-52) do not preserve the total intrinsic spin and the total orbital angular momentum of a two-nucleon system. In the presence of these terms, the two-nucleon system is invariant only in the combined space of L and S labelled by J. The dependence of the nuclear force on the two-body spin-orbit operator is expressed by fifth and sixth terms in (3-52),

$$V_{\text{spin-orbit}} = V_{LS}(r)\boldsymbol{L} \cdot \boldsymbol{S} + V_{LS\tau}(r)(\boldsymbol{L} \cdot \boldsymbol{S})(\boldsymbol{\tau}_1 \cdot \boldsymbol{\tau}_2) \tag{3-57}$$

The reason that two separate components are needed for this and subsequent terms comes from the possibility that the radial dependence of the isospin-dependent and the isospin-independent parts may be quite different from each other; for example, as the result of different mesons being exchanged. The spatial part of the two-body spin-orbit operator involves $\boldsymbol{L}$. Since it does not change sign under an inversion of the spatial coordinate system, it is an axial vector. In order to maintain parity invariance as well as rotational invariance of V, only a scalar product with another axial vector may enter here. It is easy to see that the operator $\boldsymbol{L}^2$ is not useful for this purpose, since it conserves both L and S and is therefore a part of the central force. The only other possibility is the product $\boldsymbol{L} \cdot \boldsymbol{S}$. (See Problem 3-9 for other forms.)

The two-body spin-orbit operator, however, cannot connect two states with different orbital angular momenta, *i.e.*,

$$\langle LS|\boldsymbol{L} \cdot \boldsymbol{S}|L'S'\rangle = 0 \quad \text{for} \quad L' \neq L$$

This comes from a combination of two reasons. From angular momentum coupling requirements, the matrix element $\langle LS|\boldsymbol{L} \cdot \boldsymbol{S}|L'S'\rangle$ vanishes if $|L' - L| > 1$, since the operator $\boldsymbol{L}$ carries only one unit of orbital angular momentum. On the other hand, the parity of the orbital part of the wave function of a state with angular momentum L is $(-1)^L$. Under a space reflection, the operators $\boldsymbol{L}$ and $\boldsymbol{S}$ do not change sign. The matrix element $\langle LS|\boldsymbol{L} \cdot \boldsymbol{S}|L'S'\rangle$, however, changes sign if $L' = L \pm 1$ and must therefore vanish. As a result, the spin-orbit term is non-zero only between states of the same orbital angular momentum.

The next pair (seventh and eighth) of terms in (3-52) are the tensor force terms which we have already encountered earlier in §3-4. The quadratic spin-orbit terms, $V_Q(r)Q_{12}$ and $V_{Q\tau}(r)Q_{12}\boldsymbol{\tau}_1 \cdot \boldsymbol{\tau}_2$, enter only when there is momentum dependence in the potential. The last two terms $V_{PP}(r)(\boldsymbol{\sigma}_1 \cdot \boldsymbol{p})(\boldsymbol{\sigma}_2 \cdot \boldsymbol{p})$ and $V_{PP\tau}(r)(\boldsymbol{\sigma}_1 \cdot \boldsymbol{p})(\boldsymbol{\sigma}_2 \cdot \boldsymbol{p})(\boldsymbol{\tau}_1 \cdot \boldsymbol{\tau}_2)$ are often dropped since, for elastic scattering, they can be expressed as a linear combination of other terms. Their contributions therefore cannot be determined using elastic scattering, from which most of our information on nucleon-nucleon interaction is derived.

Returning now to the deuteron, we see that if only the central force terms V_0, V_σ, V_τ, and $V_{\sigma\tau}$ are present in the nuclear potential, then both L and S are good quantum numbers. The same is true for the spin-orbit terms for reasons mentioned earlier. Among the remaining terms, the simplest one that can admix the ${}^3S_1$ and ${}^3D_1$ states is the tensor force. The presence of the ${}^3D_1$ component in the ground state wave function provides the clearest indication of the presence of such a term in the nuclear force.

§3-6 YUKAWA THEORY OF NUCLEAR INTERACTION

From a practical point of view, the meson-exchange idea introduced by Yukawa in 1934 is a useful starting point for the examination of nucleon-nucleon interaction beyond that obtained in the previous section through symmetry arguments. In the Yukawa picture, the interaction between two nucleons is mediated by the exchange of various mesons. Although it is not easy to make a quantitative connection with the underlying quark structure of the hadrons, the theory makes it possible to relate nuclear interaction with various other hadronic processes such as the strength of meson-nucleon interaction. On a more empirical level, the Yukawa idea provides us with a reasonable form of the radial dependence for a nuclear potential. Such forms may be used, for example, as the starting point for semi-empirical forms of nuclear potentials; over the years they have led to our understanding of a variety of nuclear phenomena. Our focus in this section will be mainly on the origin of the meson-exchange idea itself. We shall leave applications of the idea to the last section after we have first taken a look at the experimental information derived from nucleon-nucleon scattering.

A proper derivation of a boson exchange potential requires a relativistic quantum field theory treatment that is beyond our present scope. However, the essence may be obtained by drawing an analogy to classical electrodynamics. The electrostatic potential $\phi(\boldsymbol{r})$ in a source-free region is a solution of the Laplace's equation

$$\nabla^2\phi(\boldsymbol{r}) = 0 \tag{3-58}$$

In the presence of a point source with charge q and located at the origin, the equation takes on the form

$$\nabla^2\phi(\boldsymbol{r}) = -\left[\frac{1}{4\pi\epsilon_0}\right]4\pi q\delta(\boldsymbol{r}) \tag{3-59}$$

The solution is the familiar Coulomb potential,

$$\phi(\boldsymbol{r}) = \left[\frac{1}{4\pi\epsilon_0}\right]\frac{q}{r} \tag{3-60}$$

When the electromagnetic field is quantized, photons emerge as the field quanta and the charge becomes the source of the field.

The nuclear force is different from electromagnetic forces in several respects, the most important one being its short range. We shall see evidence in support of this in the next chapter. For now, we are concerned mainly with the question of finding an equation similar to (3-59) and its analogue in quantum field theory for a short-range nuclear potential. The equation must also be invariant under a Lorentz transformation so that it is correct in the relativistic limit as well. This rules out the Schrödinger equation which applies only in the nonrelativistic limit. The field quantum that is exchanged between the nucleons must be a boson, as only bosons can be created and annihilated singly. A fermion, on the other hand, must be created and annihilated together with its antiparticle. The Dirac equation is therefore unsuitable here, since it is an equation for spin-$\frac{1}{2}$ particles. This leaves the Klein-Gordon equation as the prime candidate.

The relativistic energy-momentum relation is given by the equation,

$$E^2 = p^2c^2 + m^2c^4$$

We can quantize this equation in the same way as in nonrelativistic quantum mechanics by replacing energy E with operator $i\hbar(\partial/\partial t)$ and momentum $\boldsymbol{p}$ with operator $-i\hbar\nabla$. This gives us a quantized version of the relativistic energy-momentum relation,

$$-\hbar^2\frac{\partial^2}{\partial t^2}\phi(\boldsymbol{r}) = (-\hbar^2c^2\nabla^2 + m^2c^4)\phi(\boldsymbol{r}) \tag{3-61}$$

where m is now the mass of the field quantum. After dividing both sides of the equation by $(\hbar c)^2$ and rearranging the terms, we obtain the familiar form of the Klein-Gordon equation,

$$\left(\nabla^2 - \frac{1}{c^2}\frac{\partial^2}{\partial t^2}\right)\phi(\boldsymbol{r}) = \frac{m^2c^2}{\hbar^2}\phi(\boldsymbol{r}) \tag{3-62}$$

This is only the analogue of (3-58), since it does not yet contain a source term for field quanta. This point may be further demonstrated by letting the mass of the field quantum m go to zero and ignoring the time dependence. The result is a quantized version of (3-58).

In order to include a source, we must find the equivalent of Poisson's equation (3-59) by adding a source term to (3-62). For simplicity, we shall consider only the static limit and ignore terms involving time derivatives. For a point source with strength g located at the origin, we have the result

$$\nabla^2\phi(\boldsymbol{r}) = \frac{m^2c^2}{\hbar^2}\phi(\boldsymbol{r}) - g\delta(\boldsymbol{r}) \tag{3-63}$$

The solution for this equation,

$$\phi(\boldsymbol{r}) = \frac{g}{4\pi r} e^{-\frac{mc}{\hbar} r} \tag{3-64}$$

has the well-known form of a Yukawa potential and reduces to (3-60) on letting $m = 0$ and $g = [(4\pi\epsilon_0)^{-1}]4\pi q$. On the other hand, if the field quantum has a finite mass, we find that the strength of the potential drops by $\sim 1/e$ at a distance $r_0 = \hbar/mc$. The quantity r_0 may be taken as a measure of the range of the force mediated by a boson of mass m. For pions ($m \sim 140$ MeV/c^2), the value of r_0 is around 1.4 fm. We shall see later that the exchange of a single pion gives a good representation of the long-range part of the nuclear potential.

§3-7 NUCLEON-NUCLEON SCATTERING PHASE SHIFTS

The form of nucleon-nucleon interaction potential given in (3-52) was obtained using properties of the deuteron and symmetries of a two-nucleon system. To make further progress, we need additional experimental information, and this is provided by the scattering of one nucleon off another at different energies.

Nucleon-nucleon scattering. In principle there are four types of scattering measurements involving two nucleons that can be carried out. The scattering of an incident proton off a proton (*pp*-scattering) is the simplest one of the four to carry out experimentally, since it is relatively easy to accelerate protons and to construct targets containing hydrogen (proton). For neutron scattering, there are two major sources of incident beams. At low energies, neutrons from nuclear reactors may be used. At higher energies, neutrons produced by proton bombardment, for instance, through a (p, n) reaction on a ^{7}Li target, are often used. However, both the intensity and the energy resolution of neutron beams obtained in this way are much more limited than those of proton beams. As a result, neutron scattering is, in general, a more difficult experiment than proton scattering. The scattering of neutrons off proton targets (*np*-scattering) is an important source of information on two-nucleon systems. A system of two protons, or two neutrons, can only be coupled together to isospin $T = 1$, whereas a neutron-proton system can be either $T = 0$ or 1. Hence *np*-scattering, and the corresponding *pn*-scattering, contain information that cannot be obtained from *pp*- and *nn*-scattering.

In addition to *pp*- and *np*-measurements, one can, in principle, carry out *pn*- and *nn*-scattering experiments as well. Here, instead of using protons as the target, a "neutron target" is used. Free neutrons are unstable, with a half-life on the order of 10 minutes. It is therefore not possible to construct a "fixed" neutron target as it is with protons. There are, in principle, two methods of getting round this limitation. One way is to carry out a "colliding beam" experiment. In place of a target fixed in the laboratory, a neutron beam is used and, instead of having an incident beam scattering from a fixed target, two beams of particles are directed toward each other. Scattering takes place when the two beams collide. To be practical, such an experiment requires high intensities in both beams and, at this moment, highly intense beams of neutrons are not easily available. One such experiment involving two colliding beams of neutrons is designed to measure the low energy neutron-neutron scattering parameters to which we shall return in the next section.

The other way to "simulate" a fixed neutron target is to use deuterium. Since the deuteron is a loosely bound system of a neutron and a proton, the desired *pn*- or *nn*-scattering results can be obtained by carrying out the corresponding *pd*- or *nd*-scattering experiments. The contribution to the scattering from protons in the deuterium target may be removed by subtracting from the measured values the corresponding results obtained in *pp*- or *np*-scattering. This procedure is correct provided that:

(1) The subtraction procedure can be carried out with enough accuracy. This requires that the corresponding scattering data on a proton target are available with comparable or better accuracy, and that the effect of deuteron binding energy can be corrected in a satisfactory manner. In general, both points are relatively easy to achieve.

(2) Three-body effects are negligible. When a nucleon interacts with a deuteron, the entire system now consists of three nucleons. If there are fundamental three-body forces, their contributions will be present in *pd*-and *nd*-scattering but not in the scattering of one nucleon off another one. Hence, nucleon-deuteron scattering contains contributions from three-body forces that cannot be removed by subtracting the contributions of the proton alone. As we have seen earlier, this may not be a problem, since three-body forces, if they exist, are expected to be weak.

The information obtained from *pn*- and *nn*-scattering may not be any different from that obtained from *np*- and *pp*-scattering. For example, the only difference between *pn*- and *np*-scattering is whether the neutron or the proton is the target. Under time-reversal invariance, these two arrangements should give identical results.

As we have seen earlier, both *pp* and *nn* are $T = 1$ systems. If nuclear force is charge independent, the results of *pp*- and *nn*-scattering can only be different by the contribution made by the Coulomb interaction. Since the latter is well known, a comparison of *pp*- and *nn*- scattering results can, in principle, test the charge independence of nuclear interaction. However, the accuracy that can be achieved with *nn*-scattering is still inadequate for such a task. In the next section we shall see that there may be such a test possible at low energies where higher accuracies are possible.

To simplify the notation, we shall use the symbol NN from now on to represent a system of two nucleons when there is no need to differentiate between neutrons and protons, and the symbol np to represent both np and pn unless further distinction is required by the occasion. Furthermore, we shall assume that contribution of the Coulomb force, where present, has already been taken out and we can therefore ignore it in the discussion.

Our primary interest here is to relate scattering data to the NN potential. A large quantity of measured values at a variety of bombarding energies and scattering angles have been accumulated over the years. Instead of relating the potential V directly to the scattering results, it is more common to reduce the experimental information to phase shifts δ_ℓ for different partial waves. The merit of a particular potential is often discussed by comparing the phase shifts calculated using the potential with those extracted from experimental data, such as the examples shown later in Fig. 3-3. For this reason we shall first briefly review the subject of partial wave analysis for NN-scattering. A more detailed discussion is given in §C-2.

Scattering cross section. The quantity measured in a scattering experiment is the number of counts registered by a detector at a certain fixed angle (θ, ϕ). The counting rate depends on the solid angle subtended by the detector at the scattering center, the intensity of the incident beam, the number of target nuclei involved, and the differential cross section $d\sigma/d\Omega$. Our primary interest is in $d\sigma/d\Omega$, which, as defined in §C-1, has the dimension of area per unit solid angle. It is a function both of the bombarding energy and the scattering angle.

The scattering of one particle off another is described in the nonrelativistic limit by the Schrödinger equation. In the center of mass, the scattering wave function is the solution of the equation,

$$-\frac{\hbar^2}{2\mu}\nabla^2\psi + (V - E)\psi = 0 \tag{3-65}$$

where μ is the reduced mass of the two-nucleon system. We shall be mainly concerned with short-range nuclear forces here (see §C-5 for Coulomb scattering) and consequently we can assume that $V = 0$ except in a very small region where scattering takes place.

In the asymptotic region, far away from the small area where V is different from zero, the wave function has the form

$$\psi(r, \theta, \phi) \xrightarrow[r\to\infty]{} e^{ikz} + f(\theta, \phi)\frac{e^{ikr}}{r} \tag{3-66}$$

where the term e^{ikz} represents the incident plane wave and the part of the incident beam that is not scattered. The scattered wave is represented by a spherical function, $r^{-1}e^{ikr}$, radiating outward from the scattering region. The probability of scattering to direction (θ, ϕ) is represented by the scattering amplitude $f(\theta, \phi)$. For simplicity, we shall consider first only elastic scattering and, as a result, the wave number k in the center of mass of the two particles has the same magnitude

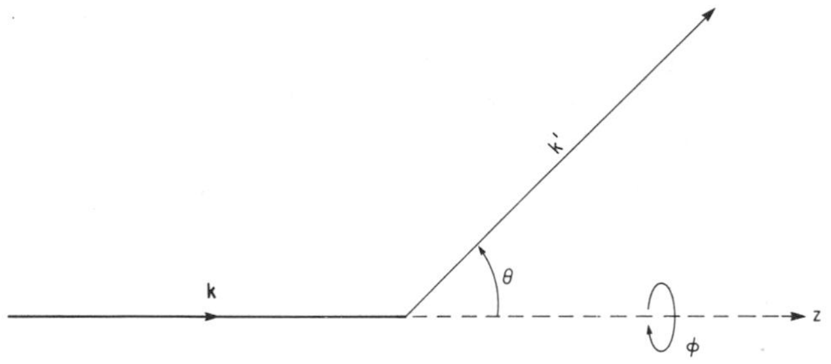

Fig. 3-2 Schematic diagram of a scattering arrangement. The vector $\boldsymbol{k}$ is along the direction of the projectile and the vector $\boldsymbol{k}'$ is along the direction of the scattered particle. The scattering angle θ is between $\boldsymbol{k}$ and $\boldsymbol{k}'$. The result of the scattering is independent of the azimuthal angle ϕ unless the orientation of the spin of one of the particles involved is known.

before and after the scattering. The differential scattering cross section at angles (θ, ϕ) is given by (C-8) as the square of the scattering amplitude,

$$\frac{d\sigma}{d\Omega}(\theta, \phi) = \left|f(\theta, \phi)\right|^2 \tag{3-67}$$

As shown in Fig. 3-2, the geometry of a scattering arrangement is such that it is convenient to place the origin of the coordinate system at the center of the scattering region and take the direction of the incident beam as the positive z-axis. The incident wave vector $\boldsymbol{k}$ and the scattered wave vector $\boldsymbol{k}'$ then define a plane, the scattering plane.

If nucleons in the incident beam and in the target are not polarized –that is, there is no preferred direction in space with which the intrinsic spins are aligned – the scattering is invariant with respect to a rotation around the z-axis. In such cases, the scattering cross section is independent of the azimuthal angle ϕ and the differential cross section is a function of the scattering angle θ only. We shall return later to the more general case where the orientation of the intrinsic spin of one or more nucleons is also detected.

Partial wave analysis. For a central potential, the relative angular momentum $\boldsymbol{\ell}$ between the two nucleons is a conserved quantity in the reaction. Under such conditions, it is useful to expand the wave function as a sum over the contributions from different partial waves,

$$\psi(r, \theta) = \sum_{\ell=0}^{\infty} a_\ell \, Y_{\ell 0}(\theta) \, R_\ell(k, r) \tag{3-68}$$

where a_ℓ are the expansion coefficients. Only spherical harmonics $Y_{\ell m}(\theta, \phi)$ with $m = 0$ appears in the expansion since, in the absence of polarization, the wave function is independent of the azimuthal angle ϕ (see also B-16). We have explicitly included the wave number k here in the argument of the radial wave function $R(k,r)$ so as to emphasize the dependence on energy.

For a free particle, $V = 0$, and the radial wave function reduces to the form

$$R_\ell(k,r) \xrightarrow{\text{free}} j_\ell(kr) \xrightarrow[r\to\infty]{} \frac{1}{kr}\sin(kr - \tfrac{1}{2}\ell\pi) \tag{3-69}$$

where $k = \hbar^{-1}\sqrt{2\mu E}$ and $j_\ell(\rho)$ is the spherical Bessel function of order ℓ. If only elastic scattering is allowed by the potential, the probability current density in each partial wave channel is conserved. The only effect the potential can have on the wave function is a change in the phase angle,

$$R_\ell(k,r) \xrightarrow[r\to\infty]{\text{scatt.}} \frac{1}{kr}\sin(kr - \tfrac{1}{2}\ell\pi + \delta_\ell) \tag{3-70}$$

where δ_ℓ is the *phase shift* in the ℓ-th partial wave channel. (For more details, see §C-2.)

In terms of δ_ℓ, the scattering amplitude may be expressed in the form

$$f(\theta) = \frac{\sqrt{4\pi}}{k}\sum_{\ell=o}^{\infty}\sqrt{2\ell+1}\,e^{i\delta_\ell}\sin\delta_\ell Y_{\ell 0}(\theta) \tag{3-71}$$

Using (3-67), the differential scattering cross section may be written in terms of the phase shifts,

$$\frac{d\sigma}{d\Omega} = \frac{4\pi}{k^2}\left|\sum_{\ell=0}^{\infty}\sqrt{2\ell+1}\,e^{i\delta_\ell}\sin\delta_\ell Y_{\ell 0}(\theta)\right|^2 \tag{C-20}$$

and the scattering cross section, the integral of $d\sigma/d\Omega$ over all solid angles, has the form

$$\sigma = \int\frac{d\sigma}{d\Omega}d\Omega = \frac{4\pi}{k^2}\sum_{\ell=0}^{\infty}(2\ell+1)\sin^2\delta_\ell(k) \tag{3-72}$$

Decomposition into partial waves is a useful way to analyze the scattering results for a given bombarding energy. In particular, only a few of the low order partial waves can contribute to the scattering at low energies as discussed in §C-2.

Nucleon-nucleon scattering phase shifts. Realistic nucleon-nucleon scattering differs in several important respects from the simple, central potential scattering discussed so far. First, we have learned earlier in §3-5 that nuclear potential in general depends on the total intrinsic spin of the two nucleons. As a result, the total angular momentum $\boldsymbol{J} = \boldsymbol{\ell} + \boldsymbol{S}$, rather than the orbital angular momentum $\boldsymbol{\ell}$, is conserved in the scattering. For two nucleons, the value of total intrinsic spin S can be either 0 or 1. In order to determine the value of S, we need to detect the orientations, or *polarizations*, of the spins of the nucleons involved. In fact, the information on NN scattering is incomplete unless polarizations are also observed. Second, with sufficient energy, scattering can excite the internal degrees of freedom of nucleons, for example, by changing one of them to a Δ-particle through such reactions as

$$p + p \rightarrow \Delta^{++} + n$$

or producing other particles such as pions from the reaction

$$p + p \rightarrow p + n + \pi^{+}$$

and baryon-antibaryon pairs through the reaction

$$p + p \rightarrow p + p + p + \overline{p}$$

These are inelastic scattering events, since part of the incident kinetic energy is now converted into excitation energies or mass of the particles created.

Since we are dealing with identical fermions, the scattering of two nucleons can take place only in a state that is totally antisymmetric with respect to a permutation of the two particles, in the same way calculated earlier for the deuteron. For *pp*-scattering, we have $T = 1$ and the two nucleons are symmetric as far as their total isospin wave function is concerned. If the intrinsic spins of the two protons are coupled together to $S = 0$ (antisymmetric state), the relative orbital wave function must be in a symmetric state and, as a result, only even ℓ-values are allowed. For $S = 0$, we have $J = \ell$, and the partial waves for the lowest two orders of *pp*-scattering are $^1S_0$ and $^1D_2$. The phase shifts extracted from the measured *pp*-scattering data for these two partial waves at bombarding energy less than 300 MeV in the laboratory are shown in Fig. 3-3(*a*) as illustrative examples. Only the real part of the phase shifts are shown. For bombarding energy less than 300 MeV in the laboratory, contributions from inelastic scattering are still relatively unimportant and the imaginary parts of the phase shifts extracted from measured scattering cross sections are still small.

By the same token, the partial waves for triplet ($S = 1$) *pp*-scattering have odd ℓ values. The lowest order in this case is *p*-wave ($\ell = 1$). When $\ell = 1$ is coupled with $S = 1$ three states, with $J = 0$, 1 and 2 are produced. The phase shifts for these three triplet states are also shown in Fig. 3-3(*a*). There is no admixture between the two $J = 0$ states, $^3P_0$ and $^1S_0$, since they are of different parity. As a result, we find that both ℓ and S are good quantum numbers here by default.

The *np*-system may be coupled together to either isospin $T = 0$ or $T = 1$. For $T = 0$, the two nucleons are antisymmetric in isospin. As a result, the $S = 0$ states must have odd ℓ values in order to be antisymmetric in the total wave function. The lowest order partial wave here is $\ell = 1$ and the phase shifts for $^1P_1$-scattering extracted from experimental data are shown in Fig. 3-3(*b*). In order for *p*-wave *np*-scattering to be in $S = 1$ state, it is necessary for the total isospin to be $T = 1$. The phase shifts in this case should be identical to those found in *pp*-scattering if nuclear force is charge independent and Coulomb effects are removed. An examination of these three sets of empirical *p*-wave phase shifts,

Fig. 3-3 Nucleon-nucleon scattering phase shifts for low order partial waves obtained from scattering and other related measurements (R.A. Arndt, J.S. Hyslop, III and L.D. Roper, *Phys. Rev. D* **35**, [1987] 128): (*a*) proton-proton scattering with contributions from the Coulomb potential removed, (*b*) isovector neutron-proton scattering and (*c*) isoscalar neutron-proton scattering. The crosses in the $^1S_0$ and $^3S_1$ phase shifts of *np*-scattering are the calculated results using a Paris potential (R. Vinh Mau, in *Mesons in Nuclei*, ed. by M. Rho and D.H. Wilkinson, North Holland, Amsterdam, 1979) and the circles are the corresponding results using a Bonn potential (R. Machleidt, K. Holinde, and Ch. Elster, *Phys. Rep.* **149** [1987] 1). Only the real part of the phase shifts are shown.

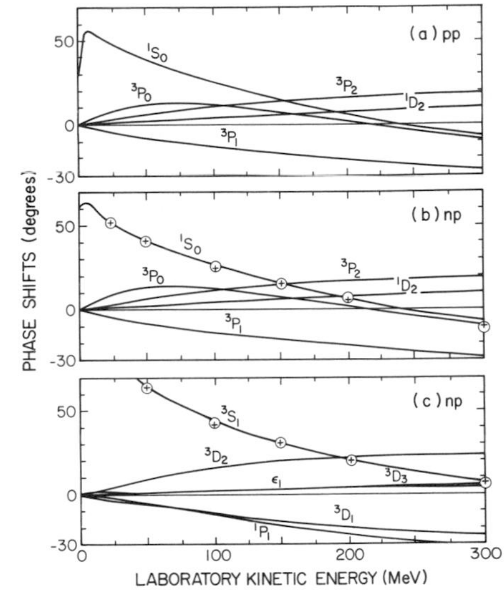

$^3P_0$, $^3P_1$, and $^3P_2$, in Fig. 3-3(*c*) shows that they are only slightly different from the corresponding values given in Fig. 3-3(*a*) for *pp*-scattering. It is not clear whether these small differences come from the way the phase shifts are extracted from experimental scattering cross sections or are indications of a weak charge dependence in the nuclear force. We shall return to this point in the next section in a discussion on the difference between the scattering lengths for *pp*- and *np*-scattering.

The other $T = 0$ phase shifts in the *np*-system, shown in Fig. 3-3(*b*), belong to triplet ($S = 1$), $\ell =$ even scattering. This is the first time we encounter a mixing of different ℓ-partial waves. Up to now, each phase shift has been characterized by a definite ℓ-value (as well as J- and S-values) even though the orbital angular momentum is not fundamentally a good quantum number. Because of parity and other invariance requirements, mixing of different ℓ-partial waves has not taken place. As in the case of the deuteron, the tensor force can cause admixture between two triplet states of the same J but different in ℓ by two units ($\ell = J \pm 1$). For a given J-value, the scattering is now specified by two (energy-dependent) phase shifts, $\delta_{J>}$ for $\ell = J - 1$ and $\delta_{J<}$ for $\ell = J + 1$, as well as an energy-dependent parameter ϵ_J to indicate the amount of mixing between the two in a physical state.

There are several ways to define the parameter ϵ_J. The usual convention used in the literature today is that of Stapp, Ypsilantis, and Metropolis (*Phys.

Rev. **105** [1957] 302). In this system of definitions, the scattering matrix (see §C-6) for a given J is written in the form

$$\{S\} = \begin{pmatrix} e^{2i\delta_{J>}} \cos 2\epsilon_J & ie^{i(\delta_{J>}+\delta_{J<})} \sin 2\epsilon_J \\ ie^{i(\delta_{J>}+\delta_{J<})} \sin 2\epsilon_J & e^{2i\delta_{J<}} \cos 2\epsilon_J \end{pmatrix} \tag{3-73}$$

In other words, for the scattering from the $\ell = J+1$ channel to the $\ell = J+1$ channel, the scattering matrix element is given by

$$e^{2i\delta_J} = e^{2i\delta_{J>}} \cos 2\epsilon_J$$

and from the $\ell = J-1$ channel to the $\ell = J-1$ channel,

$$e^{2i\delta_J} = e^{2i\delta_{J<}} \cos 2\epsilon_J$$

On the other hand, the scattering from $\ell = J-1$ to $\ell = J+1$ and from $\ell = J+1$ to $\ell = J-1$ is given by

$$e^{2i\delta_J} = e^{i(\delta_{J>}+\delta_{J<})} \sin 2\epsilon_J$$

These are generally referred to as the *nuclear bar* phase shifts. For the triplet $J = 1$ state, the values of ϵ_1 deduced from experimental data are shown as a part of Fig. 3-3(*b*).

Spin polarization in nucleon-nucleon scattering. We have seen earlier that nuclear potentials contain spin-dependent terms and, as a result, the scattering cross sections between nucleons are conditional on whether the sum of the intrinsic spins of the two nucleons is coupled to $S = 0$ or 1. It is not easy to observe the value of S directly in a scattering experiment. Usually it is the orientation of the nucleon intrinsic spins that is detected. Since each nucleon is a spin-$\frac{1}{2}$ particle, its projection on the quantization axis can either be $+\frac{1}{2}$ or $-\frac{1}{2}$. If the spins of all the nucleons in the incident beam are aligned in a particular direction, the beam is said to be a *polarized* one. Similarly, if the spins of the target nucleons are oriented along a given direction, the target is a polarized one. When the orientation of spins is taken into account, there are four possible arrangements of the two nucleons in the initial state, and in the final state, of a nucleon-nucleon scattering experiment. These may be represented in the form $|+\frac{1}{2},+\frac{1}{2}\rangle$, $|-\frac{1}{2},+\frac{1}{2}\rangle$, $|+\frac{1}{2},-\frac{1}{2}\rangle$ and $|-\frac{1}{2},-\frac{1}{2}\rangle$. Because of spin-dependent terms in the nuclear force, the orientations of the spins of the two nucleons may be changed as a result of the scattering. Since the final state may be a different one of the four possible combinations of spin orientations of the two nucleons, there are, in principle, sixteen different polarization measurements that can be carried out corresponding to starting from any one of the four possible incident spin combinations to any one of the four possible final spin combinations. Mathematically, we may write the scattering amplitude as a 4×4 matrix with each of the elements representing the probability for one of the sixteen possible arrangements for the scattering.

These sixteen quantities are not independent of each other. Because of time reversal and other symmetries inherit in the scattering, only five matrix elements are unique for each type of nucleon pairs, *i.e.,* *np*- or *pp*-scattering. This can be seen by writing the five independent scattering amplitudes in following manner:

$$\begin{aligned} f_1 &= f_{++,++} = f_{--,--} \qquad\qquad f_2 = f_{++,--} = f_{--,++} \\ f_3 &= f_{+-,+-} = f_{-+,-+} \qquad\qquad f_4 = f_{+-,-+} = f_{-+,+-} \\ f_5 &= f_{++,+-} = f_{-+,--} = f_{--,+-} = f_{-+,++} \\ &= f_{--,-+} = f_{+-,++} = f_{++,-+} = f_{+-,--} \end{aligned} \tag{3-74}$$

where we have used + and − in the subscript to stand, respectively, for $+\frac{1}{2}$ and $-\frac{1}{2}$ projections of the spins of two nucleons in the initial and in the final states.

Instead of scattering amplitudes f, it is more common to express NN scattering as the matrix element of the t-matrix operator defined in (C-98). An element of the t-matrix is related to the scattering amplitude in the following way,

$$f_{\boldsymbol{k}'\boldsymbol{k}} = -\frac{\mu}{2\pi\hbar^2}\langle \boldsymbol{k}'|t|\boldsymbol{k}\rangle \tag{3-75}$$

where $|\,\boldsymbol{k}\,\rangle$ and $|\,\boldsymbol{k}'\,\rangle$ are, respectively, the initial and final states of the two nucleons. In the place of f_1 to f_5, the t-matrix element for nucleon-nucleon interaction is often written as a function of five coefficients, A, B, C, E, and F, in the following form:

$$\begin{aligned} t_{\boldsymbol{k}'\boldsymbol{k}}(1,2) = A &+ B\boldsymbol{\sigma}_n(1)\boldsymbol{\sigma}_n(2) + C\{\boldsymbol{\sigma}_n(1) + \boldsymbol{\sigma}_n(2)\} \\ &+ E\boldsymbol{\sigma}_q(1)\boldsymbol{\sigma}_q(2) + F\boldsymbol{\sigma}_p(1)\boldsymbol{\sigma}_p(2) \end{aligned} \tag{3-76}$$

The three directional vectors $\boldsymbol{n}$, $\boldsymbol{p}$, and $\boldsymbol{q}$, along which the nucleon spin components are taken, are defined in terms $\boldsymbol{k}$ and $\boldsymbol{k}'$,

$$\hat{\boldsymbol{n}} = \frac{\boldsymbol{k}\times\boldsymbol{k}'}{|\boldsymbol{k}\times\boldsymbol{k}'|} \qquad \hat{\boldsymbol{q}} = \frac{\boldsymbol{k}'-\boldsymbol{k}}{|\boldsymbol{k}'-\boldsymbol{k}|} \qquad \hat{\boldsymbol{p}} = \hat{\boldsymbol{q}}\times\hat{\boldsymbol{n}} \tag{3-77}$$

The relation between coefficients A to F and scattering amplitudes f_1 to f_5, as well as other common ways of writing the NN-scattering t-matrix, can be found in standard references such as Bystricky, Lehar, and Winternitz (*J. de Phys.* **39** [1978] 1).

Instead of *np*- and *pp*-pairs, the decomposition of the NN-scattering amplitude into five independent quantities may be carried out in terms of $T = 0$ and $T = 1$ states of the two nucleons. The net result is that the amount of independent information obtained from scattering is greatly increased with polarization measurements. The experiments are, unfortunately, far more difficult than ordinary scattering measurements. Polarized beams are fairly common these days and it is not difficult to obtain analyzing power (A_y) measurements whereby a polarized beam is scattered from an unpolarized target and the polarization of the particles in the final state is not detected. Because of spin-dependence, the differential cross section at a fixed scattering angle θ may be different depending on whether the spin of the incident nucleon is polarized along the unit vector $\hat{n}$ defined in (3-77) or antiparallel to it. Such a difference is characterized by the analyzing power. This supplies one of the five independent quantities in the scattering. The sum of the two differential cross sections supplies the other.

For additional information, the polarization of the scattered particle must be measured. The only efficient way to detect the polarization of a nucleon is to carry out a second scattering off a target of known analyzing power. The asymmetry in the differential cross section of the second scattering gives information on the polarization direction of the nucleon after the first scattering. Since the cross section is low in general for nuclear processes, a second scattering greatly complicates the experimental setup and reduces the rate of data collection. In spite of such difficulties, more and more high-precision data involving the polarization of the scattered particle are becoming available. Valuable information can also be obtained using polarized targets. However, this requires sophisticated low temperature techniques to "freeze" the spin orientations of the nucleons in the target. Data involving such targets are, as a result, still quite rare.

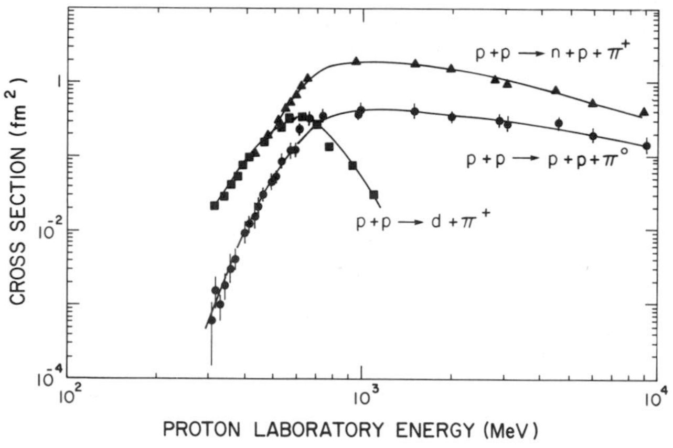

Fig. 3-4 Energy dependence of the total cross section for pion production in *pp*-scattering through reactions $p+p \rightarrow d+\pi^+$, $p+p \rightarrow p+n+\pi^+$ and $p+p \rightarrow p+p+\pi^0$. (Adapted from G. Jones, in *Pion Production and Absorption in Nuclei – 1981* ed. by R.D. Bent, AIP Conf. Proc. **79** [1982], Amer. Inst. Phys., New York, p. 15.)

Inelastic scattering. If the kinetic energy in the center of mass of the two-nucleon system is above the amount required for meson production, inelastic reactions become possible (see Fig. 3-4). Since the mass of the lightest meson, $\pi^{\pm}$, is around 140 MeV/c^2, we expect that some of the kinetic energy in the center of mass of the scattering system will be transferred to pion production once the bombarding energy is above the threshold. As the energy increases, excitation of the internal degrees of freedom of the nucleon as well as the production of other particles becomes more and more probable. Inelastic scattering represents a loss of flux from the incident channel and, as far as the incident channel is concerned, the probability amplitude is no longer conserved. Such a situation may be described by a complex scattering potential.

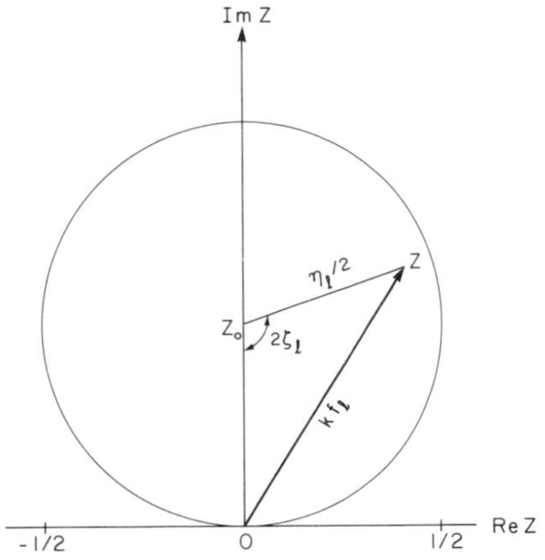

Fig. 3-5 Argand diagram showing the energy dependence of complex scattering amplitude for the ℓ-th partial wave. The length of the vector $\boldsymbol{z} = \boldsymbol{k} f_\ell$ is a function of the scattering energy and touches the unit circle centered at $z_0(0, \frac{1}{2}i)$ only in the limit of elastic scattering and η_ℓ becomes unity. The magnitude of η_ℓ is given by twice the distance $|\boldsymbol{z} - \boldsymbol{z}_0|$ and the phase shift is half the angle between $(\boldsymbol{z} - \boldsymbol{z}_0)$ and the imaginary axis.

Both the scattering amplitude and the phase shifts produced by a complex potential are generally also complex. Let us define a scattering amplitude f_ℓ for the ℓ-th partial wave by the following expression:

$$f(\theta) = \sum_{\ell=o}^{\infty} \sqrt{4\pi(2\ell+1)}\, f_\ell\, Y_{\ell 0}(\theta) \tag{3-78}$$

For purely elastic scattering,

$$f_\ell = \frac{1}{k} e^{i\delta_\ell} \sin\delta_\ell = \frac{1}{2ik}\left(e^{2i\delta_\ell} - 1\right) \tag{3-79}$$

as we have seen earlier. If inelastic scattering is taking place, the scattering amplitude becomes complex and may be expressed in terms of two real numbers: η_ℓ, the inelasticity parameter, and ζ_ℓ, defined by the relation

$$f_\ell = \frac{1}{2ik}\left(\eta_\ell e^{2i\zeta_\ell} - 1\right) \tag{3-80}$$

The energy dependence of a complex scattering amplitude is often displayed in the form of an Argand diagram in terms of the locus of the point

$$z = kf_\ell = \frac{1}{2i}\left(\eta_\ell e^{2i\zeta_\ell} - 1\right) \tag{3-81}$$

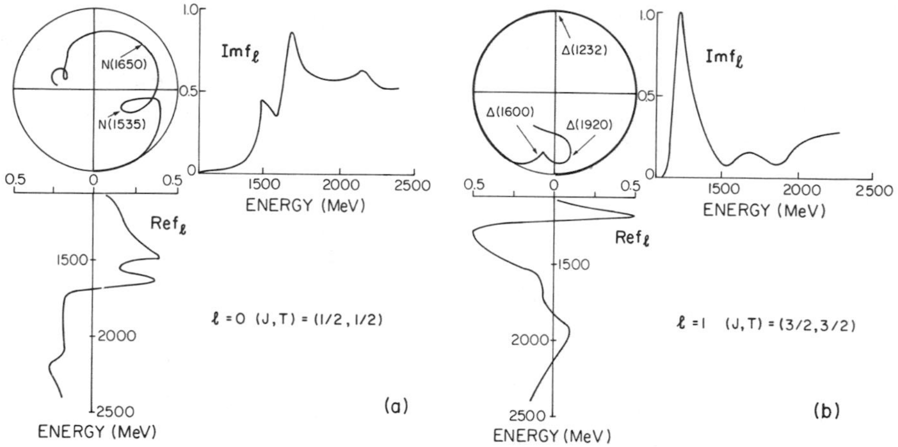

Fig. 3-6 Examples of Argand plot taken from pion-nucleon scattering; (*a*) for $\ell = 0$, $(J,T) = (\frac{1}{2}, \frac{1}{2})$ partial wave and (*b*) for $\ell = 1$, $(J,T) = (\frac{3}{2}, \frac{3}{2})$ partial wave which includes the strong P_{33}-resonance. The real and imaginary parts of the scattering amplitudes for each channel are also shown for comparison. (Adapted from Particle Data Group, *Phys. Lett.* **204B** [1988] 1.)

in the complex plan shown in Fig. 3-5. In terms of an Argand plot, the strong P_{33}-resonance in pion-nucleon scattering is indicated by a large imaginary scattering amplitude as shown in Fig. 3-6.

An examination of the values deduced from experimental nucleon-nucleon scattering data shows that the phase shifts are essentially real until the energy is above 300 MeV in the laboratory ($\sim$ 150 MeV in the center of mass). At much higher energies, the real and imaginary parts become comparable to each other as more and more inelastic channels are open. Complete lists of phase shifts up to 1 GeV in laboratory scattering energy are available, for example, from Arndt, Hyslop, and Roper (*Phys. Rev. D* **35** [1987] 128).

§3-8 LOW ENERGY SCATTERING PARAMETERS

Effective range analysis. If we make a partial wave expansion of the scattering wave function as was given in (3-68), for example, and substitute the results into the Schrödinger equation (3-65), we obtain an equation for the radial wave function of the ℓ-th partial wave,

$$-\frac{\hbar^2}{2\mu}\left\{\frac{1}{r^2}\frac{d}{dr}r^2\frac{d}{dr} - \frac{\ell(\ell+1)}{r^2}\right\}R_\ell(k,r) + V(r)R_\ell(k,r) = ER_\ell(k,r) \qquad \text{(3-82)}$$

The ℓ-dependent term, $\ell(\ell+1)/r^2$, comes from the angular part of the kinetic energy and is sometimes referred to as a centrifugal barrier since it is repulsive to an incoming particle. The "effective potential" experienced by the scattering particle is then

$$\tilde{V}(r) = V(r) + \frac{\hbar^2}{2\mu}\frac{\ell(\ell+1)}{r^2} \tag{C-26}$$

Because of the barrier, scattering at low energies is dominated by low order partial waves. In particular, for $E < 10$ MeV, nucleon-nucleon scattering is essentially given by s-wave ($\ell = 0$) alone, as can be seen from the fact that, among the observed values shown in Fig. 3-3, only $\ell = 0$ phase shifts δ_0 are significantly different from zero.

When kinetic energy $E \to 0$, the total cross section remains finite for nucleon-nucleon scattering. The limiting value is often characterized by a length parameter a defined by the relation

$$\lim_{E\to 0} \sigma = 4\pi a^2 \tag{3-83}$$

The quantity a is known as the *scattering length* and it is often convenient to discuss extremely low energy scattering in terms of it instead of the s-wave phase shift. The two quantities are related in the following way,

$$a = \lim_{k\to 0} \mathrm{Re}\left\{-\frac{1}{k}e^{i\delta_0}\sin\delta_0\right\} \tag{C-32}$$

where $k^2 = 2\mu E/\hbar^2$. The energy dependence of δ_0 at low energies is given by the *effective range* parameter r_e, defined by the relation

$$k\cot\delta_0 = -\frac{1}{a} + \frac{1}{2}r_e k^2 \tag{3-84}$$

A more detailed discussion of these parameters and their relation to the nucleon-nucleon interaction potential is given in §C-3.

Scattering length and effective range provide a useful way to parametrize information on low energy nucleon-nucleon scattering. Furthermore, these parameters may be related to observations other than NN scattering, such as deuteron binding energy. In addition, very accurate results can be obtained for the np-system by scattering slow neutrons off protons in hydrogen atoms bound in H_2 molecules. For these reasons, a great deal of attention is devoted to the measurement and understanding of these parameters.

Neutron scattering off hydrogen molecules. The hydrogen molecule, H_2, is a *homonuclear* molecule, a diatomic molecule made of two identical nuclei. Since the distance between the two atoms is large (7.8×10^{-11} m) compared with the range of nuclear force, we do not need to consider any nuclear interaction between the two protons in the H_2 molecule. On the other hand, the two protons, being identical particles, must obey the Pauli exclusion principle. Like other two-nucleon systems, the allowed states for two protons in a hydrogen molecule must be antisymmetric in the product of their orbital and spin wave functions. For this reason, the spin orientations of the two protons are correlated with their relative orbital angular momentum and such a correlation may be exploited for neutron-proton scattering length measurements.

There are two low-lying states for a hydrogen molecule. The lower one in energy is the *para*-hydrogen state in which the two protons are symmetric relative to each other in their spatial wave function. The higher energy state is the *ortho*-hydrogen state in which the two protons are antisymmetric in their spatial wave function. For an ortho-hydrogen, it is necessary that the intrinsic spins of the two protons be coupled together to $S_H = 1$ in order to satisfy the Pauli principle. Since, for this arrangement, M_S, the projection of $\boldsymbol{S}_H$ on the quantization axis, can take on three different values, -1, 0, and $+1$, there are three possible states associated with each ortho-hydrogen molecule. The total intrinsic spin of the two protons in a para-hydrogen, in contrast, is $S_H = 0$ and there is only one possible state. Because of this difference, ortho-hydrogen has three times the statistical weight of para-hydrogen in a sample in equilibrium at room temperature. On the other hand, at low temperatures, hydrogen molecules tend to go into the lowest possible energy state and, as a result, are almost completely in the lower energy, para-hydrogen state. Thus the relative amount of para- and ortho-hydrogen in a sample may be controlled by varying the temperature of the sample.

Measurements of low energy neutron scattering from hydrogen molecules may be carried out with high precision partly because of the intense neutron flux available at low energies. By lowering the energy, the wave length of incident neutrons can be made sufficiently long so that scattering off the two protons in a hydrogen molecule is a coherent one. Low energy neutrons are also useful in that very little energy is transferred to the hydrogen target. Energy received by a molecule may cause transitions from para- to ortho-hydrogen states and this decreases the accuracy that can be achieved in a measurement. For these reasons, the neutron energy is kept low, around 10 meV (10^{-3} eV), corresponding to a temperature of 100 K. At such low energies, contributions from the effective range term in (3-84) may be ignored, and the scattering is characterized by the two *np*-scattering lengths alone.

For $\ell = 0$, a neutron-proton system is either in their singlet state with $S = 0$ and $T = 1$ or triplet state with $S = 1$ and $T = 0$. The scattering length in the form of an operator may be expressed in the following way:

$$\boldsymbol{a} = \frac{1}{4}\big(3a_t + a_s\big) + \big(a_t - a_s\big)\boldsymbol{s}_n \cdot \boldsymbol{s}_p \tag{3-85}$$

where a_t and a_s are the scattering lengths for triplet and single states, respectively. The operators $\boldsymbol{s}_n = \frac{1}{2}\boldsymbol{\sigma}_n$ and $\boldsymbol{s}_p = \frac{1}{2}\boldsymbol{\sigma}_p$ act, respectively, on the intrinsic spins of the neutron and the proton wave functions. Similar to $\boldsymbol{\sigma}_1 \cdot \boldsymbol{\sigma}_2$ given in (3-56), the scalar product of the neutron and proton intrinsic spin operators, $\boldsymbol{s}_n \cdot \boldsymbol{s}_p$, is sensitive to the sum of intrinsic spins of the neutron-proton system. It is easy to check that the expectation value of $\boldsymbol{a}$ is a_t in a triplet state and a_s in a singlet state for a neutron-proton system.

Returning now to the hydrogen molecule, we may write the operator for the sum of the intrinsic spins of the two protons as

$$\boldsymbol{S}_H = \boldsymbol{s}_{p_1} + \boldsymbol{s}_{p_2} \tag{3-86}$$

In terms of $\boldsymbol{S}_H$, the scattering length for a slow neutron from these two protons may be written in a form similar to (3-85) above,

$$\boldsymbol{a}_H = \tfrac{1}{2}(3a_t + a_s) + (a_t - a_s)\boldsymbol{s}_n \cdot \boldsymbol{S}_H \tag{3-87}$$

The cross section expressed in terms of the expectation value of the square of $\boldsymbol{a}_H$ may be calculated in the following way,

$$\begin{aligned}\sigma_{\mathrm{H}} &= 4\pi\langle a_H^2\rangle \\ &= 4\pi\Big\{\tfrac{1}{4}(3a_t + a_s)^2 + (3a_t + a_s)(a_t - a_s)\langle \boldsymbol{s}_n \cdot \boldsymbol{S}_H\rangle \\ &\qquad + (a_t - a_s)^2\big\langle(\boldsymbol{s}_n \cdot \boldsymbol{S}_H)^2\big\rangle\Big\}\end{aligned} \tag{3-88}$$

For unpolarized incident neutrons, the second term on the right hand side vanishes on averaging over all possible orientations of the intrinsic spin of an incident neutron. The third term may be simplified by applying the same argument in the following way. First we expand the term explicitly as a function of the cartesian components of the intrinsic spins,

$$\begin{aligned}(\boldsymbol{s}_n \cdot \boldsymbol{S}_H)^2 =& s_{nx}^2 S_{Hx}^2 + s_{ny}^2 S_{Hy}^2 + s_{nz}^2 S_{Hz}^2 \\ &+ 2s_{nx}s_{ny}S_{Hx}S_{Hy} + 2s_{ny}s_{nz}S_{Hy}S_{Hz} + 2s_{nz}s_{nx}S_{Hz}S_{Hx}\end{aligned}$$

The expectation values of the last three terms in the expression are again zero for an unpolarized neutron beam. For the first three terms we note that, since $\boldsymbol{s}^2 = s_x^2 + s_y^2 + s_z^2$ and $\langle \boldsymbol{s}^2\rangle = \frac{3}{4}$, we have

$$\langle s_{nx}^2\rangle = \langle s_{ny}^2\rangle = \langle s_{nz}^2\rangle = \tfrac{1}{4}$$

As a result,

$$\big\langle s_{nx}^2 S_{Hx}^2 + s_{ny}^2 S_{Hy}^2 + s_{nz}^2 S_{Hz}^2\big\rangle = \tfrac{1}{4}\big\langle S_{Hx}^2 + S_{Hy}^2 + S_{Hy}^2\big\rangle = \tfrac{1}{4}\big\langle \boldsymbol{S}_H^2\big\rangle = \tfrac{1}{4}S_H(S_H + 1)$$

or,

$$\sigma_H = \pi\{(3a_t + a_s)^2 + (a_t - a_s)^2 S_H(S_H + 1)\} \tag{3-89}$$

For para-hydrogen, we have $S_H = 0$, and the cross section is

$$\sigma_{\text{para}} = \pi(3a_t + a_s)^2 \tag{3-90}$$

For ortho-hydrogen, we have $S_H = 1$, and the result is

$$\sigma_{\text{ortho}} = \pi(3a_t + a_s)^2 + 2\pi(a_t - a_s)^2 \tag{3-91}$$

From the values of σ_{para} and σ_{ortho} measured with slow neutron scattering off hydrogen molecules, the values of scattering lengths a_s and a_t may be deduced,

Table 3-3 Nucleon-nucleon scattering length and effective range.

		$S=0, T=1$	$S=1, T=0$
pp	a	-17.1 ± 0.2 fm	—
	r_e	2.794 ± 0.015 fm	—
nn	a	-16.6 ± 0.6 fm	—
	r_e	2.84 ± 0.03 fm	—
np	a	-23.715 ± 0.015 fm	5.423 ± 0.005 fm
	r_e	2.73 ± 0.03 fm	1.73 ± 0.02 fm

and these are listed in Table 3-3. Data of similar quality can also be obtained from coherent scattering of slow neutrons from protons bound in crystals, from crystal diffraction and from the reflection of slow neutron by liquid hydrocarbon mirrors.

Neutron-proton scattering length. Let us examine first the singlet scattering length for the np-system. Since this is a system with isospin $T=1$, we can compare its value with a_{pp} and a_{nn}, the scattering length for pp- and nn-scattering. The signs of all three scattering lengths are negative. This means that, using the definition for the sign given in §C-3, there is no bound state for two nucleons in $T=1$, a fact we have already examined earlier in the discussion on deuterons.

The pp-scattering length a_{pp} is easily measured from low energy proton scattering off a hydrogen target. However, since the cross section for Coulomb scattering, given later in (4-1), is inversely proportional to the square of the energy, pp-scattering at low energies is dominated by electromagnetic effects. In principle, one can subtract the contributions of the Coulomb term: in practice, the accuracy achieved in this way is rather limited, since the cross section for nuclear scattering is only a very small part of the measured quantity. For example, the value of scattering length corresponding to the measured cross section is -7.823 ± 0.01 fm and, after correction for Coulomb effects, the pp-scattering length is -17.1 ± 0.2 fm (see, *e.g.*, H.P. Noyes, *Ann. Rev. Nucl. Sci.* **22** [1972] 465 for more detail).

Measurements of the nn scattering length are complicated by the absence of fixed neutron targets. Several different types of experiment have been carried out to deduce the value of a_{nn} using either deuterons or tritons in reactions such as

$$n+d\rightarrow p+n+n \qquad n+t\rightarrow d+n+n$$
$$d+d\rightarrow p+p+n+n \qquad t+d\rightarrow {}^3\text{He}+n+n$$
$$t+t\rightarrow \alpha+n+n$$

With the availability of good quality pion beams in recent years, the reaction

$$\pi^-+d\rightarrow \gamma+n+n$$

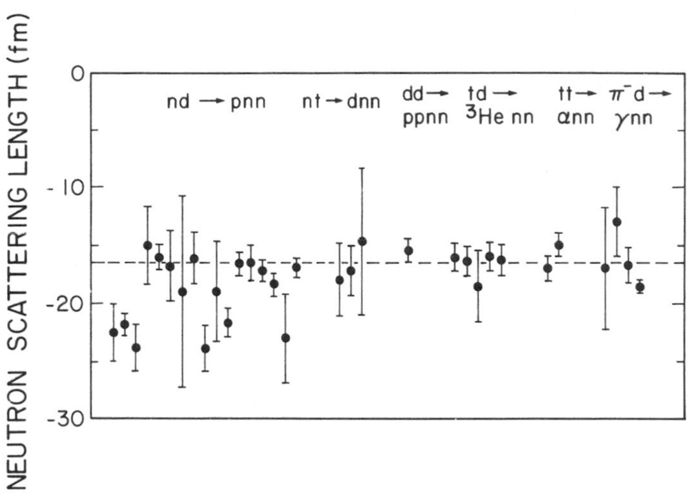

Fig. 3-7 Distribution of the measured values of nn-scattering length in chronological order from left to right. The average value of -16.6 ± 0.6 fm is indicated by the dashed line. (Taken from D.W. Glasgow et al., in *Nuclear Data for Basic and Applied Science,* ed. by G.F. Young et al., Gordon and Breach, New York, 1985).

has also been used to reduce the measured uncertainty of a_{nn}. Here, instead of relying on scattering of neutrons off neutrons, the value of a_{nn} is deduced from effects due to "final state interaction"; that is, changes in the observed reaction cross section due to the interaction between the two emerging neutrons. The values of a_{nn} in chronological order are displayed in Fig. 3-7 and the average of these values, shown as the dashed line in the figure, is listed in Table 3-3.

A comparison of the values of the three $T = 1$ scattering length is interesting. First we note that the value $a_{np} = -23.715 \pm 0.015$ fm is noticeably larger than a_{pp} and a_{nn}. This is an indication of the charge dependence of nuclear force. However, most of the difference may explained by the following argument. At low scattering energies, the two nucleons are never very close to each other and, as a result, only the long range part of the nuclear force is operating. The nuclear interaction here is dominated by the exchange of a single pion. For a pair of protons and a pair of neutrons, only a neutral pion can be exchanged. On the other hand, a charged pion can also be exchanged in the interaction between a neutron and a proton, as shown in Fig. 3-8. This can take place by the proton emitting a π^+ and changing itself into a neutron while the original neutron becomes a proton on absorbing the positive pion. Alternatively, the neutron may emit a π^- and change into a proton while the original proton converts itself into a neutron on absorbing the pion. These are the "exchange" processes. Since in quantum mechanics it is not possible to follow the trajectory of a particle as it interacts with another indistinguishable particle, there is no way to associate either one of the two nucleons in the final state with a particular one in the initial state. As a result, we cannot distinguish an exchange process from a direct process in which a neutral pion is exchanged; the contributions of both processes must be included in an np scattering. Because of the small mass difference between charged and neutral pions,

$$m_{\pi^\pm} - m_{\pi^0} = 4.6\text{MeV}$$

we expect a small difference in the interaction between a proton and a neutron from that of two identical nucleons. In this way, most of the difference between the np-scattering length in the triplet state and the scattering lengths for pp- and nn-systems can be accounted for (for more details, see, *e.g.,* D. Wilkinson, *J. Phys. Soc. Jpn.* **55** [1986] Suppl. 347). The observed difference of a_{np} from a_{pp} and a_{nn}, therefore, cannot be taken as an indication of a fundamental charge dependence in the nuclear force or in the strong interaction itself.

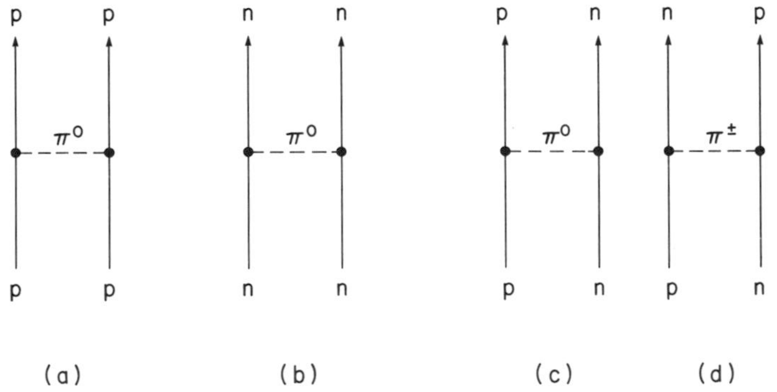

Fig. 3-8 One-pion exchange diagrams. (*a*) *pp*-interaction, (*b*) *nn*-interaction, (*c*) direct term and (*d*) exchange term for *np*-interaction.

The difference between $a_{pp} = -17.1 \pm 0.2$ fm and $a_{nn} = -16.6 \pm 0.6$ fm is not significant at this moment because of the large uncertainty surrounding the measured value of a_{nn}. Within experimental error, pp- and nn-scattering lengths are equal to each other and the results support the charge independence of nuclear force. However, it is worth noting that the last measurement of nn-scattering length by Gabioud et al. (*Phys. Lett.* **103B** [1981] 9) gave a value $a_{nn} = -18.6 \pm 0.5$ fm from $\pi^- + d \rightarrow \gamma + n + n$ reactions. Using improved techniques, the experimental uncertainty in this measurement has been reduced compared with previous results. Since the value obtained differs from a_{pp} by more than one standard deviation of experimental uncertainty, the charge independence of nuclear force may be again in question. A new type of experiment involving direct scattering of neutrons from neutrons using two intense colliding beams of neutrons has been planned at the Los Alamos National Laboratory (Glasgow et al., in *Nuclear Data for Basic and Applied Science,* ed. by Young et al., Gordon and Breach, New York, 1985) and is expected to produce results that will shed new light on the question.

For $T = 0$, the scattering length can only be measured on the triplet *np*-system. The large number of significant figures in the value, 5.423 ± 0.05 fm, is a reflection of the accuracy that can be achieved in slow neutron scattering. The positive sign indicates that there is a bound state which we have already seen as the ground state of the deuteron. The fact that the value is significantly different from that for $T = 1$ is a clear indication of the isospin dependence of nuclear force.

The values for the effective range may be obtained from low energy nucleon-nucleon scattering as well as, for example, photodisintegration of a deuteron or capture of a slow neutron by a proton. The best known values are given in Table 3-3 for comparison. Again, we find evidence for isospin dependence but there is no indication of any contradiction to the assumption of charge independence of nuclear force. The accuracy of the measured values is, however, somewhat lower than the corresponding scattering length measurements, in particular, for the *np*-system. This is not surprising since we are no longer in the extremely low energy region where high accuracy is achievable, as we have seen with scattering length measurements.

§3-9 THE NUCLEAR POTENTIAL

One-pion exchange potential. When Yukawa's idea of a simple one-pion exchange potential (OPEP) was applied to nuclear force, it was found that it could fit experimental data only for inter-nucleon distances greater than 2 fm. In retrospect this is not a surprise, since the pion mass is around 140 MeV/c^2, corresponding to a range of approximately 1.4 fm, as we have seen earlier in §3-6. At shorter distances, contributions from sources other than single-pion exchange become evident upon examination of data on nucleon-nucleon scattering. For example, the values of *s*-wave phase shifts shown in Figs. 3-3 are large and positive at low energies, indicating that the force is an attractive one. As the energy is increased to around 250 MeV in the laboratory for the $^1S_0$-channel and to just above 300 MeV for the $^3S_1$-channel, the phase shifts become negative, showing that the force is now repulsive. This is generally interpreted as evidence of a ***hard core*** in the nucleon-nucleon interaction when the two nucleons are within a distance of the order of a femtometer between their centers. From a quark picture, such a strongly repulsive, short-range term in the interaction between nucleons is to be expected. Being fermions, each one of the three quarks inside a nucleon must occupy one of the three lowest available states. When two nucleons are close together, a large fraction of their volumes overlap each other. As a result, the six quarks in a two-nucleon system can no longer be considered as two separate groups consisting of three quarks each. The action of the Pauli exclusion principle between the quarks demands that three of the six quarks must go to states above the lowest three occupied by the other three quarks. A large amount of energy is required to make this transition. From the point of view of nucleon-nucleon scattering, this additional energy shows up as a great resistance for the two nucleons to come very close to each other, almost as if there is some sort of impenetrable barrier between them.

Although it is easy to justify the existence of a hard core in terms of quarks, it is quite a different story to obtain a quantitative predication for it. To start with, given our present level of knowledge, 300 MeV is a very low energy for QCD to carry out any sort of reliable calculations. As a result, we have no way of calculating, for example, the range or the strength of the repulsive core.

At the hadron level, it is also difficult to generalize on the one-pion-exchange picture, which was designed to understand the long-range part of the force. In addition, the OPEP also has difficulties in relating the strength of nucleon-nucleon interactions to the observed size of pion-nucleon interaction. If two nucleons interact with each other through the exchange of a virtual pion, the strength of the interaction must be related to the probability of a nucleon emitting and absorbing real pions. Such probabilities are, in turn, connected to the *coupling*, or interaction, strength between a pion and a nucleon. In the language of field theory, strength is characterized by the pion-nucleon coupling constant $g_{\pi N}$, analogous to the factor g in (3-63). Models of nuclear force built solely upon the one-pion-exchange picture have found that, in general, the value of $g_{\pi N}$ obtained, for example, from pion-nucleon scattering cannot be used directly to calculate the coupling constant for nucleon-nucleon scattering. As a result, the coupling constant is often treated in practice as a parameter adjusted to fit the nucleon-nucleon data.

One-boson exchange potential. Our present view is that the nuclear force may be divided into three parts, as illustrated in Fig. 3-9. The long range part ($r > 2$ fm) is dominated by one-pion exchange. If the exchange of single pions is important, there is no reason to exclude exchanges of two or more pions and mesons heavier than pions. The range of force associated with these more massive intermediate bosons is shorter and, for this reason, the intermediate range part of the nuclear force (1 fm$< r <$ 2 fm) comes mainly from exchanges of two pions and heavier mesons. The hard core in the interaction ($r \lesssim 1$ fm) is made of heavy mesons exchanges, multi-pion exchanges as well as QCD effects.

It is helpful to use diagrams, based on Feynman diagrams in field theory, to represent the various boson exchange terms. The process of exchanging one pion between two nucleons may by represented by the diagram given in Fig. 3-10(a). An implicit assumption in the diagram is that the time axis be in the vertical direction. Two nucleons, with momenta $\boldsymbol{p}_1$ and $\boldsymbol{p}_2$, represented by the two solid lines, are moving freely until time t_1 when a pion, represented by the dashed line, is emitted by nucleon 1. The pion emission changes the momentum of nucleon 1 from $\boldsymbol{p}_1$ to $\boldsymbol{p}_1'$. At time t_2, the pion is absorbed by nucleon 2 and the momentum

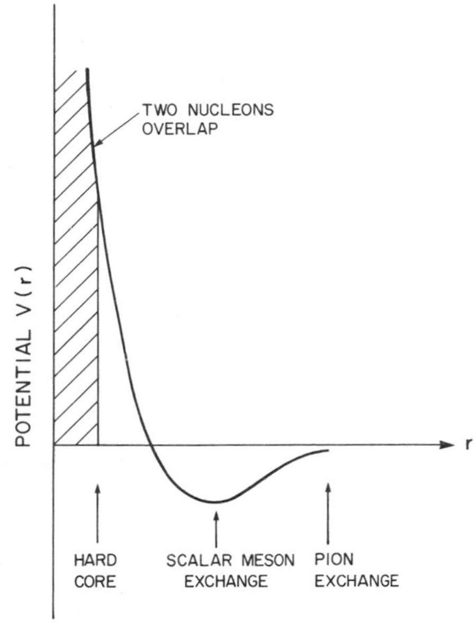

Fig. 3-9 Schematic diagram showing the different parts of nucleon-nucleon potential as a function of distance r between two nucleons. The radius of the hard core is around 0.4 fm and it takes energy in excess of 1 GeV to bring two nucleons closer than twice this distance. The main part of the attraction lies at intermediate ranges, at radius $\sim$ 1 fm, and is believed to be dominated by the exchange of scalar mesons. The long range part, starting around 2 fm or so, is due to the single-pion exchange.

of the nucleon is changed from $\boldsymbol{p}_2$ to $\boldsymbol{p}_2'$ as a result. For simplicity, the diagram is often abbreviated in the form shown in Fig. 3-10(*b*). Following the same rules, a two-pion exchange diagram may be represented by that shown in Fig. 3-10(*c*). A particular type of two-pion exchange term is given by Fig. 3-10(*d*) in which the intermediate state of one of the nucleons becomes a Δ-particle, shown as a double line, as a result of absorbing the pion. Since a ρ-meson decays into two pions with a mean life of only 4×10^{-24} s, the exchange of a ρ-meson, shown in Fig. 3-10(*e*), may be considered as a special type of two-pion exchange term. Similarly, the exchange of an ω-meson is a type of three-pion exchange (not shown) as ω decays to three pions with a mean life of 8×10^{-23} s. Fig. 3-10(*f*) is another type of two-pion exchange term where both pions are emitted before either one is absorbed. In contrast, the two pions in Fig. 3-10(*c*) are emitted and absorbed one after another. As a side interest, a three-body force (not shown) may arise, for example, as the result of a nucleon emitting a ρ-meson which decays into two pions, each of which is absorbed by a different nucleon.

Nucleon-nucleon potentials. There are two general approaches to constructing a potential that has the correct form for long, intermediate, and short range parts. The first is a phenomenological one which generalizes the one-pion exchange picture to a one-boson exchange (OBE) picture. To keep the form simple, only the exchange of a single boson is allowed. In addition to pions, heavier mesons are introduced to account for the intermediate range. To compensate for multi-meson

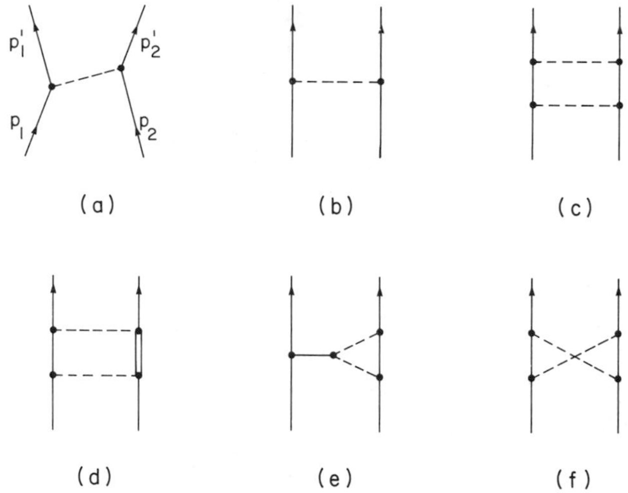

Fig. 3-10 Diagrammatic representation of meson exchange between two nucleons. Diagrams (a) and (b) represent one-pion exchange, (c) indicates two-pion exchange, (d) indicates two-pion exchange with intermediate state involving a Δ-particle, (e) shows a ρ-meson exchange and (f) is another type of two-pion exchange term where both pions are emitted before either one is absorbed.

exchanges, the strength for each type of meson exchange is left as a parameter to be determined by fitting, for example, NN-scattering data. The hard core is put in explicitly "by hand" without any reference to its source. The strength of such an approach lies in its simplicity. There are, however, several problems. The lifetimes of many of the mesons involved are sufficiently short that the validity of a picture involving the exchange of these particles without considering their decay is not very sound. Furthermore, in order to fit experimental data with a minimum number of terms, the range of each OBE term, and consequently the masses of the mesons exchanged, often become adjustable parameters as well, with little or no relation to real mesons. These "fundamental" objections to such phenomenological potentials, however, should not detract from their success achieved in a variety of applications.

A second approach in constructing a nucleon-nucleon potential is to make use of our knowledge of hadrons as much as possible and treat phenomenologically only those aspects, mainly short-range interactions, of which we have incomplete knowledge. Such a program was carried out, for example, by the Paris group (R. Vinh Mau, in ***Mesons in Nuclei***, ed. M. Rho and D.H. Wilkinson, North-Holland, Amsterdam, 1979) and the Bonn group (R. Machleidt, K. Holinde, and Ch. Elster, *Phys. Rep.* **149** [1987] 1). The one- and two-pion exchange parts of the potential are well known; here, both groups used essentially the same approach. For the less well known short-range parts, different techniques were used by each group.

It is perhaps more interesting to examine three important differences in the two potentials, in part to forecast the possible future development of nuclear force studies. The first is the treatment of three- and four-pion exchanges that form a part of the short-range interaction. Here the Paris potential used a phenomenological approach and determined some of the parameters involved by fitting them to known data. The Bonn potential made an estimate of the effect instead. It will be interesting to see whether QCD can help achieve a more satisfactory solution to this problem.

A second difference in the two potentials is in the treatment of the Δ-particle. As we have seen earlier in Fig. 2-3, a strong resonance in the scattering of π^+ off protons is found at laboratory pion energy 195 MeV. Such a strong resonance in the pion-nucleon reaction must have a profound influence on the nucleon-nucleon interaction. For example, a nucleon may be excited to become a Δ-particle in the intermediate state, as shown in Fig. 3-10(d). As a particle that is distinguishable from a nucleon, the Δ-particle is not affected by the Pauli exclusion principle with respect to nucleons in the nucleus. For applications involving energies above 300 MeV in the laboratory, the formation of a Δ-particle is expected to play a significant role and must be included as a part of the potential. On the other hand, it is not easy to incorporate such a strong inelastic channel in a potential except by putting in the resonance explicitly, an approach adopted by the Bonn group.

The interactions between antinucleons and between a nucleon and an antinucleon are also integral parts of a nucleon-nucleon potential. This is especially true if we take a fully relativistic approach where both nucleon and antinucleons appear together in a wave function. Furthermore, experimental data are available for the scattering of antinucleons off nucleons and nuclei (see, *e.g.*, Fig. 7-9). Studies of such scattering using nucleon-nucleon potential, with antinucleons incorporated as a part, can tell us far more about the two-nucleon system than the consideration of nucleons and antinucleons as totally separate entities. There are several different ways to carry out the extensions to include antinucleons and, in respect to this, the Paris and Bonn potentials differ from each other.

In spite of these differences, it is important to realize that, at low energies where most of the experimental data are taken, calculations using either potential have produced very similar results. For instance, the values of ${}^1S_0$ and ${}^3S_1$ phase shifts obtained with both potentials are shown in Fig. 3-3. The close agreement between the two sets of calculated results and with values extracted from NN-scattering data is a demonstration of the degree of understanding already achieved in nucleon-nucleon interaction.

Nucleon-nucleon interaction for bound nucleons. One of the reasons for having a nucleon-nucleon potential is to make use of it in investigations of nuclear structure and nuclear reaction. For this purpose it is not essential, in principle, to have a potential. Most of the applications involve many-body matrix elements of the nuclear interaction. A two-body force acts between two nucleons at a time, and many-body matrix elements of such an interaction can always be expressed in terms of two-body matrix elements similar in form to the nucleon-nucleon t-matrix given in (3-76). These matrix elements are, however, different from those in the two-nucleon space in two important aspects. In the first place, there may be a difference in the interaction between a pair of nucleons inside a nucleus from that between a pair of free nucleons. In this chapter we have dealt mainly with the latter category. As we shall later in §6-8, the "effective" interaction between bound nucleons is modified by the presence of other nucleons in the same nucleus and may be different from the interaction operating between free or *bare* nucleons discussed here.

A second problem is associated with the question of whether or not a nuclear potential can be specified completely within a two-nucleon system. When two free nucleons interact, energy and momentum are conserved within the two-particle system. That is, the momenta of the two nucleons are restricted by the relation,

$$\frac{p_1^2}{2\mu_1} + \frac{p_2^2}{2\mu_2} = E \tag{3-92}$$

where p_1 and p_2 are the momenta, and μ_1 and μ_2 are the reduced masses of the two particles in their center of mass. In other words, the sum of the momenta of the two nucleons is confined to lie on a spherical "shell" in momentum space with the square of the radius, $p_x^2 + p_y^2 + p_z^2 = 2\mu E$, determined by the total kinetic energy E available in the center of mass. The two-body interaction t-matrix elements under such circumstances are said to be "on the energy shell" and they are called *on-shell* matrix elements. Once nucleons are bound to a nucleus, energy-momentum conservation applies to the nucleus as a whole and the momenta of a pair of nucleons are no longer restricted by (3-92). Interaction between two nucleons under the condition that (3-92) is not satisfied, *i.e.*, the sum of the momenta is not equal to the kinetic energy of their relative motion, is said to be "off the energy shell" or *off-shell* for short. Such off-shell interactions are usually built into a nucleon-nucleon interaction potential; however, since off-shell conditions do not exist in two-nucleon systems, we have no way of testing these parts of the potential here. In this sense, the nuclear potential cannot be completely specified by a study of the two-nucleon system alone. By the same token, purely phenomenological potentials with parameters fitted to data on two-nucleon systems have no way of knowing *a priori* whether they are correct for off-shell effects.

Again there are two possible ways to solve the problem of off-shell behavior of a nuclear potential. The first is to have a theory connecting off-shell effects to those on-shell, a relation that is implicit in all the models of nuclear potential. If we have the correct association between these two types of effects, then the off-shell behavior of a potential is completely determined once the on-shell matrix elements are given. However, we have not yet arrived at this level of understanding of the nucleon-nucleon interaction. In the absence of such a theory, an alternative is to take a semi-empirical approach and to determine off-shell matrix elements by comparing them with experimental data sensitive to such effects. Unfortunately, such tests must be carried out on systems with more than two nucleons, and not too many quantities have been found that are useful for checking the off-shell matrix elements. An unambiguous determination of off-shell behavior of nuclear force is therefore still lacking.

Relation with quark-quark interaction. Although it is generally accepted that the force between nucleons is a facet of the strong interaction between quarks, a quantitative connection between nuclear force and quark-quark interaction is still lacking. The root of the problem is, as usual, the difficulty of carrying out QCD calculations at the low energies where nuclear physics operates.

In order to study nuclear interaction in a quark model, we need a system of at least six quarks. The force between the quarks must be such that it satisfies the condition of confinement; that is, unless the two nucleons are very close together the six quarks are clustered tightly into two separate groups of three quarks each. At large enough distances compared with the average value between quarks inside a nucleon, the force between these two "bags" of quarks must have a form consistent with that given successfully by meson exchange; at intermediate distances, the force must be attractive and not too different from that given by a model involving the exchange of several pions and heavier mesons. With our present level of knowledge of low energy strong interaction, it is not difficult to demonstrate such a relation between quark-quark and nucleon-nucleon interactions; however, a proper derivation of the nucleon force from QCD is still being developed.

Qualitatively we can see how nuclear force may arise from a quark-quark interaction by making an analogy with the force between chemical molecules. Charge distributions in many molecules are spherically symmetric so that there should not be any net electrostatic force left to act between any two such molecules. However, since we know that such molecules do condense into liquids and solids, there must be a residual force acting between two molecules; this is generally known as the van der Waals force. It is useful to see how such a force arises between molecules so that we may gain insight into the question of how forces between nucleons arise from the interaction between quarks.

Suppose for an instant that a neutral molecule acquires a electric dipole moment $\boldsymbol{p}$, say, as a result of fluctuation in its shape and, consequently, its charge distribution. The electrostatic potential at a point $\boldsymbol{r}$ from the center of a dipole is given by the expression

$$\phi(\boldsymbol{r}) = \left[\frac{1}{4\pi\epsilon_0}\right]\frac{\hat{\boldsymbol{r}}\cdot\boldsymbol{p}}{r^2} = \left[\frac{1}{4\pi\epsilon_0}\right]\frac{p\cos\theta}{r^2} \tag{3-93}$$

where the angle θ is between vectors $\boldsymbol{p}$ and $\hat{\boldsymbol{r}}$ ($\hat{\boldsymbol{r}} = \boldsymbol{r}/r$ and $r = |\boldsymbol{r}|$). The electric field from such a dipole,

$$\boldsymbol{E}(r,\theta) = -\nabla\phi = -\Big[\frac{1}{4\pi\epsilon_0}\Big]\Big\{\frac{\boldsymbol{p}}{r^3} - 3\Big(\frac{p\cos\theta}{r^4}\Big)\boldsymbol{r}\Big\} \tag{3-94}$$

induces a dipole moment $\boldsymbol{p}'$ in another molecule with a magnitude proportional to the polarizability χ of the molecule,

$$\boldsymbol{p}' = \chi\boldsymbol{E}$$

As a result, a dipole-dipole interaction arises between these two molecules with a strength

$$V(\boldsymbol{r}) = -\boldsymbol{p}'\cdot\boldsymbol{E} = -\Big[\Big(\frac{1}{4\pi\epsilon_0}\Big)^2\Big]\chi(1+3\cos^2\theta)\frac{p^2}{r^6} \tag{3-95}$$

Note that the interaction energy is always negative regardless of the orientation of the first dipole we have initially assumed. Consequently we have a force that is always attractive and varies as r^{-7}.

For a spherically symmetric molecule, the dipole moment is zero on the average, $\langle\boldsymbol{p}\rangle = 0$. However, because of fluctuation, the instantaneous value of $\boldsymbol{p}$ may be different from zero; in other words, we have $\langle\boldsymbol{p}^2\rangle \neq 0$. An attractive force between two nucleons therefore results from the fluctuation. The same can be expected of nucleons and baryons in general. Instead of an electrostatic force, we are dealing with the "color" force between quarks. Although strong interactions confine quarks within nucleons, a *color* van der Waals force can arise between nucleons just as a dipole-dipole force originates between molecules. In this way, we can see how a nuclear force can arise from the residual interaction between quarks in separate nucleons. Although the idea of a color van der Waals force is an attractive one, the actual force it produces has a range much longer than what is observed. At this moment, the color van der Waals force does not seem to be a correct model for nuclear interaction without further development.

Problems

3-1. Find the possible range of values for the depth of a one-dimensional square well, 3 fm wide, that has only one bound state for a nucleon.

3-2. If the surface of a deformed nucleus is given by the equation

$$x^2 + y^2 + 1.2z^2 = R^2$$

where $R = 1.2A^{1/3}$ fm, calculate classically the electric quadrupole moment of the nucleus assuming $A = 200$, $Z = 80$ and the nuclear density is uniform inside the surface and zero outside.

3-3. Carry out the angular part of the integration,

$$\int_0^{2\pi}\int_0^{\pi} (Y_{LM_L}(\theta\phi))^* Y_{20}(\theta\phi)\, Y_{LM_L}(\theta\phi) \sin\theta\, d\theta\, d\phi$$

in (3-25) for the expectation value of the quadrupole operator in the $L = 2$ and $M = 2$ state using the explicit forms of the spherical harmonics given in (B-18).

3-4. For an infinite three-dimensional harmonic oscillator potential well with oscillator frequency ω, the radial wave functions for the lowest s-state and the lowest d-state are, respectively,

$$R_{1s}(r) = 2\nu^{3/4}\pi^{-1/4}e^{-\frac{1}{2}\nu r^2} \qquad R_{1d}(r) = \frac{4}{\sqrt{15}}\nu^{7/4}\pi^{-1/4}r^2e^{-\frac{1}{2}\nu r^2}$$

where the oscillator length parameter $\nu = M\omega/\hbar$ and M is the mass of a nucleon. Find the root-mean-square radii in each of these states taking $\hbar\omega = 15$ MeV. Compare the values obtained with the measured deuteron radius. For the radial wave function given above, what is the value of the off-diagonal matrix element $\langle R_{1s}|r^2|R_{1d}\rangle$? Use this model to calculate the quadrupole moment of the deuteron assuming that the wave function is predominantly made of the ${}^3S_1$-state with a 4% admixture of the ${}^3D_1$-state.

3-5. Rewrite the right hand side of following the rank 1 spherical tensor product

$$\big(\boldsymbol{\sigma}(1)\times\boldsymbol{\sigma}(2)\big)_{1m} = \sum_{pq}\langle 1p1q|1m\rangle \boldsymbol{\sigma}_p(1)\boldsymbol{\sigma}_q(2)$$

given in (3-37) in terms of the cartesian components of Pauli matrices $\boldsymbol{\sigma}(1)$ and $\boldsymbol{\sigma}(2)$ for nucleons 1 and 2, respectively. Show that, in cartesian coordinates, it has the same form as an ordinary vector product of the same two vectors.

3-6. Show that the spherical tensor rank of the operator

$$S_{12} = \frac{3}{r^2}(\boldsymbol{\sigma}_1\cdot\boldsymbol{r})(\boldsymbol{\sigma}_2\cdot\boldsymbol{r}) - \boldsymbol{\sigma}_1\cdot\boldsymbol{\sigma}_2$$

is two in intrinsic spin space; that is, it may be written as an operator of the form $\sum_{qq'}\langle j_1qj_2q'|\lambda M\rangle\boldsymbol{\sigma}_{1q}\boldsymbol{\sigma}_{2q'}$ with $\lambda = 2$ and $j_1 = j_2 = 1$. Here $\langle j_1qj_2q'|\lambda M\rangle$ is the Clebsch-Gordan coefficient.

3-7. For a Yukawa potential $V(r) = V_0\frac{e^{-r/r_0}}{r}$, with the range $r_0 = \hbar/mc$ given in terms of the mass m of the boson exchanged, find the angular distribution for elastic scattering in first Born approximation for the potential. Show that in the limiting case of zero-mass boson, the result is identical to Rutherford scattering.

3-8. In classical electrodynamics, the scalar field $\phi(\boldsymbol{r})$ produced by an electron located at the origin is given by Poisson's equation,

$$\nabla^2\phi(\boldsymbol{r}) = -4\pi e\delta(\boldsymbol{r})$$

Show that the radial dependence of the field is given by

$$\phi(r) = \frac{e}{r}$$

For a nucleon, the scalar field satisfies the Klein-Gordon equation,

$$\left(\nabla^2 - \frac{1}{r_0^2}\right)\phi(\boldsymbol{r}) = 4\pi g\delta(\boldsymbol{r})$$

Show that the radial dependence of the field is given by

$$\phi(r) = -g\frac{e^{-r/r_0}}{r}$$

Derive that the range r_0 is given by the relation $r_0 = \hbar/mc$ using the fact that the boson, with mass m, is a virtual particle and can therefore exist only for a time Δt given by the Heisenberg uncertainty relation.

3-9. For a velocity-independent two-body potential, the only two-body scalars that can be formed of $\boldsymbol{r} = \boldsymbol{r}_1 - \boldsymbol{r}_2$, $\boldsymbol{S} = \boldsymbol{\sigma}_1 + \boldsymbol{\sigma}_2$ and $\boldsymbol{T} = \boldsymbol{\tau}_1 + \boldsymbol{\tau}_2$ are r, $\boldsymbol{\sigma}_1 \cdot \boldsymbol{\sigma}_2$, $\boldsymbol{\tau}_1 \cdot \boldsymbol{\tau}_2$, $\boldsymbol{\sigma}_1 \cdot \boldsymbol{\sigma}_2 \boldsymbol{\tau}_1 \cdot \boldsymbol{\tau}_2$ and $S_{12} \equiv 3(\boldsymbol{r} \cdot \boldsymbol{\sigma}_1)(\boldsymbol{r} \cdot \boldsymbol{\sigma}_2)/r^2 - (\boldsymbol{\sigma}_1 \cdot \boldsymbol{\sigma}_2)$. Show that the following operators

a) $\boldsymbol{S} \cdot \boldsymbol{S}$ *b)* $(\boldsymbol{r} \cdot \boldsymbol{S})^2$
c) $(\boldsymbol{r} \times \boldsymbol{S}) \cdot (\boldsymbol{r} \times \boldsymbol{S})$ *d)* $(\boldsymbol{r} \times (\boldsymbol{\sigma}_1 - \boldsymbol{\sigma}_2)) \cdot (\boldsymbol{r} \times (\boldsymbol{\sigma}_1 - \boldsymbol{\sigma}_2))$

can be reduced to functions of these scalars. Give the symmetry argument of why scalar products $\boldsymbol{r} \cdot \boldsymbol{S}$ and $\boldsymbol{r} \cdot \boldsymbol{T}$ are not allowed.

With velocity or momentum dependence, the only additional operators required is $\boldsymbol{L} \cdot \boldsymbol{S}$ where $\boldsymbol{L} = \boldsymbol{r} \times \boldsymbol{p}$ and $\boldsymbol{p} = \frac{1}{2}(\boldsymbol{p}_1 - \boldsymbol{p}_2)$. Show that the following terms do not form independent scalars either:

e) $\boldsymbol{p} \cdot \boldsymbol{p}$, $\boldsymbol{L} \cdot \boldsymbol{L}$, $(\boldsymbol{L} \cdot \boldsymbol{S})^2$ *f)* $\boldsymbol{r} \times \boldsymbol{L} \cdot \boldsymbol{p}$ *g)* $(\boldsymbol{L} \cdot \boldsymbol{S})(\boldsymbol{L} \cdot \boldsymbol{L})$
h) $(\boldsymbol{r} \cdot \boldsymbol{p})(\boldsymbol{r} \cdot \boldsymbol{S})$ *i)* $(\boldsymbol{r} \cdot \boldsymbol{p})(\boldsymbol{L} \cdot \boldsymbol{S})$

3-10. Calculate the s-wave phase shift of a neutron scattered by an attractive square-well potential of depth V_0 and width W. Obtain the scattering length a and effective range r_0 in terms of V_0 and W.

3-11. Show that the angular distribution of an s-wave scattering is isotropic. If the only non-zero phase shifts in a hypothetical scattering of a particle off another are s- and p-waves, find the angular distribution of the scattering cross section assuming that the particle is a neutron, the s-wave phase shift is $\delta_0 = 45°$, p-wave phase shift is $\delta_1 = 30°$, and the scattering takes place with laboratory energy 5 MeV. Plot the results for scattering angle in the range 0° to 180°.

3-12. If instead of the observed value of $J^\pi = 1^+$ the ground state of a deuteron is $J^\pi = 0^-$, what are now the possible values of orbital angular momentum L, sum of intrinsic spin S, and isospin T in this hypothetical state? What are the implications for nuclear force if this is true?

3-13. Show that for any function $f(r)$ of r

$$\nabla(\boldsymbol{\sigma}\cdot\nabla f) = \hat{r}(\boldsymbol{\sigma}\cdot\hat{r})\left[\frac{\partial^2 f}{\partial r^2} - \frac{1}{r}\frac{\partial f}{\partial r}\right] + \boldsymbol{\sigma}\frac{1}{r}\frac{\partial f}{\partial r}$$

where $\hat{r}$ is an unit vector and $\nabla_2 = -\nabla_1 = \nabla$.

3-14. At distance sufficiently large that overlap between their densities may be ignored, the interaction between two nucleons may be shown to be similar to that between two point dipoles,

$$V(r) \sim (\boldsymbol{\sigma}_1\cdot\nabla_1)(\boldsymbol{\sigma}_2\cdot\nabla_2)f(r)$$

Under the assumption of one-pion exchange, we may take the radial dependence to have the form

$$f(r) = \frac{e^{-r/r_0}}{r}$$

where

$$r_0 = \frac{\hbar c}{m_\pi c^2}$$

is the range. The strength of the potential may be related to the pion-nucleon coupling constant g ($g^2/\hbar c \simeq 0.081 \pm 0.002$). Except for isospin dependence, which we shall ignore here for simplicity, the potential may be written in the form

$$V(r) = -g^2 r_0^2(\boldsymbol{\sigma}_1\cdot\nabla_1)(\boldsymbol{\sigma}_2\cdot\nabla_2)\frac{e^{-r/r_0}}{r}$$

Use the result of Problem 3-13 above to show that $V(r)$ can be expressed in terms of the tensor operator S_{12} given in (3-39)

$$V(r) = \frac{g^2}{3}\left\{\left[\left(1 + \frac{3r_0}{r} + \frac{3r_0^2}{r^2}\right)S_{12} + \boldsymbol{\sigma}_1\cdot\boldsymbol{\sigma}_2\right]\frac{e^{-r/r_0}}{r} - 4\pi r_0^2\delta(r)\boldsymbol{\sigma}_1\cdot\boldsymbol{\sigma}_2\right\}$$

where $r = |\boldsymbol{r}_1 - \boldsymbol{r}_2|$.

Chapter 4

BULK PROPERTIES OF NUCLEI

The abundance of experimental information on nuclei comes partly from the large number of available nuclear species and partly from the wide variety of measurements that can be carried out. Our interest here is to see what we can learn about the atomic nucleus from these known facts.

Most of the data may be divided into four categories: energies, static moments, transition probabilities, and reaction rates. In this chapter, we shall concentrate primarily on the first two categories. By energy, we include both ground state binding energy of a nucleus as well as excitation energies of higher states. The main difference between these two types of energies is a matter of the reference point. The binding energy is the amount of energy required to separate the nucleus into its constituent parts: the excitation energy, on the other hand, is the energy of a state relative to the ground state of the nucleus. For static moments, we shall include the electromagnetic moments of a nuclear state as well as moments of nuclear matter distribution. In the latter category, we shall take nuclear size as the zeroth order moment of matter distribution.

§4-1 NUCLEAR SIZE

The fact that the size of a nucleus is far smaller compared with that of an atom was established, long before there was much else known about the nucleus, by Coulomb scattering using α-particles from radioactive elements. Before 1911, the model of the atom that embodied most of the observations known at the time consisted of a uniform distribution of positive and negative charge throughout the entire volume of radius $\sim 10^{-10}$ m occupied by the atom. If this were true, the incident α-particle would only experience Coulomb scattering with differential cross section given by the Rutherford formula,

$$\left(\frac{d\sigma}{d\Omega}\right)_{\text{Rutherford}} = \left\{\left[\frac{1}{4\pi\epsilon_0}\right]\frac{zZe^2}{4T\sin^2\left(\frac{1}{2}\theta\right)}\right\}^2 = \left\{\frac{\alpha\hbar c z Z}{4T}\frac{1}{\sin^2\left(\frac{1}{2}\theta\right)}\right\}^2 \tag{4-1}$$

where z is the charge number of the projectile, equal to 2 for an α-particle, and Z is the number of protons in the target nucleus. The kinetic energy of the α-particle in the center of mass is represented by T so as to distinguish it from the total relativistic energy E of the projectile we need later in this section. The quantity inside the square brackets converts the formula from cgs to SI units. Since the angular distribution was proportional to $\sin^{-4}(\theta/2)$, where θ is the scattering angle, most of the α-particles were expected to emerge in a small forward cone.

The experimental results were, however, quite different. When scattering was carried out on thin foils of heavy elements such as gold, many more α-particles were observed at very large scattering angles than expected. The explanation offered by Rutherford and collaborators was that the positively charged nucleus was concentrated in a very small part of the volume occupied by the atom. The experiments were carried out before the advent of particle accelerators and the energy of the α-particles used was limited to what could be obtained from naturally occurring radioactive material. For example, one of the higher energy sources used was ^{214}Po which emits α-particles of 7.68 MeV. The corresponding de Broglie wave length is around 5 fm, comparable with the nuclear radius of the heavy elements and much larger in size than the α-particle projectile.

When the kinetic energy of the projectile is increased, the de Broglie wave length decreases and the scattering become sensitive not only to the finite size of the projectile and the target, but also to the nuclear force interacting between the nucleons. Our interest at the moment is primarily in the size of the nucleus. For this purpose, where possible, it is preferable to avoid complications due to finite size of the projectile and nuclear interaction with the target. Leptons are the ideal probes here since they are point particles and they do not participate in strong interactions. Neutrinos would have been the best choice since they interact with the nucleus only through the extremely short-range weak interaction. However, the experiments are complicated by the difficulty in detecting neutrinos; the interaction cross section is small and a good neutrino beam is hard to come by. The charged leptons – electrons and muons – interact with the nucleus predominantly through the Coulomb interaction. Here, the interaction has a long range, but the reaction is well understood. For this reason, electrons and, to a lesser extent, muons become the common lepton probes used to measure nuclear sizes and charge distributions. One of the drawbacks of electromagnetic probes is that they are not sensitive to the distribution of the neutral neutrons in a nucleus. As a result, a study of nuclear matter distribution requires hadronic probes as well. We shall see later in §4-6 that this information is much harder to obtain, in part because the analyses are complicated by the nuclear interaction between the probe and the nucleus.

From electron scattering measurements it is found that, on the average, the (charge) radius of a nucleus consisting of A nucleons is well represented by the form

$$R_{\text{ch}} = r_0 A^{1/3} \tag{4-2}$$

with $r_0 = 1.2$ fm. When the energy of electrons is increased so that the de Broglie wave lengths become much smaller than the size of a nucleon, the scattering becomes sensitive to the charge distribution within each individual nucleon and the quarks begin to play a role in the measured results.

§4-2 ELECTRON SCATTERING FORM FACTOR

The Rutherford formula (4-1) is a convenient starting point for studying charged particle scattering. Before we make any further use of it, let us examine the premises used in deriving the expression so that we may be aware of the modifications required when we apply it to electron scattering off nuclei. For our purposes, the most important assumptions are the following three:

(1) Relativistic effects are ignored,

(2) Both projectile and target are considered to be spinless, *i.e.*, $J = 0$, and

(3) Both projectile and target are assumed to be point particles.

For low-energy α-particle scattering discussed in the previous section, condition 1 is satisfied. Furthermore, for α-particle scattering off even-even nuclei (ground-state spin zero), condition 2 is also fulfilled. The finite size of the nuclear charge distribution, however, does violate condition 3. This was evident even in Rutherford's experiment and will become more important as we increase the bombarding energy. This point, however, may be used as an advantage if our interest is in the charge distribution of the nucleus itself.

Scattering off point particles. Electron scattering off nuclei differs from low-energy α-particle scattering in several respects. First, an electron is a Dirac particle with intrinsic spin $\boldsymbol{s} = \frac{1}{2}$, whereas an α-particle has spin $J = 0$. Second, the rest mass of an electron is very small. As a result, electrons with de Broglie wave length smaller than or comparable with nuclear dimensions are already relativistic particles, with total energy much larger than the rest mass energy. The differential cross section for scattering relativistic electrons off point-charged particles is described in the center of mass by the Mott formula,

$$\left(\frac{d\sigma}{d\Omega}\right)_{\rm Mott} = \left\{\left[\frac{1}{4\pi\epsilon_0}\right]\frac{e^2ZE}{2p^2c^2\sin^2\left(\frac{1}{2}\theta\right)}\right\}^2\left\{1 - \frac{p^2c^2}{E^2}\sin^2(\tfrac{1}{2}\theta)\right\}$$

$$= \left\{\frac{\alpha\hbar cZE}{2p^2c^2\sin^2\left(\frac{1}{2}\theta\right)}\right\}^2\left\{1 - \beta^2\sin^2(\tfrac{1}{2}\theta)\right\} \qquad (4\text{-}3)$$

where p is the magnitude of the momentum, $E = \sqrt{(pc)^2 + (m_ec^2)^2}$ is the total relativistic energy of the incident electron, and $\beta = v/c$. In the nonrelativistic limit, we have $E \approx m_ec^2$ and the kinetic energy of the incident electron becomes $T = p^2/2m_e$. This gives us the results:

$$\frac{E}{2p^2c^2} \xrightarrow[v \ll c]{} \frac{1}{4T} \qquad\qquad \beta \xrightarrow[v \ll c]{} 0$$

and the Mott formula reduces to the Rutherford formula.

It is often useful to express the scattering result as a function of the *momentum transfer* $\hbar\boldsymbol{q}$ from the electron to the nucleus. In the nonrelativistic limit, we can define a three-momentum transfer

$$\boldsymbol{q} = \boldsymbol{k}_i - \boldsymbol{k}_f \tag{4-4}$$

where $\hbar\boldsymbol{k}_i$ is the incident electron momentum and $\hbar\boldsymbol{k}_f$ is the final electron momentum. If the electron energy is high, it is more appropriate to use the four-momentum transfer

$$t = \frac{(E_i - E_f)^2}{\hbar^2 c^2} - (\boldsymbol{k}_i - \boldsymbol{k}_f)^2 \tag{4-5}$$

which is shown in Appendix D to be a Lorentz scalar and one of the Mandelstam variables. For elastic scattering, the final electron energy E_f is equal to the initial electron energy E_i in the center of mass ($E_f = E_i = E$), and we have the result,

$$q^2 = -t = 4k^2 \sin^2(\tfrac{1}{2}\theta) \simeq \left(\frac{2E}{\hbar c}\right)^2 \sin^2(\tfrac{1}{2}\theta) \tag{4-6}$$

where we have taken $k = k_i \approx k_f$ as shown in (D-41). The magnitude of the momentum transfer is therefore a function of the scattering angle.

It is also possible to express the differential cross section in terms of q^2 instead of the solid angle. From (4-6), we have

$$d\Omega = 2\pi d(\cos\theta) = \frac{\pi}{k^2} dq^2 \tag{4-7}$$

In the limit $E \to pc = \hbar kc$, the Rutherford cross section may be approximated by the form

$$\left(\frac{d\sigma}{dq^2}\right)_{\text{Rutherford}} \approx \frac{4\pi Z^2 \alpha^2}{q^4} \tag{4-8}$$

This result demonstrates that the differential cross section is mainly a function of the momentum transfer without any explicit dependence on the energy.

Since a nucleus is much heavier than an electron, it is often more convenient to express the scattering cross section in the laboratory frame of reference in which the target is initially at rest. The difference between E_i and E_f is, then, the energy taken away by the recoil of the target particle as a result of the scattering,

$$\frac{E_f}{E_i} = \frac{1}{1 + \frac{2E_i}{Mc^2} \sin^2(\frac{1}{2}\theta)} \tag{4-9}$$

where M is the mass of the target particle. Transformations between center-of-mass and laboratory frames of reference for various quantities of interest in scattering are given in Appendix D.

In the limit that the electron rest mass may be ignored, the cross section in the laboratory for elastic scattering of unpolarized electrons off spinless ($J = 0$), point-charged particles, is given by

$$\left(\frac{d\sigma}{d\Omega}\right)_{\text{point}} = \left(\frac{Z\alpha\hbar c}{2E_i \sin^2(\frac{1}{2}\theta)}\right)^2 \frac{E_f}{E_i} \cos^2(\tfrac{1}{2}\theta) \tag{4-10}$$

For targets with a finite spin, additional contribution from the "magnetic" scattering is also present. Instead of the above expression, we have

$$\left(\frac{d\sigma}{d\Omega}\right)_{\text{Dirac}} = \left(\frac{Z\alpha\hbar c}{2E_i \sin^2(\frac{1}{2}\theta)}\right)^2 \frac{E_f}{E_i} \left\{\cos^2(\tfrac{1}{2}\theta) + \frac{(\hbar q)^2}{2(Mc)^2} \sin^2(\tfrac{1}{2}\theta)\right\} \tag{4-11}$$

This is known as the Dirac formula. In order to differentiate between the two terms in the expression, the first one is called the *electric* term and the second one the *magnetic* term. The relative contributions of these two terms may be found in the following way. From (4-6), we find that

$$\frac{(\hbar q)^2}{2(Mc)^2} \simeq 2\left(\frac{E}{Mc^2}\right)^2 \sin^2(\tfrac{1}{2}\theta)$$

is much less than unity if the electron energy is much smaller than the rest mass energy of the target particle ($E \ll Mc^2$). As a result, the magnetic scattering term in (4-11) may be ignored in elastic scattering. The exceptions are high energies and backward angles. In the latter case, the $\cos^2(\frac{1}{2}\theta)$ factor reduces contributions from the electric term relative to magnetic scattering which has a $\sin^2(\frac{1}{2}\theta)$ dependence. For inelastic scattering, the magnetic term dominates where the electric term is forbidden by selection rules.

Charge form factor. Scattering described by the Mott or the Dirac formula are for point charges. Such a picture is correct for nuclei only at very low incident energies where the de Broglie wave lengths are much larger than nuclear dimensions. As the bombarding energy increases, the charge and magnetization density distribution in a nucleus become "visible" to the incident electrons and the scattering results are sensitive to the finite distributions of charge and magnetism in the target nucleus.

Let $\Psi(\boldsymbol{r})$ be the ground state wave function of a nucleus. The charge density distribution is then given by

$$\rho_{\text{ch}}(\boldsymbol{r}) = Z\left|\Psi(\boldsymbol{r})\right|^2 \tag{4-12}$$

where it is taken that the nuclear wave function $\Psi(\boldsymbol{r})$ is normalized to unity and the charge density is normalized to Z, the number of protons in the nucleus. The Fourier transform of $\rho_{\text{ch}}(\boldsymbol{r})$, given by the integral

$$F(\boldsymbol{q}) = \int \rho_{\text{ch}}(\boldsymbol{r}) e^{i\boldsymbol{q}\cdot\boldsymbol{r}}\, dV \tag{4-13}$$

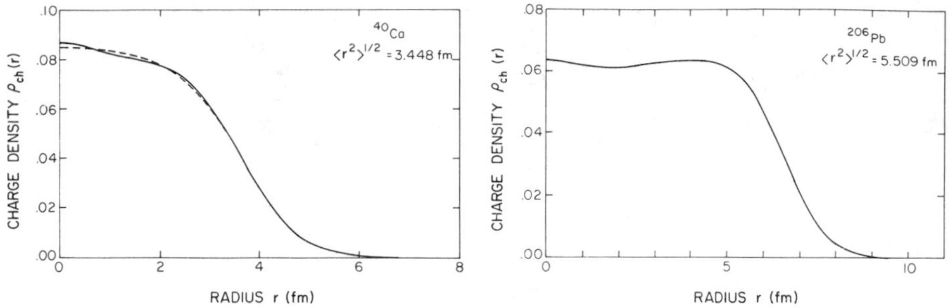

Fig. 4-1 Charge densities of ^{40}Ca and ^{206}Pb obtained from electron scattering. The solid lines are observed values reconstructed from Fourier-Bessel coefficients and the dashed line is the calculated results obtained from a three-parameter Fermi form for ^{40}Ca. The values of all the parameters are taken from de Vries, de Jager, and de Vries (*Atomic Data and Nucl. Data Tables*, **36** [1987] 495), obtained by fitting electron scattering data.

is called the charge or *longitudinal* form factor so as to distinguish from the transverse form factor to be discussed later.

For nuclear size studies, we are primarily interested in the radial dependence of the density. For this purpose, we can take an average over the angular distribution and consider only $\rho_{\rm ch}(r)$, the radial distribution of the charge density. With this simplification, the angular part of the integration in (4-13) can be carried and the (radial) charge form factor reduces to the expression

$$F(q^2) = \frac{4\pi}{q} \int \rho_{\rm ch}(r) \sin(qr) r \, dr \tag{4-14}$$

It is a function of q^2, the square of the momentum transfer given in (4-6), rather than q, since only q^2 is a proper Lorentz scalar. We shall also see later in (4-18) that only even powers of q enter into an expansion of the form factor in terms of its argument.

In terms of $F(q^2)$, the cross section for elastic scattering of electrons off a finite nucleus, with ground state spin $J = 0$, may be expressed in the form

$$\frac{d\sigma}{d\Omega} = \left(\frac{d\sigma}{d\Omega}\right)_{\rm point} |F(q^2)|^2 \tag{4-15}$$

From this expression we see that the ratio of the measured and the point particle scattering cross sections yields the square of the form factor. By applying an inverse of the transformation given in (4-13), charge density distributions are obtained from measured scattering cross sections through form factors deduced using (4-15),

$$\rho_{\rm ch}(r) = \frac{1}{2\pi^2 r} \int_0^\infty F(q^2) \sin(qr) q \, dq \tag{4-16}$$

Examples of charge density distribution obtained this way are shown in Fig. 4-1.

Transverse form factor. For elastic scattering off $J = 0$ states, only the longitudinal form factor can contribute to the cross section. This comes from the fact that this is the only operator having a spherical tensor rank zero part and, hence, allowed by angular momentum coupling requirements. More generally, instead of (4-15), the cross section is given by

$$\frac{d\sigma}{d\Omega} = \left(\frac{d\sigma}{d\Omega}\right)_{\text{point}} \left\{ |F(q^2)|^2 + \left(\tfrac{1}{2} + \tan^2\left(\tfrac{1}{2}\theta\right)\right)|F_T(q^2)|^2 \right\} \tag{4-17}$$

where the *transverse* form factor $F_T(q^2)$ can be further decomposed into electric and magnetic multipole components. The operators for these terms are related to those for electromagnetic moments given in §4-7 and for transitions between states in §5-2. Since the tensorial ranks of these operators are greater than zero, their contributions are felt only for elastic scattering involving states with $J > 0$ and for inelastic scattering (other than those with both initial and final spin zero).

§4-3 CHARGE RADIUS AND CHARGE DENSITY

Charge radius. Using the charge form factor $F(q^2)$ deduced from electron scattering experiments, we can obtain the root-mean-square (rms) radius of a nucleus in a model-independent way. For low momentum transfers, $F(q^2)$ may be expanded as an infinite series in q^2. Starting with (4-16), we can first write the sine function as a power series in terms of its argument qr. Upon integrating term by term, we obtain the result

$$\begin{aligned} F(q^2) &= \frac{4\pi}{q} \int \rho_{\text{ch}}(r)\left(qr - \tfrac{1}{3!}(qr)^3 + \cdots\right) r\,dr \\ &= \int \rho_{\text{ch}}(r) 4\pi r^2\,dr - \tfrac{1}{6} q^2 \int r^2 \rho_{\text{ch}}(r) 4\pi r^2\,dr + \cdots \\ &= Z\left\{1 - \tfrac{1}{6}\, q^2 < r^2 > + \cdots\right\} \end{aligned} \tag{4-18}$$

where the overall factor Z comes from the normalization factor used for charge densities given in (4-12). The radial integral in the second term is the expectation value of the radius squared. In this way, the q^2 dependence of $F(q^2)$ at low q gives directly the expectation value of r^2. A plot of the measured value of $\langle r^2\rangle^{1/2}/A^{1/3}$ as a function of nucleon number A is given in Fig. 4-2. Except for small values of A, we see that the ratio $\langle r^2\rangle^{1/2}/A^{1/3}$ is roughly a constant, with a value 0.97 ± 0.04 fm. This lends support to the idea that the nucleus is made of an "incompressible fluid"; as a result, the volume increases linearly with nucleon number and the radius increases with $A^{1/3}$.

It should be pointed out that $\langle r^2\rangle^{1/2}$ is not the nuclear radius R. This is simply illustrated by the example of a constant-density sphere of radius R with density given by

$$\rho(r) = \begin{cases} \rho_0 & \text{for } r \le R \\ 0 & r > R \end{cases}$$

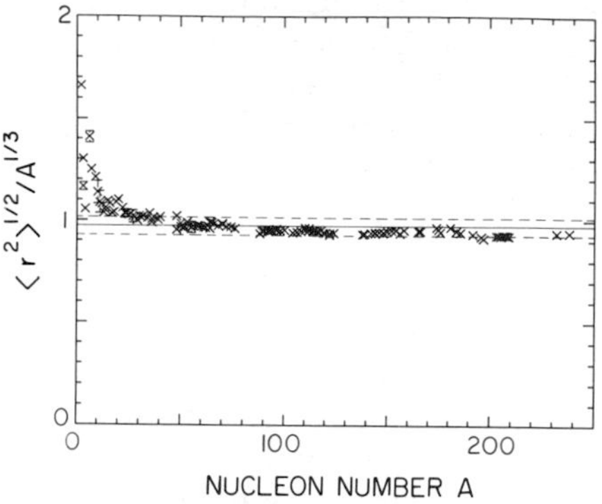

Fig. 4-2 Distribution of $\langle r^2 \rangle^{1/2} A^{-1/3}$ as a function of nucleon number A using values of $\langle r^2 \rangle^{1/2}$ deduced from electron scattering data. Except for light nuclei, we find $\langle r^2 \rangle^{1/2} = (0.97 \pm 0.04) \times A^{1/3}$ fm, indicating that the nuclear volume is essentially proportional to the number of nucleons.

The volume of the sphere is then

$$V = 4\pi \int_0^\infty \rho(r)\, r^2\, dr = \frac{4\pi}{3} R^3 \rho_0$$

and the expectation value of r^2,

$$\langle r^2 \rangle = \frac{3}{R^3 \rho_0} \int_0^\infty \rho(r)\, r^4\, dr = \frac{3}{5} R^2 \tag{4-19}$$

This gives a value of $R = \{5/3\langle r^2 \rangle\}^{1/2} = 1.29\, \langle r^2 \rangle^{1/2}$. For more realistic radial density distributions such as the ones given below, the ratio $R^2/\langle r^2 \rangle$ is slightly smaller than 5/3. This, together with the result of $\langle r^2 \rangle^{1/2} \approx 0.97 A^{1/3}$ obtained from Fig. 4-2, gives the relation $R = r_0 A^{1/3}$ with $r_0 = 1.2$ fm as given in (4-2).

Fourier-Bessel coefficients. Nuclear charge densities $\rho_{\text{ch}}(r)$, deduced from charge form factor $F(q^2)$ measured in electron scattering experiments, are often given in terms of Fourier-Bessel coefficients. By this way of parametrizing an empirical charge density distribution, the density up to some cutoff radius R_c is expanded in terms of $j_0(\xi)$, spherical Bessel function of order zero,

$$\rho_{\text{ch}}(r) = \begin{cases} \sum_k a_k j_0(k\pi \frac{r}{R_c}) & \text{for } r \le R_c \\ 0 & \text{for } r > R_c \end{cases} \tag{4-20}$$

The reason that only spherical Bessel function of order zero enters here comes from the fact that the charge density operator is a spherical tensor of rank zero and therefore carries no angular momentum. For an inelastic transition involving multipole excitation of order λ, spherical Bessel function $j_\lambda(\xi)$ comes in instead of $j_0(\xi)$.

The Fourier-Bessel coefficients may be expressed in terms of the charge density in the following way. Since

$$j_0(\xi) = \frac{\sin \xi}{\xi}$$

and

$$\int_0^1 \sin(m\pi x)\sin(n\pi x)\,dx = \tfrac{1}{2}\delta_{mn} \qquad \text{for integer } m \text{ and } n$$

the Fourier-Bessel coefficients are related to the charge density through the integral,

$$a_m = \frac{2m^2\pi^2}{R_c^3}\int_0^{R_c} \rho_{\rm ch}(r) j_0\left(m\pi\frac{r}{R_c}\right) r^2\,dr$$

In practice, the form factor $F(q^2)$ can be measured only up to some maximum momentum transfer. For this reason, the charge density $\rho_{\rm ch}(r)$ may be determined only up to a certain accuracy. This, in turn, implies that there is only a finite number of Fourier-Bessel coefficients that can be found from a given measurement. The accuracy achieved in using a finite number of Fourier-Bessel coefficients to represent a charge density depends somewhat on the choice of the cutoff radius R_c as well. Usually R_c is taken to be slightly beyond where the density essentially drops off to zero. For light nuclei, R_c is often taken to be around 8 fm and, for heavy nuclei, a value of 12 fm is used instead.

Other forms of charge density. For many practical applications, density distributions in terms of Fourier-Bessel coefficients remain too complicated. As can be seen from Fig. 4-1, the charge density as a function of the radial distance r is essentially a constant except for the central ($r \approx 0$) and the surface ($r \approx R$) regions. This is particularly true for heavy nuclei where the nucleus is large enough for the separation into various regions to be significant. In general, the central region is unimportant partly for the reason that it occupies only a small fraction of the nuclear volume and partly because most studies are sensitive primarily to the surface region.

Because of the short-range nature of nuclear force, nucleons near the surface of the nucleus are less tightly bound than those inside for the simple reason that there are fewer nucleons in the vicinity to interact with. As a result, nuclear density drops off more or less exponentially in the surface region. A density distribution with a constant central region and a diffused edge may be represented by simple analytical forms involving fewer parameters than the number of Fourier-Bessel coefficients in (4-19).

A density distribution with a diffused edge may be written in the form

$$\rho_{2pF}(r) = \frac{\rho_0}{1 + \exp\frac{r-c}{z}} \tag{4-21}$$

The two parameters, c and z, are determined, for instance, by fitting to densities derived from measured form factors and the factor ρ_0 is given by normalization condition. Eq. (4-21) is generally known as a two-parameter Fermi form or Woods-Saxon form. The meaning of c may be interpreted as the radius of the distribution at half of its central value, and z is the *diffuseness*, related to the thickness of

Table 4-1 Parameters for Fermi forms of charge density distribution.

Nucleus	$\langle r^2\rangle^{1/2}$	c	z	w
^{16}O	2.730±0.025	2.608	0.513	−0.051
^{28}Si	3.086±0.018	3.340	0.580	−0.233
^{40}Ca	3.482±0.025	3.766	0.586	−0.161
^{88}Sr	4.17 ±0.02	4.83	0.496	
^{112}Cd	4.608±0.007	5.38	0.532	
^{148}Sm	4.989±0.037	5.771	0.596	
^{184}W	5.42 ±0.07	6.51	0.535	
^{206}Pb	5.509±0.029	6.61	0.545	
^{238}U	5.84	6.805	0.605	

Three of the entries with the value of w given are for the 3pF form and six entries without the value of w are for the 2pF form. Values of $\langle r^2\rangle^{1/2}$ are included for comparison.

the surface region. Examples of the parameters extracted from observed charge densities are listed in Table 4-1 for illustration.

A better description of the observed density is provided by a modified Fermi form with an additional parameter w,

$$\rho_{\rm 3pF}(r) = \rho_0 \frac{1 + w\left(\frac{r}{c}\right)^2}{1 + \exp\frac{r-c}{z}} \tag{4-22}$$

generally known as the three-parameter Fermi form. Such a parametrization of the density is used, for example, to obtain the dashed line for ^{40}Ca in Fig. 4-1. Other forms, such as the three-parameter Gaussian,

$$\rho_{\rm 3pG} = \rho_0 \frac{1 + w\left(\frac{r}{c}\right)^2}{1 + \exp\frac{r^2-c^2}{z^2}} \tag{4-23}$$

and the harmonic-oscillator model form,

$$\rho_{\rm HO} = \rho_0 \left\{1 + z\left(\frac{r}{c}\right)^2\right\} e^{-\left(\frac{r}{c}\right)^2} \tag{4-24}$$

are also used to characterize the charge density. Tabulated values of charge density distribution for various nuclei using these forms can be found, for example, in de Vries, de Jager, and de Vries (*Atomic Data and Nucl. Data Tables* **36** [1987] 495).

§4-4 NUCLEON FORM FACTOR

At sufficiently high energies, say in the GeV region, the de Broglie wave length of an electron becomes much shorter than the size of a typical nucleus. In such cases, the scattering result is dominated by the charge distributions within individual nucleons. The primary interest of scattering at these energies shifts to the structure of nucleons rather than that of a nucleus. For this purpose, it is more convenient to consider for the moment that the target is made of isolated nucleons. Instead of nuclear form factors, we shall be concerned with the analogous nucleon form factors. We shall return at the end of the next section to the question of whether there is any difference if the scattering is off nucleons bound in nuclei.

Nucleon form factors. Since nucleons are spin-$\frac{1}{2}$ particles, both electric and magnetic scattering contribute to the electron scattering cross section. For reasons that will soon become obvious, it is more convenient to express nucleon form factors in terms of the Sachs form factors $G_E(q^2)$ and $G_M(q^2)$. This gives us the Rosenbluth formula for the differential cross section of electron scattering off nucleons,

$$\left(\frac{d\sigma}{d\Omega}\right)_{\text{lab.}} = \left(\frac{d\sigma}{d\Omega}\right)_{\text{point}} \left\{\frac{G_E^2(q^2) + \zeta G_M^2(q^2)}{1+\zeta} + 2\zeta G_M^2(q^2)\tan^2\left(\tfrac{1}{2}\theta\right)\right\} \tag{4-25}$$

where the dimensionless quantity ζ is given by

$$\zeta = \left(\frac{\hbar q}{2Mc}\right)^2$$

The relation between Sachs form factors and longitudinal and transverse form factors defined earlier may be seen by comparing the form of (4-25) with that of (4-17). The Sachs form factors have the property that, at zero momentum transfer,

$$G_E(0) = \begin{cases} 1 & \text{for proton} \\ 0 & \text{for neutron} \end{cases} \qquad G_M(0) = \begin{cases} \mu_p & \text{for proton} \\ \mu_n & \text{for neutron,} \end{cases} \tag{4-26}$$

where μ_p and μ_n are, respectively, the magnetic dipole moment of a proton and a neutron in units of nuclear magnetons (see Table 2-8).

In the place of $G_E(q^2)$ and $G_M(q^2)$, the scattering cross section may also be written in terms of Dirac and Pauli form factors, $F_1(q^2)$ and $F_2(q^2)$, defined by the following expressions:

$$G_E(q^2) = F_1(q^2) - \zeta F_2(q^2) \qquad G_M(q^2) = F_1(q^2) + F_2(q^2) \tag{4-27}$$

The main difference between these two sets of form factors is that, instead of electric and magnetic scattering, $F_1(q^2)$ and $F_2(q^2)$ are distinguished according to helicity $\boldsymbol{\sigma}\cdot\boldsymbol{p}/|\boldsymbol{p}|$, projection of the intrinsic spin $\boldsymbol{\sigma}$ of the electron along its direction of motion $\boldsymbol{p}/|\boldsymbol{p}|$. The Dirac form factor $F_1(q^2)$ represents the helicity preserving

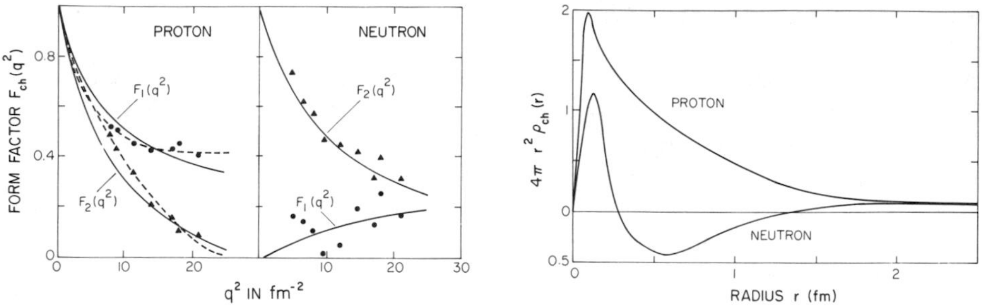

Fig. 4-3 Nucleon form factors and charge distributions. (Adopted from Hofstadter and Herman, *Phys. Rev. Lett.* **6** [1961] 293 and Littauer, Schopper, and Wilson, *Phys. Rev. Lett.* **7** [1961] 144.)

part of the scattering and the Pauli form factor $F_2(q^2)$ represents the helicity-flipping part. It should be noted that the Rosenbluth formula (4-25) was derived using first Born approximation involving only the exchange of one photon for the interaction between the electron and the nucleon. In principle, corrections due to two or more photon exchanges are needed, but comparisons with observations indicate that the formula works well to fairly high energies.

Asymptotic forms. There are two interesting points connected with nucleon form factors that are worth noting. The first is that, in the limit of large momentum transfer, the two proton form factors and the magnetic form factor of a neutron are identical to each other except for a *scaling factor* which may be deduced by examining (4-26),

$$G_E^{\mathrm{p}}(q^2) = \frac{1}{\mu_p} G_M^{\mathrm{p}}(q^2) = \frac{1}{|\mu_n|} G_M^{\mathrm{n}}(q^2) = G(q^2) \tag{4-28}$$

The function $G(q^2)$ may be described by a dipole form

$$G(q^2) = \frac{1}{\left\{1 + (q/q_0)^2\right\}^2} \tag{4-29}$$

Empirically, the parameter q_0 is found to be $\hbar q_0 = 0.84$ GeV/c.

Using the dipole form, the proton charge distribution may be expressed in the following way:

$$\rho_{\mathrm{ch}}(r) = \rho_0 e^{-q_0 r} \tag{4-30}$$

From this we obtain the square of the charge radius of a proton,

$$\langle r^2 \rangle = \frac{\int r^2 \rho_{\mathrm{ch}}(r)\, r^2\, dr}{\int \rho_{\mathrm{ch}}(r)\, r^2\, dr} = \frac{12}{q_0^2} = \left(0.81 \text{ fm}\right)^2 \tag{4-31}$$

Because of the scaling relation (4-28), the magnetic radius of a proton must also have the same value. Note that the value 0.81 fm for the rms radius of a nucleon is slightly smaller than the corresponding average value $\langle r^2 \rangle^{1/2}/A^{1/3} = 0.97 \pm 0.04$ for nucleons in nuclei.

The electric form factor of a neutron $G_E^n(q^2)$ is only known at small momentum transfers, $q < 10$ GeV/c, and is found to be much smaller in value than the corresponding magnetic form factor $G_M^n(q^2)$ at the same momentum transfer. In addition, there are two other reasons which make measurements of neutron electric form factors difficult at high q. The first is the increase in the value of ζ of (4-25) with q and, as a result, the scattering cross section at high q is dominated by magnetic form factors. The second is the absence of a fixed neutron target, so that all our experimental knowledge of neutrons must be deduced indirectly from scattering off such targets as deuteron and ^{3}He. Based on the information available, the charge distributions obtained by applying (4-16) to the form factors obtained are shown in Fig. 4-3. The small neutron charge form factor, however, may be important in understanding some of the details in nuclear charge distribution to be discussed later.

§4-5 HIGH-ENERGY LEPTON SCATTERING

Let us return again to the question of electron scattering off nuclei. At low energies, where the electron wave length is much larger than the nuclear size, the scattering cross section is essentially given by the Mott formula (4-3). As the energy of the incident electron is increased, the extended size of nuclear charge distribution comes into play and the scattering cross section is modified by the nuclear charge form factor. Our interest is still confined to elastic scattering; that is, the energy of the scattered electron E_f differs from the incident energy E_i only by the amount taken up by nuclear recoil. The energy transferred to the nucleus is then

$$\hbar\omega \equiv E_i - E_f = \frac{(\hbar q)^2}{2M} \qquad (4\text{-}32)$$

The relation between ω and q^2 given above may be taken as the definition of elastic scattering.

Quasi-elastic scattering. As we increase the incident energy further, the electron wave length will become short enough to be comparable with the size of a nucleon in the nucleus. At this point, coherence in the scattering from several nucleons at the same time is lost and the scattering takes place essentially from individual nucleons. For elastic scattering of electrons off free nucleons, the energy transferred is given by

$$\hbar\omega = \frac{(\hbar q)^2}{2M_N} \qquad (4\text{-}33)$$

This expression differs from (4-32) in that the nucleon mass M_N rather than M, the mass of the target nucleus, appears in the denominator. Since M_N is different from M, the scattering of an electron from a "bound" nucleon is no longer a true elastic scattering according to the definition given by (4-32). It is, instead, a *quasi-elastic* scattering of electrons off individual nucleons. Quasi-elastic scattering also differs from elastic scattering off free nucleon, in that nucleons in a nucleus are not stationary with respect to the nuclear center of mass. The average momentum of a nucleon may be estimated from the uncertainty relation,

$$p_{\mathrm{F}} \sim \frac{\hbar}{R} \tag{4-34}$$

where R is the size of the potential well binding a nucleon to the nucleus. This is known as the Fermi momentum of a nucleon inside a nucleus. Since R is of the order of the size of the nucleus, a few femtometers, p_{F} is of the order of 100 to 200 MeV/c. As a result, there is a spread of the order of 100 MeV in the energy transferred in quasi-elastic scattering, around 10% of the total.

Structure functions. At forward angles, the momentum transferred in a scattering is small and the cross section is dominated by elastic scattering. Since form factors decrease in value very quickly with increasing momentum transfer, the elastic scattering cross section rapidly becomes very small as momentum transfer is increased, and the importance of inelastic scattering processes becomes apparent. The reaction cross section now depends in general on the amount of energy $\hbar\omega$ as well as the momentum q transferred. The scattering result is usually expressed as a double differential cross section in these two variables,

$$\frac{d^2\sigma}{dq^2 d\omega} = \frac{4\pi Z^2\alpha^2}{q^4}\frac{\hbar E_f}{E_i Mc^2}\left\{\frac{Mc^2}{\hbar\omega}F_2(q^2,\omega)\cos^2\left(\tfrac{1}{2}\theta\right) + 2F_1(q^2,\omega)\sin^2\left(\tfrac{1}{2}\theta\right)\right\} \tag{4-35}$$

where the factor $4\pi Z^2\alpha^2/q^4$ is the familiar Rutherford scattering cross section off a point charge given in (4-8). The functions $F_1(q^2,\omega)$ and $F_2(q^2,\omega)$, related to the form factors defined earlier for (single) differential cross section $d\sigma/d\Omega$, are often referred to as the nucleon *structure functions* since they express the difference of a nucleon from a point particle.

The definition of the momentum transfer in an inelastic scattering remains the same as that given in (4-4); however, its relation to energy is slightly different. In the limit that the electron mass may be ignored, the final form of (D-37) is equivalent to

$$(\hbar cq)^2 = 4E_iE_f\sin^2\left(\tfrac{1}{2}\theta\right) \tag{4-36}$$

compared with that given by (4-6). In addition to the energy taken away by target particle recoil, some of the incident energy is also expended in exciting particles from ground states to excited states in an inelastic scattering.

Since nucleons are not "elementary" particles, quasi-elastic scattering off the constituent quarks and inelastic scattering involving excitation of a nucleon to a higher state can take place in the same way as inelastic scattering off nuclear targets: the only difference between these two types of scattering is that the energies involved in nucleon scattering are usually much higher. For high-energy electron scattering, it is customary to express the scattering in terms of two dimensionless quantities:

$$x = \frac{\hbar q^2}{2M\omega} \qquad\qquad y = \frac{\hbar\omega}{E_i} \tag{4-37}$$

instead of q^2 and ω. Let us rewrite the double differential cross section for inelastic scattering in terms of these two variables.

From (4-36), we obtain the relation,

$$\sin^2\left(\tfrac{1}{2}\theta\right) = \frac{(\hbar c q)^2}{4E_i E_f} = \frac{Mc^2}{2E_f}xy \tag{4-38}$$

and, for high-energy scattering in the forward directions,

$$\cos^2\left(\tfrac{1}{2}\theta\right) = 1 - \sin^2\left(\tfrac{1}{2}\theta\right) = 1 - \frac{(\hbar c q)^2}{4E_i E_f} \approx 1 \tag{4-39}$$

Instead of (4-36), we may express q^2 in terms of x and y,

$$q^2 = \frac{2ME_i}{\hbar^2}xy \tag{4-40}$$

Using (4-38) and (4-39), the angular dependence on the right hand side of (4-35) may be expressed in terms of x and y,

$$\frac{d^2\sigma}{dq^2 d\omega} = \frac{4\pi Z^2\alpha^2}{q^4}\frac{1}{\omega}\left\{F_2(q^2,x)(1-y) + F_1(q^2,x)xy^2\right\}$$

where we have made use of the fact that

$$\frac{E_f}{E_i} = 1 - y$$

from the definition of $\hbar\omega$ in (4-32) and y in (4-40). In terms of x and y, the double differential cross section may be written in the form usually found in the literature

$$\frac{d^2\sigma}{dx\,dy} = \frac{4\pi Z^2\alpha^2}{q^4}\frac{2ME_i}{\hbar^2}\left\{F_2(q^2,x)(1-y) + F_1(q^2,x)xy^2\right\} \tag{4-41}$$

In the derivation, the only property of an electron used is that its rest mass may be ignored; the formula can therefore be applied to describe the scattering of other charged leptons, such as muons, at high energies. For neutrino scattering, however, one must replace the factor $4\pi\alpha^2/q^4$ in Rutherford scattering with the factor $G_F^2/2\pi$, where G_F is the Fermi coupling constant for weak interactions (see §5-4). The details can be found in standard texts on high-energy physics such as Perkins (*Introduction to High Energy Physics*, 3rd ed., Addison Wesley, Menlo Park, California, 1987).

EMC effect in deep-inelastic scattering. Many different types of final states can be reached in high-energy scattering. If the cross section includes all the possible final states, the scattering is called an *inclusive* one, in contrast with exclusive scattering to a particular final state. Inclusive scattering is also referred to as *deep-inelastic* scattering. One of the interesting questions in high-energy, deep-inelastic lepton scattering off nuclei and nucleons concerns the quark substructure of nucleons. Indeed, it was the identification of point-like objects inside nucleons in lepton-nucleon scattering, known as partons at the time of discovery, that provided early experimental evidence for the existence of quarks in hadrons. The effect of quark substructure in lepton scattering may be formulated in terms of the nucleon structure functions introduced in (4-41).

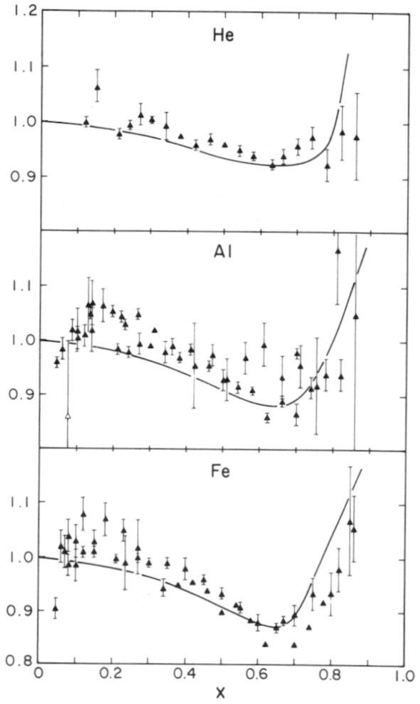

Fig. 4-4 Examples of cross section for high-energy lepton scattering off ^{4}He, ^{26}Al and ^{56}Fe as ratios to that for the deuteron. The values are essentially the same as $R(x)$, the ratio of $F_2^A(q^2,x)$, the nuclear structure function per nucleon, to $F_2^d(q^2,x)$, the nucleon structure factor for deuteron. The points are the measured values taken from Sloan, Smadja, and Voss (*Phys. Rep.* **162** [1988] 45) and the smooth curves are the calculated values from Akulinichev et al. (*Phys. Rev. Lett.* **55** [1985] 2239).

Eq. (4-41) applies equally well to scattering of leptons off nucleons as well as off nuclei. For the convenience of comparison between the measured results on nucleon and nuclear targets, we shall define the structure functions for nuclear targets in terms of their values per nucleon, $F_1^A(q^2,x)$ and $F_2^A(q^2,x)$, by taking out a factor A from $F_1(q^2,x)$ and $F_2(q^2,x)$ in (4-41),

$$\frac{d^2\sigma}{dxdy} = \frac{4\pi Z^2\alpha^2}{q^4}\frac{2ME_i}{A\hbar^2}\left\{F_2^A(q^2,x)(1-y) + F_1^A(q^2,x)xy^2\right\} \tag{4-42}$$

Let us compare the structure functions per nucleon obtained from high-energy lepton scattering off free nucleons and off bound nucleons in nuclei.

The structure functions for neutrons and protons are different from each other and may in principle be measured individually by separate experiments. However, for scattering off finite nuclei, the structure function measured will be the average between contributions from neutrons and protons, and there is no easy way to distinguish between them. Let us assume for simplicity that we have $N = Z$, true for light nuclei. For such targets, the contributions from bound neutrons and protons will have equal weight. The same relation between neutron and proton is also obtained from scattering off a deuteron target. Since the deuteron is a loosely bound system, we can treat the two nucleons in it as essentially free nucleons as we have done earlier in nucleon-nucleon scattering.

For simplicity, we shall ignore $F_1(q^2, x)$, since it contributes only a small amount to the measured cross sections. If the quark substructure of a free nucleon is the same as that of a bound nucleon in a nucleus, we expect the ratio

$$R(x) = \frac{F_2^A(q^2, x)}{F_2^d(q^2, x)} \qquad \text{(4-43)}$$

to be unity for all values of momentum transfer characterized by the dimensionless variable x. The experimental results, first reported by the European Muon Collaboration group (Aubert et al., *Phys. Lett.* **123B** [1983] 275) and later confirmed by other groups, however, showed not only that $R(x)$ differs from unity, but also the ratio changes as a function of x. Examples of such results are shown in Fig. 4-4. The apparent departure would, on the surface, imply that the internal structure of a bound nucleon, as revealed by $F_2(q^2, x)$, is somewhat different from that of free nucleon. However, because of the large uncertainties in some x regions, the available data at the moment are also compatible with calculated results (solid lines in Fig. 4-4) assuming that the only difference in bound and free nucleon structure functions is due to binding energy experienced by a nucleon in a nucleus. If future experiments can reduce the uncertainties, especially at the low-x region, we may yet be able to find a nontrivial difference in the quark substructure between bound and free nucleon. If this is true, it will be the first time that a signature of quark presence is found in nuclei.

If nucleon structure is modified when nucleons are imbedded in a nucleus, high-energy, strong-interaction probes may also be used to study the effects. Similarly, more selective exclusive reactions such as $(e, e'N)$, whereby a nucleon N is ejected from the nuclear target by inelastic electron scattering, may also be used to deduce the nucleon structure function inside a nucleus. The subject belongs to the new field of high-energy nuclear physics, to which we shall return to briefly at the end of §7-6.

§4-6 MATTER DENSITY AND CHARGE DENSITY

Since the interaction between an electron and a nucleus is primarily electromagnetic in nature, the density distribution observed through electron scattering is predominantly the distribution of charge inside a nucleus. Highly accurate charge distributions have been measured for many nuclei in this way and the results provide us with an opportunity to examine some of the related properties in detail. One question of interest is the influence of neutrons on the charge distribution.

One way to detect the influence of neutrons in charge distributions experimentally is to measure the isotopic difference, the difference in the charge distributions of nuclei with the same number of protons but different in the number of neutrons. If charge distribution is independent of neutrons in a nucleus, we expect the *isotopic shift* in the distributions for different isotopes to be negligible. The measured results indicate that, in general, the shifts are small but not zero. The same effect can also be observed in other measurements such as the energy of x-rays from muonic atoms. We shall also make use of such data in examining the influence of neutrons on the charge distribution.

Muonic atom. A muon is a lepton with properties very similar to an electron; it is therefore possible to replace one of the electrons in an atom by a muon. However, since the mass of a muon is 207 times larger than that of an electron, the radii of the muonic orbits are much smaller than those of electrons. This may be seen from the following argument.

Consider the case of a simple, hydrogen-like atom with Z protons in the nucleus and only a single electron outside. Using the Bohr model for this hydrogen-like atom, the radius of the n-th orbit is given by

$$r_n(e^-) = [4\pi\epsilon_0]\frac{n^2\hbar^2}{Zm_e e^2} = \frac{1}{\alpha\hbar c}\frac{n^2\hbar^2}{Zm_e} \tag{4-44}$$

where m_e is the mass of an electron, α is the fine structure constant, and the quantity inside the square brackets converts the formula from cgs to SI units. For a hydrogen atom ($Z = 1$), the ground state ($n = 1$) radius is the well-known Bohr radius,

$$a_0 = \frac{\hbar}{\alpha c m_e} = 5.29 \times 10^{-11} \text{ m}$$

In arriving at this result, only the electrostatic potential between a positively charged nucleus and a negatively charged electron is used; nothing else special about the electron enters into the argument. As a result, we can obtain the analogous result for a muonic atom by replacing m_e in (4-44) with the muon mass m_μ,

$$r_n(\mu^-) = \frac{1}{\alpha\hbar c}\frac{n^2\hbar^2}{Zm_\mu} = a_0\frac{n^2 m_e}{Zm_\mu} \tag{4-45}$$

Using similar arguments, we can arrive at the energy level of a muonic atom consisting of a single muon,

$$E_n = -\frac{m_\mu}{2}\frac{(Z\alpha c)^2}{n^2} \tag{4-46}$$

by starting with the energy levels of a hydrogen-like atom given in most standard textbooks on quantum mechanics.

The results of (4-45) and (4-46) apply only to a hydrogen-like atom. For atoms with $Z > 1$, this means all the electrons except one are stripped off so as to remove influences due to other electrons. Such "screening" effects are, in general, quite difficult to calculate accurately. Fortunately, for a muonic atom with only one electron replaced by a muon, we do not have this problem. Since the muonic orbits are so much smaller than the electronic orbits, there is very little chance of finding electrons between the muon and the nucleus, in particular for the low-lying orbits of interest to us. For this reason, the screening effect due to electrons in the muonic atom may be ignored for our purposes here.

For a heavy nucleus, such as ^{208}Pb with $Z = 82$, the radius of the lowest muonic orbit from (4-45) is

$$r_1(\mu^-) \simeq a_0 \frac{0.511}{82 \times 106} = 3.1 \times 10^{-15} \text{ m}$$

or 3.1 fm, using a muon mass of 106 MeV/c^2. This value is smaller than the radius of ^{208}Pb of 7.1 fm as estimated from $R = r_0 A^{1/3}$ using $r_0 = 1.2$ fm. A more elaborate calculation will show that the muon spends roughly 50% of the time inside a heavy nucleus. As a result, the actual muonic orbits are different from those given by (4-45,46), since these equations were derived assuming the nucleus as a point charge. We have seen that the low-lying muonic orbits are very close to the nuclear surface and the detailed charge distribution in the nucleus plays an important role. The modifications to the muonic orbits may be observed from the changes in the energy level positions which, in turn, affect the energy of x-rays emitted when the muonic atom decays from one level to anther.

When a muon is captured by an atom, it is likely that initially the muon will be in one of the higher levels. The muonic atom then decays to lower levels by emitting electromagnetic radiation. Since the differences in the energy levels are larger than the corresponding electronic orbits because of the greater muonic mass, as can be seen in (4-46), the wave lengths of the radiation emitted are in the x-ray range. From the energies of these x-rays, we can deduce the muonic energy levels. The differences from the values given by (4-46) provide a measure of the charge distribution in the nucleus (for more details, see, *e.g.*, Devons and Duerdoth, *Adv. Nucl. Phys.* **2** [1969] 295).

Isotopic shifts in calcium isotopes. It is useful to examine the influence of neutrons on the measured nuclear charge distributions using the even calcium isotopes as example. The isotopic shift data, obtained from electron scattering, are taken from Frosch et al. (*Phys. Rev.* **174** [1968] 1380) and summarized in Table 4-2. In addition to root-mean-square radius and parameters c, z, and w of a three-parameter Fermi distribution defined in (4-22), the values of the surface thickness t, the distance between 90% and 10% of the peak density, are also given to provide an idea of the large surface region in these nuclei.

Table 4-2 Charge distribution of Ca isotopes.

Nucleus	$\langle r^2\rangle^{1/2}$ (fm)	t (fm)	c (fm)	z (fm)	w
^{40}Ca	3.4869	2.681	3.6758	0.5851	−0.1017
^{42}Ca	3.5166	2.724	3.7278	0.5911	−0.1158
^{44}Ca	3.5149	2.630	3.7481	0.5715	−0.0948
^{48}Ca	3.4762	2.351	3.7444	0.5255	−0.03
^{48}Ti	3.5844	2.580	3.8551	0.5626	−0.0761

The difference, for example, in the values of the root-mean-square radius $\langle r^2\rangle^{1/2}$ between different isotopes given in the table is quite small. However, because of the great accuracy achieved in measuring the relative scattering cross sections, the results indicate a genuine difference in the size of the different isotopes as far as their charge distributions are concerned. Since the radius decreases by 0.01 fm in going from ^{40}Ca to ^{48}Ca, it means that the addition of neutrons to the calcium isotopes reduces the size of the charge distribution of the same 20 protons when the neutron number is increased from 20 to 28. If we take the simple view that charges were distributed evenly throughout the nuclear volume, the charge radius would have increased by 6% using the simple $R = r_0 A^{1/3}$ relation. In contrast, when we compare ^{40}Ca with ^{48}Ti, a nucleus with two more protons and six more neutrons than ^{40}Ca, we find that the size of charge distribution is increased by 0.1 fm, not far from the expectation of an $A^{1/3}$-dependence.

There are two possible explanations for the decrease in the charge radius with increasing neutron number among the even calcium isotopes. The first is that the addition of neutrons makes the protons more tightly bound and, hence, a smaller charge radius. This is, however, not true for nuclei in general and has led to the speculation that, for some other reasons, ^{48}Ca is a more tightly bound nucleus than its neighbors.

A second explanation depends on the charge distribution within a neutron. The net charge of a neutron is zero; however, as we have seen in §4-4, the charge distribution inside a neutron is not zero everywhere. One possible model for the charge distribution in a neutron is that the central part is positive and the region near the surface is negative, as shown in Fig. 4-3. The detailed charge distribution inside a neutron is not well known because of the difficulty in measuring the small charge form factor. However, a small excess of negative charge in the surface region can produce about a third of the decrease in the charge radius in going from ^{40}Ca to ^{48}Ca as suggested by Bertozzi et al. (*Phys. Lett.* **41B** [1972] 408). The other two-thirds of the decrease may be attributed to the spin-dependence (the Darwin-Foldy terms) in the interaction of protons with other nucleons in the nucleus.

Regardless of the exact cause of the isotopic shifts among the calcium isotopes, it is clear that neutrons have a definite influence on the measured charge distribution of a nucleus. Unfortunately data obtained using electromagnetic probes are not sufficient to provide us with the type of detailed information we need. In principle, strong interaction mechanisms can be used to deduce the neutron distribution. The difficulty here lies in separating out effects due to neutrons from a multitude of others — in particular, those involving aspects of strong interaction that are not yet very well known. The only exception is, perhaps, pion scattering.

Pion-nucleus scattering. We have seen earlier in §2-6 that there is a strong pion-nucleon resonance in the isospin $t = \frac{3}{2}$ (and spin $S = \frac{3}{2}$) channel at pion laboratory energy 195 MeV (see Fig. 2-3). The dominance of the P_{33}-resonance provides us with a unique way to probe the difference between neutron and proton density distributions in a nucleus.

There are six different reactions possible in pion-nucleon scattering using charged pions as the projectile:

$$\begin{array}{llll} (a) & \pi^+ + p \to \pi^+ + p & \text{(b)} & \pi^- + n \to \pi^- + n \\ (c) & \pi^+ + n \to \pi^+ + n & \text{(d)} & \pi^- + p \to \pi^- + p \\ (e) & \pi^+ + n \to \pi^0 + p & \text{(f)} & \pi^- + p \to \pi^0 + n \end{array} \tag{4-47}$$

We shall ignore the last two reactions, since the scattered neutral pions are much harder to detect than charged particles. Among the remaining four, only (a) and (b) have $|t_0| = \frac{3}{2}$ and are therefore purely isospin $\frac{3}{2}$ reactions. For reactions (c) and (d), $|t_0| = \frac{1}{2}$, and the isospin is a mixture of $\frac{1}{2}$ and $\frac{3}{2}$. From simple Clebsch-Gordan coupling of isospin, we find that

$$|\pi^+ n\rangle = \sum_t \langle 11\,\tfrac{1}{2}\,{-\tfrac{1}{2}}|t\tfrac{1}{2}\rangle\,|t,\tfrac{1}{2}\rangle = \sqrt{\tfrac{1}{3}}\,|t{=}\tfrac{3}{2},t_0{=}\tfrac{1}{2}\rangle + \sqrt{\tfrac{2}{3}}\,|t{=}\tfrac{1}{2},t_0{=}\tfrac{1}{2}\rangle$$

$$|\pi^- p\rangle = \sum_t \langle 1\,{-1}\,\tfrac{1}{2}\tfrac{1}{2}|t\,{-\tfrac{1}{2}}\rangle\,|t,-\tfrac{1}{2}\rangle = \sqrt{\tfrac{1}{3}}\,|t{=}\tfrac{3}{2},t_0{=}-\tfrac{1}{2}\rangle - \sqrt{\tfrac{2}{3}}\,|t{=}\tfrac{1}{2},t_0{=}-\tfrac{1}{2}\rangle$$

That is, only $\frac{1}{3}$ of the scattering amplitude of either one of these two reactions is in the isospin $t = \frac{3}{2}$-channel and the other $\frac{2}{3}$ is in the $t = \frac{1}{2}$-channel.

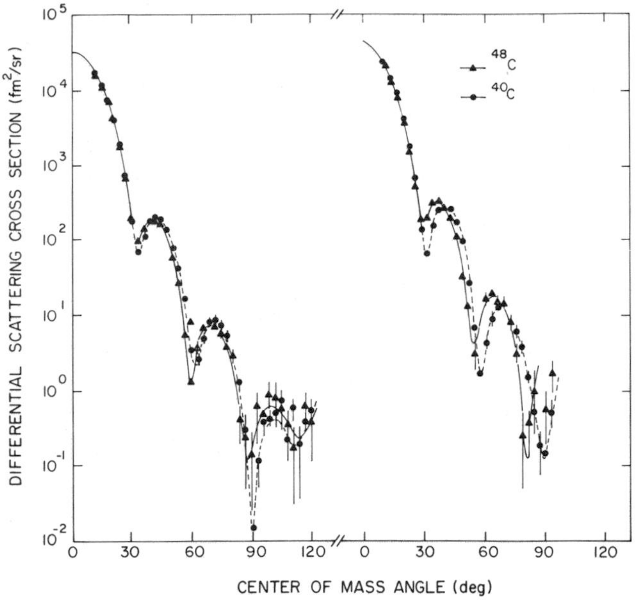

Fig. 4-5 Angular distributions of 180 MeV π^+ (left) and π^- (right) scattering off ^{40}Ca and ^{48}Ca targets. In each case, the triangles indicate data on ^{40}Ca and the circles indicate data on ^{48}Ca. The similarity in the π^+ cross sections shows that proton distributions in the surface regions of the two nuclei are essentially the same whereas the difference in the π^- cross sections shows that the neutron distributions are different. The data are taken from Corfu et al., as quoted by Ingram (*Meson-Nuclear Physics – 1979*, AIP Conf. Proc. **54**, ed. by E.V. Hungerford III, Amer. Inst. Phys., New York, 1979).

Since isospin is conserved in a reaction, the scattering amplitudes for the first four reactions in (4-47) may be decomposed in terms of isospin:

(a) $f_{\pi^+p}(\theta) = f_{t=\frac{3}{2}}(\theta)$ (b) $f_{\pi^-n}(\theta) = f_{t=\frac{3}{2}}(\theta)$

(c) $f_{\pi^+n}(\theta) = \frac{1}{3}f_{t=\frac{3}{2}}(\theta) + \frac{2}{3}f_{t=\frac{1}{2}}(\theta)$ (d) $f_{\pi^-p}(\theta) = \frac{1}{3}f_{t=\frac{3}{2}}(\theta) + \frac{2}{3}f_{t=\frac{1}{2}}(\theta)$

At energies near the P_{33}-resonance, the scattering cross section in the isospin $t = \frac{1}{2}$-channel is much smaller than that in the $t = \frac{3}{2}$-channel, and we shall ignore it here to simplify the argument. In this approximation, we obtain the ratios

$$\frac{\sigma(\pi^+p)}{\sigma(\pi^+n)} \approx \frac{\sigma(\pi^-n)}{\sigma(\pi^-p)} \approx 9 \tag{4-48}$$

from squares of the scattering amplitudes. Thus, the difference between the elastic π^+- and π^--scattering cross section off a nucleus is, to $\sim$ 10% uncertainty, the difference between neutron and proton density distributions. The method is particularly useful since pion scattering is sensitive mainly to the nuclear surface region where most of the differences in the neutron and proton densities are expected. Tests have been applied to a variety of nuclei, and the results have demonstrated the usefulness of the method.

For the even calcium isotopes discussed earlier, differential cross sections for charged pion scattering confirm the observation made with electromagnetic probes that the proton distribution is essentially unchanged as we add more neutrons. When the results of 180 MeV π^+ scattering off ^{40}Ca and ^{48}Ca are plotted on the same graph as shown in the left side of Fig. 4-5, very little difference is found. Since the energy is very close to the P_{33}-resonance and the cross sections are dominated by scattering off protons, the data give strong support to the similarity of proton distributions in ^{40}Ca and ^{48}Ca. On the other hand, a definite difference is found in the results of scattering of π^- off the same two nuclei at the same energy, as shown on the right side of Fig. 4-5. This is caused by the difference in the neutron distributions in the two nuclei, a result we expect from the eight additional neutrons in ^{48}Ca. When we examine the π^--scattering data off the same two nuclei at energies away from the P_{33}-resonance, the same type of difference is also observed except that the magnitudes are somewhat smaller, since scattering off neutrons no longer dominates over that off protons. However, the accuracy that can be achieved in pion scattering is not yet as high as that of electron scattering. As a result, it is not possible to examine the detailed differences as we have done earlier with electron scattering results.

Nucleon-nucleus scattering. As we shall see later in Chapter 7, the time spent by an intermediate energy (100 to 1000 MeV) proton in a nucleus is sufficiently short that it is unlikely to suffer multiple scattering, *i.e.*, projectile scattered more than once inside a target nucleus. In this way, the incident proton interacts only with one of the nucleons in the target and, as a result, the scattering is sensitive to the density of nucleon distribution in a fairly direct way. The incident proton interacts with both neutrons and protons in the nucleus, and there is a small difference between their interactions coming from the differences in the isospin of the reaction channels as we have seen above for pion scattering. The isospin dependence of the nuclear force is, however, not strong enough for the scattering to differentiate clearly between scattering off a proton and off a neutron as charged pions are capable of doing at the P_{33}-resonance. On the other hand, we can make use of the fact that proton density distribution in a nucleus is known through electron scattering, and any difference measured by proton scattering off nuclei may be attributed to the presence of neutrons. Neutron distributions obtained in this way depend a great deal on the model used to analyze the data, and the results are somewhat ambiguous compared with, for example, those obtained from charged pion scattering at the P_{33}-resonance.

§4-7 NUCLEAR SHAPE AND ELECTROMAGNETIC MOMENTS

Multipole expansion of charge density. In electromagnetism, the potential $\phi(\boldsymbol{r})$ due to a point charge q located at $\boldsymbol{r}'$ is given by the expression,

$$\phi(\boldsymbol{r}) = \left[\frac{1}{4\pi\epsilon_0}\right]\frac{q}{|\boldsymbol{r}-\boldsymbol{r}'|}$$

where the quantity inside the square brackets converts the expression from cgs to SI units. In the region $r > r'$, the potential may be expressed as a series in terms of spherical harmonics,

$$\phi(\boldsymbol{r}) = \left[\frac{1}{4\pi\epsilon_0}\right]\sum_{\lambda\mu}\frac{4\pi q}{2\lambda+1}\frac{(r')^\lambda}{r^{\lambda+1}}Y^*_{\lambda\mu}(\theta',\phi')Y_{\lambda\mu}(\theta,\phi)$$

where we have made use of the notation $\boldsymbol{r} = (r,\theta,\phi)$ and $\boldsymbol{r}' = (r',\theta',\phi')$. The expansion may be extended to a charge distribution in a nucleus $\rho_{\text{ch}}(\boldsymbol{r}')$,

$$\phi(\boldsymbol{r}) = \left[\frac{1}{4\pi\epsilon_0}\right]\sum_{\lambda\mu}\frac{4\pi Z}{2\lambda+1}\frac{1}{r^{\lambda+1}}\mathcal{Q}_{\lambda\mu}Y_{\lambda\mu}(\theta,\phi) \tag{4-49}$$

where the multipole coefficients

$$\mathcal{Q}_{\lambda\mu} = \frac{1}{Z}\int er'^\lambda Y^*_{\lambda\mu}(\theta',\phi')\rho_{\text{ch}}(\boldsymbol{r}')\,dV' \tag{4-50}$$

are quantities characterizing the charge density distribution. Along the z-axis, we have $\cos\theta = 1$ and (4-49) reduces to the familiar form given in many electromagnetic theory texts for the potential of an arbitrary, finite charge distribution,

$$\phi(r) = \left[\frac{1}{4\pi\epsilon_0}\right]\frac{1}{r}\sum_{\lambda\mu}\sqrt{\frac{4\pi}{2\lambda+1}}\frac{Q_{\lambda\mu}}{r^\lambda}$$

If the charge distribution is nearly spherical in shape, (4-49) is a fast convergent series and the importance of higher order terms decreases very rapidly with increasing order (λ,μ). In fact, the potential of a charge distribution can often be approximated by the contribution from the lowest order nonvanishing moment alone. We shall adopt the same philosophy here to describe the electromagnetic properties of a nucleus. Our interest is mainly in the low-order electric and magnetic multipoles, since the moments of these multipoles can be measured by a variety of experimental techniques.

For simplicity, let us start with the charge density. Using the normalization given in (4-12), the multipole coefficients may be expressed as the expectation value of an operator,

$$\mathcal{Q}_{\lambda\mu} = \langle\Psi(\boldsymbol{r})|er^\lambda Y^*_{\lambda\mu}(\theta,\phi)|\Psi(\boldsymbol{r})\rangle \tag{4-51}$$

From the expression, we can identify that the operator for electric multipole of order (λ, μ) has the form

$$O_{\lambda\mu}(E) = er^{\lambda}Y^{*}_{\lambda\mu}(\theta, \phi) \tag{4-52}$$

Note that the complex conjugation on the spherical harmonics is irrelevant in most of the subsequent discussion, since we shall be mainly using the $\mu = 0$ component. If we adopt a model that the nuclear wave function is made of products of single-particle wave functions (see §6-6), the operator may be reduced to a sum of operators each acting on an individual nucleon,

$$O_{\lambda\mu}(E) = e \sum_{\text{protons}} r_i^{\lambda}Y^{*}_{\lambda\mu}(\theta_i, \phi_i) = \sum_{i=1}^{A} e(i)r_i^{\lambda}Y^{*}_{\lambda\mu}(\theta_i, \phi_i) \tag{4-53}$$

where, in the final form, we have introduced the symbol

$$e(i) = \begin{cases} 1e & \text{for proton} \\ 0 & \text{for neutron} \end{cases}$$

so that the summation may be taken over all A nucleons in the nucleus. We will find later that the form given by (4-53) is a general one and is useful for discussing electric multipole transitions as well.

If the charge distribution of a nucleus is spherical in shape, only the zeroth order moment,

$$\mathcal{Q}_{00} = e$$

is different from zero; all higher order moments vanish, as can be seen from (4-50). A finite multipole coefficient therefore provides us with a measure of the departure from a spherical shape. However, the moments of some multipoles may also vanish because of symmetry reasons. For example, the transformation property of spherical harmonics under an inversion of the coordinate system,

$$Y_{\lambda\mu}(\pi - \theta, \pi + \phi) = (-1)^{\lambda}Y_{\lambda\mu}(\theta, \phi) \tag{B-14}$$

requires that all odd electric multipole coefficients be zero. As a result, we expect that the electric dipole moment vanishes for all nuclei and any observation of a nonvanishing value implies a violation of this symmetry. For this reason, there is great interest in measuring the electric dipole moment of a neutron. Since a neutron has no net charge ($\mathcal{Q}_{00} = 0$), it provides the best case for observing a nonzero electric dipole moment. This can happen if both time- and parity-invariance symmetries are violated. The measured value at the moment is $(-1.1 \pm 0.8) \times 10^{-25}$ e-cm, consistent with zero: however, it does not rule out the existence of a small symmetry violation contribution either. For our purposes, we shall assume that both parity- and time-invariance are exact symmetries and only even-order electric multipole moments may be different from zero.

Angular momentum coupling places a second restriction on the multipole coefficients a state can have. The multipole operator, $r^{\lambda}Y_{\lambda\mu}(\theta, \phi)$, is a spherical tensor of rank (λ, μ) and carries an angular momentum $\boldsymbol{\lambda}$. The expectation value of such an operator in a state with spin J vanishes unless the three angular momenta involved, $\boldsymbol{J}$, $\boldsymbol{\lambda}$, and $\boldsymbol{J}$, form a closed triangle. This means that all multipole coefficients vanish for a state with $2J < \lambda$. For this reason, a $J = 0$ state has no multipole moment except $\lambda = 0$, regardless of the intrinsic shape of the nucleus. This is a quantum mechanical phenomenon, as we can see from the following argument.

Classically, we can "see" the intrinsic shape of an object, for example, by taking a photograph. This is always possible, even for an object rotating at high angular velocities, if the picture is taken with sufficiently short exposure time. For a quantum mechanical object, we can also think of taking a photograph of the object in order to find out its shape. However, if the exposure time, Δt, is short, the picture does not necessarily correspond to that of an eigenstate: in general, it is a superposition of all the states in an energy interval $\Delta E = \hbar/\Delta t$, as dictated by the Heisenberg uncertainty principle. A sharp image of the intrinsic shape of an object requires extremely short exposure time and therefore corresponds to a superposition of a large number of eigenstates. To "see" the property of an object in a given eigenstate requires good energy resolution. In order to reduce ΔE, we need to take a time-exposure photograph with large Δt. A $J = 0$ state is one in which the apparent shape of the object is spherical, *i.e.*, it appears to be the same from all directions. In terms of a photograph, it is analogous to an extremely blurred image of the rotating object as the result of averaging over a long time. Although in practice we use scattering experiments to observe the shape of a quantum mechanical object rather than literally taking a photograph, the limitations imposed by the uncertainty principle on our "thought experiment" nevertheless apply.

For a state with $J > 0$, the value of a particular multipole coefficient is the expectation value of an operator and is, therefore, also a function of the M-value, the projection of the spin $\boldsymbol{J}$ on the quantization axis, of the state. The dependence on the M-value is, however, a trivial one, as we can see from the Wigner-Eckart theorem. Using (B-52), the expectation value may be reduced into a product of two parts,

$$\langle JM|Q_{\lambda\mu}|JM\rangle = (-1)^{J-M}\begin{pmatrix} J & \lambda & J \\ -M & \mu & M \end{pmatrix}\langle J\|Q_\lambda\|J\rangle$$

where $\begin{pmatrix} J & \lambda & J \\ -M & \mu & M \end{pmatrix}$ is the $3j$-coefficient, containing all the dependence of the matrix element on M. Since the reduced matrix element $\langle J\|Q_\lambda\|J\rangle$ is common to all the states differing only by their M-values, there is only a single independent quantity characterizing all the multipole coefficients of order λ for all $(2J + 1)$ substates of the same J. For this reason, it is convenient to define the *multipole moment* as the expectation value of the multipole operator in the state of maximum M, as we have done earlier for both the magnetic dipole moment of baryons and the electric quadrupole moment of the deuteron.

Electric quadrupole moment. Since the dipole moment vanishes for symmetry reasons, the lowest order electric multipole moment that can give us some idea of the "shape" of a nucleus is the quadrupole moment ($\lambda = 2$). The existence of a nonvanishing electric quadrupole moment implies that the charge distribution of the state is no longer a spherical one and the nucleus is said to be *deformed.* Usually nuclei near a closed shell are spherical in shape and have small absolute values for their quadrupole moments. On the other hand, nuclei in the middle of a major shell are deformed and have large absolute value of quadrupole moment. A more detailed discussion of deformed nuclei is given later in §6-3.

The quadrupole moment is the expectation value of the operator $r^2 Y_{20}(\theta, \phi)$ in the state of $M = J$,

$$\begin{aligned} Q &= \sqrt{\frac{16\pi}{5}} e \langle J, M=J | r^2 Y_{20}(\theta, \phi) | J, M=J \rangle \\ &= e \langle J, M=J | (3z^2 - r^2) | J, M=J \rangle \end{aligned} \qquad (4\text{-}54)$$

For a spherical nucleus, $\langle x^2 \rangle = \langle y^2 \rangle = \langle z^2 \rangle = \frac{1}{3}\langle r^2 \rangle$, and the quadrupole moment vanishes. For a deformed nucleus having an oblate shape; *i.e.,* the polar axis is shorter than the equatorial axis, Q is negative. In contrast, for a prolate shape nucleus with the polar axis longer than the equatorial axis, the quadrupole moment is positive, as we have seen earlier for the deuteron.

The next higher order, nonvanishing static electric multipole moment beyond quadrupole is the hexadecapole moment. The spherical tensor rank of the operator is $\lambda = 4$ here and the expectation value vanishes except for states with $J \geq 2$. For such a state, the quadrupole moment is usually nonzero. As a result, it is hard to measure the hexadecapole moment, as it is not easy, in general, to differentiate its effects from those due to quadrupole moment. Furthermore, since most of the nuclei are very nearly spherical in shape, (4-49) converges very fast and, for most observables, the contribution of the higher multipole moments decreases very quickly with increasing order. For this reason, observed effects due to quadrupole moment dominate over those from hexadecapole moment in most measurements. The lack of data is also due in part to the fact that static moment measurements are far more easily to be carried out on ground states and there are only very few stable nuclei with ground state spin $J \geq 2$. Most of the known values of hexadecapole moments are for excited states, deduced often in a model-dependent way from measurements such as Coulomb excitation discussed in §7-1.

Magnetic moments. In addition to charge distribution, a deformed nucleus may also possess a nonspherical "magnetic charge" distribution. Nuclear magnetism, as we have seen earlier, originates from a combination of two sources, the intrinsic magnetic dipole moment of individual nucleons and the orbital motion of protons. Analogous to an electric charge density distribution, we may define a ***magnetic charge density*** $\rho_{\rm m}(\boldsymbol{r})$ as the divergence of a magnetization density $\boldsymbol{M}(\boldsymbol{r})$,

$$\rho_m(\boldsymbol{r}) = -\nabla \cdot \boldsymbol{M}(\boldsymbol{r}) \qquad (4\text{-}55)$$

The magnetization density can, in turn, be written in terms of a magnetization current,

$$\mathcal{J}(\boldsymbol{r}) = \frac{c}{[c]}\nabla \times \boldsymbol{M}(\boldsymbol{r})$$

As far as $\mathcal{J}(\boldsymbol{r})$ is concerned, we can again adopt a model that nuclei are made of point nucleons having an intrinsic spin but no internal structure. A neutron is, then, a particle with magnetic dipole moment $\frac{1}{2}g_n$, and a proton is a particle having a magnetic dipole moment $\frac{1}{2}g_p$ as well as carrying a unit of positive charge. In such a point particle picture, the magnetization current density may be written as a sum of the contributions from all the nucleons in the nucleus,

$$\mathcal{J}(\boldsymbol{r}) = \sum_{i=1}^{A}\left\{eg_\ell(i)\frac{\boldsymbol{p}(i)}{M_N} + \frac{e\hbar}{2M_N}g_s(i)\nabla \times \boldsymbol{s}(i)\right\}\delta(\boldsymbol{r} - \boldsymbol{r}(i)) \tag{4-56}$$

where, in units of nuclear magneton μ_N,

$$g_\ell = \begin{cases} 1 & \text{for a proton} \\ 0 & \text{for a neutron} \end{cases} \qquad g_s = \begin{cases} 5.586 & \text{for a proton} \\ -3.826 & \text{for a neutron,} \end{cases}$$

and M_N is the mass of a nucleon (in practice, the proton mass is used).

Similar to a charge distribution, we may decompose a magnetization density distribution in terms of multipole coefficients given by the integral

$$\mathcal{M}_{\lambda\mu} = \int r^\lambda Y^*_{\lambda\mu}(\theta,\phi)\rho_m(\boldsymbol{r})\,dV = -\int r^\lambda Y^*_{\lambda\mu}(\theta,\phi)\,\nabla\cdot\boldsymbol{M}(\boldsymbol{r})\,dV \tag{4-57}$$

Because of the divergence operator, the parity of the integrand for magnetic multipole moment of order λ is $(-1)^{\lambda+1}$ instead of $(-1)^\lambda$, as in the case of an electric multipole moment operator of the same order. As a result, even-order magnetic multipoles vanish for the same reason as odd-order electric multipole moments.

The lowest order nonvanishing magnetic multipole for a nucleus is the dipole. From (4-57), we see that the operator is proportional to $rY_{1\mu}$. Since $rY_{1\mu}$ is proportional to $\boldsymbol{r}_\mu$, where $\boldsymbol{r}_\mu$ is a component of the vector $\boldsymbol{r}$ in spherical coordinates, the expectation value of the operator may be obtained by the following steps. Using the definition given in (4-57), the magnetic dipole coefficient may be written in the form

$$\mathcal{M}_{1\mu} = -\int rY_{1\mu}\nabla\cdot\boldsymbol{M}(\boldsymbol{r})\,dV$$

Upon integration by parts, we obtain the result

$$\mathcal{M}_{1\mu} = \int \boldsymbol{M}_\mu(\boldsymbol{r})\,dV$$

It is convenient here to discuss separately the contributions from intrinsic dipole moments of nucleons and orbital motion of protons.

From the second term of (4-56), we find that the contribution from nucleon intrinsic magnetic moment to the dipole operator is given by the following expression:

$$\boldsymbol{O}_{1\mu}(M1, s) = \frac{e\hbar[c]}{2M_N c}\sum_{i=1}^{A} g_s(i)\boldsymbol{s}_\mu(i) \tag{4-58}$$

For contributions from orbital motion, we note that

$$\mathcal{M}_{1\mu}(\ell) \equiv \int \boldsymbol{M}_\mu(\ell)\,dV = \frac{1}{2}\int \Big(\boldsymbol{r}\times(\nabla\times\boldsymbol{M}(\ell))\Big)_\mu dV = \frac{[c]}{2c}\int \big(\boldsymbol{r}\times\mathcal{J}(\ell)\big)_\mu dV$$

obtained, again, with the help of integration by parts. For a proton, the contribution to the magnetization current due to orbital motion is given by $(e/M_p)\boldsymbol{p}$. Using the fact that orbital angular momentum is the vector product of $\boldsymbol{r}$ and $\boldsymbol{p}$,

$$\boldsymbol{\ell}\hbar = \boldsymbol{r}\times\boldsymbol{p}$$

we obtain the contribution to the magnetic dipole operator due to proton orbital motion as

$$\boldsymbol{O}_{1\mu}(M1,\ell) = \frac{[c]}{2c}\Big(\boldsymbol{r}\times\sum_{i=1}^{A} e g_\ell(i)\frac{\boldsymbol{p}_i}{M_N}\Big)_\mu = \frac{e\hbar[c]}{2M_N c}\sum_{i=1}^{A} g_\ell(i)\,\boldsymbol{\ell}_\mu(i) \tag{4-59}$$

(see also Problem 2-6).

Combining the contributions given in (4-58,59), we obtain an expression for the magnetic dipole operator explicitly in terms of orbital angular momentum $\boldsymbol{\ell}(i)$ and intrinsic spin $\boldsymbol{s}(i)$ of each nucleon,

$$\begin{aligned}\boldsymbol{O}_{1\mu}(M1) &= \boldsymbol{O}_{1\mu}(M1,\ell) + \boldsymbol{O}_{1\mu}(M1,s)\\ &= \frac{e\hbar[c]}{2M_N c}\sum_{i=1}^{A}\Big\{g_\ell(i)\,\boldsymbol{\ell}_\mu(i) + g_s(i)\,\boldsymbol{s}_\mu(i)\Big\}\\ &= \mu_N\sum_{i=1}^{A}\Big\{g_\ell(i)\,\boldsymbol{\ell}_\mu(i) + g_s(i)\,\boldsymbol{s}_\mu(i)\Big\}\end{aligned} \tag{4-60}$$

The equivalent general expression for a magnetic multipole operator of arbitrary order is

$$\boldsymbol{O}_{\lambda\mu}(M\lambda) = \mu_N\sum_{i=1}^{A}\Big\{\frac{2}{\lambda+1}g_\ell(i)\boldsymbol{\ell}(i) + g_s(i)\boldsymbol{s}(i)\Big\}\cdot\nabla_i\big(r_i^\lambda Y^*_{\lambda\mu}(\theta_i,\phi_i)\big) \tag{4-61}$$

For a derivation, see, *e.g.*, Bohr and Mottelson (*Nuclear Structure,* vol. I, Benjamin, Reading, Massachusetts, 1969), or de Shalit and Talmi (*Nuclear Shell Theory,* Academic Press, New York, 1963).

§4-8 MAGNETIC DIPOLE MOMENT OF ODD NUCLEI

The magnetic dipole operator $O_{1\mu}(M1)$ given in (4-60) is an operator carrying one unit of angular momentum. For such an operator, the only nonvanishing expectation values are found in states with spin $J \geq \frac{1}{2}$. The magnetic dipole moment μ of a state is defined as the expectation value of the operator in a state with $M = J$. In units of nuclear magnetons, the result is

$$\mu = \langle J, M{=}J|\mathcal{M}_{10}|J, M{=}J\rangle = \sum_{i=1}^{A}\langle J, M{=}J|g_\ell(i)\boldsymbol{\ell}_0(i)+g_s(i)\boldsymbol{s}_0(i)|J, M{=}J\rangle \quad (4\text{-}62)$$

For a given state, in particular the nuclear ground state, this expression can be evaluated by making some simplifying assumptions about the wave function.

As we shall see in the next section, pairs of protons and of neutrons in the ground state of a nucleus tend to couple to angular momentum zero. For such a *zero-coupled pair*, contributions to the magnetic dipole moment vanish. This is fairly easy to see both from angular momentum selection rules and, more explicitly, from the following arguments. For a pair of identical nucleons, the combination of total orbital angular momentum L and sum of intrinsic spin S must be either $(L, S) = (0,0)$ or $(1,1)$ in order to satisfy the Pauli exclusion principle. The $(L, S) = (1,1)$ pair has a higher energy and is therefore of no interest to a discussion of the nuclear ground state. We are left only with the $(L, S) = (0,0)$ combination for a $J = 0$ pair. For a pair of identical nucleons coupled to $S = 0$, their intrinsic spins are aligned in the opposite direction with respect to each other and their contributions to the magnetic dipole moment from their intrinsic magnetic moments cancel each other as a result. Similarly, for a pair of protons coupled to $L = 0$, the net contribution to the magnetic dipole moment due to their orbital motion is also zero. We can see this result by imagining that the two protons are orbiting in the opposite directions with respect to each other.

In the limit that the pairing force completely dominates the nuclear ground state, all pairs of neutrons and protons will be coupled to $J = 0$. Since these nucleons do not make any contributions to the magnetic moment, we can leave them aside for our present purpose. For even-even nuclei, this means that the magnetic dipole vanishes on account of the fact that there is nothing other than zero-coupled pairs (and the ground state spin is $J = 0$). This is observed to be true for all stable even-even nuclei. For odd-mass nuclei, all nucleons except one are in zero-coupled pairs in this scheme. As a result, we need only to consider the nucleon that is not a member of a zero-coupled pair for both ground state magnetic dipole moment and spin. The expression for the magnetic dipole moment given in (4-62) for such a nucleus reduces to the expectation value of the unpaired nucleon alone,

$$\mu_{\text{s.p.}} = \mu_N \langle j, m{=}j|g_\ell \boldsymbol{\ell}_0 + g_s \boldsymbol{s}_0|j, m{=}j\rangle \quad (4\text{-}63)$$

where $|j, m\rangle$ is the single-particle wave function of the unpaired nucleon in angular moment state (j, m).

Again, using the Landé formula (B-62), as we have done earlier in the case of deuteron magnetic moment in §3-2, we obtain

$$\mu_{\text{s.p.}} = \frac{1}{j(j+1)}\langle j, m=j|(\boldsymbol{\mu}\cdot\boldsymbol{j})j_0|j, m=j\rangle \tag{4-64}$$

where, using (4-63), the magnetic dipole operator may be written in the form,

$$\boldsymbol{\mu} = g_\ell \boldsymbol{\ell} + g_s \boldsymbol{s}$$

The expectation values of $\boldsymbol{\ell}\cdot\boldsymbol{j}$ and $\boldsymbol{s}\cdot\boldsymbol{j}$ that enter here may be obtained from the relations,

$$\boldsymbol{\ell}\cdot\boldsymbol{j} = \boldsymbol{\ell}\cdot(\boldsymbol{\ell}+\boldsymbol{s}) = \boldsymbol{\ell}^2 + \tfrac{1}{2}(\boldsymbol{j}^2 - \boldsymbol{\ell}^2 - \boldsymbol{s}^2)$$

$$\boldsymbol{s}\cdot\boldsymbol{j} = \boldsymbol{s}\cdot(\boldsymbol{\ell}+\boldsymbol{s}) = \boldsymbol{s}^2 + \tfrac{1}{2}(\boldsymbol{j}^2 - \boldsymbol{\ell}^2 - \boldsymbol{s}^2)$$

This gives us the final result:

$$\mu_{\text{s.p.}} = j\left\{g_\ell \pm \frac{g_s - g_\ell}{2\ell+1}\right\} \qquad \text{for} \qquad j = \ell \pm \tfrac{1}{2} \tag{4-65}$$

In this extreme single-particle picture, the magnetic dipole moment of an odd-mass nucleus is completely specified once the values of ℓ and j of the unpaired nucleon is given.

We can make use of the same single-particle model to deduce the spin and orbital angular momentum of the unpaired nucleon as well. If all the nucleons except one are members of zero-coupled pairs, the spin of the state is also given by that of the unpaired nucleon. Thus we have $J = j$. For a given j-value there are two possible ℓ-values, $\ell = j \pm \frac{1}{2}$. The choice between them is determined by the ground state parity. Since all other nucleons are grouped in pairs occupying the same single-particle orbits, the parity of the state is given by the orbital angular momentum of the unpaired nucleon, $\pi = (-1)^\ell$. As a result, from the spin and parity of the ground state and whether the odd-mass nucleus has an odd number of neutrons or protons, we can calculate the ground state magnetic dipole moment of an odd-mass nucleus in an extreme single-particle model using (4-65). The results fall into groups characterized by the values of j and ℓ and are known as the Schmidt values. The observed values of magnetic dipole moments for odd-mass nuclei are compared with the Schmidt values in Fig. 4-6.

From the figure, we see that the observed magnetic dipole moments are within the limits given by the Schmidt values with only a small number of exceptions. This is not surprising. The main approximation used in the model is that all the nucleons except one are tied in zero-coupled pairs. A more realistic ground state wave function will include other components which may be characterized by the number of "broken" zero-coupled pairs, pairs of nucleons coupled to $J > 0$. From considerations of angular momentum coupling alone, it is unlikely that these non-zero coupled pairs can contribute to the magnetic dipole moment of the nucleus as whole in a coherent way with the unpaired nucleon so as to

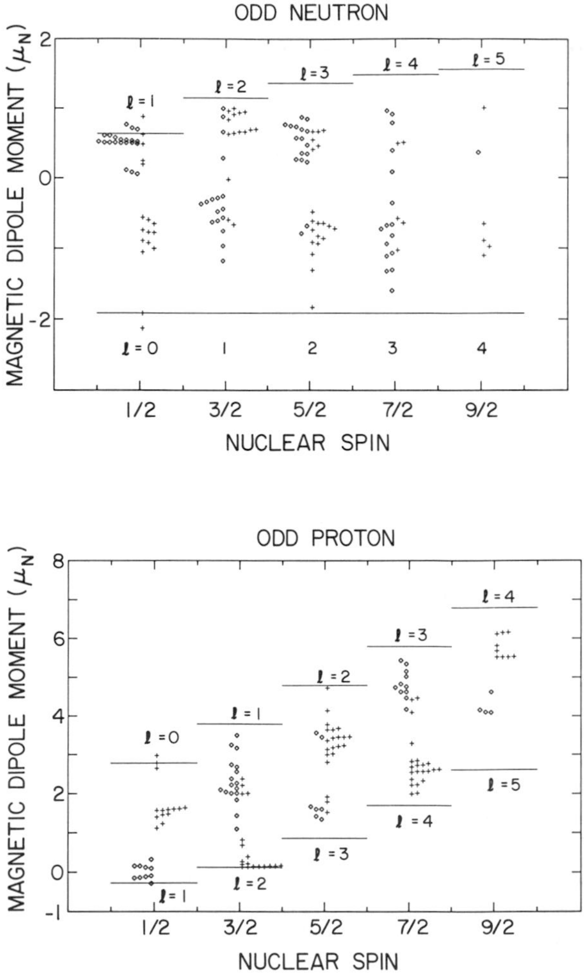

Fig. 4-6 Magnetic dipole moment of odd-mass nuclei. The solid lines are calculated results using (4-65) and the observed values are taken from C.M. Lederer, J.M. Hollander, and I. Perlman (*Table of Isotopes*, 7th edition, Wiley, New York, 1978).

increase the absolute value. The results of more quantitative calculations also tend to support this intuitive argument. As a result, components with broken pairs tend to decrease the absolute values of the magnetic dipole moment given by the single-particle model, the most likely dominant component in the ground state. For this reason, the Schmidt values form the limits of possible ground state magnetic dipole moment for odd-mass nuclei.

The model can be checked easily in the simple cases of a nucleon away from closed shell nuclei (see §6-5) such as ^{3}He and ^{15}N, which are one nucleon less than, or one hole in, the closed shell nuclei of ^{4}He and ^{16}O, respectively, and ^{17}O and ^{17}F, which have one nucleon outside ^{16}O. In such cases, the single-particle description is expected to be good to start with and corrections to the single-particle model can be calculated fairly reliably. The actual results found are, however, quite surprising. Corrections to the Schmidt values for the magnetic dipole moments of these nuclei due to terms other than single-particle contributions are found to be much larger in absolute values than required to explain the difference between observed and Schmidt values. This means that there are also other adjustments besides those coming from the nuclear wave functions adopted.

In calculating the expectation of operators, we have assumed that each nucleon in the nucleus has essentially the same properties as a free nucleon. For example, the intrinsic magnetic dipole moment of a proton is taken to be $\frac{1}{2}g_p$ and a neutron $\frac{1}{2}g_n$. Such an assumption is sometimes referred to as the *impulse approximation*, a term originated from scattering studies (see §7-4). There are two ways by which the impulse approximation may fail for magnetic dipole moment calculations of odd-mass nuclei. The first is the effect of mesons in nuclei. If charged mesons are exchanged between nucleons, their flow constitutes an electric current which must also contribute to the observed magnetic moments. The effects of such a *mesonic current* are also thought to be responsible for the discrepancies from observed values in calculations based on the impulse approximation for other nuclear properties as well. The second is that we have made the naive assumption in (4-56) that nucleons in nuclei behave like point particles carrying the same charge and magnetic dipole moments as free nucleons. Instead, effective values should be used in order to account for modifications of nucleons bound in nuclei, in the same spirit as we did earlier in our calculation of the magnetic dipole moment in the quark model for the baryon octet in §2-8. These two possibilities, mesonic current and effective nuclear operator, are related to each other and to the more general question on the modifications a nucleon experiences in the nuclear medium. In the case of magnetic dipole moment of odd-mass nuclei near closed shells, these two effects seem to cancel each other to a large extent, resulting in much closer agreement of the observed values to the Schmidt values than that expected from the size of either correction term alone. We shall return to this question in discussing other nuclear properties as well.

§4-9 GROUND STATE SPIN AND ISOSPIN

The ground state is the lowest state in energy for a nucleus. It is a special state of a system of N neutrons and Z protons by virtue of the fact that it is the most stable one against any variations on the internal degrees of freedom of the system. In addition it is, in general, the most accessible state and, as a result, often the best known and most extensively studied state in the nucleus. For these reasons we have learned a great deal about nuclear physics from investigations made of ground states of nuclei.

The ground state properties commonly observed are binding energy, spin, isospin, and static electromagnetic moments. We have already discussed the lowest order electromagnetic moments in the previous two sections and we shall return to the question of binding energy in the next section. Other observables such as transition rates to and from ground states and reactions involving ground states will be covered, respectively, in Chapters 5 and 7. In this section, we shall concentrate on the possible values for spin and isospin of the ground state of a nucleus.

Ground state spin. Since each nucleon possesses an intrinsic spin $\boldsymbol{s} = \frac{1}{2}$ and an (integer) orbital angular momentum $\boldsymbol{\ell}$, the total angular momentum or spin $\boldsymbol{j}$ carried by a nucleon in a nucleus is a half-integer quantity. As a result, the total spin, the vector sum of the spins of all the nucleons in a nucleus,

$$\boldsymbol{J} = \sum_{i=1}^{A} \boldsymbol{j}_i \tag{4-66}$$

is a half-integer quantity for odd-mass (A = odd) nuclei and an integer quantity otherwise. The same considerations apply to isospin as well,

$$\boldsymbol{T} = \sum_{i=1}^{A} \boldsymbol{t}_i \tag{4-67}$$

where $\boldsymbol{t}_i = \frac{1}{2}$ is the isospin of the i-th nucleon. Even-mass nuclei can be divided further into two categories: those with both neutron and proton numbers even (N = even, Z = even) are called even-even nuclei and those with both neutron and proton numbers odd (N = odd, Z = odd) are called odd-odd nuclei.

For even-even nuclei, the ground spin is observed to be zero without any exception. This remarkable phenomenon reflects a fundamental property of nuclear interaction generally known as *pairing*. Since ground state spins of odd-odd nuclei are observed to be nonzero in general, we can conclude further that pairing interaction is important only between two identical nucleons, two protons or two neutrons, but not between a neutron and a proton. For example, we have seen that the ground state spin of a deuteron is $J = 1$ (and $T = 0$). If there were a pairing force between a neutron and a proton, the spin would have been $J = 0$ instead. In terms of isospin decomposition, we find that pairing force reveals itself only in the $T = 1$ state of two nucleons but not in the $T = 0$ state. Because of antisymmetrization requirements, a neutron and a proton occupying the same single-particle orbit and having relative angular momentum $\ell = 0$ form an isoscalar pair $\frac{3}{4}$ of the time and an isovector pair $\frac{1}{4}$ of the time (see §3-8). From this we can deduce that the $T = 1$ pairing force cannot be very strong, or else it would have overwhelmed the contributions from the $T = 0$ components in a neutron-proton pair and pairing between a neutron and a proton (in the $T = 1$ state) would have been observed as well.

Because of the pairing force, the ground state spin of an odd-mass nucleus is given by the j-value of the unpaired nucleon. We have made some use of this argument already in the previous section. The basic idea here is that an odd-mass nucleus may be considered as a nucleon coupled to an even-even core consisting of all the neutrons and protons locked in zero-coupled pairs. The total angular momentum of such a core is zero and, as a result, the ground state spin of an odd-mass nucleus assumes that of the unpaired nucleon. In Chapter 6 we shall see that the j-value of the unpaired nucleon may be found from the single-particle energy level spectrum. In this way, the ground state spin of an odd-mass nucleus can usually be deduced from the number of neutrons and protons.

For odd-odd nuclei, it is not an easy matter to predict the ground state spin. An estimate may be made in the following way. As an extension of the idea used for deducing the ground state spin of odd-mass nuclei, we can treat an odd-odd nucleus as an even-even core plus a neutron and a proton outside. The even-even core in its lowest state has spin zero because of the pairing force and may again be ignored for our purpose here. If the single-particle spin of the proton is $\boldsymbol{j}_p$ and that of the neutron is $\boldsymbol{j}_n$, the total angular momentum of the neutron-proton pair outside the core is the vector sum of the two spins. The possible range of values is then

$$|j_{\rm p} - j_n| \leq J \leq j_{\rm p} + j_n$$

For the values of j_p, we can use the ground state spin of the neighboring odd-mass nucleus with one less neutron. Similarly, the value of j_n may be obtained from the neighboring odd-mass nucleus with one less proton. It is not possible, in general, to narrow down the possible J-value any further, but some guidance may be obtained from the empirical Nordheim rules:

$$\begin{array}{lll} \text{Strong rule,} & J = |j_p - j_n| & \text{for } \eta = 0 \\ \text{Weak rule,} & J = \text{either } |j_p - j_n| \text{ or } j_p + j_n & \text{for } \eta = \pm 1 \end{array} \qquad \text{(4-68)}$$

where $\eta = j_p - \ell_p + j_n - \ell_n$. In practice, many exceptions to the rule are found and only a general guide to the ground state spin of an odd-odd nucleus is provided.

Ground state isospin. The possible isospin of a nucleus may be deduced from the proton and neutron numbers. For a system of Z protons and N neutrons, the projection of isospin on the quantization axis is

$$T_0 = \tfrac{1}{2}(Z - N) \qquad \text{(4-69)}$$

The absolute value of T_0 gives the minimum of the possible isospin of a nucleus.

The maximum possible value is limited by the total number of nucleons. This may be seen from the following arguments. Isospin is related to the symmetry in interchanges between protons and neutrons. Since each nucleon has $|t_0| = \frac{1}{2}$, the maximum absolute value of T_0 for a system of A nucleons is $\frac{1}{2}A$, attained when all the nucleons are either protons or neutrons. This forms the upper limit of the possible value of T. The value of isospin T of a system of N neutrons and Z protons must therefore be within the range,

$$\tfrac{1}{2}|Z - N| \leq T \leq \tfrac{1}{2}(Z + N) \qquad \text{(4-70)}$$

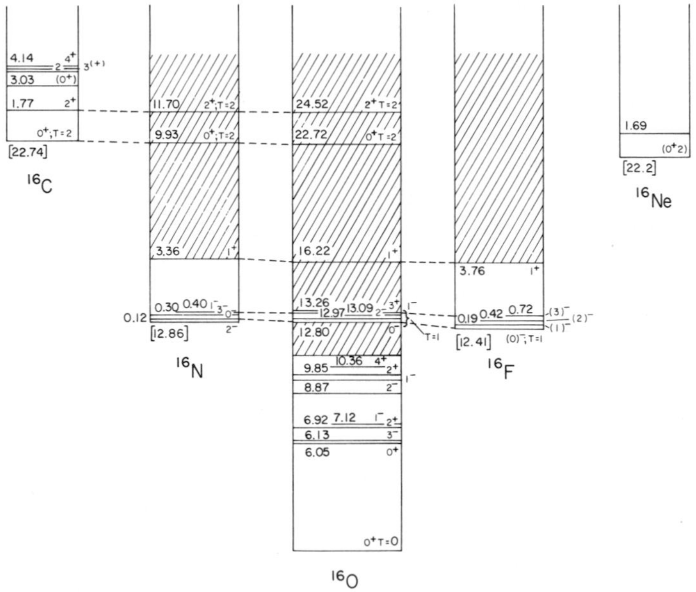

Fig. 4-7 The $A = 16$ isobar with known isobaric analogue states connected by dashed lines. The energies of ^{16}C, ^{16}N, ^{16}F, ^{16}Ne relative to the ground state of ^{16}O, corrected for Coulomb effect and neutron-proton mass difference, are shown inside square brackets.

The isospin dependence of the nuclear force is not large enough to separate states belonging to different T into isolated groups in energy. However, the lowest member of each allowed T-value is, except for odd-odd nuclei, well separated in energy from each other.

Based on the fact that there is a bound two-nucleon state for $T = 0$, the deuteron, but not for $T = 1$, we can infer that nuclear force favors the minimum value, $T = \frac{1}{2}|Z - N|$, as the isospin for the ground state. For higher isospins, the lowest member of each group usually appears at successively higher energies. For example, in ^{16}O the ground state has $T = 0$ and the lowest $T = 1$ state occurs around 13 MeV excitation and the lowest $T = 2$ state at 23 MeV, as shown in Fig. 4-7. States with $T \geq 3$ are expected to be at energies above 30 MeV. However, the density of states is too high at these excitation energies for individual states to be identified.

For odd-odd nuclei, the separation in energy between the lowest member of the two lowest possible isospins is often quite small. This is the result of competition between the T- and J-dependence of the nuclear force. Because of this, the ground state isospin is often a choice between $\frac{1}{2}|Z-N|$ and $(\frac{1}{2}|Z-N|+1)$. For example, the ground state of $^{26}_{13}$Al has $(J^\pi, T) = (5^+, 0)$ and the first excited state at 0.229 MeV has $(J^\pi, T) = (0^+, 1)$. In $^{42}_{21}$Sc, the ground state has $(J^\pi, T) = (0^+, 1)$ and the lowest state with $T = 0$ occurs at 0.6 MeV as a $J^\pi = 7^+$ state.

We have assumed from the beginning that nuclear force depends not on the charge states of the interacting nucleons, but on the isospin. However, in addition to the nuclear force, we also have the Coulomb force, which acts between protons only. (We shall not be concerned with the much weaker magnetic dipole-dipole interaction between nucleons that is also dependent on whether it is a neutron pair, a proton pair, or a proton-neutron pair). Such a charge-dependent force violates the symmetry between proton-neutron exchange. Although on the nucleon-nucleon level the Coulomb interaction is much weaker than nuclear interactions and may be ignored for most purposes, it is not true for the nucleus as a whole. As we shall see in the next section, through short-range nuclear forces, a nucleon can interact with only a few of its neighbors and, as a result, the contribution of nuclear force in a many-nucleon system increases, on the whole, only linearly with the number of nucleons. The Coulomb force, on the other hand, has a long range and its contributions increase quadratically with proton number. Consequently, the effect of the Coulomb force may become quite significant in heavy nuclei, and the nuclear many-body system is no longer invariant when we exchange a proton with a neutron and vice versa. In such cases, isospin is no longer a good quantum number.

Isospin mixing. We can also examine the question of whether isospin is a good quantum number by investigating the amount of admixture of different isospin components in an eigenstate of the nuclear Hamiltonian. Consider first two eigenstates of the isospin-conserving part of the Hamiltonian, $|\,JTx\,\rangle$ and $|\,J'T'y\,\rangle$, where x and y are labels other than spin and isospin required to specify these two states. If in the complete Hamiltonian there are isospin-breaking terms, such as the Coulomb force, these two states will no longer be the eigenstates. We can find the eigenstates $|\,\psi_1\,\rangle$ and $|\,\psi_2\,\rangle$ of the complete Hamiltonian using $|\,JTx\,\rangle$ and $|\,J'T'y\,\rangle$ as the basis states and solve the eigenvalue problem in the two-dimensional model space adopted here.

It is convenient to carry out this calculation using a matrix method. The Hamiltonian matrix may be represented in the following form,

$$\{H\} = \begin{pmatrix} H_{xx} & H_{xy} \\ H_{yx} & H_{yy} \end{pmatrix} \tag{4-71}$$

where values of the diagonal matrix elements,

$$H_{xx} \equiv \langle JTx|H|JTx\rangle \qquad\qquad H_{yy} \equiv \langle J'T'y|H|J'T'y\rangle$$

can be large since there are contributions from the nuclear interaction. In contrast, only isospin-breaking terms are effective in the off-diagonal elements of the basis we have chosen here,

$$H_{xy} \equiv \langle JTx|H|J'T'y\rangle \qquad\qquad H_{yx} \equiv \langle J'T'y|H|JTx\rangle$$

and they are, in general, much smaller than the diagonal elements. Furthermore, the Coulomb force preserves rotational symmetry and is invariant under time reversal. As a result, the Hamiltonian matrix is real and symmetric. We can therefore take

$$H_{xy} = H_{yx} = S\,\delta_{JJ'}$$

where S is the size of the off-diagonal matrix element. The eigenvectors of the Hamiltonian are linear combinations of the two basis states in the form

$$\begin{aligned} |\psi_1\rangle &= \cos\theta|JTx\rangle + \sin\theta|J'T'y\rangle \\ |\psi_2\rangle &= -\sin\theta|JTx\rangle + \cos\theta|J'T'y\rangle \end{aligned} \tag{4-72}$$

where the angle θ is given by the relation

$$\tan 2\theta = \frac{2S}{H_{xx} - H_{yy}} \tag{4-73}$$

From this expression we see explicitly that the admixture between the two basis states depends on the size of the off-diagonal matrix element as well as the difference in the values of the diagonal matrix elements.

Although in a realistic nucleus the number of states that can be admixed by isospin breaking forces may be much larger, the general features of mixing different isospin states are well illustrated by the simple, two-dimensional example given above. There are several points in the problem that are worth noting. The first is that the admixture is important only between states whose "unperturbed" locations, given by the values of the diagonal matrix elements in the example above, are close to each other in energy. This can be seen by looking at the denominator on the right hand side of (4-73). The second is that since the isospin-breaking term in the Hamiltonian is rotational invariant, the off-diagonal matrix elements vanish between states having different J-values. As a result, isospin admixture can take place only between states of the same J.

We can go one step further by examining the value of the off-diagonal matrix elements more closely. Since the Coulomb force has a long range, the matrix element is small unless the wave functions of the states involved are very similar to each other in every respect except isospin. This can most easily be seen in the limit that the isospin-breaking term has an infinite range and, as a result, the radial dependence of the term is a constant. In this case, the off-diagonal matrix elements vanish unless $x = y$; *i.e.,* the two states must be identical except for isospin. In reality, we find that isospin mixing is important only between nearby states having the same spin and a large overlap between their spatial wave functions.

In light nuclei, isospin purity of a state is maintained by a combination of two factors. First, the Coulomb force is relatively weak, since the number of protons is still small. As a result, the value of S in (4-73) is small in general. Second, the density of states is relatively low in the low-lying regions of interest to us. As a

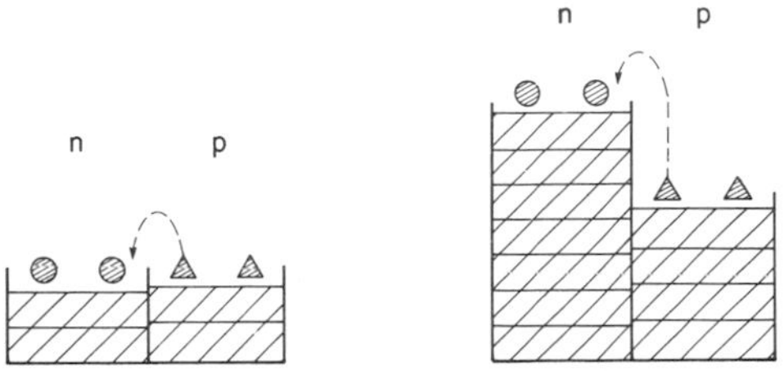

Fig. 4-8 Occupancies of neutron and proton single-particle states. The location of the isobaric analogue state to the ground state of an isobar with one more neutron and one less proton may be found by changing a proton (triangles) to a neutron (circles). Since the corresponding neutron single-particle orbits are filled, such states are much higher in energy in heavy nuclei (right) than in light nuclei (left).

result, it is difficult to find two (unperturbed) states of different isospin near each other in energy as well as having a large overlap in their wave functions.

In heavy nuclei, isospin remains pure for the ground state and a few of the low-lying states nearby for a quite different set of reasons. Because of neutron excess, the Fermi energy for neutrons is much higher than that for protons. The lowest possible isospin is $T_{\rm min} = |T_0| = \frac{1}{2}|Z - N|$ and the dominant component in the ground state wave function is given by the configuration with all the nucleons occupying the lowest available single-particle states. Admixtures of isospin will have to come from states with $T = T_{\rm min} + 1$. The location of the lowest member of such a state may be found by estimating the excitation energy of the isobaric analogue state (see below) to the ground state of a neighboring isobar with one more neutron and one less proton. The dominant configuration for the ground state of such a nucleus may be obtained by changing a proton in the nucleus with $T_0 = T_{\rm min}$ to a neutron. This takes a large amount of energy since the first empty neutron single-particle is much higher in energy. As a crude estimate, the amount of energy is given by the difference in the neutron and proton Fermi energies and is, therefore, a function of neutron excess, as can be seen from looking at Fig. 4-8. As a result, the location of the isobaric analogue state to the ground state of the $T = T_{\rm min} + 1$ nucleus is quite high for nuclei with a large neutron excess. This, in turn, means that it is difficult to have any significant isospin mixing in the low-lying states of heavy nuclei.

In a sense, isospin is no longer a meaningful symmetry in heavy nuclei, since the active neutrons and protons are occupying different single-particle states and are therefore distinguishable by the states they occupy. Furthermore, the argument for isospin purity given above applies only to the ground state region: for excited states, the strong Coulomb effect will lead to large mixing, and isospin ceases to be a meaningful quantum number altogether.

Isobaric analogue state. The importance of isospin in light nuclei can also be seen from similarities in the properties of different members of an isobar, *i.e.,* nuclei having the same A but different in Z and N. An example was seen earlier in Fig. 3-1 for the low-lying states in $A = 11$ and 21 mirror nuclei. A more interesting example can be found in the $A = 16$ isobar shown in Fig. 4-7. The nucleus ^{16}N has $N = 9$ and $Z = 7$. The ground state isospin is $T = 1$ on account of the fact that $T_0 = -1$. If we apply an isospin raising operator T_+ on the ground state wave function of ^{16}N, we obtain a state of $T = 1$ and $T_0 = 0$. This is a state in ^{16}O ($N = Z = 8$). Since the ground state of ^{16}O has isospin $T = 0$, the state produced by the T_+ operation on the ground state of ^{16}N must be an excited state in ^{16}O. The isospin raising operation does nothing except change a neutron to a proton; the structure of the state it produces is unchanged otherwise from the state it acted on. Furthermore, since the nuclear force is charge independent, this state formed by raising T_0 must also be an eigenstate of the nuclear Hamiltonian, and hence corresponds to an observable state in ^{16}O. This state must have very similar properties as the ground state of ^{16}N, since the wave functions are the same except for T_0. Two states, one in ^{16}N and the other in ^{16}O in our example, related by the isospin raising and lowering operators, are called the *isobaric analogue state* (IAS) of each other.

We can easily estimate where such an excited state in ^{16}O should lie. If the forces acting on a nucleon are completely charge independent, the excitation energy of the isobaric analogue state to the ground state of ^{16}N in ^{16}O is given by the binding energy differences of these two nuclei. From a table of binding energies, we find that

$$E_B(^{16}\text{O}) = 127.62 \text{ MeV} \qquad\qquad E_B(^{16}\text{N}) = 117.98 \text{ MeV}$$

The difference is 9.64 MeV.

Two corrections must be applied before we can compare the result with the observed excited states in ^{16}O. The first is that contributions of the Coulomb interaction to the binding energies depend on the number of protons. We may estimate the difference in the amount of Coulomb energy between ^{16}O and ^{16}N using a uniformly charged sphere model given in (1-19). The result, calculated using a radius of $R = 1.2A^{1/3} = 2.52$ fm, is 4.00 MeV. This means that, instead of 9.64 MeV, the IAS should be at excitation energy 13.64 MeV. A second correction is due to the difference in mass between a neutron and a proton (together with an atomic electron to keep the atom neutral as required in the definition of binding energy given in the next section). Since ^{16}O has one more proton and one less neutron than ^{16}N, the calculated value of the excitation energy of IAS, before considering the mass difference between a neutron and a proton, is too large by an amount equal to the neutron-proton mass difference of 0.78 MeV. This puts the calculated excited energy of the IAS to the ground state of ^{16}N in ^{16}O at $13.64 - 0.78 = 12.86$ MeV. The ground state spin and parity of ^{16}N is $J^\pi = 2^-$ and there is a $2^-, T = 1$ state at 12.97 MeV in ^{16}O, as shown in Fig. 4-7. It is also known that this state is very similar in property to the ground state of ^{16}N.

The difference of 0.11 MeV between our estimate and the observed excitation energy may be attributed to the crudeness of our Coulomb energy calculation and to a possible small difference in the wave function of the 12.97 MeV state in ^{16}O and that produced by applying T_+ to the ground state wave function of ^{16}N. In fact, a more careful examination of Fig. 4-7 will show that there is a quartet of states with $J^\pi = 2^-, 0^-, 3^-$, and 1^- in the ground state region of ^{16}N. The same quartet of $T = 1$ states is also found in ^{16}O at around 12.9 MeV; however, the sequence of the four levels is different from that in ^{16}N, showing that some small violation of the IAS idea is found in actual nuclei.

Besides $T = 1$ states, IAS for $T = 2$ are also known in the $A = 16$ isobar. The ground state of $^{16}_{6}$C ($T_0 = -2$) is 0^+ and an excited 2^+ is found at 1.77 MeV. The IAS of these two $T = 2$ states are found in ^{16}N at 9.93 MeV for the 0^+ state and at 11.90 MeV for the 2^+ state. Similarly, the same $T = 2$ pair is found in ^{16}O at 22.72 MeV (0^+) and 24.52 MeV (2^+). On the other hand, since the ground state of ^{16}F is not stable (half-life $T_{1/2} \sim 10^{-19}$ s), the level structure is not known to sufficiently high excitation energies to identify the $T = 2$ states.

§4-10 NUCLEAR BINDING ENERGY

The binding energy of a nucleus, $E_B(Z, N)$, is the energy it takes to remove all Z protons and N neutrons from the nucleus. It is given by the mass difference between the nucleus and the sum of the masses of the (free) nucleons that make up the nucleus,

$$E_B(Z, N) = \big(ZM_{\rm H} + NM_n - M(Z, N)\big)c^2 \tag{4-74}$$

where $M(Z, N)$ is the mass of the neutral atom, $M_{\rm H}$ is the mass of a hydrogen atom, and M_n is the mass of a free neutron. It is conventional to use neutral atoms as the basis for tabulating nuclear masses and binding energies. This comes from the reason that all mass measurements are carried out in practice with most, if not all, of the atomic electrons present. Because of this convention, the electron rest mass energy, $m_e c^2 = 0.5110034$ MeV, as well as the binding energies of the electrons to an atom, often enters into binding energy difference and reaction Q-value calculations as, for example, in β-decay (see §5-5).

It is sometimes convenient to measure masses in terms of atomic mass unit u (also abbreviated as amu in some books) defined using the mass of a neutral ^{12}C atom as the standard,

$$\begin{aligned} \mathrm{u} &= \frac{\text{mass of } ^{12}\text{C atom}}{12} = \frac{1 \text{ kg}}{N_A} \\ &= 1.6605402(10) \times 10^{-27} \text{ kg} = 931.49432(28) \text{ MeV}/c^2 \end{aligned} \tag{4-75}$$

where $N_A = 6.0221367(36) \times 10^{26}$ (kg mole)$^{-1}$ is the Avogadro's number. In terms of u, the masses of a free proton and a free neutron are, respectively,

$$M_p = 1.007276470(12)\ u \qquad\qquad M_n = 1.008664904(14)\ u$$

By definition, the mass of ^{12}C is exactly 12 u.

Since binding energy is a small fraction of the rest mass energy of a nucleus, atomic masses in atomic mass units are usually not very different numerically from the number of nucleons $A = N + Z$. It is therefore more convenient to express nuclear masses in terms of the *mass excess*, $\Delta(Z, N)$ (also referred to on occasion as mass defect), defined in the following manner:

$$\Delta(Z, N) \equiv \big(M(Z, N) \text{ in } u - A\big) \times 931.49432 \text{ MeV} \tag{4-76}$$

where multiplication by 931.5016 converts the mass excess from atomic mass units to energy units MeV. For a hydrogen atom, the mass excess is

$$\Delta(H) = (1.007276470 - 1) \times 931.49432 + 0.51110 = 7.2890 \text{ MeV}$$

and for a free neutron,

$$\Delta(n) = (1.008664904 - 1) \times 931.49432 = 8.0713 \text{ MeV}$$

Given the mass excess of a nucleus, the binding energy in (4-74) may be expressed in the form

$$E_B(Z, N) = Z\Delta(H) + N\Delta(n) - \Delta(Z, N) \tag{4-77}$$

A table of mass excess for all the nuclei with known binding energy is given at the end of the book.

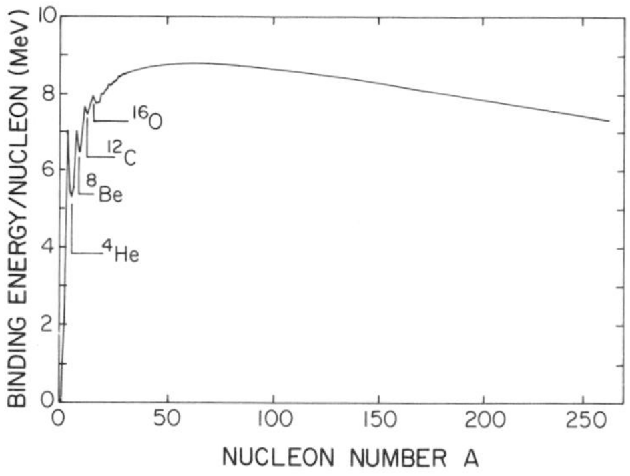

Fig. 4-9 Average binding energy per nucleon as a function of nucleon number A for the most stable nucleus of each isobar. The values are calculated from mass excess.

Saturation of nuclear force. With the exception of light nuclei, $A < 30$, the average binding energy per nucleon is fairly constant, with a value around 8 to 8.5 MeV. This is a direct indication of the short-range nature of the nuclear force. For a two-body, infinite-range interaction, the total binding energy is proportional to the number of pairs of interacting particles and will, therefore, increase more or less quadratically with the number of particles and the binding energy per nucleon increases linearly with A. The fact that the nuclear binding energy per particle is roughly a constant suggests that each nucleon interacts mainly with a few of its nearby neighbors, a characteristic of having a very limited range.

The most tightly bound nuclei are found around $A \approx 56$. For heavier nuclei, the average binding energy per nucleon decreases slowly with increasing A due to rising Coulomb repulsion among a larger number of protons. Because of the cumulative effect of electrostatic repulsion, no stable nuclei are found beyond $Z = 82$. By the same token, when a heavy nucleus undergoes fission and converts into two or more lighter nuclei, energy is released.

Among light nuclei, the binding energy per nucleon shows many sharp peaks occurring at $A = 4n$ for $n = 1, 2, \ldots$ as can be seen in Fig. 4-9. All these nuclei are even-even ones with $Z = N$, ^{4}He, ^{8}Be, ^{12}C, and ^{16}O, etc. For other nuclei nearby, the binding energy per nucleon is much less; for example, only 2.83 MeV for triton ($^3_1$H), 2.57 MeV for $^3_2$He, 5.48 MeV for ^{5}He, and 5.27 MeV for $^5_3$Li. On the other hand, for ^{4}He, the binding energy is 7.07 MeV per nucleon. The corresponding values for the other light $4n$-nuclei are also above 7 MeV. This phenomenon is known as the *saturation* of nuclear force; that is, nuclear interaction is strong among the nucleons in an α-particle-like cluster consisting of two protons and two neutrons. The interaction between α-clusters, on the other hand, is relatively weak by comparison. This is best illustrated by the example of the ground state of ^{8}Be which is unstable toward decay into two α-particles. The same saturation effect is also seen, for example, when we examine the binding energy per nucleon of ^{12}C and ^{16}O: the values, 7.68 MeV and 7.98 MeV, respectively, are not much larger than the value of 7.07 MeV for ^{4}He. A simple interpretation of the differences leads to the conclusion that, on the average, around 90% of the average binding energy is taken up by nucleons within separate α-clusters and only less than 10% is between nucleons in different clusters.

Saturation property is a reflection of a fundamental symmetry of nuclear force, known as SU_4 or Wigner supermultiplet symmetry. As the number of nucleons increases, the rise in binding energy per nucleon of $4n$-nuclei is no longer visible beyond ^{16}O as the increase in the binding among four nucleons, when a new cluster is completed, is averaged over a large number of nucleons in the entire nucleus. However, the saturation effect persists to heavy nuclei. This may be seen by the local increase in the energy required to take away a nucleon shown later in Fig. 6-14.

The removal or *separation energy* of a neutron from a nucleus is defined as the difference between the binding energy of a nucleus and that of a nearby one with one less neutron,

$$S_n(Z,N) = E_B(Z,N) - E_B(Z,N-1) \tag{4-78}$$

Similarly the proton separation energy is given by the binding energy difference between a nucleus and its neighbor with one less proton,

$$S_p(Z,N) = E_B(Z,N) - E_B(Z-1,N) \tag{4-79}$$

In addition to an increase in the separation energies for $4n$-nuclei because of the saturation property of nuclear force, we also find that the values of $S_n(Z,N)$

are larger for even N than the corresponding values for nearby odd-N nuclei. Similarly, there are variations in the value of $S_p(Z,N)$ between neighboring nuclei with odd and even proton numbers. This is the consequence of the pairing force mentioned earlier in §4-9. Additional variations of the separation energy are also found whenever N and Z is near one of the magic numbers 2, 8, 20, 50(40), 82 and 126. This is related to the shell structure in the single-particle energy term of the nuclear Hamiltonian which we shall return to in §6-5.

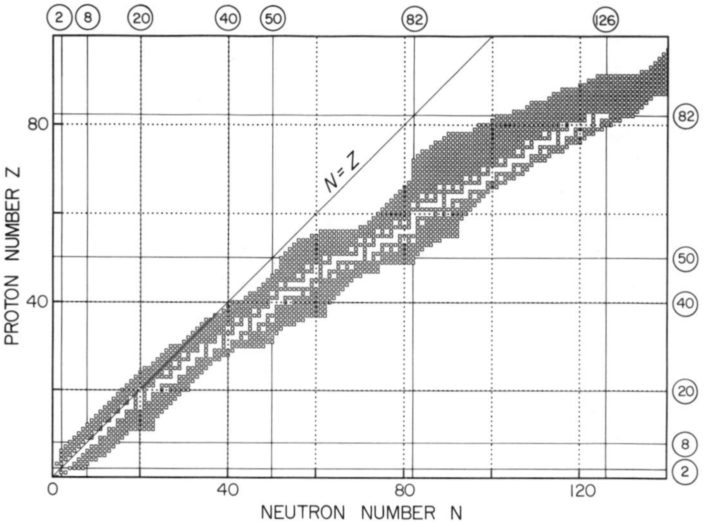

Fig. 4-10 Distribution of observed nuclei as a function of neutron number N and proton number Z. Stable nuclei are shown as empty squares and they exist between nuclei unstable against β-decay, shown as partially filled squares. Unstable nuclei may also decay by nucleon emission, α-particle emission and fission.

Valley of stability. Instead of binding energy per nucleon, we can plot the negative of the binding energy of nuclei as a function of Z and N. A schematic representation of such a two-dimensional plot is shown in Fig. 4-10. A prominent feature of this plot is that stable nuclei exist in a very narrow region along the $N \approx Z$ line. The only exception is a slight neutron excess, an increase in N over Z, toward heavy nuclei so as to overcome the increased Coulomb repulsion between a larger number of protons. In most cases, the number of stable nuclei for a given N, Z, or A is fairly small, and the lifetimes of unstable nuclei on both sides of the stable ones decrease rapidly as we move away from the central region. Such a concentration of stable nuclei in a narrow region of the (N,Z)-plane is called the *valley of stability.*

For nuclei with a few more neutrons than those in the valley of stability, β^--decay is energetically favored. Similarly, for nuclei with a few excess protons, the rates of β^+-decay determines their lifetimes. As the number of neutrons or protons becomes too large compared to those of stable nuclei in the same region, particle emission becomes a possible channel of decay and the lifetimes decrease dramatically as strong interaction processes take over as the dominant modes of decay to more stable nuclear species. The upper end (large N and Z) of the valley of stability is terminated by the fact that nuclei become unstable toward α-decay and fission.

The reason for stable nuclei to be clustered along a narrow region with $N \approx Z$ can be understood from our knowledge of the two-nucleon system discussed earlier in Chapter 3. Since nuclear force is attractive only between a neutron and a proton, the total nuclear binding energy increases with the product NZ. On the other hand, the force is repulsive between a pair of identical nucleons and the total binding energy decreases with the number of proton pairs and the number of neutron pairs, which we may take, for simplicity, as proportional to Z^2 and N^2, respectively. The net result is that nuclei with $N = Z$ are the more stable ones. This is, however, only a first order estimate. The most important modification to this result comes from the Coulomb repulsion between protons and, as a result, neutrons are favored over protons. As Z increases, more and more nuclear attractions are required to balance the repulsion in order to keep the nucleus stable. However, since the nuclear force has only a short range, the Coulomb repulsion dominates eventually as Z increases. Beyond $Z = 82$ no more stable nuclei are found in nature.

The local variations in the "width" of the valley of stability; that is, the number of stable nuclei for a given Z, N, or A, reflects the finer details of the nature of nuclear force. For example, the fact there are more even-even stable nuclei than odd-mass and odd-odd nuclei is a result of pairing interaction discussed earlier. The largest numbers of stable nuclei are found near the magic numbers.

§4-11 SEMI-EMPIRICAL MASS FORMULAE

The discussions in the previous section show that nuclear binding energy is, to a first approximation, a fairly smooth function of nucleon number and other "macroscopic" degrees of freedom, such as the number of protons and nuclear radius, that pertain to the nucleus as a whole. If we are willing to ignore small local departures, it is possible to develop simple formulae that express the binding energy $E_B(Z,N)$, or the equivalent quantity mass $M(Z,N)$, of a nucleus in terms of these bulk coordinates. On the other hand, in order to keep the formulae simple, we may not wish, for instance, to make complicated calculations to relate various terms in the expression to the underlying nucleon-nucleon interaction. As a result, semi-empirical approaches become very attractive.

Weizacker mass formula. One of the more popular and well-established approaches to obtain nuclear binding energies is based on the analogy of a nucleus to a drop of incompressible fluid. The first evidence in support of such a simple "liquid drop" model for the nucleus comes from the fact that nuclear volume increases linearly with the number of nucleons. For binding energies, we have seen earlier that the average value per nucleon is more or less a constant for stable nuclei. As a first approximation, we can write the binding energy in the form

$$E_B(Z,N) = \alpha_1 A \tag{4-80}$$

where the volume energy parameter α_1 is to be determined later from known binding energies.

Like a drop of liquid, nucleons on the surface of a nucleus are less tightly bound as there are fewer particles nearby with which to interact. As a result, we expect a decrease in the binding energy proportional to the number of nucleons on the surface. This number, in turn, depends on the surface area. Since the nuclear volume increases linearly with A, the surface area is proportional to $A^{2/3}$. A correction term can now be added to (4-80) reflecting the decrease in binding energy due to nucleons on the surface. The expression for nuclear binding energy now takes on the form

$$E_B(Z,N) = \alpha_1 A - \alpha_2 A^{2/3} \tag{4-81}$$

Again, the surface energy parameter α_2 is to be determined by fitting it to known binding energies.

The next correction term to be included comes from the repulsive electrostatic interaction between protons. From (1-19), we have

$$E_c = \left[\frac{1}{4\pi\epsilon_0}\right]\frac{3}{5}\frac{Z(Z-1)e^2}{R} = \frac{3\alpha\hbar c}{5r_0}\frac{Z(Z-1)}{A^{1/3}} = 0.72\frac{Z(Z-1)}{A^{1/3}}\ \text{MeV} \tag{4-82}$$

for the Coulomb energy of a nucleus in the form of an uniform sphere of charge of radius $R = 1.2A^{1/3}$ fm. On the other hand, a real nucleus has a diffused surface region and its shape may not necessarily be spherical. For these reasons, we expect the nuclear Coulomb energy to be somewhat different from that given by a uniform spherical charge distribution. In the absence of any other guidance, we shall maintain the form of the dependence on Z and A but leave the strength of the Coulomb energy contribution as an adjustable parameter. The formula for nuclear binding energy changes to the form

$$E_B(Z,N) = \alpha_1 A - \alpha_2 A^{2/3} - \alpha_3\frac{Z(Z-1)}{A^{1/3}} \tag{4-83}$$

with three adjustable parameters α_1, α_2, and α_3.

So far purely classical ideas have been used to obtain the first three terms of an expression for $E_B(Z,N)$. The nucleus is a quantum mechanical object and there are contributions to the binding energy that do not have classical analogues.

The first one comes from the isospin dependence of nuclear force. We have seen in earlier sections that, except for Coulomb repulsion between protons, stable nuclei prefer to have $N \approx Z$. Such an effect may be expressed by a term with a quadratic dependence on $(N - Z)$. The binding energy formula now takes on the form

$$E_B(Z,N) = \alpha_1 A - \alpha_2 A^{2/3} - \alpha_3 \frac{Z(Z-1)}{A^{1/3}} - \alpha_4 \frac{(N-Z)^2}{A} \tag{4-84}$$

where division by A in the new term is necessary since both N and Z, and hence $(N - Z)$, increase with A, whereas the isospin effect is known to be essentially constant independent of A. The new term is sometimes known as *symmetry energy* for obvious reasons.

We have also seen earlier that, as a result of pairing force, even-even nuclei are more tightly bound than their odd-odd counterparts with the same A, and odd-mass nuclei nearby have intermediate values between them. To account for pairing force, a term Δ is added. The complete binding formula now has the form

$$E_B(Z,N) = \alpha_1 A - \alpha_2 A^{2/3} - \alpha_3 \frac{Z(Z-1)}{A^{1/3}} - \alpha_4 \frac{(N-Z)^2}{A} + \Delta \tag{4-85}$$

where the pairing energy parameter,

$$\Delta = \begin{cases} \delta & \text{for even-even nuclei} \\ 0 & \text{for odd-mass nuclei} \\ -\delta & \text{for odd-odd nuclei} \end{cases}$$

with δ taken as a parameter, again to be fitted to known data.

The final form, (4-85), is often known as the Weizacker mass formula. The values of the five parameters, α_1, α_2, α_3, α_4, and δ, depend somewhat on the set of input binding energies used to find their values. A commonly used set,

$$\begin{array}{lll} \alpha_1 = 16 \text{ MeV} & \alpha_2 = 17 \text{ MeV} & \alpha_3 = 0.6 \text{ MeV} \\ \alpha_4 = 25 \text{ MeV} & \delta = \frac{25}{A} \text{ MeV} & \end{array} \tag{4-86}$$

provides some idea of the magnitude of each of the terms. The value of the pairing parameter δ is, perhaps, the least well determined quantity in the set and is found to be much larger in heavy nuclei (see Problem 4-4).

The usefulness of such a mass formula lies, for example, in obtaining fission energies for a large number of different nuclei where the primary interest is to get a global picture over a wide range of A, Z, and N values. The form given in (4-85) does not include effects such as the local increase in binding energy for nuclei near closed shells. This is known as the *shell* effect and is included in the modifications suggested by Myers and Swiatecki (*Ann. Rev. of Nucl. Part. Sci.* **32** [1982] 309) and Strutinsky (*Soviet J. Nucl. Phys.* **3** [1966] 449).

Kelson-Garvey mass formula. Since the primary aim of the Weizacker mass formula is to obtain global agreement, it does not always give good results for the binding energy differences of nearby nuclei that are important in many applications. For this purpose, the Kelson-Garvey approach (Jänecke and Masson, *Atomic Data and Nucl. Data Tables* **39** [1988] 265) is more useful. Instead of a liquid drop model, a microscopic model is used as starting point. Nuclear binding energy is considered to be the result of a sum of one- and two-nucleon interaction terms. The values of these terms may vary from one mass region to another but, in a small region differing only by a few nucleons, they must essentially be constants. We may use the known binding energies of nuclei to extract the values of these terms and the results may then be used to predict unknown binding energies in the same region.

Let us illustrate the procedure by first considering a simple example by assuming for the time being that the nuclear interaction consists of one-body terms alone. This should not be taken as a realistic assumption; it is made only for the purpose of illustrating the method. In this limit, the binding energy of a nucleus made of Z protons and N neutrons has the form

$$E_B(Z,N) = \alpha N + \beta Z \tag{4-87}$$

where the parameters α and β represent, respectively, the average values of the interaction of a neutron and a proton with the rest of the nucleons in the nucleus.

The values of α and β in a neighborhood may be obtained from the differences in the known binding energies, for example,

$$\alpha = E_B(Z,N+1) - E_B(Z,N) \qquad \beta = E_B(Z+1,N) - E_B(Z,N) \tag{4-88}$$

The difference between $E_B(Z+1,N+1)$ and $E_B(Z,N)$ is then given by the following expression:

$$\begin{aligned} E_B(Z+1,N+1) - E_B(Z,N) &= \alpha + \beta \\ &= E_B(Z,N+1) - E_B(Z,N) + E_B(Z+1,N) - E_B(Z,N) \end{aligned}$$

This relation may be expressed in the form of a difference equation relating the binding energies of four nearby nuclei,

$$E_B(Z+1,N+1) + E_B(Z,N) - E_B(Z,N+1) - E_B(Z+1,N) = 0 \tag{4-89}$$

Using this relation, the binding energy of any one of the four nuclei may be deduced from the known values of the other three. Such a simple relation is, of course, the result of our model, correct only within the limit that binding energy is made of one-body terms only. We can test whether the premise that went into deriving (4-87) is correct by calculating the actual binding energy difference between four known nuclei,

$$\Delta E_B = E_B(Z+1,N+1) + E_B(Z,N) - E_B(Z,N+1) - E_B(Z+1,N) \tag{4-90}$$

In practice, we will find that the values of ΔE_B are rather large, reflecting the fact that we have ignored two-body terms that are important in binding energy considerations.

To generalize (4-87) so as to include two-body contributions, we need three additional terms, one each for proton-proton, neutron-neutron, and neutron-proton interactions. Since the number of each type of pairs is given by

$$\text{Number of pairs} = \begin{cases} \frac{1}{2}Z(Z-1) & \text{for protons} \\ \frac{1}{2}N(N-1) & \text{for neutrons} \\ NZ & \text{for neutron-proton} \end{cases}$$

the binding energy of a nucleus may be expressed in the form

$$E_B(Z,N) = a\,N + b\,Z + c\,N(N-1) + d\,Z(Z-1) + e\,NZ \tag{4-91}$$

where the five parameters a, b, c, d, and e are to be determined from known binding energies in nuclei with very similar proton and neutron numbers.

We can follow the same approach as we have used earlier in the one-body interaction model and construct a difference equation to express the relation between binding energies of nearby nuclei in the presence of both one- and two-body terms. One such equation satisfying (4-91) may be written in the following form,

$$\begin{aligned} &E_B(Z+1,N-1) + E_B(Z-1,N) + E_B(Z,N+1) \\ &- E_B(Z,N-1) - E_B(Z+1,N) - E_B(Z-1,N+1) = 0 \end{aligned} \tag{4-92}$$

There are several ways to derive this relation from (4-91). One approach is to arrange the six binding energies into three groups, each consisting of the difference between a pair of nuclei having the same neutron number. In this way, terms depending only on neutron number do not enter in and the difference equation becomes one involving only three parameters, b, d, and e.

Starting from (4-91) and ignoring the purely neutron terms involving parameters a and c we have, for the difference in the binding energies in the pair having $(N-1)$ neutrons,

$$\begin{array}{lllll} E_B(Z+1,N-1) &= b(Z+1) &+d\,Z(Z+1) &+e\,(N-1)(Z+1) &+f(a,c,N) \\ -E_B(Z,N-1) &= b\,Z &+d\,Z(Z-1) &+e\,(N-1)Z &+f(a,c,N) \\ \hline E_B(Z+1,N-1) & & & & \\ \quad -E_B(Z,N-1) &= b &+2d\,Z &+e\,(N-1) & \end{array}$$

Similarly, for the pair involving $(N+1)$ neutrons,

$$\begin{array}{lllll} E_B(Z,N+1) &= b\,Z &+d\,Z(Z-1) &+e\,(N+1)Z &+g(a,c,N) \\ -E_B(Z-1,N+1) &= b\,(Z-1) &+d\,(Z-1)(Z-2) &+e\,(N+1)(Z-1) &+g(a,c,N) \\ \hline E_B(Z,N+1) & & & & \\ \quad -E_B(Z-1,N+1) &= b &+2d\,(Z+1) &+e\,(N+1) & \end{array}$$

It is clear that the sum of these two differences is the same as the difference between a pair with N neutrons and proton number differing by two,

$$\begin{array}{rl} E_B(Z+1,N) = & b(Z+1) + dZ(Z+1) + eN(Z+1) + h(a,c,N) \\ -E_B(Z-1,N) = & b(Z-1) + d(Z-1)(Z-2) + eN(Z-1) + h(a,c,N) \\ \hline E_B(Z+1,N) & \\ -E_B(Z-1,N) = & 2b + 2d(2Z+1) + 2eN \end{array}$$

By eliminating b, d, and e from these three equations, we obtain the result given in (4-92). The same relation can also be derived by considering (4-92) as three pairs of differences with each pair having the same proton number.

We may check how well (4-91) works by calculating from six known binding energies the quantity

$$\begin{aligned} \Delta E = & E_B(Z+1,N-1) + E_B(Z-1,N) + E_B(Z,N+1) \\ & - E_B(Z,N-1) - E_B(Z+1,N) - E_B(Z-1,N+1) \end{aligned} \tag{4-93}$$

We do not expect ΔE_B to be exactly zero for an arbitrary group of six nuclei. However, if the assumptions that went into constructing (4-91) are correct, ΔE_B should be distributed randomly around zero. This was found to be true in practice and the standard deviation is found to be

$$\frac{1}{N}\left\{\sum_{i=1}^{N}(\Delta E_i)^2\right\}^{1/2} \approx 100 \text{ keV}$$

This results means that (4-92) may be used to calculate the binding energy for any one the six nuclei from the other five with an uncertainty on average of about 100 keV.

The usefulness of a relation such as the one given in (4-91) is not limited to extrapolating binding energies for nuclei one nucleon away from known masses. If we wish to estimate the value for a nucleus far way from known ones, we can use results calculated with the Kelson-Garvey mass formula as input and extrapolate further and further away from the valley of stability. The uncertainties in the prediction, however, increase roughly as the square root of the number of steps.

One can, in principle, include contributions other than simple one- and two-body effects to reflect higher order terms that may occur in an effective nucleon-nucleon interaction. The difference equation, in this case, may involve more than six nuclei. Since binding energies are known for a large number of nuclei, it is possible, in principle, to consider such a higher order approach to extrapolate binding energies for unknown nuclei very far away from the valley of stability.

§4-12 DENSITY OF EXCITED STATES

The density of states for a nucleus, $\rho_A(E)$, is the number of states per unit energy interval. It is an important quantity in determining the rate of reactions taking place in a given energy region and in giving a rough idea of the probability of finding a particular state at a given energy. Being a statistical quantity, the concept of state density is not meaningful at low excitation energies where spacings between levels are large.

It is well known that the density of nuclear states increases rapidly with excitation energy E. This may be seen by looking at the energy level spectrum of any well-studied nucleus such as those given in Fig. 4-7. By treating nucleons in a nucleus as noninteracting fermions, Bethe (*Rev. Mod. Phys.* **9** [1937] 69) derived the relation,

$$\rho_A(E) = \frac{1}{12a^{1/4}E^{5/4}}\, e^{2\sqrt{aE}} \tag{4-94}$$

generally known nowadays as the Fermi-gas model formula. The quantity a in the expression is the *level density* parameter and is often obtained empirically by fitting (4-94) to the measured state densities. The formula gives an acceptable estimate of the state density at various energies and is used widely in many nuclear reaction calculations.

Before giving a derivation of (4-94), it is useful to make a distinction between state density and level density. For simplicity, let us ignore isospin here. Because of rotational invariance of the nuclear Hamiltonian, all states differing only in M, the projection of spin $\boldsymbol{J}$ on the quantization axis, are degenerate in energy. Thus, all $(2J+1)$ states having the same J and other quantum numbers are observed as a single "level." If the number of levels of a given J per unit energy interval at energy E is $\omega_J(E)$, the total level density $\omega_A(E)$ for a nucleus is given by the sum,

$$\omega_A(E) = \sum_J \omega_J(E) \tag{4-95}$$

The state density $\rho_A(E)$ differs from $\omega_A(E)$ by the weighting factor $(2J+1)$ for each level of spin J,

$$\rho_A(E) = \sum_J (2J+1)\,\omega_J(E) \tag{4-96}$$

The relation between $\omega_A(E)$ and $\rho_A(E)$ depends on the average value of spin at energy E and is given by the relation,

$$\omega_A(E) = \frac{\rho_A(E)}{\sqrt{2\pi\sigma_J^2(E)}} \tag{4-97}$$

where $\sigma_J^2(E)$ is the *spin cutoff* factor. We shall be dealing primarily with $\rho_A(E)$, since it is physically a more meaningful quantity.

The relation given by (4-97) and the meaning of spin cutoff factor may be seen in the following way. For a given J, the possible values of M range from $-J$ to $+J$ in integer steps and, as a result, the average of M is zero. The distribution of the values of M at a given energy may be taken to be a Gaussian function centered at $M = 0$,

$$p(M) = \frac{1}{\sqrt{2\pi\langle M^2\rangle}}\, e^{-\frac{M^2}{2\langle M\rangle^2}} \tag{4-98}$$

The justification behind this assumption is based on the central limit theorem in statistics, but we shall not go into the connection between them here. With (4-98), we can obtain $q(J)$, the distribution of states as a function of J from the difference between $p(M{=}J)$ and $p(M{=}J{+}1)$,

$$\begin{aligned} q(J) &= p(M{=}J) - p(M{=}J{+}1) \\ &\approx -\left(\frac{\partial p(M)}{\partial M}\right)_{M=J+1/2} = \frac{1}{\sqrt{2\pi\langle M^2\rangle}}\,\frac{2J+1}{2\langle M^2\rangle}\, e^{-\frac{(J+\frac{1}{2})^2}{2\langle M^2\rangle}} \end{aligned} \tag{4-99}$$

Since the degeneracy of a level with spin J is $(2J + 1)$, the average number of states per level at energy E, in the limit of high level density, is

$$\langle(2J+1)\rangle = \frac{\int (2J+1)\, q(J)\, dJ}{\int q(J)\, dJ} = \sqrt{2\pi\langle M^2\rangle} \tag{4-100}$$

where the integrals are taken over all possible J values. Comparing (4-100) with (4-96) and (4-97), we can make the identification

$$\sigma_J^2(E) = \langle M^2\rangle \tag{4-101}$$

In other words, the spin cutoff factor is the variance of the distribution of states as a function of M.

The state density given by (4-94) may be derived using an independent-particle model for the nucleus (see §6-7). In this limit, the nuclear Hamiltonian is made of one-body terms only and the energy of a nucleus consisting of A nucleons is given by the sum of energies of the occupied single-particle states. For simplicity, we shall ignore any degeneracies in the single-particle spectrum and assume that each state can accommodate at most one nucleon. Furthermore, we shall ignore for the time being any differences in the single-particle energies between neutrons and protons. The occupancy of single-particle state i may be denoted by m_i, and

$$m_i = \begin{cases} 1 & \text{if state } i \text{ occupied} \\ 0 & \text{otherwise} \end{cases}$$

We shall adopt a shorthand notation using the vector $\boldsymbol{m} \equiv (m_1, m_2, \ldots)$ to denote a particular set of occupancies, or *configuration*. Since the total number of nucleons is A, we have a condition on the possible value of $\boldsymbol{m}$,

$$\sum_i m_i = A \tag{4-102}$$

where the summation is over all the single-particle states. A further constraint is imposed by the energy of such a configuration,

$$E_{\boldsymbol{m}} = \sum_i m_i \epsilon_i \tag{4-103}$$

Since the single-particle spectrum $\{\epsilon_i\}$ has only a lower bound and no upper bound, the possible energy of a nucleus can, in principle, extend to infinity, a fact already assumed in (4-94).

The state density of a nucleus with A nucleons may be written formally in the following way,

$$\rho_A(E) = \sum_m \sum_{\boldsymbol{m}} \delta(A - m)\delta(E - E_{\boldsymbol{m}})$$

Instead of trying to obtain the state density directly from this expression, we shall make use of a standard method in statistical mechanics in dealing with similar situations. If a system has many possible energy states, the probability for it to be in a particular state with energy E_r is proportional to the Boltzmann factor $\exp\{-E_r/kT\}$, where k is the Boltzmann constant and T is the temperature of the system. The sum of the Boltzmann factor over all allowed states is the partition function,

$$z(T) = \sum_r e^{-E_r/kT}$$

If a system does not have a definite number of particles, the probability for finding the system with particle number m and energy E_r is, instead, proportional to the Gibbs factor $\exp\{(\mu m - E_r)/kT\}$, where μ is the chemical potential. In place of the partition function, we have now the grand partition function,

$$Z(\mu, T) = \sum_{m=0}^{\infty} \sum_r e^{(\mu m - E_r)/kT} \tag{4-104}$$

In statistical mechanics, the grand partition function is often the starting point for calculating many quantities of interest, and we shall follow the same approach here. In particular, we shall obtain the state density by applying an inverse Laplace transform to $Z(\mu, T)$.

Chemical potential and temperature are terms defined only for a large collection of macroscopic objects at equilibrium. It is not possible to extrapolate these concepts to subatomic physics without first identifying them with quantities that, at least, can be measured in principle in nuclei. This we shall do later. For the time being, we shall simply treat them as two parameters, α and β, and express the grand partition function for a nuclear system in the form

$$Z(\alpha, \beta) = \sum_{m\boldsymbol{m}} e^{\alpha m - \beta E_{\mathbf{m}}}$$

On substituting the values of m and E_m from (4-102) and (4-103), the expression simplifies to a product over quantities pertaining only to single-particle states,

$$Z(\alpha,\beta)=\sum_{mm}\exp\Big\{\sum_i m_i(\alpha-\beta\epsilon_i)\Big\}=\prod_i\Big\{1+e^{\alpha-\beta\epsilon_i}\Big\} \tag{4-105}$$

The final equality is obtained by noting that the occupancy m_i for the single-particle state i is either 0 or 1.

The product over single-particle states may be transformed into a summation by taking the logarithm of both sides of (4-105). The summation, in turn, may be changed into an integration if the single-particle spectrum,

$$g(\epsilon)=\sum_{i=0}^{\infty}\delta(\epsilon-\epsilon_i)$$

can be taken to be a continuous one. This is a good approximation as long as the single-particle states are closely spaced. In practice, this assumption does not seem to affect the state density obtained at the end if the number of active nucleons is sufficiently large.

The logarithm of the grand partition function now takes on the form

$$\begin{aligned}\ln Z(\alpha,\beta)&=\sum_{i=0}^{\infty}\ln\{1+e^{\alpha-\beta\epsilon_i}\}\\&\longrightarrow\int_0^{\infty}g(\epsilon)\ln\{1+e^{\alpha-\beta\epsilon}\}\,d\epsilon\end{aligned} \tag{4-106}$$

Without losing any generality, we can define the zero point of the energy scale for ϵ to be such that

$$g(\epsilon)=0 \qquad \text{for} \qquad \epsilon<0$$

Since the logarithmic factor in the integrand in the final form of (4-16) goes to zero very quickly for $\epsilon>\alpha/\beta$, the integral may be expanded in a Taylor series involving $g(\epsilon)$ and its derivatives evaluated at $\epsilon=\alpha/\beta$ (see, *e.g.*, A. Bohr and B.R. Mottelson, *Nuclear Structure,* vol. I, Benjamin, Reading, Massachusetts, 1969, p. 282, for more detail):

$$\ln Z(\alpha,\beta)=\int_0^{\alpha/\beta}g(\epsilon)(\alpha-\beta\epsilon)\,d\epsilon+\frac{\pi^2}{6\beta}g(\alpha/\beta)+\frac{7\pi^4}{360\beta^3}g''(\alpha/\beta)+\cdots \tag{4-107}$$

where the $g''(\alpha/\beta)$ is the second derivative of $g(\epsilon)$. Contributions involving the first derivative of $g(\epsilon)$ vanishes and does not enter into the expression as a result.

The state density can now be evaluated as the inverse Laplace transform of the grand partition function

$$\rho_A(E)=\left(\frac{1}{2\pi i}\right)^2\int\int_{-i\infty}^{+\infty}Z(\alpha,\beta)\,e^{-\alpha A+\beta E}\,d\alpha\,d\beta$$

One of the advantages of evaluating state density in this way is that the integral in the above expression may be carried out using the saddle-point approximation method (see, *e.g.*, K. Huang, *Statistical Mechanics,* Wiley, New York, 1963, p. 210; R.K. Pathria, *Statistical Mechanics*, Pergamon, Oxford, 1972, p. 56). Since the main contributions to the integral come from a small region around $(\alpha,\beta) = (\alpha_0,\beta_0)$ given by the conditions

$$\begin{cases} \left.\dfrac{\partial \ln Z(\alpha,\beta)}{\partial\alpha}\right|_{\substack{\alpha=\alpha_0\\ \beta=\beta_0}} - A = 0 \\[2ex] \left.\dfrac{\partial \ln Z(\alpha,\beta)}{\partial\beta}\right|_{\substack{\alpha=\alpha_0\\ \beta=\beta_0}} + E = 0 \end{cases}$$

the integral may be evaluated up to second order using the expansion of $\ln Z(\alpha,\beta)$ given in (4-107). The level density is now a function of α_0 and β_0,

$$\rho_A(E) = \frac{Z(\alpha_0,\beta_0)}{2\pi\sqrt{|D|}}\, e^{-\alpha_0 A-\beta_0 E} \tag{4-108}$$

where D is the determinant

$$D = \begin{vmatrix} \dfrac{\partial^2 \ln Z}{\partial\alpha^2} & \dfrac{\partial^2 \ln Z}{\partial\alpha\partial\beta} \\[2ex] \dfrac{\partial^2 \ln Z}{\partial\beta\partial\alpha} & \dfrac{\partial^2 \ln Z}{\partial\beta^2} \end{vmatrix}_{\substack{\alpha=\alpha_0\\ \beta=\beta_0}}$$

Note that on the right hand side of (4-108), both $Z(\alpha_0,\beta_0)$ and D depend on the energy E as well.

In an independent-particle model, the ground state of a nucleus is formed by filling single-particle states up to the Fermi energy ϵ_F. As a result, we have the relation

$$A = \int_0^{\epsilon_F} g(\epsilon)\, d\epsilon \tag{4-109}$$

Similarly, the ground state energy is given by the integral

$$E_0 = \int_0^{\epsilon_F} \epsilon g(\epsilon)\, d\epsilon \tag{4-110}$$

To make further progress, an assumption must be made of the form of the single-particle spectrum $g(\epsilon)$. The usual one used to derive (4-94) is that $g(\epsilon)$ is sufficiently constant near $\epsilon = \epsilon_F$ and, as a result, contributions from the derivatives of $g(\epsilon)$ to the various quantities of interest here may be ignored. Thus, at the saddle point, we have the results

$$A = \int_0^{\alpha_0/\beta_0} g(\epsilon)\, d\epsilon$$

$$E_0 = \int_0^{\alpha_0/\beta_0} \epsilon g(\epsilon)\, d\epsilon + \frac{\pi^2}{6\beta_0^2}\, g(\alpha_0/\beta_0) \tag{4-111}$$

Comparing (4-111) with (4-110), we can identify the Fermi energy as

$$\epsilon_F = \alpha_0/\beta_0$$

It has the same role as the chemical potential μ for macroscopic systems. Similarly, we may identify β_0 as $(kT)^{-1}$, the inverse of the product of the Boltzmann constant and the temperature by, for example, comparing (4-108) with (4-104).

Using (4-111), we can write the excitation energy in the form

$$E_x = E - E_0 = \frac{\pi^2}{6\beta_0^2} g(\alpha_0/\beta_0) \tag{4-112}$$

Similarly, the value of $Z(\alpha_0, \beta_0)$ and the determinant D may be evaluated and expressed as functions of E_x. On substituting these results into (4-108), we arrive at the expression for the state density in the form

$$\rho_A(E_x) = \frac{1}{E_x\sqrt{48}} e^{2\left(\frac{\pi^2}{6} g(\epsilon_F) E_x\right)^{1/2}} \tag{4-113}$$

In order to get to the form of (4-94), we need to redefine the zero point of our energy scale to be at the ground state of the nucleus, *i.e.*, $E_0 = 0$, so that $E_x \to E$, and to write the level density parameter as

$$a = \frac{\pi^2}{6} g(\epsilon_F) \tag{4-114}$$

The single-particle state density $g(\epsilon)$ is now a sum of both proton and neutron single-particle state densities. If the Fermi energies are different for neutrons and protons, the density $g(\epsilon_F)$ must be interpreted as the sum of neutron and proton single-particle densities at their respectively Fermi energies.

Among the assumptions made to derive (4-94), the most questionable ones are in ignoring the two-body residual interaction between nucleons and in taking the single-particle spectrum near the Fermi energy to be a constant one. Perhaps it is for these reasons that the Fermi-gas formula has some difficulty in producing results that show good agreement with observation. An example for the nucleus ^{56}Fe is shown in Fig. 4-11. A better form of the nuclear state density is given by the *back-shifted* Fermi-gas model formula,

$$\rho_A(E) = \frac{1}{12a^{1/4}(E-\Delta)^{5/4}} e^{2\sqrt{a(E-\Delta)}} \tag{4-115}$$

where both a and Δ are considered as adjustable parameters to be determined by fitting to known data. The need to shift the energy scale by an amount Δ (a negative quantity empirically) may be traced to the fact that we have ignored two-body interactions between the nucleons. Generally speaking, the ground state energy is depressed by two-body correlations in nuclei compared with the location given by an independent-particle model. Since excitation energies in a nucleus are measured from the ground state, the values given by an independent-particle

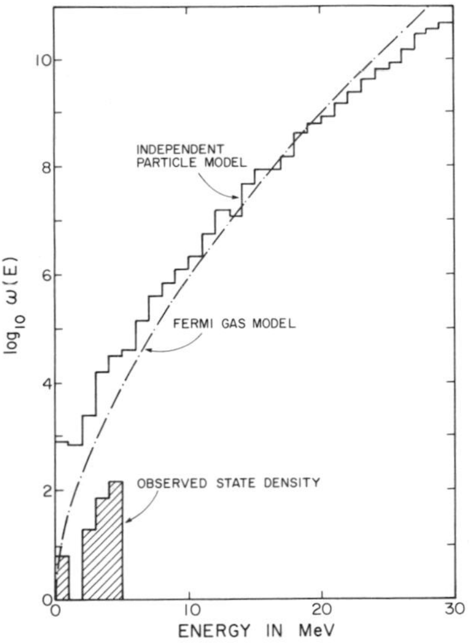

Fig. 4-11 State density of ^{56}Fe calculated with the Fermi-gas formula (4-94) (smooth curve) and an independent particle model (stair-case plot). The observed values, shown as shaded histogram, are lower than either one of the calculated values since the ground state energy is depressed by two-body correlations. This effect may be accounted for by the back-shifted Fermi-gas formula (4-115).

model are too small by an amount equal to the sum of two-body correlations in the ground state we have ignored.

The empirical values of the level density parameter a are also found to be quite different from those calculated by (4-114) using the Fermi energies of neutrons and protons. The main reason behind this failure may be related to the assumption of a constant single-particle spectrum. As we shall see in §6-5, the true single-particle density is far from a smooth one.

Experimentally, state densities can be measured directly only in two energy regions. The first is the low-lying region where individual levels are well separated from each other and all the levels are known, together with their spin and isospin assignments. In this region the level density, and hence the state density, may be obtained by direct counting. However, this is not the region of primary interest here, since the state density is not high enough to be treated as "statistical." The second region is the slow neutron resonance region, just above neutron separation energies. Here excited states, especially those with $J = \frac{1}{2}$ in odd-neutron nuclei, may be formed by absorbing a slow, s-wave neutron by a target having one less neutron. High accuracies, both in energy measurement and in level identification, can be achieved in spite of the high state densities in this region (of the order of 10^6 per MeV) because of the precision that can be achieved with slow neutron measurements. For other energy regions, the information must come indirectly from observations such as reaction rates.

§4-13 LOW-LYING EXCITED STATES

Besides ground states, low-lying excited states in most nuclei have also been extensively studied. Since the number of excited states that can be investigated in detail is quite large – tens of levels in some nuclei – a great deal of information has been accumulated as a result. There are basically two approaches we can take when faced with large quantities of information. The first is to take a statistical approach in which only the average properties of an "ensemble" of data are examined. This is the general idea behind the level density studies in the previous section. An alternative is to choose a few of the states that display features representative of nuclear states in general. Such a selection is necessary so as not to lose sight of the physics amidst the multitude of information. This is the more common approach taken in studying the low-lying excited states of nuclei.

One of the seemingly confusing features in looking at nuclear excited states in detail is that often two diametrically opposite types of behavior, namely single-particle and collective motions, coexist in the same nucleus. In the former category, we find states whose properties can best be understood in terms of a smaller number of nucleons moving in single-particle orbits. For example, some of the low-lying states in ^{17}F are well described by a proton outside a core made of ^{16}O nucleus. Electromagnetic transitions between these states occur when the proton changes the single-particle orbit it occupies. On the other hand, electromagnetic transitions between certain states in ^{20}Ne, a nearby nucleus, require the coherent or "collective" motion of many nucleons. In fact, most nuclear states display a combination of both types of behavior. In order to understand the basic physics involved, it is generally more profitable to concentrate on states that are characteristic of one type of behavior or the other.

Single-particle states, such as those mentioned above in ^{17}F, may be identified directly through one-nucleon transfer reactions. The most representative examples are the (d,n) and (d,p) reactions in which a deuteron strikes a target nucleus and a neutron or a proton is observed to emerge. Such reactions are called "stripping reactions," since a nucleon is stripped from the projectile and captured by the target nucleus without disturbing the rest of the nucleons. A state with dominant single-particle features will be strongly excited by stripping reactions. States formed by coupling a nucleon in a specific single-particle state to an eigenstate of a nucleus with one less nucleon are referred to as ***one-particle states.***

The complement of a stripping reaction is a pickup reaction in which a nucleon is removed from the target nucleus. Beside (p,d) reactions, one-nucleon pickup can also be accomplished by processes such as $(t,{}^4\text{He})$ and $({}^3\text{He},{}^4\text{He})$. If a nuclear state is predominantly formed by a nucleon in a given single-particle state coupled to a state in the neighboring $(A-1)$ nucleus, it will be strongly excited by pickup reactions. The state left in the residual nucleus is called a ***one-hole state.***

Single-particle behavior is also displayed by one-particle one-hole ($1p1h$) excitations. In this case, an excited state is formed by promoting a nucleon in an occupied single-particle state to an unoccupied one. The name $1p1h$-state is related to the fact that the state formed by a vacancy in one of the single-particle orbits is very similar to one-hole states created by pickup reactions. Similarly, the state formed by having a particle in one of the higher single-particle orbits reminds us of one-particle states left by stripping reactions. States with large $1p1h$-components are strongly excited by electron and nucleon scattering at intermediate energies. Here, since the transit time is very short, there is only the time for the projectile to interact at most with one of the nucleons in the target nucleus and promote it to a higher single-particle orbit as a result of the interaction.

A one-particle one-hole state of a slightly different nature is excited by charge exchange reactions such as (p, n), (π^+, π^0) and (π^-, π^0). In addition to promoting a nucleon to a higher orbit, the reaction may also change the charge state of the nucleon, from a neutron to a proton in the case of (p, n) and (π^+, π^0) reactions, and the other way around in the (π^-, π^0) example above. If the reaction induces only a change in the charge state without excitation to a different single-particle orbit, it is likely that the final state reached is the isobaric analogue of the initial target state.

Besides a nucleon, a pair of nucleons may also be added or removed from a nucleus through two-nucleon transfer reactions. Because of the pairing force, such reactions favor the transfer of two identical nucleons coupled to $J^\pi = 0^+$. Other studies of the correlation between a pair of nucleons in a nucleus may be carried out by double-charge exchange reactions such as (π^+, π^-) and (π^-, π^+).

Nuclear collective behaviors are characterized by many nucleons acting in unison. The most obvious examples are rotational and vibrational modes of excitation. The extreme of collective motion is reached when all the nucleons participate in the same motion — as, for example, in the rotation of a rigid body. In this case, the only degrees of freedom are rotations of the nucleus as a whole and internal excitations to change the overall shape of the nucleus. Such features may be described in terms of rotational energy, moment of inertia, and other collective coordinates. In practice, a nucleus does not quite approach a rigid rotor; the possible shapes and rotational frequencies it can attain become interesting questions.

Rather than acting as a rigid body, the behavior of the nucleus is often closer to that of a drop of liquid, a model we have already used in §4-11 in obtaining the Weizacker semi-empirical mass formula. Beside binding energies, the picture may also be extended to vibrations of the drop in describing certain excited states in nuclei. Here the nucleus oscillates between different shapes with a definite frequency, as we shall see later in §6-1 and §6-2. The energies and the possible shapes associated with the oscillations are properties of the nuclear matter and therefore form an important part of our understanding of nuclear physics.

Both rotation and vibration are concepts originated in classical mechanics. Nuclei, however, are described by quantum mechanics. Furthermore, nuclear collective behaviors cannot be observed directly as with macroscopic objects; they are revealed to us only in terms of energies, electromagnetic moments, transition rates, and reaction cross sections that can be observed in nuclei. In general, a large departure from expectations based on a single-particle picture is a signature of collective behavior. For example, high angular momenta, up to almost $100\hbar$, have been reported in heavy ion reactions. Such values are very far from those that can be achieved by the motion of one or two nucleons. In fact, some of the values observed are bordering on what is expected from classical systems and may thus be an interesting subject on their own merit.

A large amount of nuclear physics is centered around the study of low-lying states in nuclei. In addition to an interest in the phenomena displayed by these states, we also wish to understand the fundamental reasons that lead to such behaviors. In particular, we wish to give a unified description of both single-particle and collective behaviors in nuclei. Although a great deal of progress has been made, much work remains to be done.

§4-14 INFINITE NUCLEAR MATTER

As we have seen in §4-3 and §4-6, a finite nucleus has a large surface region the density of which drops gradually to zero with increasing radial distance away from the center. Even for a heavy nucleus, only a small fraction of the nucleons are in the central region where the density may be considered to be constant (see Problem 4-10). For many theoretical investigations, it is much easier if the density is uniform throughout the nuclear volume. For this reason, infinite nuclear matter is created as an idealized system of nucleons with an uniform density that approximates the interior of a heavy nucleus and having a neutron number equal to the proton number. Such a system is convenient for testing nucleon-nucleon interaction as well as techniques for solving many-body problems. Furthermore, being an infinite system, we do not have complications caused by motion of the center of mass, as in finite nuclei. Electromagnetic interactions are usually ignored in such studies, since the primary interest is in the nuclear force.

There is obviously no observed data on such an idealized system. The neutron star is as close to an infinite system of nuclear matter as we can imagine; however, experimental measurements on neutron stars having a direct interest to nuclear physics may not be forthcoming for a long time. As a result, most information concerning infinite nuclear matter must be inferred from our knowledge of finite nuclei.

Let us first try to deduce the binding energy in nuclear matter, for instance, by using the Weizacker semi-empirical mass formula discussed in §4-11. For a finite nucleus, we have used a model in which the volume is proportional to nucleon number A and the surface area to $A^{2/3}$. The ratio of the surface term to the volume term therefore varies as $A^{-1/3}$. For infinite nuclear matter, $A \rightarrow \infty$, and $A^{-1/3} \rightarrow 0$. As a result, we can ignore the surface term in (4-85). The contributions from Coulomb repulsion between protons may also be put to zero, since we do not wish to consider any electromagnetic effect here. Similarly, the symmetry energy vanishes, since we have assumed $N = Z$ to start with and the pairing effect may be ignored because A is infinite. This leaves only the volume term in the binding energy. From studies made on finite nuclei, we have the result

$$E_B/A = 16 \pm 1 \text{ MeV} \tag{4-116}$$

for the binding energy per nucleon in infinite nuclear matter. The uncertainty of 1 MeV here, in part, reflects variations in the values obtained from different ways of determining the parameter.

The density of infinite nuclear matter can be inferred from the maximum or *saturation* density in finite nuclei. The value commonly used is

$$\rho_0 = 0.16 \pm 0.02 \text{ nucleons fm}^{-3} \tag{4-117}$$

It is slightly higher than the average density of $3/(4\pi r_0^3) = 0.14$ nucleons per cubic femtometer obtained from finite nuclei using $r_0 = 1.2$ fm. The higher value comes from the fact that there is no surface region here.

The density of infinite nuclear matter is related to the Fermi momentum of nucleons in the following way. If the excitation energy is not very large – we shall come back at the end of this derivation to give an estimate of what is considered to be large – most of the low-lying single-particle states are occupied. As a result, the Pauli exclusion principle is more important than nucleon-nucleon interaction in determining the motion of nucleons inside the infinite nuclear matter. For such a system we can adopt a degenerate Fermi gas model to study the momentum distribution of the nucleons. In this approximation, nucleons are treated as non-interacting fermions filling up all the available low-lying single-particle states.

For a free particle in a cubic box of length L on each side, the wave function may be represented by a plane wave,

$$\psi(\boldsymbol{r}) = \frac{1}{L^{3/2}} e^{i\boldsymbol{k}\cdot\boldsymbol{r}} \tag{4-118}$$

The spin and isospin degrees of freedom of the nucleon have been ignored here, but we shall put them back in later. The allowed values for the wave numbers $\boldsymbol{k} = (k_x, k_y, k_z)$ are given by the condition that the wave function vanishes at the boundary of the box. As a result,

$$k_x = \frac{2\pi}{L} n_x \qquad k_y = \frac{2\pi}{L} n_y \qquad k_z = \frac{2\pi}{L} n_z \tag{4-119}$$

where n_x, n_y and n_z are integers 0, ±1, ±2,.... The number of allowed plane wave states in an volume element d^3k is given by

$$dn = 4\left(\frac{L}{2\pi}\right)^3 d^3k \tag{4-120}$$

where the factor of four comes from the fact that there are equal numbers of protons and neutrons and that each of them can either be in a state with spin-up or spin-down.

We can now relate nuclear density to the Fermi momentum. Since the total number of nucleons is A and, in the ground state, they fill all the low-lying states up to Fermi momentum k_F, we obtain the relation

$$A = \int_0^{\epsilon_F} dn = \int_0^{k_F} 4\left(\frac{L}{2\pi}\right)^3 d^3k = 4\left(\frac{L}{2\pi}\right)^3 \frac{4\pi}{3}k_F^3$$

The nucleon density is given by the number of nucleons in volume L^3. This gives us the result

$$\rho_0 = \frac{A}{L^3} = \frac{2}{3\pi^2}k_F^3 \tag{4-121}$$

On inverting the relation, we obtain the Fermi momentum in terms of the density of infinite nuclear matter,

$$k_F = \left(\frac{3\pi^2}{2}\rho_0\right)^{1/3} = 1.33 \pm 0.05 \text{ fm}^{-1} \tag{4-122}$$

where the final result is obtained using the value of ρ_0 given in (4-117).

The average kinetic energy of nucleons in infinite nuclear matter may be calculated from (4-120) using the fact that the kinetic energy of a nucleon is given by $(\hbar k)^2/2M_N$, where M_N is the mass of a nucleon. On averaging over all the nucleons, we obtain the result

$$\bar{\epsilon} = \frac{1}{A}\int_0^{k_F} \frac{(\hbar k)^2}{2M_N} 4\left(\frac{L}{2\pi}\right)^3 d^3k = \frac{3}{5}\frac{(\hbar k_F)^2}{2M_N} = \tfrac{3}{5}\epsilon_F \tag{4-123}$$

The value of Fermi energy ϵ_F can be obtained from that of k_F given in (4-122),

$$\epsilon_F = \frac{(\hbar k_F)^2}{2M_N} \approx 37 \text{ MeV} \tag{4-124}$$

The average energy of a nucleon is then $\bar{\epsilon} \approx 23$ MeV. From this result, we see that only a small fraction of the nucleons in a nucleus of nucleon number A can be excited unless the energy involved is comparable to $A\epsilon_F$, of the order of 10^3 MeV for a medium weight nucleus. This in turn justifies the use of a Fermi gas model above.

Since infinite nuclear matter is normally in a state of minimum energy, it is in an equilibrium, stable against small variations of the density. As a result, variations of the binding energy per particle as a function of Fermi momentum vanish,

$$\frac{d}{dk_F}\left(\frac{E_B}{A}\right) = 0$$

The second order derivative of the binding energy depends on the difficulty, or the *stiffness*, against variations of the density. This is measured by the compression modulus,

$$\mathcal{K} = k_F^2 \frac{d^2}{dk_F^2}\left(\frac{E_B}{A}\right)_{\rho=\rho_0} \tag{4-125}$$

the slope of the variation of binding energy per nucleon as a function of k_F. Since it is evaluated at the energy minimum, it is a positive quantity. The compression modulus is the equivalent of the bulk modulus in mechanics that characterizes variations of the volume of a material as a function of pressure applied (see Problem 4-11).

For nuclear matter, the value of $\mathcal{K}$ is generally taken to be

$$\mathcal{K} \approx 200 \text{ MeV} \tag{4-126}$$

estimated, for example, from the energy required to excite a nucleus without changing its shape. This is called a breathing mode of excitation (see §6-1) and is most easily recognized in even-even nuclei, and in particular, in closed shell nuclei. Since the ground state spin and parity of such a nucleus is 0^+, one way to excite to another 0^+ state is by a change of the density.

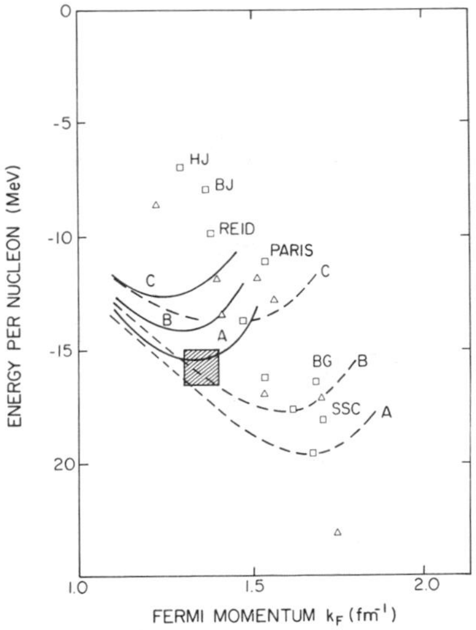

Fig. 4-12 Binding energy per nucleon as a function of nucleon Fermi momentum in infinite nuclear matter. The shaded area represents the possible values extrapolated from finite nuclei and the small squares are the results of calculations using potentials shown (HJ, Hamada-Johnston; BJ, Bethe-Johnson; BG, Bryan-Gersten; and SSC, super-soft-core). The solid lines are the results of different Dirac-Brueckner calculations and dashed lines are the results of more conventional Brueckner calculation. (R. Machleidt, in *Rel. Dynamics and Quark-Nuclear Phys.*, ed. by M.B. Johnson and A. Picklesimer, Wiley, New York, 1985, p. 71).

The aim of an infinite nuclear matter calculation is to reproduce the values of the three pieces of known "data" deduced from the properties of finite nuclei: the binding energy per nucleon (4-116), the saturation density in terms of the Fermi momentum (4-122), and (not always carried out) the compression modulus (4-126). The hope is that the simple geometry offered by the idealized system provides us with a direct and meaningful test for the nucleon-nucleon interaction and the many-body technique used. The test is a nontrivial one. For example, most of the properties of finite nuclei are not very sensitive to the hard core in nucleon-nucleon interactions. In contrast, the exact value of this term is crucial in reproducing the saturation density in infinite nuclear matter. Without the repulsion at short distances, infinite nuclear matter can gain binding energy by increasing its density. In fact, most of the calculations to date have failed to produce the correct binding energy per nucleon at the correct saturation density. The situation may be changing for some of the relativistic approaches used in nuclear many-body problems. Recent results, using a version of the Bonn potential for nucleon-nucleon interaction and a relativistic version of the most fruitful many-body technique, seem to produce the correct answers for both quantities, as shown in Fig. 4-12. We shall not go any further into these calculations, since it would require a lengthy discussion of the various techniques used to solve the many-body problem involved here.

Problems

4-1. Use a table of mass excess to:

(a) Calculate the energy released in fusing two free protons and two free neutrons into an α-particle.

(b) Show that ^{238}U is unstable toward α-decay and calculate the kinetic energy of α-particle emitted.

(c) Calculate the maximum possible energy released in the fission of ^{235}U induced by thermal neutrons. Make a reasonable assumption of the fission fragments released in the process.

4-2. The radius of a nucleus may be approximated as $R = r_0 A^{1/3}$. From the binding energy differences between odd-A mirror nuclei (*i.e.*, a pair of nuclei having the same A but with neutron and proton numbers interchanged) in the mass range $A = 11$ to 17, estimate the value of r_0 assuming a uniform spherical charge distribution for each nucleus and ignoring effects other than Coulomb.

4-3. From a table of mass excess show that the ground state of ^{8}Be is stable against β-decay and nucleon emission but unstable against decay into two α-particles. Calculate the energy released in the decay $^8\text{Be} \rightarrow \alpha + \alpha$.

4-4. Obtain the masses of members of the $A = 135$ isobar from a table of mass excess and plot the results as a function of Z. From the results deduce the value of the symmetry energy parameter α_4 in (4-85), the Weizacker semiempirical mass formula. Carry out the same calculation for members of the $A = 136$ isobar and estimate the value of the pairing parameter δ from the results.

4-5. The Weizacker semi-empirical mass formula is useful to obtain global distribution of binding energies as a function of A, Z, and N. Use this formula to show that the Q-value, *i.e.*, kinetic energy released, in fission is positive only for heavy nuclei.

4-6. Derive an expression for the form factor $F(q^2)$ assuming that the nuclear density is a uniform sphere of radius R.

4-7. The radial dependence of the density of a nucleus may be described by a two-parameter Fermi distribution,

$$\rho_{2pF}(r) = \frac{\rho_0}{1 + \exp \frac{r-c}{z}}$$

as given in (4-21). Show that the parameter c is the radius of the nucleus measured from the center to a point where the density falls to roughly half of its value at the center and the parameter $z \approx t/4.4$, where t is the radial distance between two points in the nucleus whose densities are, respectively, 10% and 90% of the maximum value.

4-8. Calculate the binding energies of members of the $A = 16$ isobar, ^{16}C, ^{16}N, ^{16}O, ^{16}F, and ^{16}Ne, using a table of mass excess. From these values calculate the position of the $T = 1$ isobaric analogue state to the ground state of ^{16}F in ^{16}O using a uniformed charge spherical of radius $R = 1.2A^{1/3}$ fm to find the Coulomb energy difference between the two nuclei. Repeat the calculation starting from the ground state of ^{16}N and compare the results obtained. Deduce also the positions of the lowest $J^\pi = 0^+$, $T = 2$ states in ^{16}N, ^{16}O, and ^{16}F from the binding energies of ^{16}C and ^{16}Ne. Compare the results with observed values shown in Fig. 4-7.

4-9. In ^{56}Fe, the level density parameter $a = 7.2$ MeV^{-1} and the spin cutoff factor $\sigma_J(E) = 20$ at excitation energy $E = 20$ MeV. Evaluate the state density of ^{56}Fe at this energy and calculate the probability of finding a $J = 0$ level in a 100 keV interval at this energy.

4-10. Use the values of the parameters given in Table 4-1 to calculate the number of nucleons in the surface region of ^{206}Pb, *i.e.*, the region where the density is 90% or less of the value at the center of the nucleus.

4-11. In the study of the property of ordinary matter, the bulk modulus B is defined as the negative of the ratio of the change in pressure applied Δp to the fractional change in the volume $\Delta V/V$,

$$B = -V\frac{\Delta p}{\Delta V}$$

where the negative sign simply means that volume shrinks when the pressure is increased. Find the relation of B to the compression modulus $\mathcal{K}$ defined in (4-125).

4-12. Show that, for high-energy inelastic scattering where the projectile mass may be ignored, the momentum transfer is given by

$$(\hbar cq)^2 = 4EE' \sin^2\left(\tfrac{1}{2}\theta\right) \tag{4-36}$$

where E is the energy of the projectile, E' the energy of the scattered particle, and θ is the scattering angle.

4-13. Use the Kelson-Garvey mass relation to find the binding energy of ^{12}O from the values of nearby nuclei calculated from a table of mass excess. Estimate the uncertainties in the value deduced. Using the value of binding energy obtained and a minimum amount of additional experimental data, find the locations of isobaric analogue states to the ground state of ^{12}O in ^{12}N and ^{12}C.

4-14. Calculate the average density of ^{238}U assuming it is a uniform sphere of radius $R = 1.2A^{1/3}$ fm. If R is the radius of ^{238}U measured from the center to a point on the surface where the density is half of that in the central region, and the nuclear density is given by a two-parameter Fermi form given in (4-21) with $z = 0.6$ fm, what is the central density of the nucleus? Compare the value obtained with the average density.

4-15. Determine the Coulomb and symmetry energy parameters of the Weizacker semi-empirical formula from the binding energies of three members of the $A = 27$ isobar $(A, Z) = (27, 12)$, $(27, 13)$, and $(27, 14)$.

4-16. The differential cross section for Rutherford scattering is proportional to $\sin^{-4}\{\frac{1}{2}\theta\}$ where θ is the scattering angle. Explain why in reality the experimental differential cross section remains finite as $\theta \to 0$.

Chapter 5

NUCLEAR EXCITATION AND DECAY

Since nucleon-nucleon interaction is the single most important factor governing the properties of a nucleus, nuclear states are usually taken as the eigenvectors of a Hamiltonian in which only the nuclear interaction potential is present. The actions of electromagnetic and weak interactions are treated as perturbations, inducing transitions from one nuclear state to another. In addition, transition between states can also take place by α-particle decay and fission. In this chapter we shall examine the rates of these transitions and see what they can tell us about the atomic nucleus. Other processes, such as nucleon emission, also change the state a nucleus. Such reactions are sufficiently fast that they must be treated differently. We shall examine them in Chapter 7.

§5-1 NUCLEAR TRANSITION MATRIX ELEMENT

Transition probability. If we have a sample of N radioactive nuclei, the probability for any one of them to decay at a given time is independent of the status of other nuclei in the sample. The number of decay taking place is therefore proportional to $N(t)$, the number of radioactive nucleus present at time t,

$$\frac{dN}{dt} = -\mathcal{W}\,N(t) \tag{5-1}$$

The constant of proportionality, $\mathcal{W}$, is the *transition probability* or *decay constant*, and its value depends on the nature of the perturbation that causes the decay and the properties of the initial and final states involved. For this reason, the decay constant is the quantity of central interest in a study of nuclear transitions.

From (5-1), we obtain the familiar exponential decay law,

$$N(t) = N_0 e^{-\mathcal{W}t} \tag{5-2}$$

where N_0 is the number of radioactive nuclei at time $t = 0$. The half-life, $T_{1/2}$, is the amount of time it takes for the activity of a sample to be reduced by half. It is inversely proportional to the transition rate,

$$T_{1/2} = \frac{\ln 2}{\mathcal{W}} = \frac{0.693}{\mathcal{W}} \tag{5-3}$$

The lifetime, or mean life, of an excited state is the average amount of time it takes for a radioactive nucleus to decay. It is connected to the transition probability and half-life by the relation:

$$\overline{T} = \frac{\int_0^\infty te^{-\mathcal{W}t}dt}{\int_0^\infty e^{-\mathcal{W}t}dt} = \frac{1}{\mathcal{W}} = \frac{T_{1/2}}{0.693} \tag{5-4}$$

These three quantities, transition probability, half-life, and lifetime, are three different ways to characterize the same physical observable.

Width. If a nucleus is in an excited state, it must discard the excess energy by undergoing a decay. It is, however, impossible to know when the decay will actually take place. As a result, there is an uncertainty in time $\Delta t = \overline{T}$ associated with the existence of the excited state. Furthermore, the limited lifetime of a state does not permit us to measure its energy to infinite precision. This conclusion is independent of the instrumental accuracy used for the measurement: for our purpose here, we can regard uncertainties introduced by the measuring apparatus as sufficiently small that they may be ignored.

In quantum mechanics, the expectation value of an observable is interpreted as the average over the measured values of a large number of identically prepared samples. In other words, if we carry out the energy measurement for a large number, N, of nuclei in the same excited state, the values obtained will not be identical to each other. If the measured value for the i-th excited nucleus is E_i, the average $\langle E\rangle$ is given by

$$\langle E\rangle = \frac{1}{N}\sum_{i=1}^{N} E_i \tag{5-5}$$

An idea of the spread in the measured values is provided by the square root of the variance,

$$\varGamma = \left\{\frac{1}{N}\sum_{i=1}^{N}\left(E_i^2 - \langle E\rangle^2\right)\right\}^{1/2} \tag{5-6}$$

Heisenberg uncertainty principle says that the product of $\varGamma$ and $\overline{T}$ is equal to $\hbar$ under the best circumstances, or

$$\varGamma = \frac{\hbar}{\overline{T}} = \hbar\mathcal{W} \tag{5-7}$$

The quantity $\varGamma$ is known as the natural line width, or width for short, of a state. It is also a way to indicate the transition probability of a state, proportional to the (inverse of) the lifetime and half-life. Since $\hbar = 6.6 \times 10^{-22}$ MeV-s, a width of 1 MeV corresponds to a mean lifetime of 6.6×10^{-22} s.

One can also relate $\varGamma$ to the probability of finding the excited state at a specific energy E. In terms of wave functions, the decay constant $\mathcal{W}$ may be defined in the following way:

$$|\Psi(\boldsymbol{r}, t)|^2 = |\Psi(\boldsymbol{r}, t{=}0)|^2 e^{-\mathcal{W}t} \tag{5-8}$$

For a stationary state, the time-dependent wave function may be separated into a product of spatial and temporal parts,

$$\Psi(\boldsymbol{r},t) = \psi(\boldsymbol{r})e^{-iEt/\hbar} \tag{5-9}$$

To carry such a separation over to a decaying (excited) state, the energy E must be changed into a complex quantity

$$E \to \langle E\rangle - \tfrac{1}{2}i\hbar\mathcal{W}$$

The time-dependent wave function now takes on the form

$$\Psi(\boldsymbol{r},t) = \psi(\boldsymbol{r})e^{-i\langle E\rangle t/\hbar - \mathcal{W}t/2} \tag{5-10}$$

so as to satisfy (5-8).

Alternatively we can use the fact that the excited state is one without a definite energy and write the wave function as a superposition of states having different energies,

$$\Psi(\boldsymbol{r},t) = \psi(\boldsymbol{r})\int a(E)e^{-iEt/\hbar}\,dE \tag{5-11}$$

where $a(E)$ is the probability amplitude for finding the state at energy E. Comparing this form with (5-10), we arrive at a relation between the decay constant $\mathcal{W}$ and probability amplitude $a(E)$,

$$e^{-\mathcal{W}t/2} = \int a(E)e^{-i(E-\langle E\rangle)t/\hbar}\,dE \tag{5-12}$$

that is, $e^{-\mathcal{W}t/2}$ is the Fourier transform of $a(E)$.

The relation can be inverted and we obtain the result,

$$a(E) = \frac{1}{2\pi\hbar}\int_0^\infty e^{\{i(E-\langle E\rangle)/\hbar - \frac{1}{2}\mathcal{W}\}t}\,dt = \frac{i}{2\pi}\frac{1}{(E-\langle E\rangle)+i\frac{1}{2}\hbar\mathcal{W}}$$

The probability for finding the excited state with energy E is given by the absolute square of the amplitude,

$$|a(E)|^2 = \frac{1}{4\pi^2}\frac{1}{(E-\langle E\rangle)^2+(\frac{1}{2}\Gamma)^2} \tag{5-13}$$

where we have replaced $\hbar\mathcal{W}$ with Γ using (5-7). The shape of such a distribution is *Lorentzian* and the width Γ may be interpreted as the full width at half maximum (FWHM) of such a distribution. Since the question of instrumental uncertainty does not enter here, the width is the "natural line width" of the distribution in energy of the excited state.

Branching ratio. A given excited state may decay to more than one final states. If the transition probability to each final state is $\mathcal{W}(i)$, the total transition probability for the initial state is the sum of transitions to all the possible final states,

$$\mathcal{W} = \sum_i \mathcal{W}(i) \tag{5-14}$$

Similarly, the total width Γ is sum of all the partial widths,

$$\Gamma = \sum_i \Gamma(i) \tag{5-15}$$

The relation between half-life $T_{1/2}$ and partial half-lives $T_{1/2}(i) = \ln 2/\mathcal{W}(i)$ is given by relation,

$$\frac{1}{T_{1/2}} = \sum_i \frac{1}{T_{1/2}(i)} \tag{5-16}$$

as evident from (5-14).

The *branching ratio* gives the partial transition probability to a particular final state as a fraction of the total transition probability from a specific initial state. For example, the mean lifetime of π^0-meson is 8.4×10^{-17} s and decays 98.8% of the time to two γ's, 1.17% of the time to a γ together with an electron-positron pair, and 2×10^{-7} of the time to an electron-positron pair alone. The branching ratios to these three decay channels are therefore 98.8%, 1.17%, and 2×10^{-7}, respectively. Among the radioactive nuclei, the ground state of the odd-odd nucleus $^{226}_{89}$Ac at the start of the actinide series has a half-life of 29 hours and can decay by emitting an electron to $^{226}_{90}$Th, transform to $^{226}_{99}$Ra by capturing one of the atomic electrons, or decay by α-particle emission to $^{222}_{87}$Fr with branching ratios 83%, 10%, and 0.06%, respectively.

Transition matrix element. The transition probability is proportional to the square of the nuclear matrix element

$$\mathcal{M}_{fi}(M_f, M_i) = \langle J_f M_f \xi | \boldsymbol{O}_{\lambda\mu} | J_i M_i \zeta \rangle \tag{5-17}$$

where $| J_i M_i \zeta \rangle$ and $| J_f M_f \xi \rangle$ are, respectively, the wave functions of the initial and final states, and $\boldsymbol{O}_{\lambda\mu}$ is the nuclear part of transition operator (see §5-2 and §5-5) with spherical tensor rank (λ, μ). The labels ζ and ξ here denote quantum numbers other than angular momentum required to specify the nuclear states uniquely. Since the transition may also involve the emission of a particle such as an electron or a nucleon, the initial and final states are not necessarily in the same nucleus: the matrix element $\mathcal{M}_{fi}$ expresses factors pertaining to the nuclear states in the transition. Since the transition itself may be induced by the interaction of the nucleus with an external field, the exact relation between the transition probability $\mathcal{W}$ and the nuclear matrix element $\mathcal{M}_{fi}$ depends also on factors related to the external field. We shall treat these factors separately for each type of transition in later sections.

The dependence of the matrix element $\mathcal{M}_{fi}$ on M_i and M_f, projections of the initial and final total angular momentum, respectively, on the quantization axis, may be factored out using the Wigner-Eckart theorem,

$$\mathcal{M}_{fi}(M_f, M_i) = (-1)^{J_f - M_f} \begin{pmatrix} J_f & \lambda & J_i \\ -M_f & \mu & M_i \end{pmatrix} \langle J_f \xi \| O_\lambda \| J_i \zeta \rangle \tag{5-18}$$

where $\begin{pmatrix} J_f & \lambda & J_i \\ -M_f & \mu & M_i \end{pmatrix}$ is the $3j$-symbol and $\langle J_f \xi \| O_\lambda \| J_i \zeta \rangle$ is the reduced matrix element defined in (B-52). Our main interest will be in $\langle J_f \xi \| O_\lambda \| J_i \zeta \rangle$, since it is a quantity invariant under a rotation of the coordinate system.

If the measurement is not sensitive to the orientation of the spin in the final state, the transition includes all the possible final states differing only by the value of M_f. Furthermore, if the transition operator is not restricted to any specific direction in space, all the allowed values of μ must be included when considering the transition. Under these conditions, the square of the transition matrix element reduces to the form

$$\begin{aligned} |\mathcal{M}_{fi}|^2 &= \sum_{\mu M_f} \left| (-1)^{J_f - M_f} \begin{pmatrix} J_f & \lambda & J_i \\ -M_f & \mu & M_i \end{pmatrix} \langle J_f \xi \| O_\lambda \| J_i \zeta \rangle \right|^2 \\ &= \left| \langle J_f \xi \| O_\lambda \| J_i \zeta \rangle \right|^2 \sum_{\mu M_f} \left| \begin{pmatrix} J_f & \lambda & J_i \\ -M_f & \mu & M_i \end{pmatrix} \right|^2 \\ &= \frac{\Delta(J_f, \lambda, J_i)}{2J_i + 1} \left| \langle J_f \xi \| O_\lambda \| J_i \zeta \rangle \right|^2 \end{aligned} \tag{5-19}$$

In arriving at the final form, we have made use of the orthogonality relation between $3j$-symbols,

$$\sum_{m_1 m_2} \begin{pmatrix} j_1 & j_2 & j_3 \\ m_1 & m_2 & m_3 \end{pmatrix} \begin{pmatrix} j_1 & j_2 & j'_3 \\ m_1 & m_2 & m'_3 \end{pmatrix} = \frac{\Delta(j_1, j_2, j_3)}{2j_3 + 1} \delta_{j_3, j'_3} \delta_{m_3, m'_3} \tag{B-35}$$

given in §B-3. The factor,

$$\Delta(J_f, \lambda, J_i) = \begin{cases} 1 & \text{for } \boldsymbol{J}_f = \boldsymbol{\lambda} + \boldsymbol{J}_i \\ 0 & \text{otherwise,} \end{cases}$$

expresses the angular momentum selection rule forbidding transitions where the triangular relation among the three angular momentum vectors $\boldsymbol{J}_f$, $\boldsymbol{\lambda}$, and $\boldsymbol{J}_i$ is not satisfied. Note also that $|\mathcal{M}_{fi}|^2$ defined in (5-19) is independent of M_i.

So far the discussion on the nuclear transition matrix element is a general one without any reference to the nature of the operator O_λ. To make further progress, we must now separate the discussion according to the type of transition that takes place.

§5-2 ELECTROMAGNETIC TRANSITION

In this section we shall deal mainly with nuclear decay through the emission of a γ-ray. The cause of such transitions is the interaction between the nucleus and an external electromagnetic field. For our purpose here, we may regard the nucleus as made of point nucleons, each carrying a magnetic dipole moment and, in the case of protons, a net charge as well. The charge distribution can interact with the external field, causing "electric" transitions. Alternatively, the intrinsic magnetism of each nucleon together with that generated by current loops set up as a result of proton orbital motion induce "magnetic" transitions in nuclei.

Electromagnetic transitions form the dominant mode of decay for low-lying excited states in nuclei, particularly for the light ones. The main reason is that nucleon emission, a much faster process than γ-decay, is forbidden until the excitation energy is above nucleon separation energies. As we can see later in Fig. 6-14, neutron separation energies are of the order of 8 to 10 MeV and the corresponding values for protons are somewhat lower because of the Coulomb repulsion. Other possible decay modes are β-decay, α-particle emission, and fission. Generally speaking, these are slower processes than γ-decay and their Q-value considerations, that is, the energy involved in the decay, are also different, as we shall see in later sections.

Besides γ-ray emission, nuclei can also de-excite through electromagnetic interaction by internal conversion whereby one of the atomic electrons is ejected. This is usually more important for heavy nuclei, where the nuclear electromagnetic fields are strong and the orbits of the inner shell electrons are close to the nucleus. Instead of a γ-ray, an electron-positron pair may also be created as a result of the de-excitation process. The probability for such internal pair creation is, in general, much smaller than γ-ray emission and becomes important where γ-ray emission is forbidden by angular momentum considerations. This happens in the case of transitions from an initial state with $J^\pi = 0^+$ to a final state that is also 0^+.

The first step in a discussion on electromagnetic transitions is to establish a connection between the transition probability $\mathcal{W}$ of (5-1) and the nuclear matrix element $\mathcal{M}_{if}$ of (5-17). This may be achieved using first-order time-dependent perturbation theory, reviewed in Appendix E. The general result, known as Fermi's golden rule, is given by (E-15) in the form

$$\mathcal{W} = \frac{2\pi}{\hbar}\left|\langle\phi_{\boldsymbol{k}}(\boldsymbol{r})|H'|\phi_0(\boldsymbol{r})\rangle\right|^2 \rho(E_f) \tag{5-20}$$

where H' represents the perturbation due to coupling between nuclear and electromagnetic fields. The density of final states $\rho(E_f)$ here is a product of the number of nuclear and electromagnetic states per energy interval at E_f. Similarly, the initial and final wave functions, $\phi_0(\boldsymbol{r})$ and $\phi_{\boldsymbol{k}}(\boldsymbol{r})$, respectively, are products of nuclear and electromagnetic parts. In keeping with the custom used by most workers in the field, all the electromagnetic calculations in this section will be carried out in cgs units.

Coupling to electromagnetic field. Our primary interest is in the nuclear part of the matrix element of H'. For this purpose, we shall first separate H' into a product of two operators, one acting only on the nuclear wave function and the other on the wave function for the external electromagnetic field. Since nuclei are described by quantum mechanical wave functions each with a definite angular momentum, it is necessary that the external electromagnetic field is also quantized and decomposed by a multipole expansion into components with definite spherical tensor ranks. As we shall see later, the decomposition is an important one, since often the lowest order multipole dominates the nuclear transition. Both quantization and multipole expansion of the electromagnetic field are fairly straightforward but tedious procedures. We shall attempt here only a brief outline of the steps involved and leave the proper derivations to more advanced textbooks such as Blatt and Weisskopf (*Theoretical Nuclear Physics*, Wiley, New York, 1952) and Sakurai (*Advanced Quantum Mechanics*, Addison-Wesley, Reading, Massachusetts, 1973)

The form of the electromagnetic perturbation in a nucleus may be visualized in the following way. Consider first the simple case of a point particle carrying a charge q but no magnetic moment for the time being. In the absence of any external electromagnetic field, the particle is free and the Hamiltonian consists of the kinetic energy term only,

$$H_0 = \frac{1}{2m}\boldsymbol{p}^2 \tag{5-21}$$

where $\boldsymbol{p}$ is the momentum. In the presence of an electromagnetic field, the momentum conjugate to $\boldsymbol{r}$ is modified from that for a free particle,

$$\boldsymbol{p} \to \boldsymbol{p} + \frac{q}{c}\boldsymbol{A}$$

and the Hamiltonian for the charged particle now takes on the form,

$$H = \frac{1}{2m}\left(\boldsymbol{p} - \frac{q}{c}\boldsymbol{A}\right)^2 \tag{5-22}$$

For simplicity we have not included here the Hamiltonian for the external electromagnetic field nor the effect of any electrostatic potential that may be present to interact with the charged particles. We shall return to this point later.

The Hamiltonian in (5-22) may be interpreted as the sum of a free particle Hamiltonian term H_0 given by (5-21) and a term H' expressing the coupling with the external electromagnetic field,

$$H = H_0 + H' \tag{5-23}$$

Comparing (5-22) with (5-21), we can make the identification

$$H' = -\frac{q}{2mc}\left(\boldsymbol{p}\cdot\boldsymbol{A} + \boldsymbol{A}\cdot\boldsymbol{p}\right) + \frac{q^2}{2mc^2}\boldsymbol{A}\cdot\boldsymbol{A} = -\frac{q}{mc}\boldsymbol{A}\cdot\boldsymbol{p} + \frac{q^2}{2mc^2}\boldsymbol{A}\cdot\boldsymbol{A} \tag{5-24}$$

where we have replaced $\boldsymbol{p}\cdot\boldsymbol{A}$ by $\boldsymbol{A}\cdot\boldsymbol{p}$ using the condition that an electromagnetic field can have only transverse components, *i.e.*, $\boldsymbol{p} = -i\hbar\nabla$ and $\nabla\cdot\boldsymbol{A} = 0$, generally known as the transversality condition. The quadratic term in $\boldsymbol{A}$ involves two photons at the same time and may therefore be ignored in the lowest order consideration we are interested in here.

In general, H' may be written in a more convenient form by expressing the momentum of the charged particle in terms of a current density,

$$\mathcal{J} = q\boldsymbol{v} = q\frac{\boldsymbol{p}}{m} \tag{5-25}$$

The first order term in (5-24) now takes on the form

$$H' = -\frac{1}{c}\boldsymbol{A}\cdot\mathcal{J} \tag{5-26}$$

For a nucleus, in addition to the electric charge carried by protons, intrinsic magnetic dipole moment of the nucleons can also interact with the external electromagnetic field. We have seen earlier in (4-56) in the discussion of static magnetic moments that the intrinsic magnetic dipole moment of a nucleon may also be expressed in the form of a current. Consequently it is unnecessary to change the form of H' given in (5-26) to include the effect of nucleon magnetic moments except a more general definition for the current density $\mathcal{J}$ than that given in (5-25) is needed. The most general form of H' must also include the possibility for the charge distribution to interact with an external electrostatic field. Such a perturbing Hamiltonian is most conveniently written in four-component notations,

$$H' = -\frac{1}{c}\sum_{\mu=1}^{4} A_\mu J_\mu \tag{5-27}$$

where $A_\mu = (\boldsymbol{A}, iV)$ contains a scale potential V and $J_\mu = (\mathcal{J}, i\rho c)$ includes a charge distribution ρ as well. The fourth component is usually not important in nuclear transitions.

External electromagnetic field. The form of the external electromagnetic field is given by the solution to Maxwell's equations. In a region outside any charge and current distributions, the four potential is the solution of the time-dependent partial differential equation,

$$\left(\nabla^2 - \frac{1}{c^2}\frac{\partial^2}{\partial t^2}\right)A_\mu(\boldsymbol{r}, t) = 0 \tag{5-28}$$

Our primary interest will be in the vector potential $\boldsymbol{A}(\boldsymbol{r}, t)$, the first three components of A_μ. The time dependence may be removed from the equation by expanding $\boldsymbol{A}(\boldsymbol{r}, t)$ in terms of components with definite wave number $\boldsymbol{k}$,

$$\boldsymbol{A}(\boldsymbol{r}, t) = \sum_{\boldsymbol{k}} \boldsymbol{A}_{\boldsymbol{k}}(\boldsymbol{r})e^{-i\omega t} \tag{5-29}$$

where $\omega = kc$ and k is the magnitude of $\boldsymbol{k}$. The spatial dependence of $\boldsymbol{A}$ is then given by the equation,

$$(\nabla^2 + k^2)\boldsymbol{A}_{\boldsymbol{k}}(\boldsymbol{r}) = 0 \tag{5-30}$$

The solution of this second order differential equation has the familiar form,

$$A_{\boldsymbol{k}}(\boldsymbol{r}) = B_{\boldsymbol{k}} e^{i\boldsymbol{k}\cdot\boldsymbol{r}} + C_{\boldsymbol{k}} e^{-i\boldsymbol{k}\cdot\boldsymbol{r}}$$

where $\boldsymbol{B_k}$ and $\boldsymbol{C_k}$ are constants to be determined by boundary conditions. Among other things, these conditions must simulate the effect of the source of the electromagnetic field. Substituting the spatial dependence of the vector potential into (5-29), we obtain the expression,

$$\boldsymbol{A}(\boldsymbol{r},t) = \frac{1}{N}\sum_{\boldsymbol{k}}\sum_{\eta}\left\{\boldsymbol{b}_{\boldsymbol{k}\eta}\boldsymbol{\epsilon}_{\eta} e^{i(\boldsymbol{k}\cdot\boldsymbol{r}-\omega t)} + \boldsymbol{b}^{\dagger}_{\boldsymbol{k}\eta}\boldsymbol{\epsilon}_{\eta} e^{-i(\boldsymbol{k}\cdot\boldsymbol{r}+\omega t)}\right\} \tag{5-31}$$

where N is a normalization constant. Being the quantum of a vector field, each γ-ray carries one unit of angular momentum. However, because of the transversality condition $\nabla\cdot\boldsymbol{A} = 0$, only two of the three components of the vector field are independent quantities. These two components may be represented by two unit vectors $\boldsymbol{\epsilon}_\eta$, with $\eta = 1, 2$. This is similar to the possibility of expressing ordinary light waves as a linear combination of horizontal and vertical polarizations. At this stage, the factors $\boldsymbol{b}_{\boldsymbol{k}\eta}$ and $\boldsymbol{b}^{\dagger}_{\boldsymbol{k}\eta}$ remain as constants related to $\boldsymbol{B_k}$ and $\boldsymbol{C_k}$ and must be determined using boundary conditions.

Up to now the solution we have obtained to Maxwell's equation is purely classical in nature. Since the form of (5-30) is identical to the equation for an harmonic oscillator, we may think of the electromagnetic field as a collection of harmonic oscillators, one for each frequency (or wave number $\boldsymbol{k}$) and polarization direction η. The separation in energy between harmonic oscillator states is in units of $\hbar\omega = (\hbar ck)$. It is now quite straightforward to have a quantum mechanical description of the electromagnetic field. The quantity $\hbar\omega$ may be taken as the energy of a field quantum for a given wave number and the electromagnetic field is now characterized by the number of quanta for each $\boldsymbol{k}$ and η. We may also interpret $\boldsymbol{b}^{\dagger}_{\boldsymbol{k}\eta}$ and $\boldsymbol{b}_{\boldsymbol{k}\eta}$ in (5-31) as the creation and annihilation operators, respectively, of a photon with labels $(\boldsymbol{k},\eta)$. In this way we see that the lowest order term of H' given in (5-27) involves a linear combination of $\boldsymbol{b}^{\dagger}_{\boldsymbol{k}\eta}$ and $\boldsymbol{b}_{\boldsymbol{k}\eta}$. The physical interpretation of H' is now quite clear: the coupling between the nuclear and electromagnetic field makes it possible for the nucleus to create a photon when it decays from a higher state to a lower one and to absorb a photon when it is excited to a higher energy state. A proper derivation of the quantized electromagnetic field will require us to demonstrate that $\boldsymbol{b}^{\dagger}_{\boldsymbol{k}\eta}$ and $\boldsymbol{b}_{\boldsymbol{k}\eta}$ are indeed creation and annihilation operators of photons by showing that they have the proper commutation relations, and so on. We shall dispense with this important step here to keep the discussion focused on the concerns of nuclear physics.

Multipole expansion of the electromagnetic field. The expansion of $\boldsymbol{A}(\boldsymbol{r},t)$ in (5-31) is made implicitly in cartesian coordinates. For applications to problems with rotational symmetry, for example, as in the case of nuclear transitions, it is more convenient to express $\boldsymbol{A}(\boldsymbol{r},t)$ in terms of operators having definite spherical tensor ranks, as we have done on many earlier occasions. The advantage here is quite obvious. We have already seen that the nuclear current density $\mathcal{J}$ may be written as a sum over terms, each carrying a definite amount of angular momentum. Since H' is a scalar, only multipoles of the same order in both $\mathcal{J}$ and $\boldsymbol{A}$ can be coupled together to form an angular momentum zero operator. For this purpose, we shall first rewrite the radiation field in terms of eigenfunctions of angular momentum operators,

$$\boldsymbol{A}(\boldsymbol{r},t) = \sum_{\lambda\mu} \boldsymbol{A}_{\lambda\mu}(\boldsymbol{r},t) \tag{5-32}$$

where the vector functions of spherical tensor ranks (λ, μ) satisfy the relations

$$\boldsymbol{J}^2 \boldsymbol{A}_{\lambda\mu}(\boldsymbol{r},t) = \lambda(\lambda+1)\boldsymbol{A}_{\lambda\mu}(\boldsymbol{r},t) \qquad \boldsymbol{J}_0 \boldsymbol{A}_{\lambda\mu}(\boldsymbol{r},t) = \mu\, \boldsymbol{A}_{\lambda\mu}(\boldsymbol{r},t) \tag{5-33}$$

The time dependence, given by (5-29), is sufficiently simple that we shall drop it from now on in order to simplify the notation. The functions $\boldsymbol{A}_{\lambda\mu}(\boldsymbol{r},t)$ are, however, different from spherical harmonics, satisfying (B-6), in that they are vector functions having spherical tensor ranks (λ, μ). They may be expressed in terms of *vector spherical harmonics* which are vector functions constructed from (scalar) spherical harmonics $Y_{\ell m}(\theta, \phi)$.

Instead of the two polarization directions allowed for $\boldsymbol{A}(\boldsymbol{r},t)$ in (5-31), we have now two different types of multipole fields satisfying (5-30). These are called *electric* multipole transitions, indicated hereafter as $E\lambda$ in the argument, and *magnetic* multipole transitions, indicated as $M\lambda$. In terms of spherical harmonics, they may be written in the following forms:

$$\boldsymbol{A}_{\lambda\mu}(E\lambda, \boldsymbol{r}) = \frac{-i}{k}\nabla \times \big(\boldsymbol{r} \times \nabla\big)\big(j_\lambda(kr)Y_{\lambda\mu}(\theta,\phi)\big)$$

$$\boldsymbol{A}_{\lambda\mu}(M\lambda, \boldsymbol{r}) = \big(\boldsymbol{r} \times \nabla\big)\big(j_\lambda(kr)Y_{\lambda\mu}(\theta,\phi)\big) \tag{5-34}$$

where $j_\lambda(kr)$ is spherical Bessel function of order λ. A general solution of $\boldsymbol{A}(\boldsymbol{r},t)$ is a linear combination of both types of terms, with time dependence given by (5-29). Again we shall refer the reader to standard references, such as Morse and Feshbach (*Methods of Theoretical Physics*, McGraw-Hill, New York, 1953) for a demonstration that $\boldsymbol{A}_{\lambda\mu}(E\lambda, \boldsymbol{r})$ and $\boldsymbol{A}_{\lambda\mu}(M\lambda, \boldsymbol{r})$ satisfy (5-33).

Electromagnetic multipole transition operators. Using the operators given in (5-34), we can write the multipole (λ, μ) part of the perturbing Hamiltonian H' of (5-27) in the following forms:

$$O_{\lambda\mu}(E\lambda) = -\frac{i(2\lambda+1)!!}{ck^{\lambda+1}(\lambda+1)}\mathcal{J}(\boldsymbol{r})\cdot\nabla\times(\boldsymbol{r}\times\nabla)\big(j_\lambda(kr)Y_{\lambda\mu}(\theta,\phi)\big)$$

$$O_{\lambda\mu}(M\lambda) = -\frac{(2\lambda+1)!!}{ck^{\lambda}(\lambda+1)}\mathcal{J}(\boldsymbol{r})\cdot(\boldsymbol{r}\times\nabla)\big(j_\lambda(kr)Y_{\lambda\mu}(\theta,\phi)\big) \tag{5-35}$$

where $(2\lambda+1)!! = 1\cdot3\cdot5\cdots(2\lambda+1)$. The normalization used in the definitions of these operators are such that they reduce to those used for the static moments in (4-53) and (5-61) in the limit $k \to 0$. For the moment, we shall not be concerned with the multipole expansion of the nuclear current density $\mathcal{J}(\boldsymbol{r})$. Since both $O_{\lambda\mu}(E\lambda)$ and $O_{\lambda\mu}(M\lambda)$ are scalar operators in the combined nuclear and electromagnetic fields, only the $(\lambda, -\mu)$-multipole part of $\mathcal{J}(\boldsymbol{r})$ can make a nonvanishing contribution in the transition.

The spherical Bessel function may expanded in a power series,

$$j_\lambda(kr) \approx \frac{(kr)^\lambda}{(2\lambda+1)!!}\left(1 - \frac{1}{2}\frac{(kr)^2}{2\lambda+3} + \cdots\right) \tag{5-36}$$

The γ-rays involved in nuclear transitions have energies E_γ typically less than 10 MeV, corresponding to wave numbers of the order $k = E_\gamma/\hbar c \approx 1/20\ \text{fm}^{-1}$ or less. Since the multipole operators act on the nuclear wave function, they cannot have contributions coming from regions outside the nucleus. The dimension of a nucleus is characterized by the nuclear radius R, and even for a heavy nucleus, such as ^{208}Pb ($R = r_0A^{1/3} \sim 7$ fm), it is less than 10 fm. As a result, the dimensionless argument kr of the spherical Bessel function is usually much less than unity for typical γ-rays emitted in nuclear decays. The series (5-36) is, then, a fast convergent one and $j_\lambda(kr)$ may be approximated by its first term in the expansion alone. This is called the *long wave length* limit. Physically, it comes from the observation that for γ-rays of these energies, the wave length is $2\pi\hbar c/E_\gamma$, of the order of 10^2 fm, much larger than the nuclear dimension. As a result, these γ-rays cannot be sensitive to the details of the nuclear radial wave functions. Under such conditions, the expectation value of $j_\lambda(kr)$ is simply proportional to its leading order term $(kr)^\lambda$.

We are now in a position to calculate the contribution of each multipole order to the transition probability given in (5-20). Upon including the density of final states and using the multipole operator given in (5-35) for H', we can express the transition probability for multipole λ from an initial nuclear state $|\,J_iM_i\zeta\,\rangle$ to a final nuclear state $|\,J_fM_f\xi\,\rangle$ in the form

$$\mathcal{W}(\lambda; J_i\zeta \to J_f\xi) = \frac{8\pi(\lambda+1)}{\lambda[(2\lambda+1)!!]^2}\frac{k^{2\lambda+1}}{\hbar}B(\lambda; J_i\zeta \to J_f\xi) \tag{5-37}$$

where the reduced transition probability $B(\lambda; J_i\zeta \rightarrow J_f\xi)$ may be written in terms of the reduced matrix element of the multipole operator for either the electric or magnetic transition in the same way as done in (5-19),

$$B(\lambda; J_i\zeta \rightarrow J_f\xi) = \sum_{\mu M_f} \left|\langle J_f M_f \xi | O_{\lambda\mu} | J_i M_i \zeta\rangle\right|^2$$
$$= \frac{1}{2J_i+1} \left|\langle J_f \xi \| O_\lambda \| J_i \zeta\rangle\right|^2 \tag{5-38}$$

It is worth noting here that the reduced transition probabilities are dimensioned quantities. For electric transitions, $B(E\lambda)$ are in units of $e^2\text{fm}^{2\lambda}$ and for magnetic transitions, $B(M\lambda)$ are in units of $\mu_N^2\text{fm}^{2\lambda-2}$. The transition rate $\mathcal{W}$ is the number of decays per unit time. In relating the numerical values of $\mathcal{W}$ and $B(\lambda)$, one must be careful with the factors e^2 for $B(E\lambda)$ and μ_N^2 for $B(M\lambda)$. For example, the values in Table 5-1 are obtained using the following relations:

$$\mathcal{W}(\lambda) = \begin{cases} \alpha\hbar c \frac{8\pi(\lambda+1)}{\lambda[(2\lambda+1)!!]^2} \frac{1}{\hbar} \left(\frac{1}{\hbar c}\right)^{2\lambda+1} E_\gamma^{2\lambda+1}\, B(\lambda \text{ in } e^2\text{fm}^{2\lambda}) \\ \alpha\hbar c \left(\frac{\hbar c}{2M_p c^2}\right)^2 \frac{8\pi(\lambda+1)}{\lambda[(2\lambda+1)!!]^2} \frac{1}{\hbar} \left(\frac{1}{\hbar c}\right)^{2\lambda+1} E_\gamma^{2\lambda+1}\, B(\lambda \text{ in } \mu_N^2\text{fm}^{2\lambda-2}) \end{cases}$$

where we have used the relation $e^2 = \alpha\hbar c$ in cgs units to obtain the numerical values for e^2 and μ_N^2.

Table 5-1

Electromagnetic transition probability for the lowest four multipoles.

$\mathcal{W}(E1) = 1.59 \times 10^{15}\ E_\gamma^3\ B(E1)$	$\mathcal{W}(M1) = 1.76 \times 10^{13}\ E_\gamma^3\ B(M1)$
$\mathcal{W}(E2) = 1.23 \times 10^{9}\ E_\gamma^5\ B(E2)$	$\mathcal{W}(M2) = 1.35 \times 10^{7}\ E_\gamma^5\ B(M2)$
$\mathcal{W}(E3) = 5.71 \times 10^{2}\ E_\gamma^7\ B(E3)$	$\mathcal{W}(M3) = 6.31 \times 10^{0}\ E_\gamma^7\ B(M3)$
$\mathcal{W}(E4) = 1.70 \times 10^{-4}\ E_\gamma^9\ B(E4)$	$\mathcal{W}(M4) = 1.88 \times 10^{-6}\ E_\gamma^9\ B(M4)$

E_γ are in MeV, $B(E\lambda)$ in units of $e^2\text{fm}^{2\lambda}$ and $B(M\lambda)$ in units of $\mu_N^2\text{fm}^{(2\lambda-2)}$.

If we take as given that the electric charge in a nucleus consists of point charges carried by individual protons and the magnetization currents are due to the magnetic dipole moments of individual nucleons and the orbital motion of protons, the electric and magnetic multipole operators in the long wave length limit simplify to the following forms:

$$O_{\lambda\mu}(E\lambda) = \sum_{i=1}^{A} e(i) r_i^\lambda Y_{\lambda\mu}(\theta_i, \phi_i) \tag{5-39}$$

$$O_{\lambda\mu}(M\lambda) = \sum_{i=1}^{A}\Big\{g_s(i)\boldsymbol{s}_i + g_\ell(i)\frac{2\boldsymbol{\ell}_i}{\lambda+1}\Big\}\cdot\nabla_i\big(r_i^\lambda Y_{\lambda\mu}(\theta_i,\phi_i)\big)$$

$$= \sqrt{\lambda(2\lambda+1)}\sum_{i=1}^{A} r_i^{(\lambda-1)}\Big\{\Big(g_s(i) - \frac{2g_\ell(i)}{\lambda+1}\Big)\big(Y_{\lambda-1}(\theta_i,\phi_i)\times\boldsymbol{s}_i\big)_{\lambda\mu}$$

$$+\frac{2g_\ell(i)}{\lambda+1}\big(Y_{\lambda-1}(\theta_i,\phi_i)\times\boldsymbol{j}_i\big)_{\lambda\mu}\Big\} \qquad (5\text{-}40)$$

where $\boldsymbol{j}_i = \boldsymbol{\ell}_i + \boldsymbol{s}_i$ and

$$e(i) = \begin{cases} 1e \\ 0 \end{cases} \qquad g_\ell(i) = \begin{cases} 1\mu_N \\ 0 \end{cases} \qquad g_s(i) = \begin{cases} 5.586\mu_N & \text{for a proton} \\ -3.826\mu_N & \text{for a neutron,} \end{cases}$$

the same values as given in §4-7. The multiplication symbols here indicate angular momentum coupled products defined in (B-32). We have omitted the derivation from (5-35) to (5-37) and (5-39,40) since it involves a large number of angular momentum recoupling and properties of vector spherical harmonics. For a proper treatment, the reader is directed to references such as Blatt and Weisskopf (*Theoretical Nuclear Physics*, Wiley, New York, 1952). Note that for historical reasons the definitions of the operators for the static moments of a state differs from those for $O_{\lambda\mu}$ by constants, for example, as in the case of the quadrupole moment operator given in (4-54).

Dimensional check. Let us first make a dimensional analysis of (5-38) to ensure that the transition probability $\mathcal{W}(\lambda)$ has the correct units. From (5-1), it is obvious that $\mathcal{W}$ is measured in number of transitions per unit time interval. Let us examine first electric multipole transitions. Since the transition operator $O_{\lambda\mu}(E\lambda)$ given in (5-39) is proportional er^λ, the reduced transition probability $B(E\lambda)$ of multipolarity λ is in units of charge squared times length to the power 2λ. It is customary to measure charge in units of e, the magnitude of charge on a proton, and length in units of femtometer (fm). Electric multipole operators are therefore in units of $e\,\text{fm}^\lambda$. For magnetic transitions, we have the nuclear magneton μ_N in the place of electric charge e. The power of length is reduced by one compared with the corresponding electric transition of the same order because of the gradient operator. As a result, operator $O_{\lambda\mu}(M\lambda)$ in (5-40) is in units of $\mu_N\text{fm}^{\lambda-1}$. Following (5-38), the units for the reduced transition probabilities are

Electric multipole λ	$B(E\lambda)$:	$e^2\text{fm}^{2\lambda}$
Magnetic multipole λ	$B(M\lambda)$:	$\mu_N^2\text{fm}^{2\lambda-2}$

Note that since the units for reduced transition probabilities involve e^2 or μ_N^2, the explicit values of these two quantities in conventional units must be put into the expression in order to evaluate the transition probability $\mathcal{W}$.

Besides the reduced transition probability, the only other dimensioned quantities in the expression for $\mathcal{W}(\lambda)$ given in (5-37) is the ratio $k^{2\lambda+1}/\hbar$. Since k is in units of inverse length and $\hbar$ is in units of length multiplied by time, the units of electric transition probability is

$$\mathcal{W}(E\lambda): \quad \frac{[\mathrm{fm}^{-1}]^{2\lambda+1}}{[\mathrm{MeV\text{-}s}]} \times e^2[\mathrm{fm}]^{2\lambda}$$

Using the relation $e^2 = \alpha\hbar c$, and the fact that the fine structure constant α is a dimensionless quantity and $\hbar c$ is in units MeV-fm, we obtained the correct result

$$\mathcal{W}(E\lambda): \quad \frac{[\mathrm{fm}^{-1}]^{2\lambda+1}}{[\mathrm{MeV\text{-}s}]} \times [\mathrm{MeV\text{-}fm}][\mathrm{fm}]^{2\lambda} = [\mathrm{s}]^{-1}$$

as expected. For magnetic multipole transitions, we need only to examine the difference in the dimension between μ_N^2 and e^2. Since

$$\mu_N = \frac{e\hbar}{2M_p c}$$

in cgs units, the dimension of μ_N^2 is

$$\mu_N^2: \quad e^2[\mathrm{fm}]^2$$

From the fact that the units for $B(M\lambda)$ is $\mu_N^2\mathrm{fm}^{2\lambda-2}$, we see that $\mathcal{W}(M\lambda)$ also has the correct dimension of inverse time.

Selection rules. We stated earlier that, for γ-rays with energy of the order of a few MeV, transitions of different multipolarities are quite different in their rates. This comes from the energy dependence of $\mathcal{W}(\lambda)$ in the form $k^{2\lambda+1}$. Because of this, the ratio between two multipole transitions λ and $(\lambda+1)$ is

$$\mathcal{R} = \frac{\mathcal{W}(\lambda+1)}{\mathcal{W}(\lambda)} \sim (kr)^2 \tag{5-41}$$

We have included a length factor r in the calculation so as to make the ratio a dimensionless quantity. Since nuclear size is of the order of a few femtometers, we can take r to be 1 fm for the purpose of making an estimate. Using this, we obtain the result that, for a 1 MeV γ-ray, $\mathcal{R}$ is of the order 3×10^{-5}. A more precise calculation of the factor relating transition probabilities with reduced probabilities for both electric and magnetic transitions produces the results shown in Table 5-1. Note that the difference in numerical factor between electric and magnetic transitions of the same multipolarity is $(\hbar/2M_p c)^2$ fm^2, coming from the difference in the units for $B(E\lambda)$ and $B(M\lambda)$.

Because of the large reduction in the probability with increasing multipolarity order, the transition between an initial nuclear state with spin-parity $J_i^{\pi_i}$ and a final state $J_f^{\pi_f}$ is usually dominated by the lowest order allowed by angular momentum and parity selection rules. For transition of order λ, the spherical tensor rank of the operator is λ and carries λ units of angular momentum. As a result, the transition vanishes unless $\boldsymbol{J}_f = \boldsymbol{\lambda} + \boldsymbol{J}_i$. The angular momentum selection rule for the λ-th multipole electromagnetic transition is therefore given by the relation,

$$|J_f - J_i| \leq \lambda \leq J_f + J_i \tag{5-42}$$

The same condition is expressed also by the symbol $\Delta(J_f, \lambda, J_i)$ in (5-19) and is implicit in the reduced matrix element of (5-38). Together with the energy dependence in $\mathcal{W}(\lambda)$, we have the multipolarity selection rule that, for allowed values of λ,

$$\mathcal{W}(E\lambda) \gg \mathcal{W}(E(\lambda+1)) \qquad \mathcal{W}(M\lambda) \gg \mathcal{W}(M(\lambda+1)) \tag{5-43}$$

We shall not make a comparison between electric and magnetic transitions until later, since the nature of these two operators is quite different.

The operator for an $E\lambda$-transition is proportional to spherical harmonics $Y_{\lambda\mu}(\theta, \phi)$, as can be seen from (5-39). Under an inversion of the coordinate system, the transformation property of $\boldsymbol{O}_{\lambda\mu}(E\lambda)$ follows that for spherical harmonics of order (λ, μ). As a result, we obtain the transformation property under parity for $E\lambda$ transition in the form

$$\boldsymbol{O}_{\lambda\mu}(E\lambda) \xrightarrow[p]{} (-1)^{\lambda} \boldsymbol{O}_{\lambda\mu}(E\lambda) \tag{5-44}$$

The magnetic operator, on the other hand, involves $\nabla(r^{\lambda} Y_{\lambda\mu}(\theta, \phi))$. The addition of the ∇ operator introduces an extra minus sign under a parity transformation, and the net result is

$$\boldsymbol{O}_{\lambda\mu}(M\lambda) \xrightarrow[p]{} (-1)^{\lambda+1} \boldsymbol{O}_{\lambda\mu}(M\lambda) \tag{5-45}$$

Eqs. (5-44,45) give the parity selection rule,

$$\pi_i \pi_f = (-1)^{\lambda} \quad \text{for} \quad E\lambda \qquad \pi_i \pi_f = (-1)^{\lambda+1} \quad \text{for} \quad M\lambda \tag{5-46}$$

where π_i and π_f are, respectively, the parities of the initial and final states. Because of this selection rule, $E\lambda$ and $M\lambda$ transitions of the same multipolarity cannot occur between the same pair of nuclear states. For example, in a $2^+ \to 0^+$ transition, only $E2$ can take place, whereas in a $2^- \to 0^+$ transition, only $M2$ is allowed.

The difference in the nature of electric and magnetic transition operators also plays a role in determining the dominant mode of transition between nuclear states. However, the selection rule is not an exact one. In general we find that magnetic transitions are weaker than electric transitions. It is not easy to make a direct comparison here, since electric and magnetic transitions of the same multipolarity cannot occur between the same pair of nuclear states as a result of the parity selection rule above. The comparison may be made instead in the following way. For a given pair of nuclear states, if both $E\lambda$ and $M(\lambda+1)$ are allowed by angular momentum and parity selection rules, the $E\lambda$ mode usually dominates the transition by a large factor. On the other hand, if both $M(\lambda)$ and $E(\lambda+1)$

transitions are allowed, the higher multipole order electric transition may be competitive in terms of transition rates with the magnetic transition in spite of the hindrance factor due to energy dependence. In fact, such a mixture often occurs between a pair of states, such as in the case of 2^+ to 1^+ transition where both $M1$ and $E2$ transitions are allowed. The mixture is characterized by the *mixing ratio* δ, defined by the relation

$$\delta^2 = \frac{\mathcal{W}(E(\lambda+1); J_i\zeta \to J_f\xi)}{\mathcal{W}(M\lambda; J_i\zeta \to J_f\xi)} \tag{5-47}$$

The sign of δ is given by the relative sign of the reduced matrix elements of the two transitions where such a sign can be determined.

The reason for the dominance of electric transitions over the magnetic transitions may be inferred from the fact that the operators for magnetic transition differ from those for electric transitions by a gradient ∇ which, in general, reduces the size of a matrix element. Similar differences are also found in other electromagnetic processes.

Internal conversion and internal pair production. Besides γ-ray emission, electromagnetic decay of an excited nuclear state can also be accomplished through internal conversion and internal pair production processes. In *internal conversion*, an atomic electron is ejected in the place of a γ-ray being emitted from the nucleus. The kinetic energy of the electron is equal to the de-excitation energy $E_i - E_f (= E_\gamma)$ minus the (atomic) binding energy of the electron. As a result, electrons emitted from internal conversion processes have discrete energies and can therefore be distinguished from the continuous spectrum of electrons emitted in β^--decays to be discussed later in §5-5. Since both types of decay can take place from the same excited state in medium and heavy nuclei, the difference makes it possible to distinguish between them.

Internal conversion may be visualized in the following way. When a nucleus de-excites, say, either by a nucleon jumping from one single-particle orbit to another or by a change of the angular momentum of the nucleus as a whole, a sudden disturbance is sent to the surrounding electromagnetic field. Atomic electrons, especially those in the innermost orbits, such as the K- and L-orbits, spend a large fraction of the time in the vicinity of the nucleus, the source of the electromagnetic field of interest here. It is therefore probable for the disturbance in the electromagnetic field to transfer all the excess energy in the nucleus to one of the electrons and eject it from the atomic orbit. This is similar to the atomic *Auger* effect where, instead of emitting a photon when an atomic electron de-excites from a higher to a lower energy orbit, one of the atomic electrons is ejected.

Internal conversion is important in heavier nuclei for two reasons. In the first place, the radii of atomic electron orbits are smaller because of the stronger Coulomb fields provided by heavy nuclei. The probability of transition is increased as a result of the large overlap between the wave functions of the nucleus and the inner shell atomic electrons. For this reason, the electrons ejected come mainly from the innermost shells. In the second place, the stronger Coulomb field in heavy nuclei exerts a larger influence on the surrounding. For these reasons, the importance of internal conversion increases roughly as Z^3 and becomes competitive with γ-ray emission for medium and heavy nuclei.

In *internal pair production*, an electron-positron pair is emitted in the place of a γ-ray when an excited nucleus decay through electromagnetic processes. As long as the energy of decay is greater than $2 \times m_e c^2 \approx 1.02$ MeV, pair production is possible in principle. However, the process is not an efficient one and is usually several orders of magnitude retarded compared with allowed γ-ray emissions. Pair production therefore becomes important only when γ-ray emission is forbidden. For example, a $0^+ \rightarrow 0^+$ transition is not allowed by γ-ray emission, as a γ-ray must carry away at least one unit of angular momentum. In such cases, pair production (and internal conversion for heavy nuclei) becomes the dominant mode of the decay.

The inverse of γ-ray emission is Coulomb excitation. Here, the nucleus is excited to a higher state as a result of changes in the surrounding electromagnetic field. This can take place, for example, as the result of a charged particle passing nearby. We shall see in §7-1 that the nuclear transition matrix elements involved are identical to the γ-ray transitions we have been discussing here.

§5-3 SINGLE-PARTICLE VALUE

It is useful to have an estimate of the reduced transition probability between a pair of states by making a few reasonable assumptions that can greatly simplify the calculations involved. There are also two further purposes for having such estimates. Up to now, we have decomposed electromagnetic transition rates $\mathcal{W}$ in nuclei according to multipoles and expressed each order λ in terms of reduced transition probabilities $B(\lambda)$. Our first goal in this section is to make an estimate of the sizes of $B(E\lambda)$ and $B(M\lambda)$ that can be expected on the average. A second motivation is to provide a basis with which we can compare observed and calculated values. As we shall see later, this is an important function. Since transition probability $\mathcal{W}(\lambda)$ itself is dominated by the energy-dependent factor $k^{2\lambda+1}$, it is difficult to obtain an idea of the size of the nuclear matrix element involved directly by looking at the numerical value of $\mathcal{W}(\lambda)$. The reduced transition probabilities $B(\lambda)$ are more intimately related to the nuclear transition matrix elements. However, because of the way the transition operators are defined, it is difficult to visualize whether a given value for a particular multipole is large or small on physical grounds. This is made even more complicated in practice when observed values for certain multipoles can differ from each other in a nucleus by several orders of magnitude — as, for example, in the case of $E2$-transitions.

$E\lambda$-transition. It is a common practice to compare measured and calculated transition rates with the Weisskopf estimates. In fact, these estimates are so widely used that they are often regarded as the "standard," or units, for measuring transition rates against.

A calculation of the reduced transition probability requires a knowledge of both the initial and final wave functions. As the first step toward establishing an average, we shall make some assumptions about the wave functions so that a reasonable estimate may be made without reference to the specific states involved in the transition. Again, we shall adopt an extreme independent-particle picture and consider nuclear transitions to be taking place by a nucleon moving from one single-particle orbit to another without affecting the rest of the nucleons in the nucleus. In the case of $E\lambda$-transitions, this means that a proton moves from an initial single-particle state $|j_i m_i\rangle$ to a final one $|j_f m_f\rangle$. In this limit, the matrix element of $\boldsymbol{O}_{\lambda\mu}(E\lambda)$ between many-body nuclear wave functions $|J_i M_i \zeta\rangle$ and $|J_f M_f \xi\rangle$ reduces to a single-particle matrix element,

$$\langle J_f M_f \xi | \sum_{k=1}^{A} e(k) r_k^\lambda Y_{\lambda\mu}(\theta_k, \phi_k) | J_i M_i \zeta \rangle = \langle j_f m_f | e r^\lambda Y_{\lambda\mu}(\theta, \phi) | j_i m_i \rangle \tag{5-48}$$

where we have made use of the explicit form of $\boldsymbol{O}_{\lambda\mu}(E\lambda)$ given in (5-39) to express the $E\lambda$-transition operator as a sum of single-particle operators.

In cases where the wave function of each nucleon has a definite orbital angular momentum ℓ, a single-particle wave function may be decomposed into a product of three parts: a radial wave function $R_{n\ell}(r)$, an orbital angular momentum part given by spherical harmonics $Y_{\ell m}(\theta, \phi)$, and an intrinsic spin part $\chi_{1/2}$. By coupling $Y_{\ell m}(\theta, \phi)$ with $\chi_{1/2}$ to angular momentum (j, m), a single-particle wave function may be expressed in the form

$$|jm\rangle = R_{n\ell}(r) \big(Y_\ell(\theta, \phi) \times \chi_{1/2} \big)_{jm} \tag{5-49}$$

where n is the principal quantum number. The single-particle matrix element on the right hand side of (5-48) can now be written in terms of a product of a radial integral and a matrix element in angular momentum space,

$$\begin{aligned} \langle j_f m_f | e r^\lambda Y_{\lambda\mu} | j_i m_i \rangle = & \int_0^\infty R^*_{n_f \ell_f}(r)\, r^\lambda R_{n_i \ell_i}(r)\, r^2\, dr \\ & \big\langle \big(Y_{\ell_f}(\theta, \phi) \times \chi_{1/2} \big)_{j_f m_f} \big| Y_{\lambda\mu}(\theta, \phi) \big| \big(Y_{\ell_i}(\theta, \phi) \times \chi_{1/2} \big)_{j_i m_i} \big\rangle \end{aligned} \tag{5-50}$$

We shall next make some further simplifying assumptions so that both factors on the right hand side may be evaluated without having to specify the particular single-particle states involved.

All the radial dependence of (5-50) is contained in the integral,

$$\langle r^\lambda \rangle = \int R^*_{n_f \ell_f}(r)\, r^\lambda R_{n_i \ell_i}(r)\, r^2\, dr$$

where $R_{n_f\ell_f}(r)$ and $R_{n_i\ell_i}(r)$ are, respectively, the normalized radial parts of the initial and final single-particle wave functions. The exact value of the integral depends on the shapes of the radial dependence. However, to a first approximation, it is determined by the power λ and the size of the nucleus. For the purpose of an estimate, we can simplify the situation greatly by assuming that the nucleus is a sphere of uniform density with the radius given by $R = r_0 A^{1/3}$. In this approximation, the radial integral reduces to

$$\langle r^\lambda \rangle = \frac{3}{\lambda+3} r_0^\lambda A^{\lambda/3} \tag{5-51}$$

For r_0, we can use the value of 1.2 fm from electron scattering. (See Problem 5-5 for values obtained with the more realistic harmonic oscillator radial wave functions.)

The estimate for the reduced transition probability of an $E\lambda$-transition now takes on the form,

$$\begin{aligned} B_{\text{est.}}(E\lambda) &= \sum_{\mu M_f} |\langle J_f M_f \xi | O_{\lambda\mu}(E\lambda) | J_i M_i \zeta \rangle|^2 \\ &= e^2 \langle r^\lambda \rangle^2 \sum_{m_f \mu} \langle (Y_{\ell_f}(\theta,\phi) \times \chi_{1/2})_{j_f m_f} | Y_{\lambda\mu}(\theta,\phi) | (Y_{\ell_i}(\theta,\phi) \times \chi_{1/2})_{j_i m_i} \rangle^2 \end{aligned} \tag{5-52}$$

The remaining matrix element involves only angular momentum and can therefore be evaluated using standard techniques of spherical tensors. However, there is very little point to doing this. Since the total solid angle about a point is 4π steradian, an average of any angular dependence must be near the value $1/4\pi$. Hence, for the purpose of an estimate, we can take

$$B_{\text{est.}}(E\lambda) = e^2 \langle r^\lambda \rangle^2 \frac{1}{4\pi}$$

On substituting the value of the radial integral given in (5-51) and using 1.2 fm for the value of r_0, we obtain the *Weisskopf single-particle estimate* for the λ-th multipole reduced electric transition probability,

$$B_{\text{W}}(E\lambda) = \frac{1}{4\pi} \left(\frac{3}{\lambda+3} \right)^2 (1.2)^{2\lambda} A^{2\lambda/3} e^2 \text{fm}^{2\lambda} \tag{5-53}$$

As mentioned earlier, this value is often used as the unit for $E\lambda$-transition and is also called the Weisskopf unit for reduced transition probability for this reason.

***M*λ-transition.** Estimates for magnetic transitions are slightly more complicated since we have contributions from both nucleon intrinsic spin and proton orbital motion. We may proceed essentially along the same line as we have done above for electric transitions. This involves the adoption of an extreme single-particle model and making use of the last form of $\boldsymbol{O}_{\lambda\mu}(M\lambda)$ given in (5-40) to reduce the nuclear matrix element to a single-particle matrix element. Parallel with (5-52), we have the expression

$$\begin{aligned}B_{\text{est.}}(M\lambda) &= \sum_{\mu M_f} |\langle J_f M_f \xi | \boldsymbol{O}_{\lambda\mu}(E\lambda) | J_i M_i \zeta\rangle|^2 \\ &= \lambda(2\lambda+1)\langle r^{(\lambda-1)}\rangle^2 \sum_{m_f \mu} \Big\{ \Big(g_s - \frac{2g_\ell}{\lambda+1}\Big) \\ &\times \langle (Y_{\ell_f}(\theta,\phi)\times\chi_{1/2})_{j_f m_f} | (Y_{(\lambda-1)}(\theta,\phi)\times \boldsymbol{s})_{\lambda\mu} | (Y_{\ell_i}(\theta,\phi)\times\chi_{1/2})_{j_i m_i}\rangle \\ &+ \frac{2g_\ell}{\lambda+1} \langle (Y_{\ell_f}(\theta,\phi)\times\chi_{1/2})_{j_f m_f} | (Y_{(\lambda-1)}(\theta,\phi)\times \boldsymbol{j})_{\lambda\mu} | (Y_{\ell_i}(\theta,\phi)\times\chi_{1/2})_{j_i m_i}\rangle \Big\}^2\end{aligned} \tag{5-54}$$

Again, we shall take the nucleus to be a constant density sphere of radius R and average over the angular dependence. The result of the radial integral may be taken to be

$$\langle r^{(\lambda-1)}\rangle = \frac{3}{\lambda+3} r_0^{(\lambda-1)} A^{(\lambda-1)/3}$$

the same as (5-51). (Strictly speaking, the factor in front should instead be $3/(\lambda+2)$ but is kept in the form given to conform with that for $E\lambda$-transitions.)

For the purpose of an estimate, it is adequate to evaluate either one of the two terms inside the curly brackets in (5-54) and multiply the result by two (before taking the square). Factors related to the gyromagnetic ratios in the first term may be replaced by a reasonable average value, and this is generally taken to be

$$\lambda(2\lambda+1)\Big(g_s - \frac{2g_\ell}{\lambda+1}\Big)^2 \approx 10$$

For the average of the square of the angular part, we can use again take the value $1/4\pi$ used earlier for $E\lambda$-transitions. The final form of the Weisskopf estimate for reduced magnetic multipole transition probability is given by

$$B_{\text{W}}(M\lambda) = \frac{10}{\pi}\Big(\frac{3}{\lambda+3}\Big)^2 (1.2)^{2\lambda-2} A^{(2\lambda-2)/3} \mu_N^2 \text{fm}^{2\lambda-2} \tag{5-55}$$

The results of (5-52) and (5-55) may be substituted into (5-37) to produce the Weisskopf units for transition probability

$$\begin{aligned}\mathcal{W}_{\text{W}}(E\lambda) &= \alpha\hbar c \frac{8\pi(\lambda+1)}{\lambda[(2\lambda+1)!!]^2} \frac{k^{2\lambda+1}}{\hbar} \frac{1}{4\pi} \Big(\frac{3}{\lambda+3}\Big)^2 (1.2)^{2\lambda} A^{2\lambda/3} \\ \mathcal{W}_{\text{W}}(M\lambda) &= \alpha\hbar c \Big(\frac{\hbar}{2M_p c}\Big)^2 \frac{8\pi(\lambda+1)}{\lambda[(2\lambda+1)!!]^2} \frac{k^{2\lambda+1}}{\hbar} \frac{10}{\pi} \Big(\frac{3}{\lambda+3}\Big)^2 (1.2)^{2\lambda-2} A^{(2\lambda-2)/3}\end{aligned} \tag{5-56}$$

Explicit values in terms of nucleon number A and transition energy E_γ in MeV are listed in Table 5-2.

Table 5-2

Weisskopf single-particle estimates for $E\lambda$- and $M\lambda$-transition probabilities and widths.

Multipole λ	$E\lambda$ $\mathcal{W}(s^{-1})$	$E\lambda$ Γ(MeV)		$M\lambda$ $\mathcal{W}(s^{-1})$	$M\lambda$ Γ(MeV)	
1	1.02×10^{14}	6.75×10^{-8}	$A^{2/3}E_\gamma^3$	3.15×10^{13}	2.07×10^{-8}	E_γ^3
2	7.28×10^{7}	4.79×10^{-14}	$A^{4/3}E_\gamma^5$	2.24×10^{7}	1.47×10^{-14}	$A^{2/3}E_\gamma^5$
3	3.39×10	2.23×10^{-20}	$A^{2}E_\gamma^7$	1.04×10	6.85×10^{-21}	$A^{4/3}E_\gamma^7$
4	1.07×10^{-5}	7.02×10^{-27}	$A^{8/3}E_\gamma^9$	3.27×10^{-6}	2.16×10^{-27}	$A^{2}E_\gamma^9$
5	2.40×10^{-12}	1.58×10^{-33}	$A^{10/3}E_\gamma^{11}$	7.36×10^{-13}	4.84×10^{-34}	$A^{8/3}E_\gamma^{11}$

E_γ in MeV. The E_γ- and A-dependent factors are common to both $\mathcal{W}$ and Γ.

In terms of Weisskopf units, the measured reduced rates have been observed to vary by orders of magnitude in different nuclei and sometimes even for transitions within the same nucleus. This shows the richness of physics contained in electromagnetic transitions between nuclear states. In order for a transition to be enhanced by an order of magnitude or more over the single-particle values we have estimated, many nucleons must be acting together in a coherent manner. As we shall see in Chapter 6, this leads to the concept of collective motion in the form of nuclear vibrations and rotations.

§5-4 WEAK INTERACTION AND BETA DECAY

Nuclear β-decay is a facet of the weak interaction. Besides transitions between nuclear states, a variety of other phenomena involving both hadrons and leptons share the same origin. In most cases, these weak processes cannot be observed, since they are slower by several orders of magnitude compared with competing reactions induced by electromagnetic and strong interactions. As a result, studies of weak interactions must be made in cases where these faster processes are either forbidden or hindered by selection rules.

The basic weak reaction in nuclei may be characterized by the decay of a free neutron and a bound proton,

$$n \rightarrow p + e^- + \bar{\nu}_e \tag{5-57}$$

$$p_{\text{bound}} \rightarrow n + e^+ + \nu_e \tag{5-58}$$

introduced earlier in (2-1) and (2-2). These transitions are examples of a general class of decay taking place in other hadrons as well. For example,

$$\begin{aligned}\pi^+ &\rightarrow \mu^+ + \nu_\mu \\ &\rightarrow e^+ + \nu_e \end{aligned} \tag{5-59}$$

$$\begin{aligned}\pi^- &\rightarrow \mu^- + \overline{\nu}_\mu \\ &\rightarrow e^- + \overline{\nu}_e \end{aligned} \tag{5-60}$$

$$\Sigma^- \rightarrow n + \pi^- \tag{5-61}$$

$$\begin{aligned}K^+ &\rightarrow \pi^+ + \pi^0 \\ &\rightarrow \pi^+ + \pi^- + \pi^+ \end{aligned} \tag{5-62}$$

Reactions such as those given by (5-57,58) and (5-59,60) are referred to as semi-leptonic processes, since both hadrons and leptons are involved. Examples given by (5-61,62) are non-leptonic processes, as leptons do not enter either in the initial or the final states. There are also purely leptonic processes, such as the decay of muons,

$$\mu^+ \rightarrow e^+ + \nu_e + \nu_\mu \tag{5-63}$$

that are also an integral part of weak interaction processes. Our main concern will be the semi-leptonic modes, as nuclear β-decay is a part of them.

Universal weak interaction. Weak interaction processes are often said to be *universal* since the strength of the basic interaction is the same for all three different types of reactions described in the previous paragraph. This point is illustrated by the fact that the coupling constant G_F, generally known as the Fermi coupling constant, has the same value

$$G_F = 1.43584(3) \times 10^{-62}\ \text{J-m}^3 = 1.16637(2) \times 10^{-11} (\hbar c)^3\ \text{MeV}^{-2} \tag{5-64}$$

regardless of whether it is measured through superallowed β-decay in nuclei (see next section and Problem 5-2), from the decay of muons shown in (5-64) or other weak interaction processes.

Weak interactions are mediated by vector bosons $W^\pm$ and Z^0, in the same way as electromagnetic interactions are carried by photons. However, because of their large masses,

$$m_W c^2 = 80.9 \pm 1.4\ \text{GeV} \qquad\qquad m_Z c^2 = 91.9 \pm 1.8\ \text{GeV}$$

the range of weak interactions is extremely short ($r_0 = \hbar/mc \sim 10^{-3}$ fm), about three orders of magnitude smaller than the long-range part of nuclear force. For

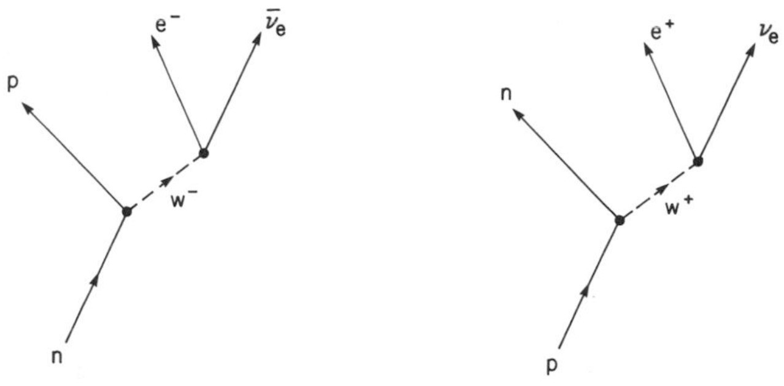

Fig. 5-1 Diagram showing the β^--decay of a neutron into a proton through the emission of a W^--boson, and the β^+-decay of a proton into a neutron through the emission of a W^+-boson. In both cases, the W-boson decays into a pair of leptons, an electron and an anti-electron neutrino in the former case, and a positron and an electron neutrino in the latter case.

this reason, weak interactions for many purposes may be considered as a zero-range or "contact" interactions.

Since bosons $W^\pm$ carry net charges, they can change the charge state of a particle as, for example, in the reactions given in (5-57,58). Pictorially, these reactions may be represented by the diagrams shown in Fig. 5-1. Most of the weak decay processes are mediated by the charged bosons, as illustrated by the examples given in (5-59) to (5-63). The neutral boson Z^0 is the source of *neutral weak current* and is responsible for reactions such as neutrino-electron scattering:

$$\nu + e^- \rightarrow \nu + e^- \tag{5-65}$$

In spite of its small cross section, such processes are important, for example, in trapping the energy inside the outer shell of a supernova, thus preventing large amounts of energy being carried away by the neutrinos immediately after the explosion.

On a more fundamental level, β-decay of hadrons may be viewed as the transformation of one type of quark to another through the exchange of charged weak currents. As we have seen in Chapter 2, the flavor of quarks is conserved in strong interactions. However, through weak interactions, it is possible for quarks to change flavor, for example, by transforming from a d-quark to a u-quark,

$$d \rightarrow u + e^- + \overline{\nu}_e \tag{5-66}$$

This is what takes place in the β^--decay of a neutron; in terms of quarks, (5-57) may be written in the form

$$(udd) \rightarrow (uud) + e^- + \overline{\nu}_e$$

Similarly, the β^+-decay of a bound proton to a neutron involves the transformation of a u-quark to a d-quark,

$$u \rightarrow d + e^+ + \nu_e \tag{5-67}$$

Diagrammatically, the processes given by (5-66,67) may be represented by Fig. 5-2(a) and (b). The other weak transitions given in (5-59) to (5-63) are represented by Fig. 5-2(c) to (e).

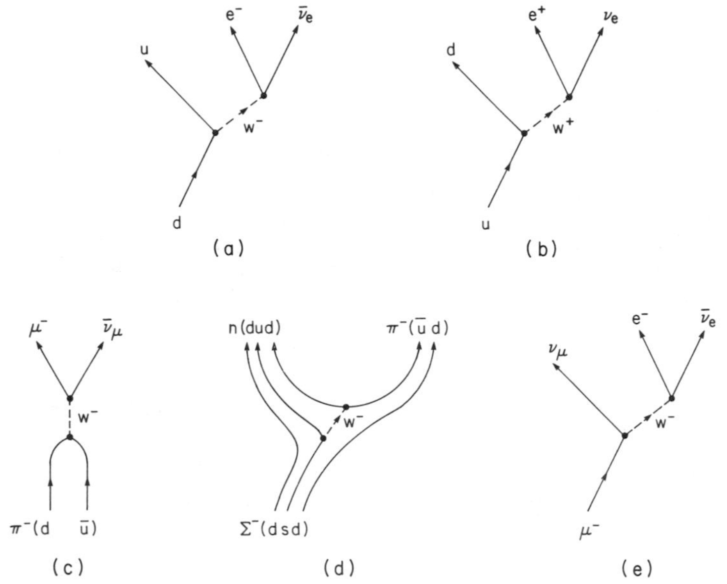

Fig. 5-2 Weak decay of quarks. In diagram (a) a d-quark becomes a u-quark by emitting a W^--boson in the same way as a neutron decays to a proton. In diagram (b), a u-quark is changed into a d-quark in the same way as a bound proton decays to a neutron. Diagram (c) is an example of weak decay of a hadron into final states involving only leptons as shown in (5-60). Diagram (d) is an example of non-leptonic weak decay shown in (5-61) and diagram (e) is an example of purely leptonic decay given by (5-63).

When a quark decays, it does not necessarily have to result in a quark of definite flavor. For the simple case of weak decay among four quarks, u, d, s, and c, the flavor mixing in the decay product may be expressed in terms of a single parameter, the Cabibbo angle θ_c,

$$\begin{aligned} u &\longrightarrow d' = d\cos\theta_c + s\sin\theta_c \\ c &\longrightarrow s' = -d\sin\theta_c + s\cos\theta_c \end{aligned} \tag{5-68}$$

This is reminiscent of what we had in §2-7 on SU_3(flavor) symmetry mixing. There the observed pairs of neutral mesons, (η, η') and (ρ, ω), are mixtures of the SU_3(flavor) symmetry conserving pairs (η_0, η_8) and (ϕ_0, ϕ_8), respectively.

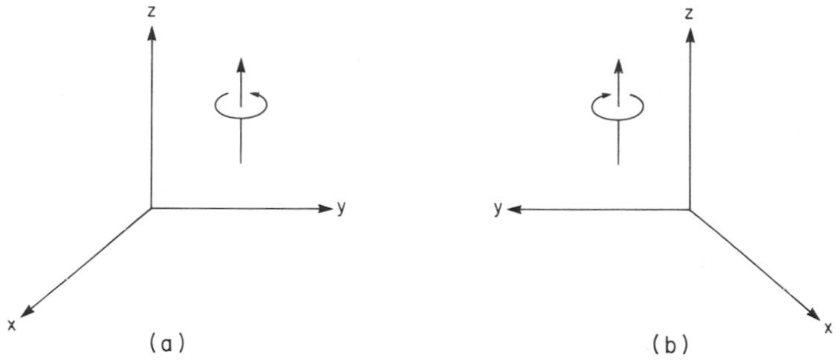

Fig. 5-3 Parity and direction of spin. Under a parity transformation, an object is changed into its mirror image. As a result, a right-handed rotation in (a) is changed into a left-handed rotation in (b). The change in the coordinate axes shown is accomplished by the operation $(x, y, z) \rightarrow (-x, -y, -z)$ followed a rotation of 180° around the y-axis.

The observed weak transitions are, however, between quarks of definite flavor; for example, u- and d-quarks as in the decay of a neutron to a proton. The relation given in (5-68) implies that the observed β-decay strength in reactions is not the fundamental weak interaction coupling constant G_F itself, but a value modified by the mixing angle.

It is customary to express the transformation (5-68) in the form of a charged current,

$$\mathcal{J}^{+}_{\text{weak}} = (\overline{u}\ \overline{c}) \begin{pmatrix} \cos\theta_c & \sin\theta_c \\ -\sin\theta_c & \cos\theta_c \end{pmatrix} \begin{pmatrix} d \\ s \end{pmatrix} \tag{5-69}$$

In terms of such a current, the more general case involving all six quarks may be written in the following form:

$$\mathcal{J}^{+}_{\text{weak}} = (\overline{u}\ \overline{c}\ \overline{t}) \begin{pmatrix} M_{11} & M_{12} & M_{13} \\ M_{21} & M_{22} & M_{23} \\ M_{31} & M_{32} & M_{33} \end{pmatrix} \begin{pmatrix} d \\ s \\ b \end{pmatrix} \tag{5-70}$$

where the 3×3 matrix is known as the Kobayashi-Maskawa matrix. The nine matrix elements are functions of three mixing angles and a phase factor. A complete determination of all the independent matrix elements involve weak decays of heavy quarks as well.

For nuclear β-decay, we are solely concerned with the transition between u- and d-quarks. As a result, only the product between the Fermi coupling constant G_F and $\cos\theta_c$ enters into the process. The mixing angle is sufficiently small that we can ignore it for most of our purposes. In order to simplify the notation and to avoid any possible confusion, we shall use the symbol G_V, the vector coupling constant, to represent the product and omit the explicit presence of the mixing angle. However, we must be aware of the difference, for example, when we compare the value of G_F with the measured strength of weak decays in nuclei, as is done in Problem 5-2.

Parity nonconservation. One of the more remarkable properties of weak decay is that parity is not conserved. It is important that we have an understanding of this point, since it is closely linked to the form of the nuclear β-decay operator. As described in §A-1, parity transformation is the operation which inverts the spatial coordinates,

$$(x, y, z) \xrightarrow[P]{} (-x, -y, -z) \tag{A-1}$$

It is often described in terms of taking a mirror image of the coordinate system, as can be seen from Fig. 5-3. Under a parity operation a usual scalar (S) is unchanged, but a usual vector (V) changes sign. In order to distinguish from axial vectors to be defined next, the usual vectors are called *polar vectors* where such a distinction is necessary. Examples of polar vectors are spatial location $\boldsymbol{r}$ and momentum $\boldsymbol{p}$.

We can also construct vectors that do not change sign under a parity transformation. For example, the angular momentum vector, $\boldsymbol{\ell}\hbar = \boldsymbol{r} \times \boldsymbol{p}$, does not change sign under a parity operation, since both $\boldsymbol{r}$ and $\boldsymbol{p}$ reverse signs. Vectors that do not change sign under an inversion of the coordinate system are called *axial vectors* (A). All angular momentum operators, including intrinsic spin operators, are axial vectors. The scalar product of an axial vector and a polar vector is a scalar that changes sign under a parity operation; such scalars are called *pseudoscalars* (P). There is also a fifth category of quantities called *tensors* (not to be confused with spherical tensors we use for angular momentum algebra) that behave differently from S, V, A, and P under a parity transformation, but we shall not be concerned with them here.

An operator made of a mixture of scalars and pseudoscalars, or a mixture of vectors and axial vectors, does not have a definite parity and, as a result, parity is not conserved under its action. For strong and electromagnetic interactions, parity is strictly conserved; that is, all processes are invariant under spatial inversion. However, this is not true for weak interaction processes.

The suspicion of parity violation in weak interaction originated from observations made of the decay of K^+-meson. Two different modes have been observed, one having two pions and the other having three pions in the final state, as shown in (5-62). Since these two modes have different parities, parity nonconservation in weak decays was proposed as the possible resolution. The confirmation of this suggestion came from observing the β^--decay of ^{60}Co (C.S. Wu et al., *Phys. Rev.* **105** [1957] 1413),

$$^{60}\text{Co} \rightarrow {}^{60}\text{Ni} + e^- + \overline{\nu}_e$$

The ground state of the odd-odd nucleus $^{60}_{27}$Co has spin-parity $J^\pi = 5^+$, as shown in Fig. 5-4. A nonzero spin makes it possible for the nucleus to be polarized along a magnetic field applied externally. The ground state decays predominantly (99%

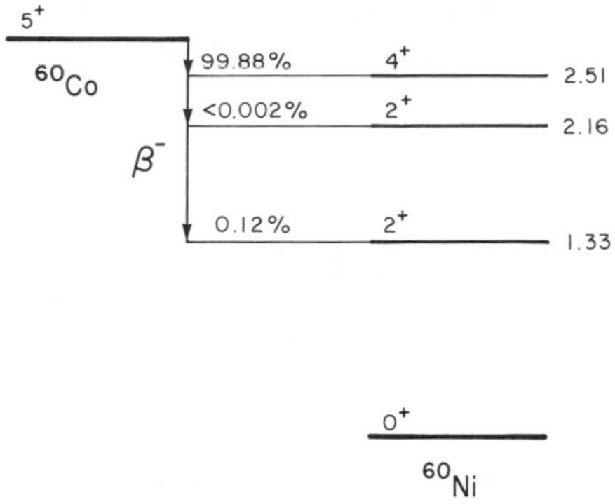

Fig. 5-4 Decay scheme of ^{60}Co to ^{60}Ni. The main branch leading to the 4^+-state at 2.51 MeV in ^{60}Ni is a Gamow-Teller transition used in one of the first experiments to demonstrate parity nonconservation in β-decay.

of the time) to the $J^\pi = 4^+$ state of $^{60}_{28}$Ni at excitation energy 2.5 MeV. The decay is purely of the Gamow-Teller type (see next section).

If the spins of all the ^{60}Co nuclei are aligned, we have a fixed direction in space that is defined in a natural way by the experimental setup. This direction may be indicated by a unit vector $\boldsymbol{\sigma}$, parallel to the alignment of the ^{60}Co ground state spin $\boldsymbol{J}$. The angular distribution of electrons emitted with momentum $\boldsymbol{p}$ and energy E may be expressed in the following form (for a derivation, see, *e.g.,* M. Morita, *Beta Decay and Muon Capture* Benjamin, Reading, Massachusetts, 1973, p. 67; J.M. Eisenberg and W. Greiner, *Excitation Mechanisms of the Nucleus*, North Holland, Amsterdam, 1970, p. 290),

$$W(\theta) = 1 + a\frac{\boldsymbol{\sigma}\cdot\boldsymbol{p}c}{E} = 1 + a\frac{v}{c}\cos\theta \tag{5-71}$$

where θ is the angle with respect to $\boldsymbol{J}$ through which the electron is emitted, E is the total (relativistic) energy of the electron, and the parameter a indicates the intensity of angular dependence. Under a parity operation $\boldsymbol{p}$, being a polar vector, changes sign and $\boldsymbol{\sigma}$, being a axial vector, does not change sign. The product $\boldsymbol{\sigma}\cdot\boldsymbol{p}$ in the second term of (5-71) is then a pseudoscalar and changes sign under an inversion of the spatial coordinate system. On the other hand, the first term (unity), being a scalar, remains invariant. If parity is conserved in the decay, the second term of (5-71) must vanish on account of the fact that $\boldsymbol{\sigma}\cdot\boldsymbol{p} \xrightarrow[P]{} -\boldsymbol{\sigma}\cdot\boldsymbol{p}$. As a result, we expect $a = 0$ and the angular distribution of electrons emitted to be isotropic. Experimentally, a turned out to be -1, indicating a maximum parity violation. The same conclusion is later confirmed by other measurements such as the decay of pions and muons.

The result $a = -1$ may also be examined from the point of view of the helicities of the leptons involved. The *helicity* of a particle is defined as the projection of $\boldsymbol{\sigma}$, twice of the intrinsic spin $\boldsymbol{s}$, along its direction of motion

$$h = \frac{\boldsymbol{\sigma}\cdot\boldsymbol{p}}{|\boldsymbol{p}|} \tag{5-72}$$

For a massless particle, the eigenvalues of h can only be ± 1. An example of a particle with only two possible orientations is provided by the photon which, as we have seen earlier, can only have two linearly independent transverse polarization directions. For electrons and other particles with nonzero rest mass, the helicity takes on values $\pm v/c$. Particles with positive helicity are often referred to as "right-handed" particles and negative helicity particles are often called "left-handed" particles. If the neutrinos are massless, they will behave in ways similar to photons and h can have values ± 1. Experimentally the helicity of neutrino was first determined by Goldhaber, Grodzins, and Sunyan (*Phys. Rev.* **109** [1958] 1015) through electron capture (see next section) in the 0^+ ground state of ^{152}Eu leading to the 1^- excited state of ^{152}Sm at 963 keV,

$$e^- + {}^{152}\text{Eu} \longrightarrow \nu_e + {}^{152}\text{Sm}^*(1^-) \longrightarrow \nu_e + {}^{152}\text{Sm}(0^+) + \gamma$$

By measuring the polarization of the γ-ray emerging from the decay of ^{152}Sm to its 0^+ ground state, the helicity of ν_e emitted in the electron capture process was determined to be -1. Other nuclear β-decays have put the helicity of $\overline{\nu}_e$ to be $+1$, as well as $h = -v/c$ for electrons and $h = +v/c$ for positrons emitted.

Two important consequences follow from these experimental observations. The first comes from the fact that all the leptons emitted in β-decays are observed to be left-handed ($h < 0$) and all antilepton right-handed ($h > 0$). For reasons we shall not go into here, operators that are scalars, pseudoscalars, and tensors under a parity transformation produce leptons (as well as antileptons) of both helicities. Only vector operators V and axial-vector operators A can accommodate the observed result that all leptons are of one helicity and antileptons the other value. Furthermore, since V and A are of different parity, a linear combination of V and A is required as the operator for β-decay. This leads to the $(V - A)$-theory of β-decay. (The minus sign is related to the fact that $a = -1$ in (5-71) rather than $+1$.)

A second consequence of the fact that neutrinos are found only with helicity $h = -1$ and antineutrinos with $h = +1$ is that neutrinos may be described by a two-component wave function. In Dirac theory, wave functions of spin-$\frac{1}{2}$ particles have four components so as to accommodate at the same time both particle and antiparticle each with projections of spin $\pm\frac{1}{2}$ along the quantization axis. If neutrinos are always of one helicity and antineutrinos always of the opposite helicity, then a two-component theory will be adequate, since particles have $h = -1$ and antiparticles $h = +1$. However, such a simplification also implies that neutrinos are massless, and an experimental determination whether a neutrino has a nonzero rest mass is of fundamental interest here as well.

Fermi and Gamow-Teller operators. Since β-decay contains both a vector part and an axial-vector part, we expect that there are two independent operators, each with its own strength and its own radial dependence. As far as nuclear β-decay is concerned, the situation is simplified for two reasons. The first comes from the fact that weak interaction has very short range, much smaller than nuclear dimensions, as we saw earlier. For this reason, the radial dependence of the operators may be approximated by a delta function. This leaves only the strengths, or coupling constants, of each of the two operators to be specified.

The two coupling constants for nuclear β-decay may be put in the form of a vector coupling constant G_V for the vector part of the operator and a Gamow-Teller coupling constant G_A for the axial-vector part. We have already seen that G_V is related to G_F, a coupling constant common to all weak interaction processes. The second simplifying feature of nuclear β-decay processes is that G_A is related to G_V. This results from the belief that the difference between G_A and G_V is due to the modification of the axial-vector operator in the presence of strong interaction. The vector current, which may be indicated by a four-vector V_μ, is known to be a conserved quantity, *i.e.*,

$$\sum_{\mu=1}^{4} \frac{\partial V_\mu}{\partial x_\mu} = 0 \tag{5-73}$$

This is generally known as the *conserved vector current* (CVC) hypothesis and (5-73) is analogous to the continuity equation in electromagnetism.

On the other hand, the axial-vector current A_μ does not obey such a relation, *i.e.*, the divergence of A_μ does not vanish. (This is related to the decay of pions which is a pseudoscalar particle.) Since A_μ is an axial vector, its divergence is a pseudoscalar. As we have seen in §2-7, the pion is a pseudoscalar particle and therefore is described by a pseudoscalar field. This leads to the *partially conserved axial-vector current* (PCAC) hypothesis,

$$\sum_{\mu=1}^{4} \frac{\partial A_\mu}{\partial x_\mu} = a\phi_\pi \tag{5-74}$$

where ϕ_π represents the pion field and a is a constant. In other words, the axial current is not conserved, but its divergence is proportional to the pion field ϕ_π. The weak axial-vector current is now related to a strong interaction field through PCAC. (For further details, see, *e.g.*, A. de Shalit and H. Feshbach, *Theoretical Nuclear Physics*, Wiley, New York, 1974; and T.D. Lee, *Particle Physics and Introduction to Field Theory*, Harwood, Chur, Switzerland, 1981).

A connection between the two weak coupling constants G_A and G_V can be made in a similar way. This is known as the Goldberger-Trieman relation which, for our purposes, may be stated in the form

$$g_A \equiv \frac{G_A}{G_V} = \frac{f_\pi g_{\pi N}}{M_N c^2} \tag{5-75}$$

where M_N is the nucleon mass. The quantity f_π is measured to be

$$f_\pi = \frac{F_\pi}{\sqrt{2}} \approx 93 \text{ MeV}$$

where F_π is known as the pion-decay constant and the quantity $g_{\pi N}$ is the pion-nucleon coupling constant. It is known empirically to have the value

$$\frac{|g_{\pi N}|^2}{4\pi} \approx 14$$

From these values, (5-75) gives the result

$$|g_A| \approx 1.31$$

The measured value from nuclear β-decay is $g_A = -1.259 \pm 0.004$ (see also Problem 5-2) in agreement with the result of the Goldberger-Trieman relation. This in turn confirms PCAC.

§5-5 NUCLEAR BETA DECAY

Nuclear β-decay is the process by which a nucleus made of Z protons and N neutrons decays to a nucleus of the same nucleon number A but with $(Z\pm1, N\mp1)$. A β^--decay,

$$A(Z,N) \longrightarrow A(Z+1, N-1) + e^- + \overline{\nu}_e \tag{5-76}$$

may be regarded as the transformation of one of the neutrons in the nucleus to a proton, and a β^+-decay,

$$A(Z,N) \longrightarrow A(Z-1, N+1) + e^+ + \nu_e \tag{5-77}$$

the transformation of a proton to a neutron.

Analogous to internal conversion in electromagnetic decays, an atomic electron may be captured by the nucleus instead of emitting a positron in β^+-decay,

$$e^- + A(Z,N) \longrightarrow A(Z-1, N+1) + \nu_e \tag{5-78}$$

Except for a small difference in the energies involved, which we shall return to later, such an *electron capture* process has the same selection rule as β^+-decay and is usually in competition with it. Again, the probability of electron capture increases as Z^3 because of the increase in the strength of the nuclear Coulomb field and the decrease in the radii of electronic orbits in atoms with increasing proton number.

Q-values. Some care is needed in calculating the Q-value for nuclear β-decay and electron capture. The Q-value of a reaction is defined as the difference in the total kinetic energy of the system before and after a reaction,

$$Q = T_f - T_i$$

For a nuclear β-decay, the parent nucleus may be assumed to be at rest in the laboratory, and the total initial kinetic energy T_i in the system is zero. In order for a β-decay to take place, the Q-value must be positive. Since either an electron or a positron is emitted in the process, the Q-value is not simply the difference between the energies of the initial and final nuclear states. (The neutrino mass is too small, if nonzero, to play a significant role in the considerations here.)

A further complication comes from the fact that the mass and binding energy of a nucleus are defined in terms of those for a neutral atom as we have seen earlier. That is, the mass difference between the parent and daughter nuclei in a β-decay,

$$\Delta M_{\beta\pm} = M(Z, N) - M(Z \mp 1, N \pm 1) \tag{5-79}$$

includes the mass and the binding energy of an atomic electron as well. For this reason, the Q-value of β^--decay is given by the following expression:

$$Q_{\beta^-} = \big(M(Z, N) - M(Z+1, N-1)\big)c^2 \tag{5-80}$$

since the electron emitted in the decay may be used, as far as energy calculations are concerned, to compensate for the additional electron required to make the daughter atom neutral. On the other hand, the Q-value for β^+-decay is given by

$$Q_{\beta^+} = \big(M(Z, N) - M(Z-1, N+1)\big)c^2 - 2m_e c^2 \tag{5-81}$$

since an amount of energy equivalent to that required to create two electrons must be supplied by the decay, one for the positron emitted and the other for the atomic electron that must be ejected in going from a neutral atom of Z electrons to one with $(Z-1)$ electrons.

In contrast, for electron capture, we have the relation

$$Q_{\rm EC} = \big(M(Z, N) - M(Z-1, N+1)\big)c^2 - B_e \tag{5-82}$$

where B_e is the ionization energy of the atomic electron captured. Since B_e is of the order of 10 eV, we may ignore it unless we are concerned with accuracies of such order as, for example, in the case of neutrino mass measurements. The difference of $2m_e c^2$ in the Q-values between β^+-decay and electron capture is, however, important. For example, the mass difference between ^{7}Be and ^{7}Li is 0.86 MeV/c^2, less than $2m_e c^2$. As a result, β^+-decay from ^{7}Be to ^{7}Li is impossible, and the transition goes purely by electron capture with a half-life of 53.4 days. Only a neutrino emerges from an electron capture. Because of the difficulty in detecting neutrinos, the most prominent signature of electron capture processes is the x-ray emitted when atomic electrons in higher orbitals decay to the lower ones left empty when an inner shell electron is absorbed by the nucleus.

In terms of binding energies, the Q-values above are given by the following expressions:

$$\begin{aligned} Q_{\beta^-} &= E_B(Z+1, N-1) - E_B(Z,N) + 0.782 \text{ MeV} \\ Q_{\beta^+} &= E_B(Z-1, N+1) - E_B(Z,N) - 2m_ec^2 - 0.782 \text{ MeV} \\ Q_{\text{EC}} &= E_B(Z-1, N+1) - E_B(Z,N) - B_e - 0.782 \text{ MeV} \end{aligned} \tag{5-83}$$

where the amount of energy of 0.782 MeV comes from the mass difference between a neutron and a neutral hydrogen atom.

Transition rates for β-decay. In order to relate the transition probability $\mathcal{W}$ for a β-decay with the nuclear matrix element involved, we shall follow a procedure that closely resembles the one used earlier in electromagnetic transitions. To simplify the discussion, we shall ignore electron capture. Again, we start with Fermi's golden rule,

$$\mathcal{W} = \frac{2\pi}{\hbar}\left|\langle\phi_{\boldsymbol{k}}(\boldsymbol{r})|H'|\phi_0(\boldsymbol{r})\rangle\right|^2\rho(E_f) \tag{5-84}$$

obtained from time-dependent perturbation theory in Appendix E. The initial state is simple, involving only the parent nuclear state,

$$|\phi_0(\boldsymbol{r})\rangle = |J_iM_i\zeta\rangle \tag{5-85}$$

which we shall assume to be stationary in the laboratory.

The final state consists of three particles, a neutral lepton, a charged lepton, and the daughter nucleus. For simplicity, we shall begin by ignoring any Coulomb effect between the charged lepton and the daughter nucleus. In this limit, both leptons are free particles and are described by plane waves travelling with wave numbers $\boldsymbol{k}_e$ and $\boldsymbol{k}_\nu$. The final state wave function is then a product of three parts,

$$|\phi_{\boldsymbol{k}}(\boldsymbol{r})\rangle = \frac{1}{\sqrt{V}}e^{i\boldsymbol{k}_e\cdot\boldsymbol{r}}\frac{1}{\sqrt{V}}e^{i\boldsymbol{k}_\nu\cdot\boldsymbol{r}}|J_fM_f\xi\rangle \tag{5-86}$$

where $|J_fM_f\xi\rangle$ is the wave function of the daughter nuclear state. The two factors of $V^{-1/2}$ are required to normalize the two lepton wave functions. We may expand the plane waves in terms of spherical harmonics, as done in §C-1,

$$e^{i\boldsymbol{k}\cdot\boldsymbol{r}} = \sum_{\lambda=0}^{\infty}\sqrt{4\pi(2\lambda+1)}\,i^\lambda j_\lambda(kr)\,Y_{\lambda 0}(\theta,0) \tag{5-87}$$

where $k = |\boldsymbol{k}| = |\boldsymbol{k}_e + \boldsymbol{k}_\nu|$ and θ is the angle between $\boldsymbol{k}$ and $\boldsymbol{r}$. The spherical harmonics $Y_{\lambda\mu}(\theta,\phi)$ is independent of the azimuthal angle ϕ for $\mu = 0$.

We can again make use of the long wave length approximation, since the Q-value of the transition is typically of the order of a few MeV. In this limit, we only need to retain the first term in the expansion of the spherical Bessel function $j_\lambda(kr)$, as we did earlier in (5-36),

$$j_\lambda(kr) \approx \frac{(kr)^\lambda}{(2\lambda+1)!!}$$

The final state wave function may now be written in the form,

$$|\phi_{\boldsymbol{k}}(\boldsymbol{r})\rangle = \frac{1}{V}\left\{1 + i\sqrt{\frac{4\pi}{3}}\,(kr)Y_{10}(\theta,0) \;+ O\big((kr)^2\big)\right\}|J_f M_f \xi\rangle \tag{5-88}$$

This is very similar to what we have taken earlier in the multipole expansion of electromagnetic transition matrix elements. The only difference, as we shall see later, is that the higher order terms are retarded by even larger factors in β-decay than the corresponding reduction between successive higher orders in electromagnetic transitions.

Nuclear transition matrix elements. Let us first examine the possible forms of the nuclear part of the β-decay operator before we proceed to find the transition matrix element. Since a neutron is transformed into a proton in β^--decay and the other way around in β^+-decay, the nuclear operator must be one-body in nature; *i.e.*, only one nucleon is involved at a time, and must involve the single-particle isospin raising or lowering operator $\tau_\pm$. Furthermore, according to the $(V - A)$ theory, there are two terms in the weak interaction, a polar vector part with coupling constant G_V and an axial-vector part with coupling constant G_A. In the nonrelativistic limit, the vector part may be represented by the unity operator times $\tau_\pm$ and the axial-vector part by a product of the intrinsic spin operator $\boldsymbol{\sigma}$ and $\tau_\pm$. A proper derivation of this result requires manipulations with Dirac wave functions and γ-matrices. We shall not carry out the discussions here, since they can be found in standard references such as Morita (*Beta Decay and Muon Capture*, Benjamin, Reading, Massachusetts, 1973).

Putting this result for the operator together with those we have obtained earlier for the wave functions in (5-85) and (5-88), we can write the $\beta^\pm$-decay transition matrix element of (5-84) in the following form:

$$\begin{aligned}\langle\phi_{\boldsymbol{k}}(\boldsymbol{r})|H'|\phi_0(\boldsymbol{r})\rangle =& \frac{1}{V}\langle J_f M_f \xi|\sum_{j=1}^{A}\{G_V \tau_\mp(j) + G_A \boldsymbol{\sigma}(j)\tau_\mp(j)\} \\ &\times \left\{1 - i\sqrt{\frac{4\pi}{3}}(kr)Y_{10}(\theta,0) + O\big((kr)^2\big)\right\}|J_i M_i \zeta\rangle \end{aligned} \tag{5-89}$$

We shall be mainly concerned with the two leading order terms in the expression, generally known as the operators for the "allowed transitions." The higher order terms involve spherical harmonics of orders greater than zero, and these induce the "forbidden decays."

For the allowed decays, the nuclear part of the $\beta^{\pm}$-decay transition operator has the form,

$$O_{\lambda\mu}(\beta) = G_V \sum_{j=1}^{A} \tau_{\mp}(j) + G_A \sum_{j=1}^{A} \boldsymbol{\sigma}(j)\tau_{\mp}(j) \tag{5-90}$$

The angular momentum carried by the first term is $\lambda = 0$ and the second term is $\lambda = 1$. The transition matrix element for this form of the β-decay operator is

$$\langle\phi_{\boldsymbol{k}}(\boldsymbol{r})|H'|\phi_0(\boldsymbol{r})\rangle \approx \frac{G_V}{V} \sum_{\mu M_f} \Big\{ \langle J_f M_f \xi | \sum_{j=1}^{A} \tau_{\mp}(j) | J_i M_i \zeta\rangle + g_A \big\langle J_f M_f \xi \big| \sum_{j=1}^{A} \boldsymbol{\sigma}(j)\tau_{\mp}(j) \big| J_i M_i \zeta \big\rangle \Big\} \tag{5-91}$$

where $g_A = G_A/G_V$. The first term here is usually referred to as *Fermi decay* and the second term as *Gamow-Teller decay.* Transitions matrix elements of operators with $\lambda > 1$ are usually much smaller in values, since they come from the higher order terms in (5-89): their contributions are important only in transitions where the two lowest order terms are forbidden by angular momentum and parity selection rules.

Density of final states. The density of states in nuclear β-decay is somewhat complicated by the three-body final state. In a two-body case, the energy and momentum of a particle is restricted by the values taken up by the other particle due to conservation laws. For this reason, the two-body problem simplifies to an equivalent one-body one. In a nuclear β-decay, the available kinetic energy, after taking care of the nuclear recoil, is shared between the neutrino and the electron (or positron). As a result, continuous energy spectra of the charged lepton and the neutrino are produced subject only to the condition that their sum, together with the nuclear recoil, satisfies energy-momentum conservation for the decay. Furthermore, the charged lepton is emitted in the Coulomb field of the daughter nucleus and its wave function is "distorted" as a result of electromagnetic interaction. This also has an effect on the density of final states available to the charged lepton.

Since a neutrino hardly interacts with its surroundings, it may be considered as a free particle once it is created. For such a particle, the number of states with momentum $p_\nu(=\hbar k_\nu)$, without any regard to the direction in which the particle is moving, is given in statistical mechanics to be

$$dn_\nu = \frac{V}{2\pi^2\hbar^3} p_\nu^2 dp_\nu, \tag{5-92}$$

where V is the same volume as that used for normalizing a three-dimensional plane wave in (5-85). If the rest mass of the neutrino is m_ν, its total energy is given by the relativistic relation

$$E_\nu^2 = (m_\nu c^2)^2 + p_\nu^2 c^2 \tag{5-93}$$

The amount E_ν is a part of the energy released by the nucleus in going from the initial to the final state; the rest of the energy is taken up by the charged lepton and the daughter nucleus.

Instead of the Q-value, it is customary to express the energies involved in β-decays in terms of the maximum kinetic energy of the charged lepton emitted. The reason for this is a practical one, as the electron or positron energy is a quantity that can be observed directly. The maximum kinetic energy, E_0, is generally referred to as the *end-point* energy in the sense that it is the point, in a plot of the number of charged leptons observed as a function of the kinetic energy, beyond which no more particle is detected. In terms of the end-point energy, we have the relation,

$$E_\nu = E_0 - E_e \tag{5-94}$$

where E_e is the kinetic energy of the charged lepton. For simplicity, we have ignored variations of the end-point energy due to small differences in the recoil energy of the daughter nucleus in a three-body final state. Since nuclear mass is much larger than those of leptons, the nuclear recoil needs to be accounted for only where high precision is required. In terms of E_0 and E_e, the density of neutrino states in (5-92) may be expressed in the form

$$dn_\nu = \frac{V}{2\pi^2\hbar^3}\frac{(E_0 - E_e)}{c^3}\left\{(E_0 - E_e)^2 - (m_\nu c^2)^2\right\}^{1/2} dE_e \tag{5-95}$$

where we have made use of the fact that $p_\nu c = \sqrt{E_\nu^2 - (m_\nu c^2)^2}$ from (5-93).

The charged leptons emitted in the decay cannot be treated as free particles, since the decay is in the Coulomb field of the daughter nucleus. A good approximation may be obtained by starting from the free particle form and folding in a distortion factor $F(Z, E_e)$ to correct for Coulomb effects. Analogous to (5-92), the density of charged lepton states may be written in the form

$$dn_e = \frac{V}{2\pi^2\hbar^3} F(Z, E_e) p_e^2 dp_e \tag{5-96}$$

The correction factor $F(Z, E_e)$ is known as the Fermi function. In the nonrelativistic limit, with the velocity of the charged lepton $v \ll c$, the function is related to the absolute square of the Coulomb wave function at the origin and has the approximate form

$$F(Z, E_e) = \frac{x}{1 - e^{-x}} \tag{5-97}$$

where $x = \mp 2\pi\alpha Zc/v$ for $\beta^\pm$-decay and α is the fine structure constant. The general form of the function is a much more complicated and does not have a simple analytical representation. Extensive tabulated values are available and they are the ones usually used in practical applications. A full discussion of the Fermi function can be found in Morita (*Beta Decay and Muon Capture*, Benjamin, Reading, Massachusetts, 1973).

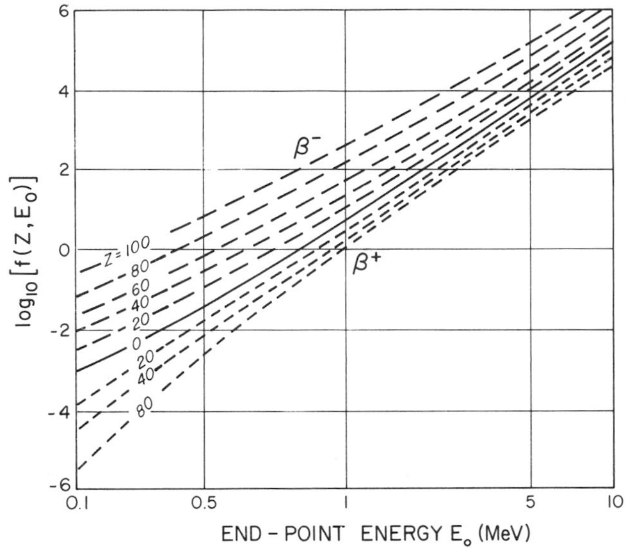

Fig. 5-5 The Fermi integral $f(Z, E_0)$, integral of $F(Z, E_e)$ over energy E_e, as a function of the end-point energy E_0 for different proton number Z of the daughter nucleus. The long dashed curves in the upper half of the diagram are for β^--decay and the short dashed curves in the lower half of the diagram are for β^+-decay.

The results of (5-91), (5-95), and (5-96) can now be put into (5-84), and we arrive at the transition probability for an electron or positron emitted with momentum $p_e(= |\boldsymbol{p}_e|)$,

$$\mathcal{W}(p_e) = \frac{1}{2\pi^3\hbar^7c^3}\sum_{\mu M_f}\left|\langle J_fM_f\xi|\boldsymbol{O}_{\lambda\mu}(\beta)|J_iM_i\zeta\rangle\right|^2 \times F(Z,E_e)p_e^2(E_0-E_e)\big\{(E_0-E_e)^2-(m_\nu c^2)^2\big\}^{1/2} \quad (5\text{-}98)$$

where all factors related to V, the (arbitrary) normalization volume, cancel each other, and the summation over M_f takes care of the requirement to include all the possible nuclear final states.

Let us ignore the small neutrino mass for the time being. The expression for $\mathcal{W}(p_e)$, then, simplifies to the form

$$\mathcal{W}(p_e) = \frac{1}{2\pi^3\hbar^7c^3}\sum_{\mu M_f}\left|\langle J_fM_f\xi|\boldsymbol{O}_{\lambda\mu}(\beta)|J_iM_i\zeta\rangle\right|^2 F(Z,E_e)p_e^2(E_0-E_e)^2 \quad (5\text{-}99)$$

The approximation affects mainly the region where E_e is very close to the end-point energy E_0 and the effect of m_ν is most evident. From (5-99), we see that $\{\mathcal{W}(p_e)/p_e^2F(Z,E_e)\}^{1/2}$ is proportional to $(E_0 - E_e)$ and a plot of the former

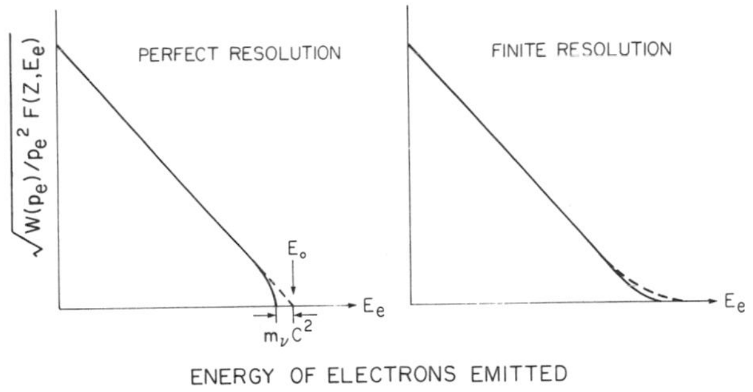

Fig. 5-6 Schematic diagram of a Kurie plot, variations of the square root of the number of electrons or positrons with momentum p_e divided by $p_e^2 F(Z, E_e)$ as a function of E_e, the energy of the electron or positron emitted in a β-decay. The solid lines are for the case of finite neutrino mass m_ν and the dashed lines are for $m_\nu = 0$. With perfect resolution the plot, as shown in (a), is a straight line intersecting the horizontal axis at the end-point energy E_0 if $m_\nu = 0$. With finite resolution of the detector, the region near the end-point energy is modified in the way shown in (b).

quantity as a function of the latter produces a straight line (except in the region of the end-point energy) with a slope proportional to the nuclear matrix element. Such a graph, represented schematically in Fig. 5-6, is called a Kurie plot.

Neutrino mass measurement. It is convenient to digress here into a few remarks on the measurement of m_ν, the mass of an electron neutrino. All evidence to date indicates that m_ν is small, of the order of tens of electron-volts, although the possibility of $m_\nu = 0$ is not ruled out either. The masses of the other two neutrinos, ν_μ and ν_τ, are expected to be much larger: the present upper limits are, respectively, 0.5 and 70 MeV/c^2.

Most of the direct measurements of m_ν, the mass of an electron neutrino, make use of the β^--decay of the triton, made of one proton and two neutrons. There are many reasons for favoring this reaction. The decay,

$$t \to {}^3\text{He} + e^- + \overline{\nu}_e$$

has a half-life of 12.3 years. The Q-value is low, 18.6 keV, so that the influence of a small m_ν will stand out more prominently than otherwise, for example, in a Kurie plot. The excited states of the daughter nucleus ^{3}He are very high in energy, so that the ground state is the only possible final nuclear state for the decay. The radioactive tritium source is relatively easy to prepare and, as a result, high precision may be achieved. There are, however, several difficulties associated with the measurement that are hard to overcome. The first one arises mainly from the low counting rate near the end-point, a common problem in all nuclear β-decays. The second one comes from the fact that the rest mass energy of the neutrino is comparable to the energies encountered in atoms. As a result, atomic effects, which are seldom a problem in nuclear measurements, become important considerations here. For example, there are two possible final atomic states for ^{3}He in the decay, and the relative probability of forming them must be known fairly well in order to obtain a reliable final answer on m_ν. At this moment, there is still no agreement between the measured values from different laboratories. The most often quoted value of $\sim$ 30 eV/c^2 came from Lubimov et al. (*Phys. Lett.* **94B** [1980] 266), and the results are shown in Fig. 5-7. Observations of neutrinos from the supernova SN 1987a put a constraint on the possible mass of an electron neutrino, but it has not changed the uncertainties around the measured value of m_ν at this time.

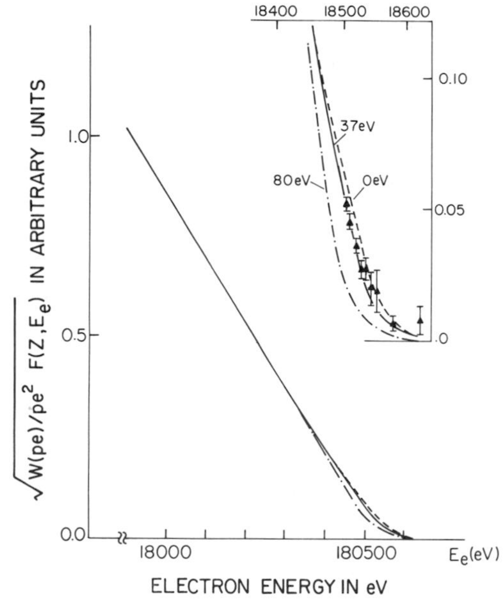

Fig. 5-7 Kurie plot near the end-point energy for tritium β-decay. The process is widely used for neutrino mass measurements. The dashed curves are for $m_\nu = 0$ eV/c^2 and the dash-dot curve for $m_\nu =$ 80 eV/c^2. The best-fit value is $m_\nu = 37$ eV/c^2, shown by the solid curve. (Adapted from Lubimov et al. (*Phys. Lett.* **94B** [1980] 266).

Total transition probability. Let us return to the question of nuclear β-decay in general. If we are not interested in the distribution of charged leptons emitted as a function of E_e, we can obtain the total transition probability $\mathcal{W}$ by integrating $\mathcal{W}(p_e)$ in (5-99) over all possible values of momentum p_e,

$$\mathcal{W} = \int \mathcal{W}(p_e)\,dp_e = \frac{m_e^5 c^4}{2\pi^3\hbar^7} f(Z, E_0) \Big| \sum_{\mu M_f} \langle J_f M_f \xi | O_{\lambda\mu}(\beta) | J_i M_i \zeta \rangle \Big|^2 \tag{5-100}$$

where the dimensionless function

$$\begin{aligned} f(Z, E_0) &= \int F(Z, E_e) \left(\frac{p_e}{m_e c}\right)^2 \left(\frac{E_0 - E_e}{m_e c^2}\right)^2 \frac{dp_e}{m_e c} \\ &= \frac{1}{m_e^5 c^7} \int F(Z, E_e)\, p_e^2 (E_0 - E_e)^2\, dp_e \end{aligned} \tag{5-101}$$

is known as the Fermi integral. Except in the trivial case of $Z = 1$ for the daughter nucleus, the integral must be evaluated numerically. Extensive tables of calculated values are available (*Tables for the Analysis of Beta Spectra*, Applied Mathematics Series 13, National Bureau of Standards, Washington, 1951) and some of the typical values are plotted in Fig. 5-5.

From the transition probability, we obtain the expression for β-decay half-life,

$$T_{1/2} = \frac{\ln 2}{\mathcal{W}} = \frac{1}{f(Z,E_0)} \frac{2\pi^3\hbar^7}{m_e^5 c^4} \frac{\ln 2}{\left|\sum_{\mu M_f} \langle J_f M_f \xi | \boldsymbol{O}_{\lambda\mu}(\beta) | J_i M_i \zeta\rangle\right|^2} \tag{5-102}$$

Instead of half-lives, nuclear β-decay rates are often quoted in terms of ft-values, the product of Fermi integral $f(Z, E_0)$ and $T_{1/2}$,

$$ft \equiv f(Z,E_0)T_{1/2} = \frac{2\pi^3\hbar^7}{m_e^5 c^4} \frac{\ln 2}{\left|\sum_{\mu M_f} \langle J_f M_f \xi | \boldsymbol{O}_{\lambda\mu}(\beta) | J_i M_i \zeta\rangle\right|^2} \tag{5-103}$$

As we can see from the definition, the ft-value is a more meaningful physical quantity in nuclear β-decay studies, since it is directly related to the square of the nuclear transition matrix element. The half-life, on the other hand, involves $f(Z, E_0)$ that depends in a complicated way on the proton number of the daughter nucleus and the end-point energy. We have already encountered a similar problem in the study of electromagnetic decays. There, half-lives are dominated by energy dependence and the quantities that are more directly related to nuclear physics are the reduced transition probabilities $B(\lambda)$ or transition rates measured in Weisskopf units. The ft-value in β-decay plays a similar role as $B(\lambda)$ in electromagnetic transitions.

The measured ft-values are found to vary over many orders of magnitude if we include both allowed and forbidden decays. As a result, it is often more convenient to use $\log ft$-values, the logarithm to the base 10 of the ft-values. A compilation of the distribution of measured $\log ft$ values is shown in Fig. 5-8.

Allowed β-decay. Let us return to the operator for allowed decays given in (5-90). The Fermi term involves only the isospin raising or lowering operator. As a result, it is possible to carry out the summation over all the nucleons explicitly,

$$\sum_{j=1}^{A} \boldsymbol{\tau}_{\mp}(j) = \boldsymbol{T}_{\mp},$$

where $T_{\mp}$ lowers or raises the third component of the isospin of the nucleus as a whole. The value of the matrix element of the Fermi operator can now be evaluated without having to know explicitly the wave functions involved,

$$\begin{aligned} &\langle J_f M_f T_f T_{0f} | \sum_{j=1}^{A} \boldsymbol{\tau}_{\mp}(j) | J_i M_i T_i T_{0i}\rangle \\ &\qquad = \sqrt{T_i(T_i+1) - T_{0i}(T_{0i} \mp 1)}\, \delta_{J_f J_i} \delta_{M_f M_i} \delta_{T_f T_i} \delta_{T_{0f}(T_{0i}\mp 1)} \end{aligned} \tag{5-104}$$

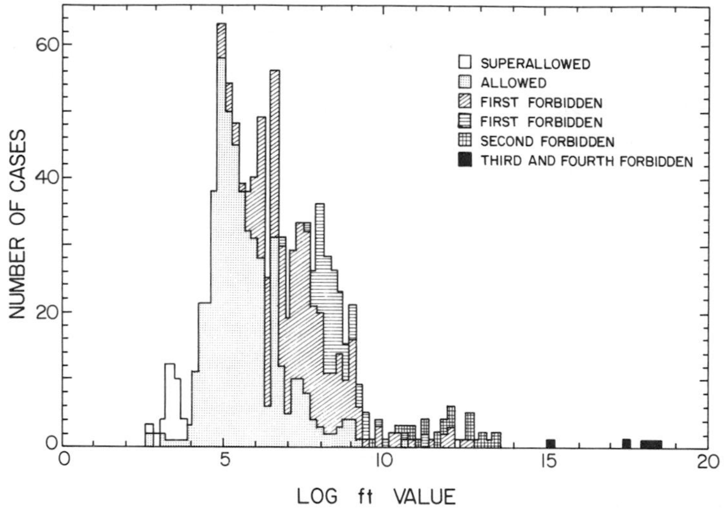

Fig. 5-8 Systematics of observed log ft-values for different types of β-decay. (Adapted from G.E. Gleit et al., *Nucl. Data Sheets* **5** [1963] set 5.)

This result is derived under the assumption that isospin is an exact quantum number. As we have seen earlier, both the Coulomb force and the difference in the mass between charged and neutral pions violate isospin symmetry and, consequently, affect the actual value of the Fermi matrix element. In practice, it is found that the necessary correction factors are quite small for light nuclei and can be evaluated to sufficient accuracy so that the final results may be reliable to an uncertainty of 0.1% or less.

From the results given in (5-104), we find that the angular momentum and isospin selection rules of Fermi-type of $\beta^{\pm}$-decay are the following:

$$\begin{array}{ll} J_f = J_i & (\Delta J = 0) \\ T_f = T_i \neq 0 & (\Delta T = 0, \text{ but } T_i = 0 \rightarrow T_f = 0 \text{ forbidden}) \\ T_{0f} = T_{0i} \mp 1 & (\Delta T_0 = 1) \\ \Delta\pi = 0 & \text{no parity change} \end{array} \tag{5-105}$$

In other words, Fermi decay goes primarily between isobaric analogue states where the only difference between the initial and the final states is the replacement of a proton by a neutron or vice versa.

For the Gamow-Teller operator, $\sum_j \boldsymbol{\sigma}(j)\tau_{\mp}(j)$, the summation over nucleons cannot be carried out explicitly, since both spin and isospin of a nucleon are acted upon at the same time. Unlike Fermi decays, matrix elements for the Gamow-Teller operator cannot be evaluated unless both the initial and the final wave functions are given. The angular momentum and isospin selection rules may be deduced from the properties of the operator itself. Since the spherical tensor ranks of the operator are unity in both spin and isospin spaces, it is necessary that the initial and final states are related in the following ways:

$$\begin{aligned}
&\Delta J = 0, 1 && \text{but } J_i = 0 \to J_f = 0 \text{ forbidden} \\
&\Delta T = 0, 1 && \text{but } T_i = 0 \to T_f = 0 \text{ forbidden} \\
&T_{0f} = T_{0i} \mp 1 && (\Delta T_0 = 1) \\
&\Delta\pi = 0 && \text{no parity change}
\end{aligned} \qquad (5\text{-}106)$$

The last point, on the parity selection rule, comes from the fact that $\boldsymbol{\sigma}$ is an axial-vector operator and, consequently, cannot change the parity between initial and final states. The absolute values of Gamow-Teller matrix elements are generally smaller than those for Fermi transitions, since both spin and isospin are involved simultaneously.

For allowed decays, the square of the nuclear transition matrix element may be written in the form

$$\begin{aligned}
\sum_{\mu M_f} \Big| \langle J_f M_f \xi | \boldsymbol{O}_{\lambda\mu}(\beta) | J_i M_i \zeta \rangle \Big|^2
&= G_V^2 \Big\{ \sum_{M_f} \Big| \langle J_f M_f \xi | T_{\mp} | J_i M_i \zeta \rangle \Big|^2 + g_A^2 \sum_{\mu M_f} \Big| \langle J_f M_f \xi | \sum_{j=1}^{A} \boldsymbol{\sigma}_\mu(j) \boldsymbol{\tau}_{\mp}(j) | J_i M_i \zeta \rangle \Big|^2 \Big\} \\
&\equiv G_V^2 \{ \langle F \rangle^2 + g_A^2 \langle GT \rangle^2 \}
\end{aligned} \qquad (5\text{-}107)$$

There is no cross term between the Fermi and Gamow-Teller matrix elements, since the latter vanishes on summing over all possible projections on the quantization axis. To simplify the notation, we shall use the abbreviations for the matrix elements adopted in the final form of (5-107).

For allowed β-decays, the ft-value may be written in the following form

$$ft = \frac{K}{\langle F \rangle^2 + g_A^2 \langle GT \rangle^2} \qquad (5\text{-}108)$$

where the vector coupling constant G_V, as well as other universal constants, are absorbed into the definition of the constant

$$K = \frac{2\pi^3 \hbar^7 \ln 2}{m_e^5 c^4 G_V^2} = 6141.2 \pm 3.2 \text{ s} \qquad (5\text{-}109)$$

Among the constants, the value of G_V is perhaps the least well known one. A determination of K is then one way to deduce the absolute value of the vector coupling constant G_V. The best measured value of K, at the moment, is 6141.2 ± 3.2 s, obtained from superallowed decays after applying such corrections as the finite size and the charge distribution of the nucleus (A. Sirlin, *Phys. Rev. D* **35** [1987] 3423). The value of the vector coupling constant obtained in this way is $G_V = 1.41549 \times 10^{-49}$ erg-cm^3 or, in its more commonly quoted form,

$G_V/(\hbar c)^3 = 1.1493 \times 10^{-11}$ MeV^{-2}. In order to obtain the Fermi coupling constant of $G_F/(\hbar c)^3 = 1.16637(2) \times 10^{-11}$ MeV^2, further correction factors are required (see, *e.g.*, R.J. Blin-Stoyle, *Fundamental Interactions and the Nucleus*, North-Holland, Amsterdam, 1973).

Superallowed β-decay. Transitions from an initial nuclear state with $J_i^\pi = 0^+$ to a final state with $J_f^\pi = 0^+$ form a special class of β-decay, since the Gamow-Teller term does not contribute anything here. The transitions are purely Fermi and, as a result, are least sensitive to the details of nuclear wave functions. Such decays are useful, for example, in determining the value of K, and hence G_V, as mentioned in the previous paragraph. Light nuclei are preferred here, since isospin breaking effects are at a minimum; however, the number of available cases is limited. Superallowed β^--decay is often forbidden by Q-value considerations, since Coulomb energy is higher for nuclei with one more proton. Most of the examples found are positron emitters, such as the reaction

$$^{14}\mathrm{O} \rightarrow {}^{14}\mathrm{N} + e^+ + \nu_e$$

leading to the 0^+ first excited state of ^{14}N at 2.311 MeV. The half-life of the ^{14}O is 74 s and the Q-value of the reaction is 1.12 MeV. The ft-value is 3109 sec, among the smallest known. If the initial and final nuclear states are truly isobaric analogue states of each other, the value of the Fermi matrix element may be obtained using (5-104) without referring explicitly to the nuclear wave functions.

The determination of g_A, the ratio between axial-vector and vector coupling constants, must be deduced from nuclear Gamow-Teller decay. The best-known value is

$$g_A = \frac{G_A}{G_V} = -1.259 \pm 0.004 \tag{5-110}$$

In principle, β^--decay of neutrons is the ideal reaction to use here, since only the intrinsic spin wave function of a free neutron enters into the calculations. In this case, the Gamow-Teller matrix element can be evaluated using the relation

$$\sum_{\mu m_f} \left| \langle \chi_{m_f} | \sigma_\mu | \chi_{m_i} \rangle \right|^2 = 3$$

However, we are limited here by our knowledge on the half-life of neutrons. The value that is quoted nowadays is $T_{1/2} = 621 \pm 7$ s. Unfortunately, it belongs to the class of data that changes with time as newer and better measurements are carried out.

Table 5-3 Typical log ft-values for nuclear β-decay.

Decay type	$\log_{10} ft_{1/2}$
Superallowed	2.9 - 3.7
Allowed	4.4 - 6.0
First forbidden	6 - 10
Second forbidden	10 - 13
Third forbidden	> 15

Forbidden decay. From the selection rules given in (5-105) and (5-106) we see that, for allowed β-decays, the spins of the initial and final states can be different at most by unity and the parities must be same. Transitions between states of different parity and $\Delta J > 1$ are also known to take place, albeit with much larger ft-values (*i.e.,* smaller probabilities). These are referred to as *forbidden decays.* As we can see from (5-88), the operators for forbidden decays involve spherical harmonics of order greater than zero.

Forbidden decays are classified into different groups by the ℓ-value of the spherical harmonics involved. For a given order ℓ, the possible operators with definite spherical tensor ranks are $Y_{\ell m}(\theta, \phi)$ and $(Y_\ell(\theta, \phi) \times \boldsymbol{\sigma})_{\lambda\mu}$. The angular momentum and parity selection rules for the ℓ-th order forbidden transition are then

$$\Delta J = \ell \text{ or } (\ell \pm 1) \qquad\qquad \Delta\pi = (-1)^\ell$$

The isospin selection rule remains the same as for allowed decays,

$$\Delta T_0 = 1 \qquad \Delta T = 0 \text{ or } 1 \qquad \text{but } T_i = 0 \to T_f = 0 \text{ transitions forbidden}$$

since nothing is different in the isospin structure of the operator for forbidden decays from that for allowed decays. Thus for first-order forbidden transitions, the operators are $r\, Y_{1\mu}(\theta, \phi)$ (proportional to $\boldsymbol{r}$) and $(\boldsymbol{\sigma} \times rY_1(\theta, \phi))_{\lambda\mu}$ with $\lambda = 0$, 1 and 2. Since the parity of $Y_{1\mu}(\theta, \phi)$ is -1, a parity change between the initial and final states is necessary.

The reason for the large ft-values in forbidden β-decays comes from the fact that, with $\ell > 0$, there is an angular momentum barrier inhibiting the emission of leptons. This results in a reduction in the size of the nuclear transition matrix element and hence an increase in the ft-values. Typical values for the various order decays are given in Table 5-3. However, as can be seen from Fig. 5-8, the separation in the observed ft-values for different order transitions is not as clearly separated as shown in Table 5-3. In general, it is quite difficult to calculate the nuclear matrix elements for forbidden β-decays with any reliability and, as a result, very few theoretical investigations are found in the literature.

Charge exchange reactions. Charge exchange reactions can also replace a proton in a nucleus by a neutron or the other way around. Although the process primarily involves nuclear interactions, the matrix elements that enter into the reaction rates are essentially the same as those in β-decay induced by weak interactions. Relations between β-decay and charge-exchange reactions are therefore of interest both from the point of making connection between these two interactions and in studying the nuclear matrix elements involved.

A typical charge exchange reaction may be illustrated by a (p,n) or (n,p) reaction. In the former case, a nucleus A is bombarded by a beam of protons. Among the different reactions that take place, we are interested here in the one where the proton is absorbed by the nucleus and a neutron is emitted in exchange. The residual nucleus B has one more proton and one less neutron than the initial target nucleus A. Such a process may be represented in the form

$$A(Z,N) + p \longrightarrow B(Z+1, N-1) + n \tag{5-111}$$

or $A(p,n)B$ for short. The nuclear structure part of this reaction bears strong resemblance to β^--decay shown in (5-76). Apart from the dynamics of scattering, the main difference between them is that the (p,n) reaction is not restricted by Q-value considerations to final states that are lower in energy, as given by (5-83) for the β^--decay.

Analogous to the relation between (p,n) reaction and β^--decay, the (n,p) reaction,

$$A(Z,N) + n \longrightarrow B(Z-1, N+1) + p \tag{5-112}$$

is the complement of β^+-decay. From a practical point of view the reaction is a difficult one to study because of the scarcity of energetic neutron beams. A combination of (p,n) and (n,p) reactions, however, allows a whole range of interesting questions to be investigated.

Charge exchange processes may also be induced by reactions involving nuclei as the projectile and scattered particles, such as $(^3\text{He},t)$, $(^6\text{Li},^6\text{He})$, and their inverses. The use of ^{3}He and heavier ions has the complication that both the incident and the scattered particle, as well as the target nucleus, may be excited in the process. This makes the analysis of the reaction more complicated. In addition, (π^+,π^0) and (π^-,π^0) reactions are also used to study charge exchange processes. At the moment, pion reactions lack the good energy resolution that can be achieved, for example, with (p,n) reaction, but the situation may change with progress in detection apparatus.

With pions and heavy ions, it is also possible to study double-charge exchange reactions, such as (π^-,π^+) and (π^+,π^-). In such reactions, a pair of nucleons change their charge states at the same time. These reactions are sensitive to two-body correlations in nuclei, a question of importance in nuclear structure studies. The information is closely related to that obtained through two-nucleon transfer reactions described in §7-2 and is also related to double β-decay described below.

Double β-decay. Double β-decay is the process by which two electrons or two positrons are emitted,

$$A(Z,N) \rightarrow A(Z+2,N-2) + 2e^- + 2\overline{\nu}_e$$
$$A(Z,N) \rightarrow A(Z-2,N+2) + 2e^+ + 2\nu_e \qquad (5\text{-}113)$$

Such processes are caused by second-order perturbations induced by weak interactions and are therefore much slower than β-decays in which only a single charged lepton is emitted. As a result, double β-decays are expected to be long-lived, with typical half-lives of the order of 10^{20} years. Processes with such long half-lives may be observed only in nuclei where ordinary β-decay and other faster reactions are forbidden by Q-value considerations. In spite of this limitation, a number of such cases are known as we can see by comparing the binding energies of neighboring nuclei. For example, ${}^{82}_{34}\text{Se}$ is stable against β^--decay to ${}^{82}_{35}\text{Br}$ since the Q-value is -0.90 MeV. However, it is unstable against double β^--decay to ${}^{82}_{36}\text{Kr}$ with a Q-value of $+3.00$ MeV.

It is not surprising that a large number of nuclei can, in principle, undergo double β-decay. In general, these are even-even nuclei near, but not at, the bottom of the valley of stability. Because of pairing energy, they are more tightly bound compared with neighboring odd-odd nuclei (see §4-11). On the other hand, a neighboring even-even nucleus with two more protons and two less neutrons may be even more tightly bound because of a larger symmetry energy. As we have seen earlier in the discussion of binding energies, this term is proportional to $(N-Z)^2$. Since most nuclei in the medium to heavy range have a neutron excess, an isobar with two neutrons less can often be more tightly bound as a result. For this reason, more nuclei are known to be capable of double β^--decay than the number of nuclei which can decay by emitting two positrons. Double β^+-decay is possible, for example, in the decay of ${}^{106}_{48}\text{Cd}$ to ${}^{106}_{46}\text{Pd}$: the Q-value of 0.7 MeV is, however, smaller than the typical double β^--decay values of two to three MeV.

As mentioned in Chapter 1, one of the primary interests in nuclear double β-decay is the question of whether the process can take place without emitting neutrinos. If neutrinos are Majorana fermions, with no distinction between particles and antiparticles, we can imagine that the neutrino from the first β-decay in a double β-decay process is absorbed in the intermediate state and that this absorption induces the emission of the second charged lepton. In such cases, no neutrino will emerge from the decay. On the other hand, such "neutrinoless" double β-decay processes are strictly forbidden if the neutrinos are Dirac particles with particles distinct from their antiparticles.

The fact that neutrinos and antineutrinos are different particles was confirmed by an experiment using the reaction,

$$\text{neutrino} + {}^{37}\text{Cl} \rightarrow e^- + {}^{37}\text{Ar}$$

The source of the neutrino for this classic experiment by R. Davis in 1955 (*Phys. Rev.* **97** [1955] 766) was a reactor which produces mainly $\overline{\nu}_e$. The observed cross section for this reaction was much smaller than one expected for Majorana neutrinos. The observation forms a proof that neutrinos are Dirac particles.

However, in a neutrinoless double β-decay, the neutrinos are virtual particles and may be different from real neutrinos observed in Davis' experiment. If virtual neutrinos are Majorana particles, then double β-decay can take place without emitting any physical neutrinos and can therefore proceed on a much faster scale, perhaps by as much as six orders of magnitude. An important fact in support of the faster rate, among others, is that the phase space available for the final states of a neutrinoless double β-decay is much larger than the competing two-neutrino mode.

One way to distinguish between the two types of double β-decay is the spectrum of the electrons emitted. If no neutrinos are emitted, the sum of the energies of the two electrons is equal to the Q-value of the decay (again ignoring the small amount of energy associated with nuclear recoil). On the other hand, if two neutrinos are also emitted, the sum of the energies of the two electrons will have a continuous distribution given by energy-momentum conservation of the five-body final state. One of the recent measurements of the double β^- decay of ^{82}Se to ^{82}Kr by Elliot, Hahn, and Moe (*Phys. Rev. Lett.* **59** [1987] 1649) gives a limit of the half-life of the decay to be 4.4×10^{20} years and an energy spectrum of the two electrons emitted consistent with the two-neutrino mode.

Measured long lifetimes alone do not necessarily rule out the possibility of neutrinoless double β-decay. As we have seen in the case of single β-decays, there is a large spread in the $\log ft$ values, especially for the allowed decays. Such a divergence in the rate is due primarily to the differences in the values of the nuclear transition matrix element involved. The same may also be true for double β-decays. If the nuclear matrix elements in double β-decays are much smaller than expected, the lifetimes of 10^{20} years could even be an underestimate of the two-neutrino mode. Consequently, long measured half-lives by themselves do not rule out the neutrinoless mode. In this sense, the energy spectrum of the experiment of Elliot, Hahn, and Moe is more conclusive evidence against the neutrinoless mode than lifetime measurements.

Since we are considering very slow processes, there are also other possibilities for double β-decay in addition to two-neutrino and neutrinoless modes. One is the weak decay of a Δ-particle to a nucleon with the emission of two charged leptons. The normal decay mode of Δ is to a pion plus a nucleon via strong interaction, but a weak branch involving leptons cannot be ruled out, especially when the Δ-particle is a part of a nucleus. The other possibility is that, instead of two neutrinos, a boson, given the name "Majaron," may be emitted. The detection of any such events requires measurements involving half-lives of the order of 10^{20} year, and these are not easy experiments to carry out.

§5-6 ALPHA-PARTICLE DECAY

Barrier for α-particle emission. In light nuclei the threshold for α-particle emission is comparable with those for nucleon emission, as can be seen from the separation energies shown in Fig. 6-14. As a result, α-decay is not energetically possible for the low-lying excited states until $A > 150$ or so. However, even for heavy nuclei, lifetimes are very long by strong interaction time scales. Furthermore, the kinetic energies of α-particles emitted are very similar, usually in the range of 4 to 9 MeV, but the half-lives associated with these emissions differ by a wide range. For this reason, α-decay by natural occurring radioactive nuclei, as mentioned in Chapter 1, was a puzzle in the early days of modern physics, one which eventually led to the discovery of quantum mechanical tunnelling.

It is perhaps easier to visualize the barrier an α-particle must tunnel through by considering first the inverse process of an α-particle approaching a heavy nucleus from large distances. Outside the nuclear surface, the interaction is purely Coulomb and the repulsive potential may be represented in the form,

$$V_c(r) = \left[\frac{1}{4\pi\epsilon_0}\right]\frac{2Ze^2}{r} = \frac{2Z\alpha\hbar c}{r} = \frac{2.88Z}{r}\ \text{MeV} \tag{5-114}$$

where r is measured in femtometers in the last equality.

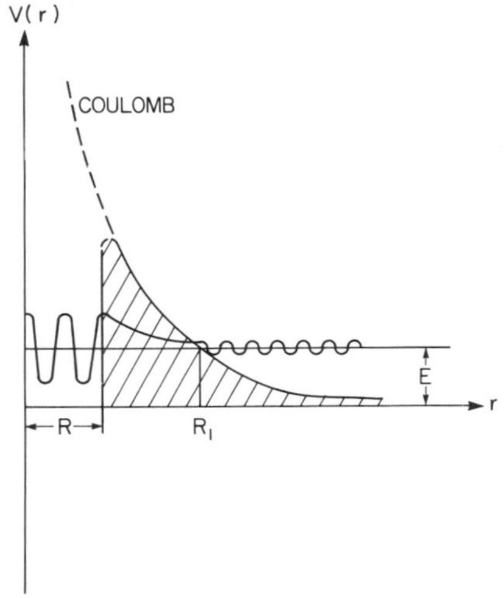

Fig. 5-9 Schematic diagram showing the wave function of an α-particle tunnelling through a potential barrier. Inside the potential well, $r \leq R$, the wave function is sinusoidal. In the region $R < r < R_1$, the energy of the particle is less than the barrier height and the amplitude of the wave decreases more or less exponentially. Once outside, $r > R_1$, the α-particle is essentially free except for Coulomb interaction with the residual nucleus.

Once the α-particle is inside the nuclear surface, short-range nuclear forces become effective. Since particles are bound to the nucleus at distances $r < R$, the combined result of Coulomb and nuclear forces must produce a minimum in the potential at these short distances. For simplicity, we shall take the shape of this part of the potential to be an attractive square well, as shown schematically in Fig. 5-9. In this approximation, the height of the barrier that retains the α-particle inside the nucleus may be estimated from the amount of work required to overcome Coulomb repulsion in bringing an α-particle with two units of charge to the surface of a heavy nucleus such as ${}^{238}_{92}$U,

$$E_c = \left[\frac{1}{4\pi\epsilon_0}\right]\frac{2Ze^2}{R} = \frac{2Z\alpha\hbar c}{r_0 A^{1/3}} \approx 35 \text{ MeV} \tag{5-115}$$

Better estimates put the Coulomb energy of an α-particle in this region of mass number to be just below 30 MeV.

Classically, in order for an α-particle to be emitted from a nucleus, it must somehow acquire enough energy to reach the top of the potential barrier. Once it is there, it can leave the nucleus carrying with it all the energy it acquired. Consequently, when the α-particle is far away from the nucleus, it should have a kinetic energy at least equal to the barrier height. The observed kinetic energies of α-particles emitted are, however, much smaller than this value and, consequently, a different mechanism must be operating here. The quantum mechanical explanation is that the α-particle does not have to go over the top of the potential barrier before being emitted: instead, it tunnels through it. The basic reason for such a possibility comes from the fact that the amplitude of the α-particle wave function does not vanish inside a barrier of finite height and, as a result, there is a finite probability of finding the particle outside the nucleus, as shown schematically in Fig. 5-9.

Table 5-4 Examples of α-decay half-lives.

α emitter	E_α(MeV)	$\tau_{1/2}$	α emitter	E_α(MeV)	$\tau_{1/2}$
^{206}Po	5.22	8.8d	^{236}U	4.49	2.39×10^7y
^{208}Po	5.11	2.90y	^{238}U	4.20	4.51×10^9y
^{210}Po	5.31	138d	^{238}Pu	5.50	86y
^{212}Po	8.78	0.30μs	^{240}Pu	5.17	6.58×10^3y
^{214}Po	7.68	164μs	^{242}Pu	4.90	3.79×10^5y
^{216}Po	6.78	0.15s	^{244}Pu	4.66	8×10^7y
^{228}U	6.69	9.1m	^{240}Cm	6.29	26.8d
^{230}U	5.89	20.8d	^{242}Cm	6.12	163d
^{232}U	5.32	72y	^{244}Cm	5.80	17.6y
^{234}U	4.77	2.47×10^5y	^{246}Cm	5.39	5.5×10^3y

The reason that α-particle emission is an important channel of decay for heavy nuclei comes from a combination of two reasons. The first is the saturation of nuclear force which turns α-clusters into tightly bound groups of four nucleons inside a nucleus. As we have seen earlier, the average binding energy per nucleon between α-clusters is much less than that between nucleons inside such a cluster.

This reduces the energy required for an α-cluster to be separated from the rest of the nucleus. The second reason is the increase in Coulomb repulsion in heavy nuclei due to the larger number of protons present. The combined effect of both reasons enables the Q-value for α-emission to become positive (negative in terms of the separation energy) for $A > 150$, as can be seen from Fig. 6-14. Examples of Q-values for some of the heavy nuclei as measured by the kinetic energy of the α-particles emitted are given in Table 5-4, together with the half-life associated with each.

Decay probability. Even before there were any theoretical models constructed to explain the decay probability, it was found empirically that the large range of α-decay half-lives, from μs to 10^{17} years, may be related to the square root of the kinetic energy E_α of the α-particles emitted in the following way:

$$\log_{10} \mathcal{W} = C - \frac{D}{\sqrt{E_\alpha}} \tag{5-116}$$

where $\mathcal{W}$ is the decay probability. The parameters C and D are weakly dependent on Z but not on the neutron number, as can be seen from the plot in Fig. 5-10. This is known as the Geiger-Nuttall law. A simple, one-body theory of α-particle decay described below provides the foundation for this observed relation.

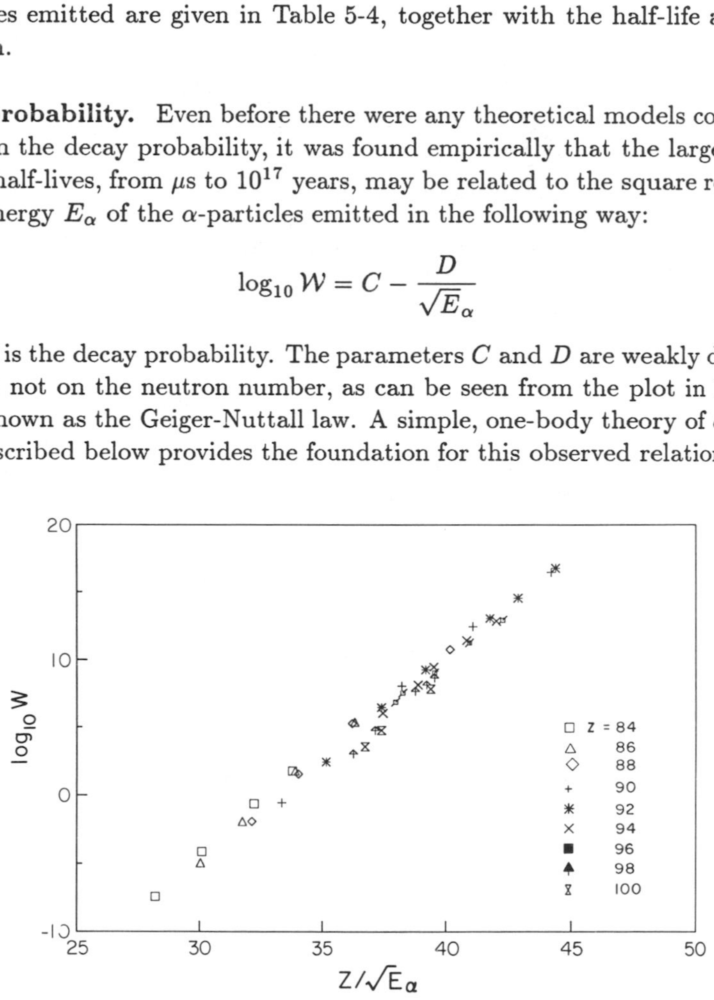

Fig. 5-10 Semi-log plot of the transition rate $\mathcal{W}$ as a function of the ratio $Z/\sqrt{E_\alpha}$ for α-decay of heavy nuclei with proton number Z and α-particle kinetic energy E_α. The clustering of data along a straight line, described by (5-116), is known as the Geiger-Nuttall law.

We do not expect α-particles to exist as entities that can be readily identified inside a heavy nucleus. On the other hand, there is a finite probability of finding two protons and two neutrons to be correlated in such a way that they have an α-particle-like structure. Let us call such an object an α-cluster. In this way, the probability $\mathcal{W}$ for α-particle emission from a heavy nucleus by tunnelling may be expressed as a product of three factors,

$$\mathcal{W} = p_\alpha \nu \mathcal{T} \tag{5-117}$$

where p_α is the probability of finding an α-cluster inside a heavy nucleus, ν is the frequency of the α-cluster appearing at the inside edge of the potential barrier where it has can leak out, and $\mathcal{T}$ is the transmission coefficient for the α-cluster to tunnel through the barrier.

Once an α-cluster is formed inside a nucleus, the frequency ν with which it appears at the edge of the potential well depends on the velocity v it travels and the size of the potential well. A reasonable way to estimate ν is to take the well size as twice the nuclear radius R. With this assumption we obtain the result,

$$\nu = \frac{v}{2R} = \frac{\sqrt{2K/M_\alpha}}{2R} \tag{5-118}$$

where K is the kinetic energy of the α-cluster inside the well and M_α is the mass. The precise value of K depends on the depth of the potential well and is not well known. It is reasonable to expect that K is of the same order of magnitude as E_α, the kinetic energy of the α-particle emitted. For our simple model, we shall take $K = E_\alpha$. This leads to the result

$$\nu = \frac{\sqrt{2E_\alpha/M_\alpha}}{2R} = \frac{\sqrt{E_\alpha}}{A^{1/3}} \times 2.9 \times 10^{21}\ \mathrm{s}^{-1} \tag{5-119}$$

where E_α is measured in MeV and, as usual, we have taken $R = r_0 A^{1/3}$ with $r_0 = 1.2$ fm. From this we obtain, for example, a value of $\nu = 10^{21}$ sec^{-1} for ^{238}U with $E_\alpha = 5.6$ MeV. It is about an order of magnitude larger than the best values deduced from measurements. Part of the reason for the poor agreement comes from the fact that heavy nuclei do not have the simple spherical shape assumed here. Furthermore, the replacement of K by E_α may also have caused some loss of accuracy.

Transmission coefficient. The transmission coefficient $\mathcal{T}$ for a particle to tunnel through a one-dimensional square potential barrier of height V_0 and width b is given in standard quantum mechanics texts as

$$\mathcal{T} = \left\{1 + \frac{V_0^2}{4E(V_0 - E)} \sinh^2 \kappa b\right\}^{-1} \tag{5-120}$$

where

$$\kappa = \frac{1}{\hbar}\sqrt{2m(V_0 - E)} \tag{5-121}$$

and E is the kinetic energy ($E < V_0$) and m is the mass of the particle. Outside the potential barrier, the particle is free and the wave function is sinusoidal. Inside the barrier, the wave function decays exponentially, since the kinetic energy of the particle is less than the barrier height, as shown schematically in Fig. 5-9. A limiting case with $V_0 \gg E$ is particularly simple to calculate. In this limit, $\kappa b \to \infty$, and $\sinh \kappa b \longrightarrow e^{\kappa b}$. The transmission coefficient in (5-120) simplifies to the form

$$\mathcal{T} \longrightarrow e^{-2\kappa b} \tag{5-122}$$

The factor $e^{-\kappa b}$ expresses the attenuation of the amplitude of the wave in going through the barrier and it is quite reasonable to expect that the transmission coefficient is essentially given by the square of this factor. For our case of $V_0 \approx$ 30 MeV and E_α in the range 4 to 9 MeV, the condition of $V_0 \gg E_\alpha$ is adequately satisfied for the accuracies we need. As a result, we can use the approximate form of $\mathcal{T}$ given in (5-122) for the rest of the discussion.

The true potential barrier experienced by an α-particle in heavy nuclei is far more complicated than the square well example in the previous paragraph. However, the results of the one-dimensional treatment remains valid on the whole as long as the potential well is spherically symmetric. In this case, the radial part of the Schrödinger equation has the form

$$\frac{d^2u(r)}{dr^2} + \frac{2\mu}{\hbar^2}\left\{(E_\alpha - V(r)) - \frac{\ell(\ell+1)\hbar^2}{2\mu r^2}\right\}u(r) = 0 \tag{5-123}$$

where μ is the reduced mass of the α-particle inside the barrier and $u(r)$ is the radial wave function divided by r.

Since we are interested here in the region just outside the range of nuclear force, the potential $V(r)$ may be taken as purely Coulomb, having the form given by (5-114). All the angular dependence is contained in the $\ell(\ell+1)/r^2$ term, which may be taken as a part of the potential barrier. We may define an "effective" potential barrier,

$$V_b(r) = \left[\frac{1}{4\pi\epsilon_0}\right]\frac{2Ze^2}{r} + \frac{\ell(\ell+1)\hbar^2}{2\mu r^2} = \frac{2Z\alpha\hbar c}{r} + \frac{\ell(\ell+1)\hbar^2}{2\mu r^2} \tag{5-124}$$

The radial equation reduces to the form

$$\frac{d^2u(r)}{dr^2} + \frac{2\mu}{\hbar^2}\left\{E_\alpha - V_b(r)\right\}u(r) = 0 \tag{5-125}$$

This result has essentially the same form as a square well used for our simple model above; the major difference is that the barrier height is now a function of r. The equation must now be solved by better techniques such as the WKB method.

The form of the solution, however, remains very similar to that given in (5-122) if we make the replacement

$$\kappa b \longrightarrow \int_R^{R_1} \sqrt{\tfrac{2\mu}{\hbar^2}}\left\{V_b(r) - E_\alpha\right\}^{1/2} dr \tag{5-126}$$

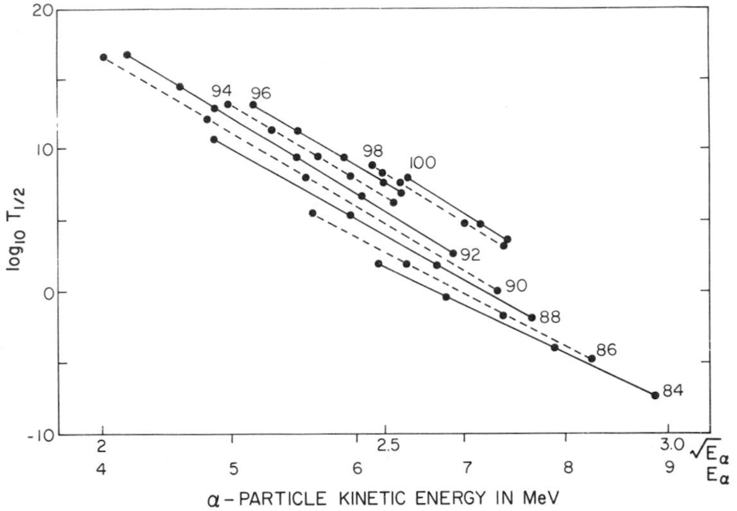

Fig. 5-11 Semi-log plot of the transition rate $\mathcal{W}$ for α-decay as a function of the square root of E_α, the kinetic energy of α-particle emitted. The observed values for the different isotopes of each element, labelled by proton number Z, are closer to straight lines than the plot given in Fig. 5-10, a result expected from the arguments leading to (5-129).

The integration is taken from $r = R$ at the nuclear surface to $r = R_1$ where R_1 is given by the relation

$$V_b(R_1) - E_\alpha = 0$$

For $\ell = 0$, R_1 has the simple form

$$R_1 = \left[\frac{1}{4\pi\epsilon_0}\right]\frac{2Ze^2}{E_\alpha} = \frac{2Z\alpha\hbar c}{E_\alpha}$$

In such cases, the integral can be carried out explicitly, and the result for the transmission coefficient becomes

$$\ln \mathcal{T} = -2\kappa b = -\frac{2R_1}{\hbar}\sqrt{2\mu E_\alpha}\left\{\cos^{-1}\sqrt{\tfrac{R}{R_1}} - \sqrt{\tfrac{R}{R_1}\left(1 - \tfrac{R}{R_1}\right)}\right\} \qquad (5\text{-}127)$$

Since we can take $R \ll R_1$ for the accuracies we need, the arc cosine term may be approximated by $(\pi/2 - \sqrt{R/R_1})$ and R/R_1 may be dropped in the last term. The result simplifies to the form

$$\ln \mathcal{T} = -\frac{2R_1}{\hbar}\sqrt{2\mu E_\alpha}\left(\frac{\pi}{2} - 2\sqrt{\tfrac{R}{R_1}}\right) = 3.26\sqrt{ZA^{1/3}} - 3.97\frac{Z}{\sqrt{E_\alpha}} \qquad (5\text{-}128)$$

where E_α in the final form of the expression is measured in MeV.

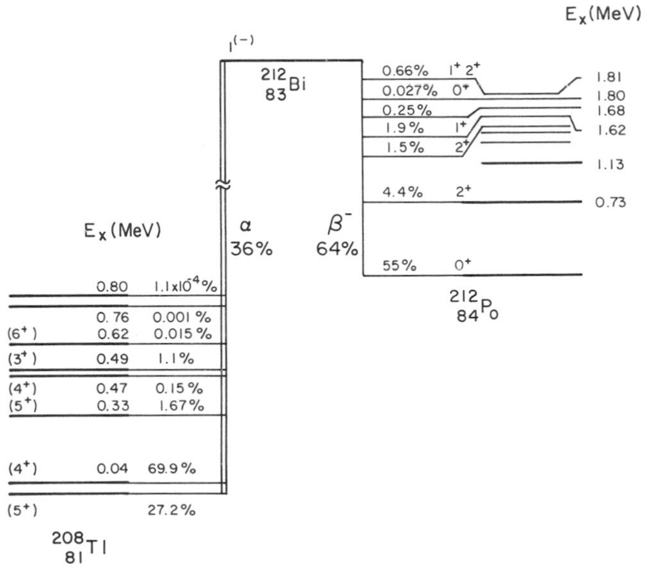

Fig. 5-12 Energy level diagram showing the decay of ^{212}Bi. The Q-value for α-decay is 6.206 MeV leading to the ground state of ^{208}Tl and the Q-value for β-decay is 2.246 MeV leading to the ground state of ^{212}Po. Decays leading to the excited states of both daughter nuclei are also observed.

It is much harder to make a simple estimate for p_α, the probability of finding an α-cluster inside a nucleus. Since this factor depends on the wave function of a nuclear state, we can expect it to be somewhat different from nucleus to nucleus. However, the value of p_α must be essentially of the same order of magnitude for all heavy nuclei, as there are only small fractional differences in their masses. For an estimate, we shall take $p_\alpha = 0.1$ for all the heavy, α-radioactive nuclei. In view of the uncertainties in calculating ν and $\mathcal{T}$, the assumption of a constant p_α for all heavy nuclei is not out of line.

Energy and mass dependence. We now have reasonable estimates for all three factors in (5-117), p_α, ν and $\mathcal{T}$. Since the Geiger-Nuttall law is a relation between the logarithm of the rate of α-particle emission as a function of the α-particle kinetic energy E_α, we need to convert the expression for $\mathcal{W}$ derived above to logarithm in the base 10, and we obtain the result

$$\begin{aligned}\log_{10}\mathcal{W} &= \log_{10}p_\alpha + \log_{10}\nu + \log_{10}\mathcal{T}\\ &= 20.46 + \log_{10}\frac{\sqrt{E_\alpha}}{A^{1/3}} + 1.42\sqrt{ZA^{1/3}} - 1.72\frac{Z}{\sqrt{E_\alpha}}\end{aligned} \qquad (5\text{-}129)$$

The dominant energy dependence comes from the last term, in agreement with the empirical result of Geiger-Nuttall law.

In addition to energy dependence, there is also a dependence of $\log_{10} \mathcal{W}$ on Z and, to a lesser extent on A, in (5-129). To show the Z dependence, we can plot $\log_{10} \mathcal{W}$ as a function of $\sqrt{E_\alpha}$ separately for each element. The experimental values are now clustered much closer to straight lines than we have seen earlier in Fig. 5-10. The curves belonging to different Z values are running parallel to each other, as shown in Fig. 5-11. The remaining deviations in the distribution of data points from straight lines, which are usually too small to be noticed in such a plot, are due to nuclear structure effects which can only be explained by a proper account of p_α. For our purposes, we shall be satisfied with the success of a relatively simple theory to explain almost 30 orders of magnitude difference in the half-lives.

We have implicitly assumed in the above discussion that there is only a single kinetic energy E_α for all the α-particles emerging from a nucleus. In fact, it is common to find several different groups of α-particles emitted by the same parent nucleus. The example of ^{212}Bi shown in Fig. 5-12 has more than 14 different decays known and each one leaves the residual nucleus in a different state. In addition to naturally occurring α-radioactive nuclei, there are also man-made α-activities to enrich the variety of samples that can be used in the study.

§5-7 NUCLEAR FISSION

In addition to α-decay, heavy nuclei can also increase the binding energy of the system of nucleons as a whole by fission into two or more fragments. Nuclear fission can take place spontaneously or be induced by another reaction. In the latter case, the fission is the result of stimulation from energy supplied by an external source such as the capture of a neutron. In general, spontaneous fission reactions are rare events, and most of our knowledge on the subject of fission is derived from induced fission.

Fission involves the movement of many nucleons at the same time and is therefore an example of the collective degrees of freedom in nuclei. We may visualize the process in the following way. For a heavy nucleus, it may be energetically more favorable to assume a shape so that the nucleons are divided into two overlapping groups separated by some distance d between their centers of mass. Since the range of nuclear force is short and the saturation property favors interaction among a few nearby nucleons, the binding energy due to nuclear interactions is not reduced in any significant way by such a separation. In terms of the Weizacker semi-empirical mass formula, the reduction comes mainly from a small increase in the surface area as a result of the deformation. On the other hand, the repulsive Coulomb energy is decreased by the larger average distance between protons. In terms of a liquid drop model (see §4-11), the tendency to deform the nuclear shape for a heavy nucleus may be viewed as sacrificing some surface energy in favor of recovering a larger amount of energy from a decrease in Coulomb repulsion. The net result is that the binding energy of the system is increased and the equilibrium shape of heavy nuclei becomes deformed.

For such nuclei, ***spontaneous fission*** is possible in principle, since the system may gain even more binding energy by separating into two or more fragments. The only thing preventing spontaneous fission from happening more often is the fission barrier through which the fragments must tunnel. We shall return to the source of this barrier later. Since the condition for large deformation required for fission is found only in extremely heavy nuclei, spontaneous fission does not become an important decay channel until $A > 240$. For example, the partial half-life for spontaneous fission in ^{232}U is around 10^{14} years, whereas the value for α-decay is only 2.85 years. On the other hand, by the time we get to $^{254}_{98}$Cf, the half-life reduces to 60 days with a branching ratio of 99% for spontaneous fission.

Induced fission. Nuclear fission was actually first discovered through ***induced fission*** before spontaneous fission was known. Induced fission may be defined as any reaction of the type $A(a, b)B$ with final products b and B being nuclei of roughly comparable mass. The best-known example is the fission of $^{235}_{92}$U induced by thermal neutrons. A compound nucleus ^{236}U* (where the asterisk in the superscript indicates that the nucleus is in an excited state) forms the intermediate state for the reaction. For all practical purposes we may regard the kinetic energy of the incident thermal neutron to be zero. In this limit, the excitation energy of the compound nucleus is simply the excess of the binding energy of ^{235}U, together with a "free" neutron, over that of ^{236}U. The value, which may be calculated from a table of mass excess, is 6.5 MeV. The ground state of ^{236}U is unstable toward α-particle emission, with a half-life of 2.4×10^7 years. However, the excitation energy brought along by the neutron capture sets the entire nucleus into vibration and many fission channels are opened as a result. A typical exit channel is the reaction

$$^{235}_{92}\text{U} + n \rightarrow {}^{92}_{36}\text{Kr} + {}^{142}_{56}\text{Ba} + 2n \tag{5-130}$$

The energy liberated in this example is around 180 MeV. The two neutrons emitted in the process are called ***prompted neutrons*** since they are released as a part of the fission process. However, both ^{92}Kr and ^{142}Ba are neutron unstable. This we can see, for example, from the fact that the heaviest stable isotopes of Kr and Ba are ^{86}Kr and ^{138}Ba, respectively. Among the possible decays of the two nuclei, neutrons may be released either directly from ^{92}Kr and ^{142}Ba or as a result of other unstable nuclei formed from their decays. These neutrons are called ***delayed neutrons*** since they emerge after some delays — as, for example, due to intervening β- and γ-ray decays.

A preference for asymmetric fission, fission into two fragments of unequal mass, may be understood from the large neutron excess carried by heavy nuclei. When such a nucleus undergoes fission, say into two medium-weight nuclei with nucleon number roughly $\frac{1}{2}A$ each, the fragments would have to be nuclei very far away from the valley of stability. As a result, many prompt neutrons would have

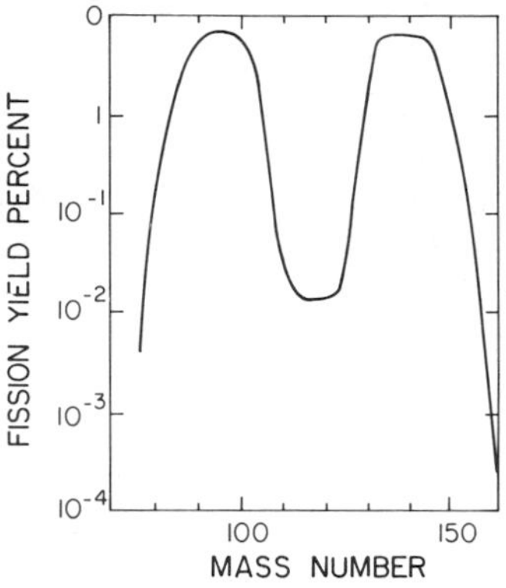

Fig. 5-13 Schematic diagram showing the distribution of fission fragments as a function of nucleon number. The reason that a bimodal distribution is favored comes from the fact that the ratio of neutron to proton numbers N/Z for heavy nuclei is larger than those for medium weight nuclei constituting the fragments.

to be emitted and this, in turn, increases the number of independent entities in the final state. Phase space considerations, however, strongly favor exit channels with the minimum number of products. It is therefore more economical to have one heavier fragment so as to reduce the number of prompt neutrons that have to be emitted. The net result, shown schematically in Fig. 5-13, is that the fission fragments has a bimodal distribution as a function of mass. Besides binary fission involving two final nuclei (plus prompt neutrons), ternary fission consisting of three final nuclei is also commonly observed. In principle larger numbers of fragments are also possible, but, again, phase space considerations reduce their probability.

The reason for using thermal neutrons to induce ^{235}U fission in most nuclear reactors is related to the neutron absorption cross section. As can be seen from Fig. 5-14, the probability for ^{235}U to form a compound nucleus decreases almost exponentially with increasing neutron energy. The only exceptions are small resonances, related to the single-particle states of ^{236}U at neutron threshold energy. For our purpose here, we may ignore them since their combined contribution to the fission cross section is not significant. In contrast, the neutron induced fission cross section for ^{238}U, the dominant component in uranium ore, is negligible until the neutron energy is above 1 MeV, or 10^{10} K in terms of temperature. Consequently, fast neutrons are more suitable to induce fission in ^{238}U, the principle behind breeder reactors.

Fission barrier. A crude model of the fission barrier that inhibits fission may be constructed from a liquid drop model using the Weizacker semi-empirical mass formula given in (4-85) as the starting point. For simplicity, we shall take a hypothetical nucleus of $A = 300$ and $Z = 100$ and take symmetric fission into two equal fragments of $A = 150$ and $Z = 50$ each for our calculation. The volume energy term in (4-85) is unchanged when the nucleus is split into two separate pieces and may be omitted from consideration here. Similarly, changes in the

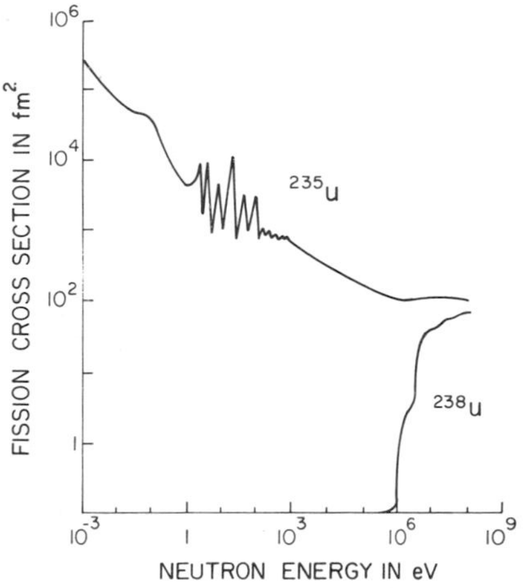

Fig. 5-14 Schematic diagram showing the cross sections for neutron induced fission on ^{235}U and ^{238}U as a function of neutron energy. Since the cross section is large at low neutron energies for ^{235}U, thermal neutrons are more effective to induce fission. In contrast, fission of ^{238}U requires neutrons of much higher energies, since the cross section is insignificant until neutron energy is above 1 MeV, corresponding to a temperature of 10^{10} K.

pairing and symmetry energy terms are too small for us to be concerned with here. Only two terms, the surface energy term $\alpha_2 A^{2/3}$ and the Coulomb energy term $\alpha_3 Z(Z-1)/A^{1/3}$ remain and they form the main contributions to the fission barrier. Since both terms enter the binding energy equation with negative signs, a decrease in either term will increase the binding energy of the system as a whole.

When the hypothetical nucleus undergoes a transformation from $A = 300$ to two fragments of $A = 150$ each, the loss in binding energy due to changes in the surface energy is given by

$$E_{\text{surface}} = \alpha_2 \left\{ 2 \times \left(\tfrac{1}{2}A\right)^{2/3} - A^{2/3} \right\} = 0.26\alpha_2 A^{2/3} \approx 200 \text{ MeV} \tag{5-131}$$

where we have used the value $\alpha_2 = 17$ MeV from §4-11. Such a change in the binding energy of the system takes place between the situation when all 300 nucleons are in a single nucleus and when they separate into two clusters sufficiently far away from each other that there is no more nuclear interaction remaining between them. We may model the separation between the two fragments in terms of a distance d between the two centers of mass. Before the commencement of fission we shall take $d = 0$, and after the fission d must be greater than twice the radius of each of the fragments. For simplicity, we shall take the radius of each fragment to be $R_f = r_0(150)^{1/3} \approx 6.4$ fm. It is not easy to calculate the contribution from the surface energy at distances between $d = 0$ and $d \approx 13$ fm. Schematically, it may take on a form as shown by the dashed line in Fig. 5-15(a).

Let us now consider the contribution to fission from the Coulomb energy term. From (4-85), the Coulomb energy of an $A = 300$ nucleus is

$$\alpha_3 \frac{Z(Z-1)}{A^{1/3}} \approx 900 \text{ MeV} \tag{5-132}$$

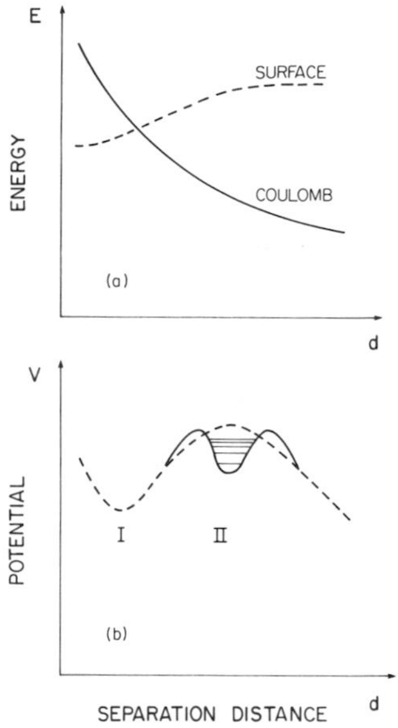

Fig. 5-15 Schematic diagram of fission barrier as a function of the distance d between two fragments. The barrier is a result of the competition between surface and Coulomb energies, represented respectively by the dashed and solid curves in (a). A combination of the two effects produces a potential shown by the dashed curve in (b). Realistic potential barriers for fission is more complicated as, for example, shown by the solid curve in (b), including the possibility of a second well, indicated as II in the diagram, in addition to the primary well indicated as I.

using $\alpha_3 = 0.6$ MeV. For each of the fragments the Coulomb energy is found to be roughly 275 MeV by the same method. There is therefore a gain of 350 MeV in the binding energy from the Coulomb term as a result of splitting the nucleus into two. Such a gain in the Coulomb energy assumes that the two fragments are at infinite distance apart at the end. For values of d of the order of nuclear radius for the $A = 300$ system to $d = \infty$, the Coulomb energy decreases according to a $1/r$ dependence if we ignore any distortion of the fragments. Schematically, the Coulomb energy as a function of d is shown by the solid line in Fig. 5-15(a).

The net gain in binding energy as a result of fission in our simple model is the difference between the gain in the Coulomb energy and the loss in the surface energy, as outlined above. The value, $350 - 200 = 150$ MeV, is of the correct order of magnitude as, for example, compared with the energy released through the reaction given in (5-130). The fission barrier is, then, a sum of the two curves shown in Fig. 5-15(a) and plotted as the dashed curve in Fig. 5-15(b). For lighter nuclei, the difference between the contributions from surface and Coulomb terms is not as clearly marked as in our example and the resulting barrier is much broader; spontaneous fission is suppressed as a result. For extremely heavy nuclei, the opposite is true and the nucleus become unstable toward fission.

Deformation, however, greatly complicates the details of fission barrier even though the main features are given correctly by our simple model. The most interesting aspect due to deformation and other finer considerations is that the detailed shape of the fission barrier may be quite different, as shown schematically by the solid line in Fig. 5-15(b). Sometimes a local minimum in the potential, marked as II in the figure, may develop. The evidence for such a secondary potential well is obtained from narrow resonances found in the fission cross section that cannot come from levels in the main well, marked as I in the figure. Such a potential barrier is often called a double-hump potential for obvious reasons. With such a potential, it is possible that the nucleus may be trapped in the excited states of well II. Such states, especially the low-lying ones, may prefer to de-excite by fission rather than returning to the main well by γ-ray emission. In some cases the rate of spontaneous fission for some of the low-lying states in well II may be hindered by a sufficiently large factor that a fission isomer may develop as a result.

Partly because of its importance to applied work, fission of heavy nuclei has been investigated extensively in many experiments. Theoretical studies that try to understand the equilibrium shape and the process leading to fission are, however, complicated, and progress seems to be rather slow.

Problems

5-1. The first excited state of ^{20}Ne at 1.634 MeV has $J^\pi = 2^+$ and decays to the 0^+ ground state with a half-life of 0.655 ps. Find the reduced transition probability $B(E2)$ in units of e^2fm^4 and the transition rate $\mathcal{W}$ in Weisskopf units.

5-2. The following list of corrected superallowed ft values in seconds are taken from A. Sirlin, *Phys. Rev. D* **35** [1987] 3423:

$^{14}\text{O} \rightarrow {}^{14}\text{N}$	3074.0 ± 3.9	$^{26m}\text{Al} \rightarrow {}^{26}\text{Mg}$	3068.1 ± 3.7
$^{34}\text{Cl} \rightarrow {}^{34}\text{S}$	3069.0 ± 4.7	$^{38m}\text{K} \rightarrow {}^{38}\text{Ar}$	3066.6 ± 4.6
$^{42}\text{Sc} \rightarrow {}^{42}\text{Ca}$	3077.5 ± 7.5	$^{46}\text{V} \rightarrow {}^{46}\text{Ti}$	3074.7 ± 4.3
$^{50}\text{Mn} \rightarrow {}^{50}\text{Cr}$	3069.6 ± 4.4	$^{54}\text{Co} \rightarrow {}^{54}\text{Fe}$	3069.0 ± 1.6

Find the vector coupling constant G_V from this list of results. From the value G_V obtained, find the ratio $|G_A/G_V|$ using the value of 621 s for the half-life of neutron β^--decay and an estimate of the value of $f(Z, E_0)$ from Fig. 5-5.

5-3. Since the end-point energy E_0 is difficult to measure precisely, the Q-value of a nuclear β-decay is often determined from the corresponding (p, n) or (n, p) reaction. Calculate the Q-value of the superallowed β^+-decay of ^{26m}Al ($E_x = 0.229 \pm 0.003$ MeV) to the ground state of ^{26}Mg given that the measured Q-value for the ^{26}Mg$(p, n)^{26}$Al reaction leading to the ground state of ^{26}Al is -4.786 ± 0.002 MeV.

5-4. The $7/2^+$ state at 1.72 MeV in ^{21}Ne has a half-life of 48 fs and decays 94% of the time to the 0.33 MeV $5/2^+$ state with a mixing ratio $\delta = 0.14 \pm 0.02$, and 6% of the time to the $3/2^+$ ground state. Find the $B(E2)$ and $B(M1)$ values for the transitions involved.

5-5. A more realistic radial wave function for nucleons than the uniform sphere model used in §5-3 for calculating the Weisskopf single-particle estimates of electromagnetic transitions is the spherical harmonic oscillator radial wave function. Use the explicit forms given in Table 6-2 to evaluate the matrix element $\langle r^2 \rangle$ and compare the results with values given by (5-51).

5-6. The orbital angular momentum part of the single-particle wave function is given by spherical harmonics $Y_{\ell m}(\theta, \phi)$. Use this together with the radial integrals evaluated above and calculate the single-particle values for $E2$-transitions.

5-7. The nucleus ${}^{12}_{7}\text{N}$ decays to ${}^{12}_{6}\text{C}$ with a Q-value of 16.38 MeV. Calculate the maximum recoil energy of the daughter nucleus. If the probability of emitting leptons with a given momentum p_ν, up to $p \sim Q/c$, is $\sim p^2 dp$, given purely by phase space conditions, calculate the distribution of the number of positrons emitted as a function of energy. Ignore Coulomb corrections to the charged lepton emitted.

5-8. Assume that a binary fission of ${}^{236}\text{U}^*$ is accompanied by two prompt neutrons, and calculate the optimum neutron and proton numbers of the two fragments using the Weizacker semi-empirical mass formula (4-85).

5-9. Calculate the excitation energy of ${}^{236}\text{U}$ formed by the capture of a thermal neutron to the ground state of ${}^{235}\text{U}$.

Chapter 6

NUCLEAR STRUCTURE

The experimental information outlined in the previous two chapters on energy level positions, static moments, transition rates, and reaction cross sections provides us with the foundation for nuclear structure studies. Partly because of the complexity of quantum-mechanical many-body systems, model construction is a necessary step toward a proper theory for understanding nuclei. In this chapter we shall examine two broad classes of nuclear structure models, microscopic models and macroscopic models. Since a nucleus is made up of nucleons, we can use a *microscopic* many-body Hamiltonian based on individual nucleons interacting with each other through a two-body potential. The eigenfunctions obtained by solving the Schrödinger equation may then be used to calculate observables and the results compared with experiments. In principle, such a calculation is possible once the nucleon-nucleon interaction is given. In practice, special techniques must be designed to solve the many-body problems involved, and we shall examine a few of the more basic ones used in nuclear physics.

On other hand, many of the observed properties of a nucleus involve the motion of many nucleons "collectively." It is more appropriate to describe these phenomena using a Hamiltonian expressed in terms of the bulk or *macroscopic* coordinates of the system such as mass, radius, and volume. We shall begin with two popular collective models describing the vibrational and rotational motions of a nucleus.

§6-1 VIBRATIONAL MODEL

We have seen earlier in the discussion of nuclear binding energies in §4-10 that in many ways the nucleus may be looked on as a drop of fluid. A large number of the observed properties can be understood from the interplay between the surface tension and the volume energy of the drop. Here, we shall extend the analogy to nuclear excitations due to *vibrational* motion.

For simplicity we shall say that, at equilibrium, the shape of a nucleus is spherical, *i.e.,* the potential energy is minimum when the nucleus assumes a spherical shape. (This is purely a simplifying assumption for our discussion here. It is made in part for the reason that spherical nuclei do not have rotational degrees of freedom and, as a result, vibrational motion stands out more clearly.) In practice, for many nuclei the most stable shape is a deformed one, as we shall see later in §6-3, and vibrational motions built upon deformed shapes are also a commonly observed feature in nuclei.

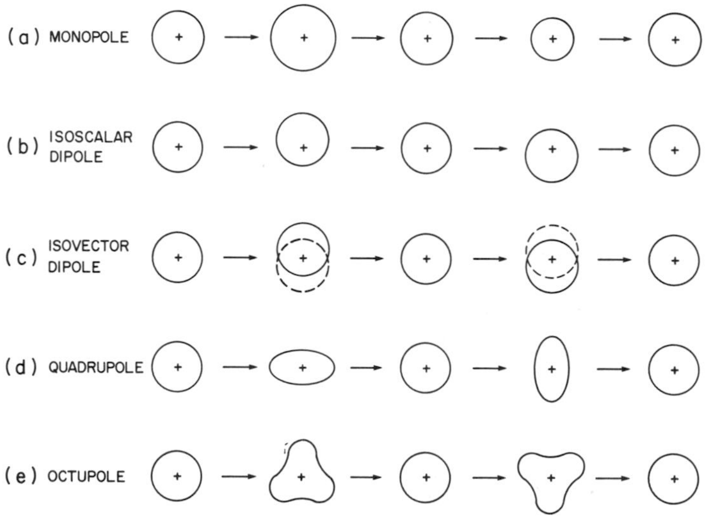

Fig. 6-1 Time evolution of low-order vibrational modes. The monopole oscillation shown in (*a*) involves variations in the size without changing the overall shape. For isoscalar dipole vibrations, shown in (*b*), the nucleus moves as a whole. An isovector dipole vibration is set up if neutrons and protons are oscillating in opposite phase as shown in (*c*). In quadrupole vibrations the nucleus changes shape from prolate to oblate and back again as shown in (*d*). A similar situation occurs for octupole vibrations shown in (*e*).

Breathing mode. When a nucleus acquires an excess of energy, for example, from Coulomb excitation due to a charged particle passing nearby, it can be set into vibration around its equilibrium. We can envisage several different types of vibration for a nucleus. For example, the nucleus may change its size without changing its shape, as shown in Fig. 6-1(*a*). Since the volume is now changing while the total amount of nuclear matter remains constant, the motion involves an oscillation in the density. Such a density vibration is similar to the motion of human respiration and is called a *breathing* mode.

For an even-even, spherical nucleus, the ground state spin and parity are 0^+. To preserve the nuclear shape, breathing mode excitation in this case generates states that are also $J^\pi = 0^+$. In Fig. 6-2, we see that in the case of doubly magic nuclei of ^{16}O, ^{40}Ca, ^{90}Zr, and ^{208}Pb, a low-lying, $J^\pi = 0^+$ state is found among the first few excited states. Such low-energy states are often the result of exciting collective degrees of freedom and may be identified as a breathing mode state. On the other hand, nuclear matter is rather stiff against compression and one expects the main part of the breathing mode strength to be much higher in energy. The observed value depends on the number of nucleons in the nucleus, and the peak location is usually found at around $80A^{-1/3}$ MeV.

Shape vibration. A second and more commonly observed type of vibration is an oscillation in the shape of the nucleus without changing the density. This is very similar to a drop of liquid suspended from a water facet. If the liquid drop is disturbed very gently, it starts to vibrate. Since the amount of energy imparted to the drop is usually too small to change its density, the motion simply involves a change in the shape.

For a drop of fluid, departures from spherical shapes without density change may be described in terms of a set of *shape* parameters $\alpha_{\lambda\mu}$ defined in the following way:

$$R(\theta, \phi) = R_0\Big\{1 + \sum_{\lambda\mu} \alpha_{\lambda\mu}(t) Y_{\lambda\mu}(\theta, \phi)\Big\} \tag{6-1}$$

where $R(\theta, \phi)$ is the distance from the center of the nucleus to the surface at angles (θ, ϕ) and R_0 is the radius of an equivalent sphere. Each mode of order λ has $(2\lambda + 1)$ parameters corresponding to $\mu = -\lambda, -\lambda + 1, \ldots, \lambda$. However, because of rotational invariance, these parameters are not independent of each other.

The $\lambda = 1$ mode corresponds to an oscillation around some fixed point in the laboratory, as shown in Fig. 6-1(*b*). If all the nucleons are moving together in unison, the internal structure of the nucleus is not changed by such an "excitation" process and the vibration corresponds to a motion of the center of mass of the nucleus. This is the isoscalar ($T = 0$) dipole mode and is of no interest to the study of the internal dynamics of a nucleus. On the other hand, the isovector ($T = 1$) mode, as we shall see in the next section, corresponds to a dipole oscillation of neutrons and protons in opposite directions, as shown in Fig. 6-1(*c*). This is the cause of giant dipole resonances observed in a variety of nuclei. The $\lambda = 2$ mode describes a quadrupole oscillation of the nucleus. A positive quadrupole deformation means that the nuclear shape is a prolate one, with its polar radius longer than its equatorial radius. On the other hand, a negative quadrupole deformation is one in which the nucleus has an oblate shape, with the equatorial radius longer than the polar radius. A quadrupole vibration, then, corresponds to the situation that the nucleus changes its shape continuously from spherical to prolate, back to spherical and then to oblate, and then back again to spherical, as shown in Fig. 6-1(*d*). Similarly, an octupole ($\lambda = 3$) vibration is depicted in Fig. 6-1(*e*).

The energy associated with vibrational motion may be discussed in terms of the shape parameters $\alpha_{\lambda\mu}$ as functions of time. The shape changes involve both kinetic energy, in transporting nucleons from one location to another, and potential energy, in forcing the nucleus away from its equilibrium shape. For this reason the shape parameters are the more appropriate canonical variables to describe the vibration rather than, for example, coordinates specifying the position of each nucleon in the nucleus.

For small amplitude vibrations, the kinetic energy may be expressed in terms of the rate of change in the shape parameters,

$$T = \frac{1}{2}\sum_{\lambda\mu} D_\lambda \left|\frac{d\alpha_{\lambda\mu}}{dt}\right|^2 \tag{6-2}$$

where D_λ is a quantity having the equivalent role as mass in ordinary (nonrelativistic) kinetic energy in mechanics. For a classical irrotational flow, D_λ is related to the mass density ρ and equilibrium radius R_0 of the nucleus in a liquid drop model,

$$D_\lambda = \frac{\rho R_0^5}{\lambda} \tag{6-3}$$

Similarly, the potential energy may be expressed as

$$V = \frac{1}{2}\sum_{\lambda\mu} C_\lambda |\alpha_{\lambda\mu}|^2 \tag{6-4}$$

Such a form follows naturally from the fact that the minimum is chosen to be at $\alpha_{\lambda\mu} = 0$. As a result, there is no linear dependence of V on $\alpha_{\lambda\mu}$; the leading order is then the quadratic term. For small amplitude vibrations, terms depending on the higher powers of $\alpha_{\lambda\mu}$ may be ignored and we are led to (6-4). The quantity C_λ may be related to the surface and Coulomb energies of the fluid in a liquid-drop model of the nucleus (see A. Bohr and B.R. Mottelson, *Nuclear Structure,* vol. II, Benjamin, 1975, Reading, Massachusetts, p. 660),

$$C_\lambda = \frac{1}{4\pi}(\lambda-1)(\lambda+2)\,\alpha_2 A^{2/3} - \frac{5}{2\pi}\frac{\lambda-1}{2\lambda+1}\,\alpha_3\frac{Z(Z-1)}{A^{1/3}} \tag{6-5}$$

where α_2 and α_3 are the surface and Coulomb energy parameters defined in (4-85).

In terms of C_λ and D_λ, the Hamiltonian for vibrational excitation of order λ may be written in the form

$$H_\lambda = \frac{1}{2}C_\lambda \sum_\mu |\alpha_{\lambda\mu}|^2 + \frac{1}{2}D_\lambda \sum_\mu \left|\frac{d\alpha_{\lambda\mu}}{dt}\right|^2 \tag{6-6}$$

If different modes of excitation are decoupled from each other, and with any other degrees of freedom a nucleus may have, H_λ, C_λ, and D_λ are constants of motion. Under these conditions, we can differentiate (6-6) with respect to time and obtain the equation of motion,

$$D_\lambda \frac{d^2\alpha_{\lambda\mu}}{dt^2} + C_\lambda \alpha_{\lambda\mu} = 0 \tag{6-7}$$

Comparing this expression with that for an harmonic oscillator,

$$\frac{d^2x}{dt^2} + \omega^2 x = 0$$

we obtain the result that, for small oscillations, the amplitude $\alpha_{\lambda\mu}$ undergoes harmonic oscillation with frequency

$$\omega_\lambda = \left(\frac{C_\lambda}{D_\lambda}\right)^{1/2} \tag{6-8}$$

and $\hbar\omega_\lambda$ is a quantum of vibrational energy for multipole λ.

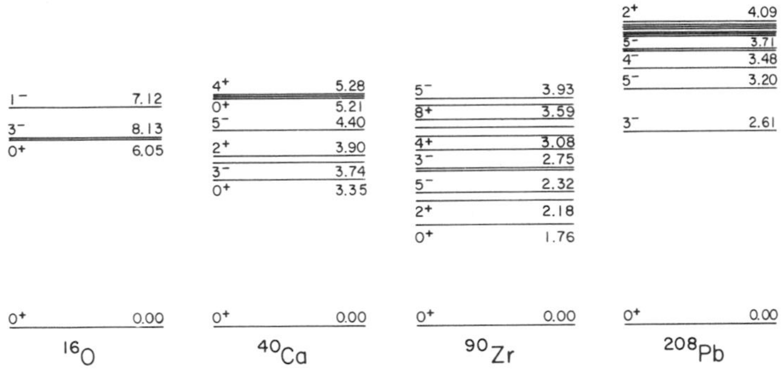

Fig. 6-2 Level structure of the low-lying spectra of doubly-magic nuclei of ^{16}O, ^{40}Ca, ^{90}Zr and ^{208}Pb showing the location of 0^+ breathing mode and 3^- octupole vibrational states.

Quadrupole and octupole vibrations. A vibrational quantum of energy is called a *phonon* since it is a form of "mechanical" energy, reminiscent of the way sound wave energy propagates through a medium. Each phonon is a boson carrying $\lambda\hbar$ units of angular momentum and having a parity $\pi = (-1)^\lambda$. For an even-even nucleus, the ground state may be regarded as the zero-phonon state. The lowest vibrational state built upon such a $J^\pi = 0^+$ state has $J = \lambda$ and $\pi = (-1)^\lambda$ because of angular momentum coupling. Examples of one-phonon octupole excitations are found in the form of a low-lying 3^- state in all the closed shell nuclei from ^{16}O to ^{208}Pb, as shown in Fig. 6-2. In terms of a single-particle picture, an excited state may be formed by promoting a particle from an occupied orbit below the Fermi surface to an empty one above. Since orbits below and above the Fermi surface near a close shell generally have opposite parities (see §6-5), a negative parity state is formed from one-particle, one-hole excitations. We shall see later in §6-5 that the typical energy for exciting a nucleon up one major shell is $41A^{-1/3}$ MeV, about 16 MeV in ^{16}O and 7 MeV in ^{208}Pb. As shown in Fig. 6-2, the observed 3^- vibrational states are much lower in energy than this value. This

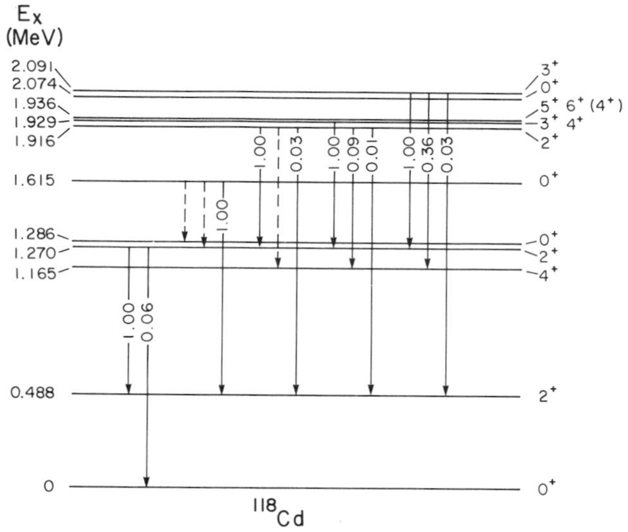

Fig. 6-3 The observed low-lying energy levels of ^{118}Cd showing quadrupole vibrational states up to three-phonon excitations. The spin and parity of the 1.929 MeV state may be either 3^+ or 4^+ and the 1.936 MeV state may be 5^+ or 6^+ with the possibility of 4^+ not ruled out. The 0^+ state at 1.615 MeV may not be a member of the vibrational spectrum. Vertical arrows indicate $B(E2)$ values relative to the observed strongest transition from each state and the dashed lines indicate transitions with only upper limited known. (Based on A. Aprahamian et al., *Phys. Rev. Lett.* **59** [1987] 535.)

is evidence that many nucleons are acting in an coherent or "collective" manner to lower the energy required to excite the state.

In nuclei such as the even-even cadmium (Cd) and tin (Sn) isotopes, the first excited state above the $J^\pi = 0^+$ ground state is inevitably a 2^+ state and at roughly twice the excitation energy there is often a triplet of levels with $J^\pi = 0^+$, 2^+, and 4^+. Such behavior is typical of nuclei excited by quadrupole vibration. The first excited state is the one-phonon state, having $J^\pi = 2^+$ of a quadrupole phonon. When two quadrupole phonons are coupled together in a two-phonon excitation, the possible angular momenta range from 0 to 4 ($= 2\lambda$). However, symmetry requirements between the two identical phonons eliminates the possibilities 1^+ and 3^+ (see Problem 6-1), and only states with $J^\pi = 0^+$, 2^+, and 4^+ are left. The splitting between the $0^+ - 2^+ - 4^+$ triplet is outside the scope of a vibrational model, since there is no mechanism in this simple picture to account for any J-dependence among states having the same number of excitation phonons. The observed fact that the sequence between these three levels is different in different nuclei implies that the nature of the J-dependence may be a complicated one.

With three quadrupole-phonons there are five possible excited levels, 0^+, 2^+, 3^+, 4^+, and 6^+. Since these states lie high in excitation energy where the density of states is large, admixture with states formed by other modes of excitation becomes important. As a result, it is not easy to identify a complete set of three-phonon excited states. One such example, shown in Fig. 6-3, is found recently in ^{118}Cd.

Electromagnetic transitions. Besides energy level positions, the vibrational model also predicts electromagnetic transition rates between states consisting of different numbers of excitation phonons. Since vibrational states have the same structure as harmonic oscillator states, we can make use of the result that, for harmonic oscillator states, the transition between an n-phonon state to an $(n-1)$-phonon state takes place by the emission of one quantum of energy. If nuclear vibrations are purely harmonic in nature, electric transition operator $\boldsymbol{O}_{\lambda\mu}(E\lambda)$ for vibrational mode of order λ must be proportional to the annihilation operator $\boldsymbol{b}_{\lambda\mu}$ for a phonon of multipolarity (λ, μ)

$$\boldsymbol{O}_{\lambda\mu}(E\lambda) \propto \boldsymbol{b}_{\lambda\mu}$$

Because of its collective nature, nuclear excitations induced predominantly by quadrupole vibrations have large $E2$-transition rates from the one-phonon state to the ground state compared with Weisskopf single-particle estimates given in §5-3. Similarly, strong $E3$-transition strength is also observed in states excited by octupole vibrations.

The matrix element of a phonon annihilation operator $\boldsymbol{b}$ between two harmonic oscillator states is given in §F-1 as

$$\langle n'|\boldsymbol{b}|n\rangle = \sqrt{n}\delta_{n',n-1} \tag{F-23}$$

Since the reduced transition probability is proportional to the square of the transition matrix element, we find that its value between n- and $(n-1)$-phonon states is proportional to n, the number of phonons in the initial state of the decay,

$$B(E\lambda, n \rightarrow n-1) \propto n \tag{6-9}$$

Because of this relation, we expect the transition probability from a two-phonon state to a one-phonon state also to be enhanced and roughly equal to twice that of the transition from a one-phonon state to a zero-phonon state in the same nucleus. Transitions between states differing by more than one phonon are higher order processes, as they involve simultaneous emission of more than one phonon. Since the probability for such processes is much lower than that for single-phonon emissions, the corresponding transition rates are expected to be small. Both points are observed to be essentially correct in vibrational nuclei, as can be seen from the examples given in Table 6-1.

Implicit in our discussion is the assumption that the vibration is an axially symmetric one; *i.e.*, variations along the x- and y-directions are always equal to each other, only their ratio to that along the z-axis is changing as a function of

Table 6-1 Quadrupole moment and $B(E2)$ values of vibrational nuclei.

Nucleus	$B(E2;2_1^+\to 0_1^+)$		$B(E2;4_1^+\to 2_2^+)$		$\frac{B(E2;4^+\to 2_1^+)}{B(E2;2_1^+\to 0^+)}$	$B(E2;2_2^+\to 0_1^+)$		Q
	10^2e^2fm^4	W.u.	10^2e^2fm^4	W.u.		10^2e^2fm^4	W.u.	$e\text{fm}^2$
^{60}Ni	1.86	13.3	8.2	59	4.4	0.03	0.22	3.0
^{62}Ni	1.8	12	2.6	18	1.5	0.09	0.6	8.8
^{102}Ru	12.4	43.9	19	66	1.5	0.31	1.09	−68
^{110}Cd	8.58	27.4	14	46	1.7	0.42	1.34	−39
^{112}Cd	9.69	30.2	20.0	62.4	2.1	0.21	0.65	−37
^{114}Cd	10	31	19	58	1.9	1.8	5.4	−36
^{116}Cd	10.6	31.6	19.4	57.8	1.8	0.37	1.1	−42
Vib. model	large		large		2.0	small		0*

* For spherical nuclei.

time. This type of vibration is generally known as β-vibration. More generally, we can also have γ-vibrations in which the nucleus deforms into an ellipsoidal shape in the equatorial direction. In other words, a section of the nucleus in xy-plan at any instant of time is an ellipse rather than a circle, as in the case of β-vibration. In addition to purely harmonic vibrational motion, anharmonic terms may be present in a nucleus. Furthermore, vibrations may also be coupled to other modes of excitation in realistic situations.

If the amplitude of vibration is large, the above treatment will no longer apply. In fact, if the vibration is energetic enough, a "drop" of nuclear matter may dissociate into two or more droplets. Such ideas are used with success in fission studies. However, in order for the nucleus to develop toward a shape that is convenient for splitting up into two or more fragments, there must be a superposition of many different vibrational modes. Furthermore, the various modes must be strongly coupled to each other so that energy can flow from one mode to another. The mathematical problem involved here is not simple, but the basic physical idea is a sound one. However, we shall not investigate this topic here.

§6-2 GIANT RESONANCE

Giant resonance is a term generally used to describe broad resonances at energies tens of MeV above the ground state. The reason that these excitations are called "giant" resonances comes from the fact that both their total strengths and their widths are much larger than typical resonances built upon single-particle (non-collective) excitations. In the energy region where giant resonances appear, the density of states is sufficiently high and the number of open decay channels sufficiently large that states in a given energy region cannot be very different from each other in character. As a result, we expect only smooth variations of reaction cross sections when the nucleus is excited. The presence of large strength localized in the region of a few MeV is therefore of interest, since it must be related to some special feature of the nuclear system particular to the energy region.

For most giant resonances, the excitation strengths are found to be essentially independent of the probe used to excite the nucleus, γ-ray, electron, proton, α-particle, and ^{16}O ion. Furthermore, both the width and the location of the peak of the strength distributions in different nuclei are found to vary smoothly as a function of nucleon number A without any significant dependence on the detailed structure of the nucleus involved. For example, the location of the isovector giant dipole resonance in different nuclei is well described by the relation

$$E_1 \approx 78A^{-1/3} \tag{6-10}$$

Prominent dipole resonances, as well as other types of giant resonances, are found in almost all the nuclei studied, from ^{16}O to ^{208}Pb, as can be seen in Figs. 6-5 and 6-6.

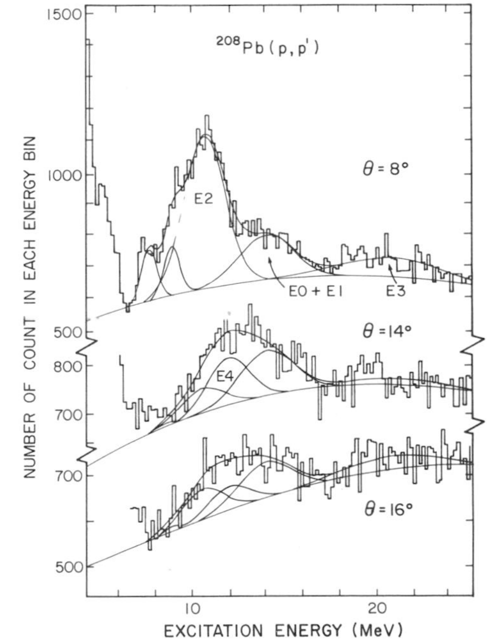

Fig. 6-4 Differential cross section of ^{208}Pb(p, p') reaction with 200 MeV protons at different scattering angles, showing the angular dependence of various giant resonances excited in the reaction. (Taken from F.E. Bertrand, in *Studying Nuclei with Medium energy Protons*, TRIUMF report [1983] TR-83-3.)

The cause of giant resonances is the collective excitation of nucleons. The energy gap between two adjacent major shells, as mentioned in the previous section, is roughly $41A^{-1/3}$ MeV and the parity of states produced by $1p1h$-excitations up one major shell is negative in general. Negative parity giant resonances are therefore predominantly caused by one-major shell excitations. For positive parity excitations there are two possibilities, rearrangement of the particles in the same major

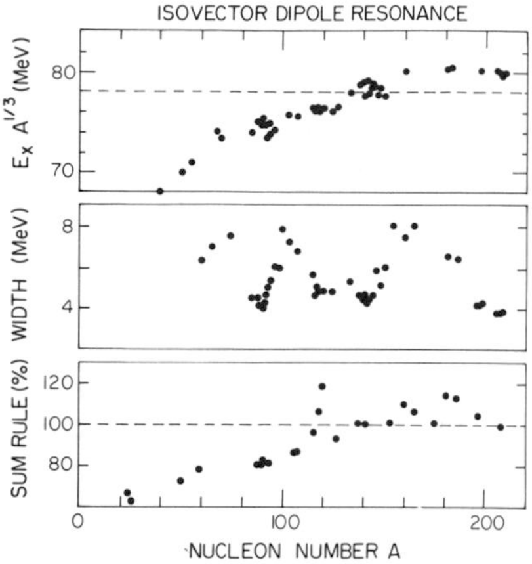

Fig. 6-5 Variations of peak location (a), width (b) and total strength (c) of isovector giant dipole resonance as a function of nucleon number. (Taken from F.E. Bertrand, *Nucl. Phys.* **A354** [1981] 129c.)

shell ($0\hbar\omega$-excitation) or elevating a particle up two major shells ($2\hbar\omega$-excitation). Other possibilities, such as excitations by four major shells ($4\hbar\omega$-excitation) for positive parity resonances and three major shells ($3\hbar\omega$-excitation) for negative parity resonances are less likely because of the larger amounts of energy involved.

Giant dipole resonance. Isovector giant dipole resonances have been studied since the late 1940s. They are excitations of the $J^\pi = 1^-$ strength by promoting nucleons up one major shell. In light nuclei, the observed peaks of strength occur around 25 MeV in excitation energy and, in heavy nuclei, the values are lower, just below 14 MeV in ^{208}Pb. The variation with nucleon number A, as can be seen in Fig. 6-5(a), is fairly well described by the form given in (6-10). The peak position is higher than that expected of a simple one $\hbar\omega$ excitation process of $41A^{-1/3}$ MeV. The difference is caused by the residual interaction between the nucleons which pushes the isovector excitation to a higher energy. The width of the resonance is found to be around 6 MeV without any noticeable dependence on the nucleon number as can be seen in Fig. 6-5(b).

An explanation of giant dipole resonances based on the collective motion of nucleons is provided by the Goldhaber-Teller model. In this model, both neutrons and protons in a nucleus vibrate collectively but in opposite phase with respect to each other, as shown schematically in Fig. 6-1(c). In the dipole mode, the neutrons are moving in one direction along the quantization axis while the protons are going in the opposite direction. The opposite phase is essential here to keep the center of mass of the entire nucleus stationary. Since neutrons and protons are "out of phase" with respect to each other, it is an isovector mode of excitation. In contrast, if the neutrons and protons are in phase, *i.e.,* an isoscalar dipole vibration, all the nucleons will be moving in the same direction and the net result is that the entire nucleus is oscillating around some equilibrium position in the laboratory. Such a motion constitutes a "spurious" state and is of no interest to the study of the nucleus, since it does not correspond to an excited state of the nucleus involving nuclear degrees of freedom.

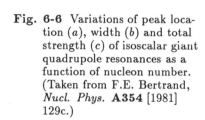

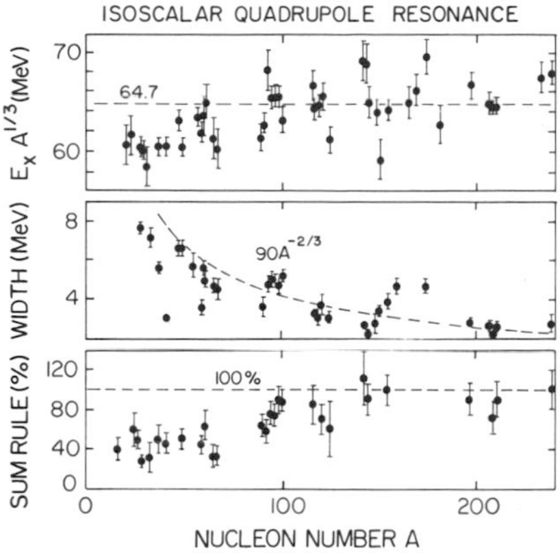

Fig. 6-6 Variations of peak location (*a*), width (*b*) and total strength (*c*) of isoscalar giant quadrupole resonances as a function of nucleon number. (Taken from F.E. Bertrand, *Nucl. Phys.* **A354** [1981] 129c.)

Sum rule quantities. Another useful way to study giant resonances is to find out how much of the total possible transition strength is represented by the observed cross section. The value may be estimated by calculating the related sum rules. The simplest quantity to examine is the total transition strength of a given multipolarity to all the possible final states. The starting state is usually chosen to be the ground state, since this is only type that can be measured directly. The (non-energy-weighted) sum rule is then

$$S = \int_0^\infty \sigma(E)\, dE \tag{6-11}$$

where $\sigma(E)$ is the cross section of the process at excitation energy E. Since an integration is carried out over all the final states, the resulting quantity is a function of the initial state only. For transition strength starting from the ground state, S is then the ground expectation value of an operator related to the transition operator. An example is given later for the case of Gamow-Teller giant resonance. Other types of sum rule quantity, such as energy-weighted ones, have also been studied; we shall, however, restrict ourselves to the simplest one defined in (6-11).

For isovector dipole transitions, the total strength S can be evaluated in a straightforward way under simplifying assumptions that there is no velocity-dependent term in the nucleon-nucleon interaction and the effect due to antisymmetrization among all the nucleons can be ignored (see, *e.g.*, de-Shalit and Feshbach, *Theoretical Nuclear Physics*, Wiley, New York, 1974, pp. 709-713). The result is known as the Thomas-Reiche-Kuhn (TRK) sum rule,

$$\int_0^\infty \sigma(E)\, dE = \frac{2\pi^2\hbar^2\alpha}{M_p}\frac{NZ}{A} \approx 6.0\frac{NZ}{A}\text{MeV-fm}^2 \tag{6-12}$$

The main correction to this simple result comes from antisymmetrization effects and is often accounted for by including a multiplicative factor $(1+\eta)$ in the final form. The value of η is estimated to around 0.5, depending on the model used to simulate the effect of antisymmetrization.

For isovector dipole transitions, the total strength is known experimentally up to ~ 30 MeV in many nuclei. The results are compared in Fig. 6-5(c) with the TRK sum rule evaluated with $\eta = 0$, *i.e.*, no correction for antisymmetry effects. As long as the actual corrections to the TRK sum rules are not too different from the generally accepted value of around 0.5 for η, we see that the measured giant dipole cross sections exhaust large fractions of the total possible strengths. Furthermore, the result is essentially independent of the particular nucleus from which the strength sum is taken. Since the sample of nuclei represented in Fig. 6-5 is quite large, a large variety of ground state wave functions must be represented. The fact that the value of S is essentially given by the TRK sum rule, which makes no specific reference to the ground state wave function of any of the nuclei involved, may be taken as another evidence of the collective nature of the excitation process itself.

Besides isovector dipole excitations, other giant resonances have also been observed in recent years. Both giant quadrupole ($E2$) and giant octupole ($E3$) resonances have been extensively studied in a variety of nuclei. The results for the former are shown in Fig. 6-6 as an example.

Gamow-Teller resonance. In addition to γ-ray excitation, giant resonances have also been observed in connection with β-decay, or more precisely, charge exchange reactions. For example, in the neutron spectra observed in the reaction $^{90}\text{Zr}(p,n)^{90}\text{Nb}$ induced by 45 MeV protons shown in Fig. 6-7, we see that a sharp peak is found leading to the $(J^\pi, T) = (0^+, 5)$ state in ^{90}Nb at 5.1 MeV excitation. The concentration of strength here is expected from the fact that the final state in ^{90}Nb is the isobaric analogue of the ground state of ^{90}Zr. The transition operator is similar to that in Fermi β-decay mediated by the isospin raising operator T_+. However, since the strength is concentrated in one state, the distribution is almost a delta function; the Fermi type of charge exchange strength, therefore, does not necessarily fit into the stereotype of a giant resonance.

Unlike the Fermi operator, the Gamow-Teller strength is shared by a number of adjacent states. Only a small part of the total strength is actually observed in β-decays, since a decay is allowed if the initial state is higher in energy than the final state. The main portion usually lies relatively high in excitation energy and is seen only in charge exchange reactions. Part of the strength in the $^{90}\text{Zr}(p,n)^{90}\text{Nb}$ reaction appears as a "giant resonance" in the neutron spectra shown in Fig. 6-7 at neutron energies just below the IAS peak.

Let us evaluate the sum of Gamow-Teller transition strength in a charge exchange reaction as an illustration. From (5-90), we find that the operator for the axial-vector transition has the form

$$\boldsymbol{O}_{\text{GT}}(\beta^\pm) = G_A \sum_{k=1}^{A} \sum_{\mu} \boldsymbol{\sigma}_\mu(k)\boldsymbol{\tau}_\mp(k) \tag{6-13}$$

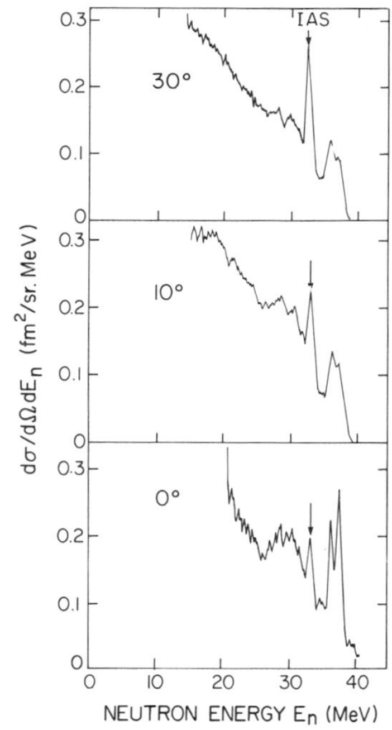

Fig. 6-7 Neutron spectra at different scattering angles from the reaction $^{90}Zr(p,n)^{90}Nb$ induced by 45 MeV protons. The results show the angular dependence of giant Gamow-Teller resonances and isobaric analogue strength excited in a (p,n) reaction. (Taken from Galonsky, A., in *The (p,n) Reaction and the Nucleon-Nucleon Force*, ed. by C.D. Goodman et al., Plenum Press, New York, 1980.)

Following (6-11), we may define the sum rule strength in the following way:

$$S_\pm = G_A^{-2} \sum_f \left|\langle f|O_{\rm GT}(\beta^\pm)|i\rangle\right|^2 \tag{6-14}$$

where $|\,i\,\rangle$ and $|\,f\,\rangle$ are the initial and final nuclear states, respectively. We have also removed the axial-vector coupling constant G_A from the definition to simplify the form of the final result. Since we are summing over all the final states, $S_\pm$ may be transformed into an expectation value using the closure relation,

$$\begin{aligned} S_\pm &= G_A^{-2} \sum_f \langle f|O_{\rm GT}(\beta^\pm)|i\rangle^* \langle f|O_{\rm GT}(\beta^\pm)|i\rangle \\ &= G_A^{-2} \sum_f \langle i|O_{\rm GT}^\dagger(\beta^\pm)|f\rangle\langle f|O_{\rm GT}(\beta^\pm)|i\rangle \\ &= G_A^{-2} \langle i|O_{\rm GT}^\dagger(\beta^\pm) O_{\rm GT}(\beta^\pm)|i\rangle \end{aligned} \tag{6-15}$$

Components of the operators involved here have the following properties:

$$\boldsymbol{\sigma}_\mu^\dagger = (-1)^\mu \boldsymbol{\sigma}_{-\mu} \qquad\qquad \boldsymbol{\tau}_\mp^\dagger = \boldsymbol{\tau}_\pm$$

as can be seen for $\boldsymbol{\sigma}_\mu$ in (3-32) and for $\boldsymbol{\tau}_\pm$ in (2-20). On substituting the explicit form of the Gamow-Teller operator into (6-15), we obtain the sum for the β^+-transitions,

$$\begin{aligned} S_+ &= \langle i|\sum_{k=1}^{A}\sum_{\mu}(-1)^\mu \boldsymbol{\sigma}_{-\mu}(k)\boldsymbol{\tau}_+(k)\boldsymbol{\sigma}_\mu(k)\boldsymbol{\tau}_-(k)|i\rangle \\ &= \langle i|\sum_{k=1}^{A}\boldsymbol{\sigma}^2(k)\boldsymbol{\tau}_+(k)\boldsymbol{\tau}_-(k)|i\rangle \end{aligned} \tag{6-16}$$

where we have made use of (B-58) to obtain $\boldsymbol{\sigma}^2$ from $\sum(-1)^\mu \boldsymbol{\sigma}_{-\mu}\boldsymbol{\sigma}\mu$. Similarly for the β^--transition,

$$S_- = \langle i|\sum_{k=1}^{A}\boldsymbol{\sigma}^2(k)\boldsymbol{\tau}_-(k)\boldsymbol{\tau}_+(k)|i\rangle \tag{6-17}$$

Now, since $\boldsymbol{\tau}_+|n\rangle = |p\rangle$, $\boldsymbol{\tau}_-|p\rangle = |n\rangle$ and $\boldsymbol{\tau}_+|p\rangle = \boldsymbol{\tau}_-|n\rangle = 0$, where $|p\rangle$ is the isospin wave function of a proton and $|n\rangle$ is that for a neutron, we have the results:

$$\begin{aligned} \boldsymbol{\tau}_+\boldsymbol{\tau}_-|p\rangle &= |p\rangle \qquad & \boldsymbol{\tau}_+\boldsymbol{\tau}_-|n\rangle &= 0 \\ \boldsymbol{\tau}_-\boldsymbol{\tau}_+|n\rangle &= |n\rangle \qquad & \boldsymbol{\tau}_-\boldsymbol{\tau}_+|p\rangle &= 0 \end{aligned} \tag{6-18}$$

In other words, we can treat $\boldsymbol{\tau}_+\boldsymbol{\tau}_-$ as the projection operator for protons and $\boldsymbol{\tau}_-\boldsymbol{\tau}_+$ as the projection operator for neutrons.

Using these results, we can now write

$$S_+ = \langle i|\sum_{k=1}^{Z}\boldsymbol{\sigma}^2(k)|i\rangle = 3Z \tag{6-19}$$

where Z is the number of protons in the initial state. The summation is restricted to be over the protons in the target nucleus as a result of the projection operator $\boldsymbol{\tau}_+\boldsymbol{\tau}_-$. In obtaining the final result, we have made use of the fact that for a single nucleon $\boldsymbol{s} = \frac{1}{2}\boldsymbol{\sigma}$ and the expectation value of $\boldsymbol{\sigma}^2$ is 3. By the same token, (6-17) can be simplified to

$$S_- = 3N \tag{6-20}$$

where N is the number of neutrons in the target.

Eqs. (6-19) and (6-20) are not very useful sum rules since they represent, respectively, the total strength if all the protons and all the neutrons are excited by the reaction. Such an excitation involves extremely high energy components and cannot be achieved in practice. Experimentally, only nucleons near the Fermi surface can be excited and there is no easy way to estimate what the numbers of such nucleons may be. However, the difference between the two sum rules

$$S_- - S_+ = 3(N - Z) \tag{6-21}$$

may not depend on how high in energy the excitation strengths are measured and may therefore be tested against observation.

A departure from (6-21) may indicate the presence of particles other than nucleons in the nucleus, such as Δs resulting from the excitation of nucleons. Such a component in the intermediate state has been conjectured as a possibility in many other reactions at intermediate energies. If these excitations can be clearly identified, it will lead to a new understanding of the structure of nuclei. As a result, there is a great amount of interest in measuring the difference in strength between (p,n) and (n,p) reactions. However, the experiments are difficult to carry out and, at this moment, the results are still too preliminary to draw any conclusion.

The strength of Gamow-Teller excitation is related to the spin-isospin term $V_{\sigma\tau}(r)\boldsymbol{\sigma}(1)\cdot\boldsymbol{\sigma}(2)\boldsymbol{\tau}(1)\cdot\boldsymbol{\tau}(2)$ in nucleon-nucleon interaction. A good knowledge of the giant Gamow-Teller resonance will therefore also shed light on this important term in the interaction between nucleons inside a nucleus. The same is true of other giant resonances as well, since each may be shown to be dependent predominantly on a particular term in the interaction.

§6-3 ROTATIONAL MODEL

Deformation. In the previous two sections we have assumed, for the convenience of discussion, that the basic shape of a nucleus is spherical and excitations are built upon the equilibrium configuration in the form of small vibrations. There is no compelling reason why the nucleus in its lowest energy state cannot be different from a sphere. The interplay between short-range nuclear force, long-range repulsive Coulomb force, and centrifugal force acting on a rotating nucleus may well favor a nonspherical or *deformed* equilibrium shape.

In general, spherical nuclei are found around closed shells. This is easy to understand. As we shall see later in §6-5, the single-particle spectrum for nucleons is not a uniform one; the single-particle states are separated into groups, with energy differences between states within a group smaller than those between groups. The presence of energy gaps between groups of single-particle states makes it energetically more favorable for nucleons to fill up each group before occupying those in the next group. A closed shell nucleus is formed when all the single-particle states in a group are fully occupied. Furthermore, in order to maintain rotational invariance, such a grouping of states usually includes all $(2\ell+1)$ magnetic substates with $m = -\ell,\ -\ell+1, \ldots,\ \ell$ for a given single-particle orbital angular momentum ℓ. If all the available m-states are filled, the total M-value of the nuclear state is zero. As a result, the total angular momentum, $\boldsymbol{L} = \sum_i \boldsymbol{\ell}_i$, cannot take on any value other than zero. Such an object is then

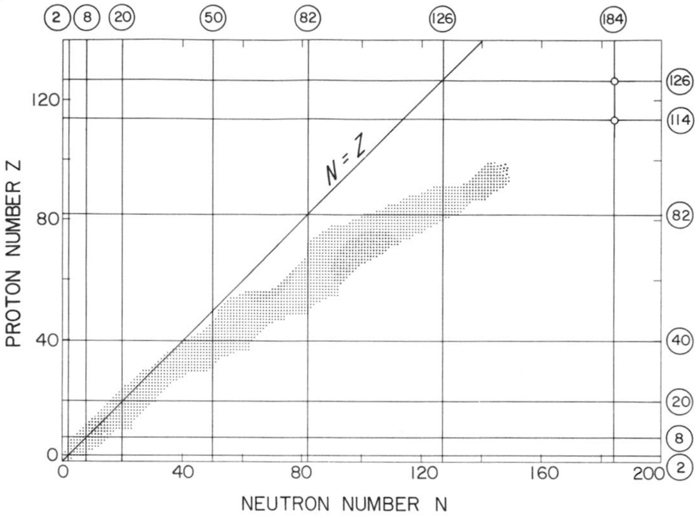

Fig. 6-8 Chart of nuclei. Deformed nuclei, indicated by the shaded areas, lie in regions between closed shell nuclei with neutron and proton numbers equal to one of the magic number given in circles. Large deformation is also found in very heavy nuclei beyond $^{208}_{82}$Pb.

invariant under any arbitrary rotations of the coordinate system and must be spherical in shape.

On the other hand, many single-particle states are available for nuclei between closed shells, and they are near each other in energy. It is quite possible for the nucleons to fill up these single-particle states in such a way that the minimum configuration has a nonspherical shape. In general, the nuclear shape tends to be prolate, *i.e.*, elongated along the z-axis, at the beginning of a major shell, and oblate, *i.e.*, flatted at the poles, toward the end of a major shell. This is caused by a preference, related to the pairing term in the nuclear force, for nucleons to fill first single-particle states with largest absolute m-values, starting from $m = \pm j$. As a result, the probability of finding nucleons increases in the polar regions at the beginning of a shell. For example, among the light nuclei in the ds-shell, between ^{16}O and ^{40}Ca, we find that signatures of deformation in the form of rotational spectra begin to appear in $A = 19$ nuclei (^{19}Ne and ^{19}Na), having only three nucleons outside the closed shell at ^{16}O. The deformation is known to be positive, for example, from the observed positive values of quadrupole moments. At the middle of the major shell, around ^{28}Si, the deformation changes sign, as can be seen from the fact that quadrupole moments are, in general, negative for nuclei with $A > 28$.

For stable nuclei, and those not too far from the valley of stability, departure from spherical equilibrium shape is generally small in the ground state region. Regions of relatively large deformation exist, for example, in medium-heavy nuclei with $150 \lesssim A \lesssim 180$ and heavy nuclei with $220 \lesssim A \lesssim 250$, as shown in Fig. 6-8. The largest deformations, or "super-deformations" as they are now known in the literature, are found in the excited states of medium-weight nuclei created when two heavy ions collide with each other.

Quantum mechanically, there cannot be a rotational degree of freedom associated with a spherical object. For a sphere, the square of its wave function is, by definition, independent of angles; that is, it appears to be the same from all directions. As a result, there is no way to distinguish the wave functions before and after a rotation. Rotation is therefore not a quantity that can be observed in this case and, consequently, cannot correspond to a degree of freedom in the system with energy associated with it. In contrast, rotational motion of a deformed object, such as an ellipsoid, may be detected, for example, by observing the changes in the orientation of the axis of symmetry with time.

Rotational Hamiltonian. Classically, the angular momentum $\boldsymbol{J}$ of a rotating object is proportional to the angular velocity $\boldsymbol{\omega}$,

$$\boldsymbol{J} = \mathcal{I}\boldsymbol{\omega}$$

with the ratio between $\boldsymbol{J}$ and $\boldsymbol{\omega}$ given by the moment of inertia $\mathcal{I}$. The rotational energy E_J is given by the square of the angular frequency and is proportional to $\boldsymbol{J}^2$ as a result,

$$E_J = \frac{1}{2}\mathcal{I}\omega^2 = \frac{1}{2\mathcal{I}}J^2$$

Quantum mechanically, the rotational Hamiltonian may be written by analogy in the form

$$H = \sum_{i=1}^{3} \frac{\hbar^2}{2\mathcal{I}_i} J_i^2 \tag{6-22}$$

where $\mathcal{I}_i$ is the moment of inertia along the i-th axis. For an axially symmetric object, the moment of inertia along a body-fixed or intrinsic set of coordinate axes, 1, 2, and 3, has the property,

$$\mathcal{I}_1 = \mathcal{I}_2 \equiv \mathcal{I} \tag{6-23}$$

(and $\mathcal{I}_3 \neq \mathcal{I}$, or else it is spherical). The Hamiltonian in this case may be written in the form,

$$H = \frac{\hbar^2}{2\mathcal{I}}(\boldsymbol{J}^2 - \boldsymbol{J}_3^2) + \frac{\hbar^2}{2\mathcal{I}_3}\boldsymbol{J}_3^2 \tag{6-24}$$

The expectation value of the Hamiltonian in the body-fixed system is therefore a function of $J(J+1)$, the expectation value of $\boldsymbol{J}^2$, and K, the eigenvalue of $\boldsymbol{J}_3$.

We have two different sets of coordinate systems here. The intrinsic coordinate system, with frame of reference fixed to the rotating body, is convenient for describing the symmetry of the object itself. On the other hand, the nucleus is rotating in the laboratory and the rotational energy is more conveniently described by a coordinate system that is fixed in the laboratory. Each system is more suitable for a different purpose, and we shall make use of both of them here. Following general convention, the intrinsic coordinate axes are labelled by subscripts 1, 2, and 3 to distinguish them from the laboratory coordinates, which are labelled by subscripts x, y, and z.

In classical mechanics, a rotating body requires three Euler angles (α, β, γ) to specify its orientation in space. The analogous situation in quantum mechanics requires three independent labels or quantum numbers to describe the rotational state. For two of these three labels we can use the constants of motion J, related to the eigenvalue of $\boldsymbol{J}^2$, and M, the projection of $\boldsymbol{J}$ along the quantization axis in the laboratory. For the third label, we can use K, the projection of $\boldsymbol{J}$ on the 3-axis in the intrinsic frame. In the body-fixed system, K takes the same role as M in the laboratory system and there are $(2J+1)$ possible values of K ranging from $-J$ to $+J$ in integer steps.

Rotational wave function. For the convenience of discussing rotational motion, we shall divide the wave function of a nuclear state into two parts, an intrinsic part describing the shape and other properties pertaining to the structure of the state, and a rotational part describing the motion of the nucleus in the laboratory. Our main concern here is in the rotational wave function, labelled by J, M, and K. Since it is a function of the Euler's angles only, it must be proportional to the $\mathcal{D}$-function, $\mathcal{D}^J_{MK}(\alpha, \beta, \gamma)$ defined in (B-22), relating wave functions of the same object in two coordinate systems rotated with respect to each other by Euler angles (α, β, γ). In terms of spherical harmonics, the function $\mathcal{D}^J_{MK}(\alpha, \beta, \gamma)$ may be defined by the relation,

$$Y_{JK}(\theta', \phi') = \sum_M \mathcal{D}^J_{MK}(\alpha, \beta, \gamma) Y_{JM}(\theta, \phi) \tag{6-25}$$

where $Y_{JK}(\theta', \phi')$ are spherical harmonics of order J in a coordinate system rotated by Euler angle α, β, γ with respect to the unprimed system.

The transformation property of $\mathcal{D}$-function under an inversion of the coordinate system is given by

$$\mathcal{D}^J_{MK}(\alpha, \beta, \gamma) \xrightarrow[P]{} (-1)^{J+K} \mathcal{D}^J_{M\,-K}(\alpha, \beta, \gamma) \tag{6-26}$$

An arbitrary $\mathcal{D}$-function, therefore, does not have a definite parity since, in addition to the phase factor, the sign of K is also changed under a reflection of the coordinate system. To construct a wave function of definite parity, a linear combination of $\mathcal{D}$-functions with both positive and negative K is required. As a result, the rotational wave function takes on the form,

$$|JMK\rangle_{\text{rot.}} = \sqrt{\frac{2J+1}{16\pi^2(1+\delta_{K0})}} \left\{\mathcal{D}^J_{MK}(\alpha, \beta, \gamma) \pm (-1)^{J+K} \mathcal{D}^J_{M\,-K}(\alpha, \beta, \gamma)\right\} \tag{6-27}$$

where the $\pm$ sign is for positive and negative parity, respectively. Since both $+K$ and $-K$ appear on the right hand side of (6-27), only $K \geq 0$ can be used to label a rotational wave function in such cases. The label K itself is no longer a good quantum number, but the absolute value of K remains a constant of motion for axially symmetric nuclei. In the more general tri-axial case with $\mathcal{I}_1 \neq \mathcal{I}_2 \neq \mathcal{I}_3$, a linear combination of $|JMK\rangle$ with different K values is required to describe nuclear rotation. Under such conditions, only J and M remain as good quantum numbers.

The complete wave function of the nucleus must also contain a part describing the object that is rotating in the laboratory frame of reference. This is the role of the intrinsic wave function. Depending on the energy and other parameters involved, a nucleus can take on different shapes, described by different intrinsic wave functions, and as a result, there may be more than one rotational band, each with a different moment of inertia in a single nucleus. For the axial symmetric case, the constant of motion K is often used as a label to identify the intrinsic state a deformed nucleus is in.

Rotational band. A nucleus in a given intrinsic state can rotate with different angular velocities in the laboratory. A group of states, each with a different total angular momentum J but sharing the same intrinsic state, forms a rotational band. Since the only difference between these states is in their rotational motion, members of a band are related to each other in energy, static moments, and electromagnetic transition rates. In fact, the signature of a rotational band is given by these relations.

The parity of a rotational state is given by (6-27). For $K = 0$, the wave function for a positive parity state vanishes if the J value is odd; as a result, only states with even J values are allowed for a $K = 0^+$ band. Similarly, there are no states with even J values in a $K = 0^-$ band. The results may summarized the following way:

$$J = \begin{cases} 0,\ 2,\ 4, \ldots & \text{for } K^\pi = 0^+ \\ 1,\ 3,\ 5, \ldots & \text{for } K^\pi = 0^- \end{cases} \tag{6-28}$$

For $K > 0$, the only restriction on the allowed spin in a band is $J \geq K$, arising from the fact that K is the projection of J on the body-fixed quantization axis or the 3-axis. The possible spins are then

$$J = K,\ K+1,\ K+2, \ldots \quad \text{for} \quad K > 0 \tag{6-29}$$

For a simple rotational Hamiltonian, such as that given in (6-24), the energy of a rotational state is given by the expression

$$E_J = \frac{\hbar^2}{2\mathcal{I}} J(J+1) + E_K \tag{6-30}$$

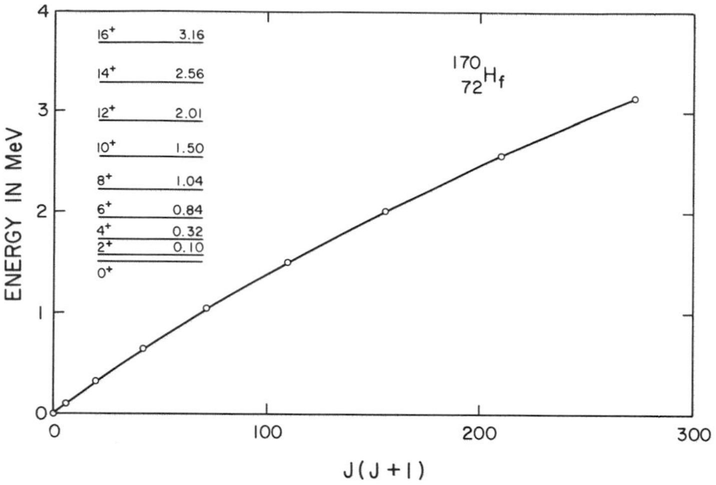

Fig. 6-9 Rotational levels in $^{170}_{72}$Hf. If the moment of inertia $\mathcal{I}$ is a constant, the relation between E_J and $J(J+1)$ would have been a straight line with slope related to $\mathcal{I}$. The small curvature found in the plot indicates that $\mathcal{I}$ increases slightly with large J, a result of centrifugal stretching of the nucleus with increasing angular velocity.

where E_K represents contributions from the intrinsic part of the wave function. An example of such a band is shown in Fig. 6-9 for ^{170}Hf.

From (6-30) we see that the relative position of members of a rotational band is given by $J(J+1)$, with the constant of proportionality related to the momentum of inertia $\mathcal{I}$. The quantity E_k enters only in the location of the band head, the position where the band starts. Different bands are distinguished by their moments of inertia and by the positions of their band heads. Both features, in turn, depend on the structure of the intrinsic state which describes the deformed nucleus that is rotating in the laboratory frame of reference.

Quadrupole moment. Besides energy level positions, the static moments of and transition rates between members of a rotational band are also given by the rotational model. The calculations here depend on the fact that all members share the same intrinsic state and differ only in their rotational motion. Let us start with the quadrupole moment. For an axially deformed object, the quadrupole moment is given by the integral,

$$Q_0 = \int (3z^2 - r^2)\rho(\boldsymbol{r})dV \tag{6-31}$$

where $\rho(\boldsymbol{r})$ is the nuclear density distribution. Since it is related to the shape of the intrinsic state, Q_0 is known as the intrinsic quadrupole moment. For an axially symmetric object, it is related to the difference in the polar and equatorial radii characterized by the parameter

$$\delta = \frac{3}{2}\frac{R_3^2 - R_\perp^2}{R_3^2 + 2R_\perp^2} \approx \frac{R_3 - R_\perp}{R} \tag{6-32}$$

where R_3 is the radius of the nucleus along the body-fixed symmetric (3-)axis, $R_\perp$ is the radius in the direction perpendicular to it, and R is the mean radius. In terms of δ,

$$Q_0 = \frac{4}{3}\Big\langle\sum_{i=1}^{A} r_i^2\Big\rangle\,\delta \tag{6-33}$$

The quantity Q_0 defined here is the "mass" quadrupole moment of the nucleus, since the density distribution $\rho(\boldsymbol{r})$ in (6-31) involves all A nucleons. The usual quantity measured in an experiment, for example, by scattering charged particles from a nucleus, is the "charge" quadrupole moment, differing from the expression above by the fact that the summation is restricted to protons only.

The observed quadrupole moment of a state, as defined in (4-54), is the expectation value of the electric quadrupole operator $\boldsymbol{Q}$ in the state $M = J$. We shall represent this quantity here as Q_{JK} for reasons that will soon become clear. The value of Q_{JK} differs from Q_0 by the fact that the former is measured in the laboratory frame of reference and the latter in the body-fixed frame. The relation between them is given by the transformation from the intrinsic coordinate system to the laboratory system. Since this is achieved by a $\mathcal{D}$-function, the result depends on both J and K. Inserting the explicit value of the $\mathcal{D}$-function for the $M = J$ case, we obtain the relation

$$Q_{JK} = \frac{3K^2 - J(J+1)}{(J+1)(2J+3)} Q_0 \tag{6-34}$$

In practice, direct measurements of quadrupole moments are possible only for the ground state of nuclei. For excited states, the quadrupole moment can sometimes be deduced indirectly through reactions such as Coulomb excitation (see §7-1).

In order to compare the calculated values of (6-34) with experimental data, we need a knowledge of the intrinsic quadrupole moment Q_0 as well as the value of K for the band. The latter may be found from the minimum value of total angular momentum J in the band. If the value of Q_{JK} is known only for one state, it may be taken as a measurement of Q_0 itself. In order to verify whether quadrupole moments of members of a rotational band are related to each other according to (6-34), values of Q_{JK} must be known for more than one member of the band. This is usually quite difficult to achieve through direct measurements. The alternative is make use of electric quadrupole transition rates, as we shall see next.

Electromagnetic transitions. In the rotational model, electromagnetic transitions between members of a band take place by changes in the angular momenta, and hence rotational frequencies, without any modifications to the intrinsic state. In principle, transitions of all multipolarity can be studied in this way; we shall avoid a general discussion here and concentrate on electric quadrupole ($E2$) and magnetic dipole ($M1$) transitions, as these are the most commonly observed ones. Since only a change in the angular momentum is involved, the transition matrix element may be related to a Clebsch-Gordan coefficient. For an $E2$-transition, the size of the matrix element must also be proportional to the deformation of the intrinsic state, represented by the quantity Q_0 given by (6-31). We shall not attempt a derivation of here (see, *e.g.,* A. Bohr and B.R. Mottelson, *Nuclear Structure,* vol. II, Benjamin, 1975, Reading, Massachusetts). The reduced transition probability from J_i to J_f is given by the relation

$$B(E2; J_i \to J_f) = \frac{5}{16\pi} e^2 Q_0^2 \langle J_i K 20 | J_f K \rangle^2 \tag{6-35}$$

For $K = 0$, $J_i = J$, and $J_f = J - 2$, the square of the Clebsch-Gordan coefficient simplifies to

$$\langle J020 | (J-2)\, 0 \rangle^2 = \frac{3J(J-1)}{2(2J+1)(2J-1)}$$

with the help of the identities given in Table B-1. The reduced transition rate for decay between adjacent members of a $K = 0$ band becomes

$$B(E2; J \to J-2) = \frac{15}{32\pi} e^2 Q_0^2 \frac{J(J-1)}{(2J+1)(2J-1)} \tag{6-36}$$

Alternatively, for electromagnetic excitation from J to $J + 2$,

$$B(E2; J \to J+2) = \frac{15}{32\pi} e^2 Q_0^2 \frac{(J+1)(J+2)}{(2J+1)(2J+3)} \tag{6-37}$$

a form more useful, for example, in Coulomb excitation measurements.

Magnetic dipole transitions may be studied in the same way as $E2$-transitions in the rotational model. For this purpose, the $K = 0$ band is not suitable, since the J-values of members of the band differ by at least two units and $M1$-transitions are, therefore, forbidden by angular momentum selection rule. For this reason, we shall consider rotational bands with $K > 0$.

The form of magnetic transition operator is given in (5-40) in terms of single-nucleon gyromagnetic ratios g_ℓ for orbital angular momentum and g_s for intrinsic spin. Here, we are dealing with collective degrees of freedom. Instead of g_ℓ and g_s, it is more appropriate to express the $M1$-transition operator in terms of g_R and g_K, respectively, the gyromagnetic ratio for rotational motion, and the intrinsic state of a deformed nucleus. In terms of these two ratios, the magnetic dipole operator retains a simple form, similar to that given by (4-60), if we restrict ourselves to $K > \frac{1}{2}$ bands,

$$\boldsymbol{\mu} = g_R \boldsymbol{J} + (g_K - g_R)\frac{\boldsymbol{K}^2}{J+1} \tag{6-38}$$

where $\boldsymbol{K} = \boldsymbol{J}_3$, the operator measuring the projection of $\boldsymbol{J}$ on the 3-axis in the intrinsic frame. For a symmetric rotor, the expectation value of $\boldsymbol{K}$ is K.

In the same spirit as (6-35) for $E2$ transitions, the $B(M1)$ value in the rotational model is given by

$$B(M1, J_i \to J_f{=}J_i{\pm}1) = \frac{3}{4\pi}(g_K - g_R)^2 K^2 \langle J_i K 10 | J_f K\rangle^2 \tag{6-39}$$

in units of μ_N^2, nuclear magneton squared. From (6-35) and (6-39) the mixing ratio between $E2$ and $M1$ rates between two adjacent members of a $K > 0$ band can be calculated. The quantity relates the intrinsic quadrupole moment Q_0 with gyromagnetic ratios g_R and g_K and provides another check of the model against experimental data.

Transitions between members of different rotation bands, or interband transitions, involve changes in the intrinsic shape of the nucleus in addition to angular momentum recoupling discussed above for intraband transitions. The main interest of interband transitions concerns the intrinsic wave function, but we shall not be going into this more complicated subject here.

Corrections to the basic model. On closer examination, the energy level positions of members of a rotational band often differ from the simple $J(J+1)$-dependence given by (6-30). Similarly, the relations between transition rates are not exactly those given by (6-35) and (6-39). There are many possible reasons for deviations from the predications of a simple rotational model. The main ones may be summarized as follows:

- We have seen that K is a constant of motion for a symmetric rotor. However, rotational wave functions are linear combinations of both $+K$ and $-K$ components in order to be invariant under a parity transformation. It is therefore possible to have a term in the Hamiltonian that couples between $\pm K$ without violating any of these conditions. Such a term is analogous to the Coriolis force in classical rotation. The size of the coupling may depend on both J and K in general but is observed to be negligible except for $K = \frac{1}{2}$. This gives rise to the decoupling term in $K = \frac{1}{2}$ bands described below.
- The moment of inertia, which gives the slope in a plot of E_J versus $J(J+1)$, may not be a constant for states of different J. This is expected on the ground that the nucleus is not a rigid body and the centrifugal force can modify slightly the intrinsic shape of the nucleus as the angular velocity is increased. Centrifugal stretching is observed at the upper end of many rotational bands. In general, departures from a simple $J(J+1)$

spectrum due to slight and gradual changes in the moment of inertia may be accounted for by adding a $J^2(J+1)^2$-dependent term in the rotational Hamiltonian.

- Rotational bands have been observed with members having very high spin values, for example, $J \approx 40\hbar$ and beyond. Such high spin states must lie quite high in energy with respect to the ground state of the nucleus. As a result, it may be energetically more favorable for the underlying intrinsic shape to adjust itself slightly and change itself to a more stable configuration as excitation energy is increased. Such changes are likely to be quite sudden, similar to a phase change in chemical reactions. Compared with centrifugal stretching, which is a smooth variation of the moment of inertia $\mathcal{I}$ with increasing J, the readjustment of the intrinsic shape takes place within a region of a few adjacent members of a rotational band. This gives rise to the phenomenon of "backbending," to be discussed later.

In practice, departures from a $J(J+1)$ spectrum for rotational excitation are small except for the decoupling term in $K=\frac{1}{2}$ bands. As a result, the $J(J+1)$-level spacing remains, for most purposes, a signature of rotational band.

Decoupling parameter. For odd-mass nuclei, rotational bands have half-integer K-values. In the case of $K=\frac{1}{2}$, the band starts with $J=\frac{1}{2}$ and has additional members with $J=\frac{3}{2}, \frac{5}{2}, \frac{7}{2}, \cdots$. If the energy level positions of the band members are given by the simple rotational Hamiltonian of (6-24), we expect that, for example, the $J=\frac{5}{2}$-member will be above the $J=\frac{3}{2}$-member in energy by an amount larger than the difference between the $J=\frac{3}{2}$ and $J=\frac{1}{2}$ members. The observed level sequence is, however, quite different and, in many cases, is more similar to the example of the $K=\frac{1}{2}$ band in ^{169}Tm shown in Fig. 6-10. Instead of a simple $J(J+1)$-sequence, we find the $\frac{3}{2}$-member of the band is depressed in energy and is located just above the $\frac{1}{2}$-member, the $\frac{7}{2}$-member is just above the $\frac{5}{2}$-member, and so on.

The special case of $K=\frac{1}{2}$ bands may be understood by an additional term $H'(\Delta K)$ to the basic rotational Hamiltonian given in (6-24). Such a term connects two components of a rotational wave function different in K by ΔK for $K \neq 0$. The contribution of this term to the rotational energy may be represented, to a first approximation, by the expectation value of $H'(\Delta K)$ in the state given by the wave function of (6-27),

$$
\begin{aligned}
&\langle JMK|H'(\Delta K)|JMK\rangle_{\text{rot.}} \\
&\qquad \longrightarrow \Big\{\langle \mathcal{D}^J_{MK}|H'(\Delta K)|\mathcal{D}^J_{MK}\rangle + \langle \mathcal{D}^J_{M\,-K}|H'(\Delta K)|\mathcal{D}^J_{M\,-K}\rangle \\
&\qquad\quad + \langle \mathcal{D}^J_{MK}|H'(\Delta K)|\mathcal{D}^J_{M\,-K}\rangle + \langle \mathcal{D}^J_{M\,-K}|H'(\Delta K)|\mathcal{D}^J_{MK}\rangle\Big\}
\end{aligned}
$$

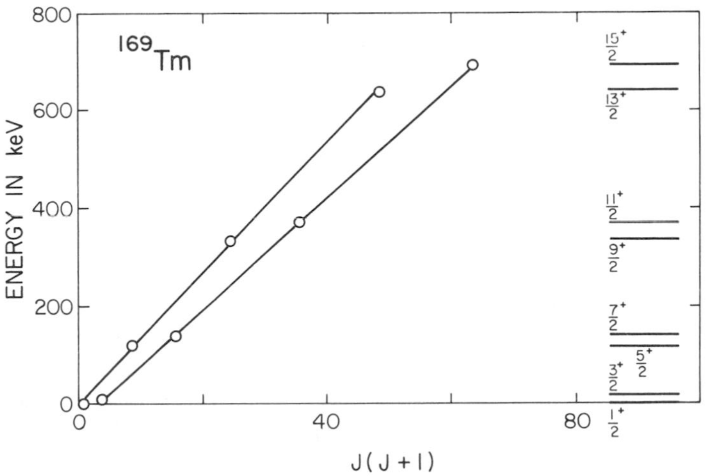

Fig. 6-10 Rotational spectrum for the $K^{\pi} = \frac{1}{2}^{+}$ band in ^{169}Tm, showing the effect of the decoupling parameter a of (6-40).

The first two terms vanish since, by definition, $H'(\Delta K)$ cannot connect two wave functions having the same K value. For $\Delta K = 1$, the last two terms are nonzero if only $K = \frac{1}{2}$.

A term in the Hamiltonian that operates only between wave functions different in K value by unity may be written in the form

$$H'(\Delta K = 1) \sim \omega_1 \boldsymbol{J}_1 = \tfrac{1}{2}\omega_1(\boldsymbol{J}_+ + \boldsymbol{J}_-)$$

where $\boldsymbol{J}_1$ is the component of the angular momentum operator $\boldsymbol{J}$ along the body-fixed 1-axis and ω_1 is the angular frequency. Such an operator may be expressed in terms of angular momentum raising and lowering operators in the intrinsic coordinate system. The analogue of such a term is the Coriolis force in classical mechanics responsible, for example, for deflecting north-south movement of air mass on earth to a counterclockwise direction in the northern hemisphere and clockwise in the south hemisphere as a result of the earth spinning on its own axis. We can view a similar situation in an odd nucleus by considering a single nucleon moving in the average potential of an even-even core. Since the core is rotating, an additional force is felt by the nucleon, and this term does not preserve K in the intrinsic frame of reference.

The decoupling term given in this way is effective only for the $K = \frac{1}{2}$ band. Because of $H'(\Delta K)$, the rotational energy of a member of the $K = \frac{1}{2}$ band, instead of (6-30), becomes

$$E_J(K = \tfrac{1}{2}) = \frac{\hbar^2}{2\mathcal{I}}\left\{J(J+1) + a(-1)^{J+1/2}(J + \tfrac{1}{2})\right\} + E_k \qquad (6\text{-}40)$$

where a is the strength of the decoupling term. Instead of a $J(J+1)$-spectrum, each level is now moved up or down from its location given by (6-30) by an amount depending on whether $J+\frac{1}{2}$ of the level is even or odd. In cases where the absolute value of the decoupling parameter a is large, a higher spin level may appear below one with spin one unit less, as seen in the ^{19}F example in Problem 6-4. The signature of a rotational band can still be recognized by the fact that one half of the members, $J = \frac{1}{2}, \frac{5}{2}, \frac{9}{2}, \cdots$, possess a E_J versus $J(J+1)$ relation with one (almost constant) slope, and the other half have a different slope, as can be seen from (6-40).

Yrast levels and backbending. The lowest level in energy for each spin is called an *yrast* level. Being the level with the highest spin in the region, yrast levels with very large spin values have been identified at high excitation energies. Most of the energy in such a state is in the form of rotational energy and, as a result, the intrinsic shape of the nucleus may remain rather "cold," near or at its equilibrium value in the ground state region. As a result, various yrast levels of a nucleus essentially share the same intrinsic state and differ only in their rotational motion. For a deformed nucleus, these levels form a rotational band, known as an yrast band. Because of the special position of yrast levels, it is very common for a very highly excited nucleus, with large amounts of angular momentum imparted to it by the excitation process, to decay first to an yrast level, for example, by particle emission. Once the excited nucleus is in such a level, it preferentially decays to the yrast level below it by γ-ray emission. The process continues successively to the next lower state, and cascades of γ-rays starting from very high spin states have been observed in a variety of nuclei. In this way, yrast bands have provided us with long observed sequences of rotational levels with both energy and transition rates measured to high accuracies.

One of the interesting observations made on yrast bands is the presence of small and sudden changes in the moment of inertia. On a plot of E_J as a function of $J(J+1)$, the sudden changes are usually too insignificant to be noticeable. However, if the moment of inertia is plotted against the square of the frequency of the rotation, a local variation in $\mathcal{I}$ is greatly amplified and appears as a Z-shaped curve, as shown in Fig. 6-11, and hence the name "backbending." In order to make such a plot, we need to calculate the local values of the rotational frequency ω and moment of inertia. The former is not a quantity that can be measured but may be inferred by analogy with classical rotational frequency through the relation

$$\hbar\omega = \frac{dE}{d\sqrt{J(J+1)}} \tag{6-41}$$

For a $K = 0$ band, the usual case for the yrast band in even-even nuclei, the value may be approximated by the difference between E_J and E_{J-2},

$$\hbar\omega \approx \left.\frac{\Delta E}{\Delta\sqrt{J(J+1)}}\right|_{J-2}^{J} \xrightarrow[J\to\infty]{} \tfrac{1}{2}(E_J - E_{J-2})$$

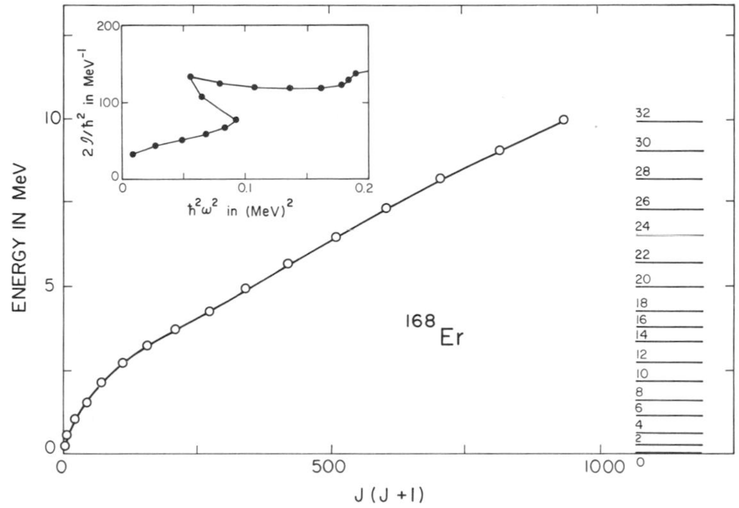

Fig. 6-11 Backbending in ^{168}Er. When the energy of each level in the yrast band is plotted against $J(J+1)$, a typical rotational spectrum is found with a slow variation of the momentum of inertia due to centrifugal stretching. However, when the relation is examined in more detail in the insert by plotting $2\mathcal{I}/\hbar^2$ against $(\hbar\omega)^2$, a sudden change is observed around $J = 14$, indicating a shift in the intrinsic structure of the nucleus.

The local value of the moment of inertia,

$$\frac{2\mathcal{I}}{\hbar^2} = \frac{4J-2}{E_J - E_{J-2}} \tag{6-42}$$

is found from the energy difference between two adjacent members using (6-30).

The explanation for backbending phenomena is that the nucleus has two rotational bands, A and B, with slightly different moments of inertia and band head positions. Such small variations may be caused by a small difference in the structure between their intrinsic states. In the absence of any coupling between them, the two bands may cross each other at some J-value, say J_c, since they have slightly different slopes in their relations of E_J as a function of $J(J+1)$. Below J_c, members of band A are lower in energy because of a lower band head and they form the yrast levels. Above J_c, band B comes down lower in energy because of a larger moment of inertia. The higher-J members of the yrast band come from B. Since the structures of these two bands are different, the interaction between them is very weak and they are, essentially, decoupled from each other except near the transition region. Here the "unperturbed" positions between members of the two bands are sufficiently close together that even a small perturbation due to a weak coupling between the two bands can cause some mixing between them, resulting in a large local variation of $\mathcal{I}$ near J_c. Other experimental observations, such as in the changes of the single-particle occupancies around the transition region, also confirm this view. More complicated band structure involving more than one backbending and different observed relations between $2\mathcal{I}/\hbar^2$ and $(\hbar\omega)^2$ have also been observed. The basic reasons behind the behavior is essentially the same as that given above for simple backbending phenomenon.

The basic concept behind rotation models is the classical rotor. Quantum mechanics enters in two places, a trivial one in the discrete (rather than continuous) distribution of energy and angular momentum, and a more important one in the evaluation of the moment of inertia. The latter is a complicated and interesting question, as we can see from the following consideration.

The equilibrium shape of a nucleus may be deduced from such measurements as the quadrupole moment. At the same time, the moment of inertia can be calculated, for example, by considering the nucleus as a rigid body,

$$\mathcal{I}_{\rm rig} = \frac{2}{5} M R_0^2 (1 + \frac{1}{3}\delta)$$

where M is the mass of the nucleus and R_0, its mean radius. The quantity δ may expressed in terms of Q_0 using (6-32). Compared with observations, the rigid body value obtained in this way turns out to be roughly a factor of two too large. Furthermore, the observed value of $\mathcal{I}$ for different nuclei changes systematically from being fairly small near closed shell nuclei, increasing toward the region in between, and decreasing once again toward the next set of magic numbers. An understanding of this question will therefore require a knowledge of the equilibrium shape of nuclei under the action of the forces between nucleons as well as the behavior of the nuclear matter under rotation. Neither of these two points is well understood yet, and the questions are under active investigation.

§6-4 INTERACTING BOSON APPROXIMATION

The importance of pairing and quadrupole interactions has been quite evident in a variety of nuclear properties examined earlier. For many nuclear states, the main features are often given by these two terms alone. For this reason, it is advantageous for certain nuclear studies to approximate the nuclear interaction to these two types only. Furthermore, we have the advantage here that analytical solutions may be possible under certain conditions. We shall examine only one representative model in this category, the interaction boson approximation (IBA).

Boson operators. A good starting point in understanding IBA is to follow the philosophy behind vibrational models and treat the possible excitation modes in the model as the canonical variables. In the interacting boson approximation, two types of excitation quanta, or *bosons*, can be constructed: a $J = 0$ quantum or s-boson and a $J = 2$ quantum or d-boson. Both types of boson are thought to be made of pairs of identical nucleons coupled to $J = 0$ and 2, respectively. Such a realization of the bosons in terms of nucleons is important if one wishes to establish a microscopic foundation for the model. However, it is not essential for the appreciation of the many successes of the model in accounting for the observed nuclear properties by very simple calculations.

Let $\boldsymbol{s}^\dagger$ be the operator that creates an s-boson and $\boldsymbol{d}^\dagger_\mu$ be the corresponding operator for a d-boson. Since a d-boson carries two units of angular momentum, it has five components distinguished by the projections of the angular momentum on the quantization axis, $\mu = -2, -1, 0, 1$, and 2. Corresponding to each of these boson creation operators, we have the conjugate annihilation operators $\boldsymbol{s}$ and $\boldsymbol{d}_\mu$. To complete the definition of these operators we need to specify the commutation relations between them:

$$\begin{aligned} &[\boldsymbol{s}^\dagger, \boldsymbol{s}] = 1 \qquad\qquad &&[\boldsymbol{s}^\dagger, \boldsymbol{s}^\dagger] = [\boldsymbol{s}, \boldsymbol{s}] = 0 \\ &[\boldsymbol{d}^\dagger_\mu, \boldsymbol{d}_\nu] = \delta_{\mu\nu} &&[\boldsymbol{d}^\dagger, \boldsymbol{d}^\dagger] = [\boldsymbol{d}, \boldsymbol{d}] = 0 \\ &[\boldsymbol{s}^\dagger, \boldsymbol{d}_\mu] = [\boldsymbol{s}, \boldsymbol{d}^\dagger_\mu] = [\boldsymbol{s}, \boldsymbol{d}_\mu] = [\boldsymbol{s}^\dagger, \boldsymbol{d}^\dagger_\mu] = 0 \end{aligned} \tag{6-43}$$

All other operators necessary to calculate nuclear properties in the model can be expressed in terms of these s- and d-boson creation and annihilation operators.

In terms of $\boldsymbol{s}^\dagger$, $\boldsymbol{s}$, $\boldsymbol{d}^\dagger_\mu$, and $\boldsymbol{d}_\mu$, the number operators for the s- and d-bosons are, respectively,

$$\boldsymbol{n}_s = \boldsymbol{s}^\dagger \cdot \overline{\boldsymbol{s}}, \qquad \boldsymbol{n}_d = \boldsymbol{d}^\dagger \cdot \overline{\boldsymbol{d}} = \sum_\mu (-1)^\mu \boldsymbol{d}^\dagger_\mu \overline{\boldsymbol{d}}_{-\mu} = \sum_\mu \boldsymbol{d}^\dagger_\mu \boldsymbol{d}_\mu \tag{6-44}$$

where the bar on top indicate the (spherical tensor) adjoint of $\boldsymbol{d}_\mu$ and $\boldsymbol{s}$,

$$\overline{\boldsymbol{d}}_\mu = (-1)^{2+\mu} \boldsymbol{d}_{-\mu} \qquad\qquad \overline{\boldsymbol{s}} = \boldsymbol{s} \tag{6-45}$$

as shown in (B-31). In addition, we can construct five additional irreducible spherical tensors made of products of two boson operators

$$\begin{aligned} \boldsymbol{P} &= \tfrac{1}{2}\{\overline{\boldsymbol{d}} \cdot \overline{\boldsymbol{d}} - \overline{\boldsymbol{s}} \cdot \overline{\boldsymbol{s}}\} \\ \boldsymbol{L} &= \sqrt{10}\big(\boldsymbol{d}^\dagger \times \overline{\boldsymbol{d}}\big)_1 \\ \boldsymbol{Q} &= \big(\boldsymbol{d}^\dagger \times \overline{\boldsymbol{s}}\big)_2 + \big(\boldsymbol{s}^\dagger \times \overline{\boldsymbol{d}}\big)_2 - \sqrt{\tfrac{7}{4}}\big(\boldsymbol{d}^\dagger \times \overline{\boldsymbol{d}}\big)_2 \\ \boldsymbol{T}_3 &= \big(\boldsymbol{d}^\dagger \times \overline{\boldsymbol{d}}\big)_3 \\ \boldsymbol{T}_4 &= \big(\boldsymbol{d}^\dagger \times \overline{\boldsymbol{d}}\big)_4 \end{aligned} \tag{6-46}$$

where the multiplication symbol $\times$ stands for the angular momentum coupled product

$$(\boldsymbol{A}_r \times \boldsymbol{B}_s)_{tm} = \sum_{pq} \langle rpsq|tm\rangle \boldsymbol{A}_{rp}\boldsymbol{B}_{sq}$$

defined in (B-32).

The simple model. If we restrict ourselves to the simple case of having either active neutrons or protons, the most general IBA Hamiltonian, known as IBA-1, may be expressed as a linear combination of the five operators given in (6-46), together with the boson number operator. In this way, there are six parameters in the IBA-1 Hamiltonian,

$$H_{\text{IBA-1}} = \epsilon \boldsymbol{n}_d + a_0 \boldsymbol{P}^\dagger \cdot \boldsymbol{P} + a_1 \boldsymbol{L}\cdot\boldsymbol{L} + a_2 \boldsymbol{Q}\cdot\boldsymbol{Q} + a_3 \boldsymbol{T}_3\cdot\boldsymbol{T}_3 + a_4\boldsymbol{T}_4\cdot\boldsymbol{T}_4 \tag{6-47}$$

where ϵ is the energy difference between a d- and an s-boson and a_J for $J = 0$ to 4 are the strengths of the other components in the expression. The dot between two spherical tensor operators in (6-46) represents a scalar product in the sense defined in (B-58). The number operator $\boldsymbol{n}_s$ for s-bosons does not enter into the expression, since the energy associated with it is taken to be zero and used for defining the zero point of the energy scale. In the absence of a microscopic connection to the nucleon degrees of freedom, these six parameters can be found, for example, by fitting results calculated with the Hamiltonian to known data.

In addition to energy, operators corresponding to other observables in the space span by the s- and d-bosons can also be expressed in terms of tensor products of the boson creation and annihilation operators. For example, the possible electromagnetic transition operators in the space take on the following forms:

$$\begin{aligned}
\boldsymbol{O}_0(E0) &= \beta_0\big(\boldsymbol{d}^\dagger \times \overline{\boldsymbol{d}}\big)_0 + \gamma_0\big(\boldsymbol{s}^\dagger \times \overline{\boldsymbol{s}}\big)_0 \\
\boldsymbol{O}_{1\mu}(M1) &= \beta_1\big(\boldsymbol{d}^\dagger \times \overline{\boldsymbol{d}}\big)_{1\mu} \\
\boldsymbol{O}_{2\mu}(E2) &= \alpha_2\big\{\big(\boldsymbol{s}^\dagger \times \overline{\boldsymbol{d}}\big)_{2\mu} + \big(\boldsymbol{d}^\dagger \times \overline{\boldsymbol{s}}\big)_{2\mu}\big\} + \beta_2\big(\boldsymbol{d}^\dagger \times \overline{\boldsymbol{d}}\big)_{2\mu} \\
\boldsymbol{O}_{3\mu}(M3) &= \beta_3\big(\boldsymbol{d}^\dagger \times \overline{\boldsymbol{d}}\big)_{3\mu} \\
\boldsymbol{O}_{4\mu}(E4) &= \beta_4\big(\boldsymbol{d}^\dagger \times \overline{\boldsymbol{d}}\big)_{4\mu}
\end{aligned} \tag{6-48}$$

where α_2, γ_0 and the β's are adjustable parameters.

Since the IBA-1 Hamiltonian contained in it many of the features of simpler Hamiltonians constructed for the specific purpose of understanding particular aspects of nuclear structure, it is possible to use the IBA-1 to investigate a variety of problems. One of the interesting features is that it has an underlying group structure and, as a result, powerful mathematical techniques may be applied to find the solutions. The communication relations among the boson creation and annihilation operators expressed in (6-43) implies that the operators form a group, the U_6 group, a unitary group in six dimensions. The energy of a state corresponding to one of the irreducible representations of this group may be expressed as a function

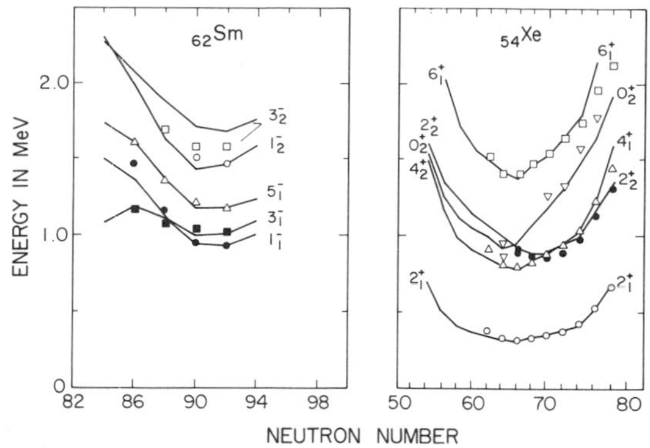

Fig. 6-12 Comparison of experimental (squares, circles and triangles) and calculated level spectra (lines) in the interacting boson approximation for octupole states in samarium isotopes (left) and states in xenon isotopes (right). (Taken from A. Arima and F. Iachello, *Adv. Nucl. Phys.* **13** [1984] 139).

of the six parameters in the Hamiltonian. If the values of these parameters are known, a large number of energy levels can be calculated. Examples of results for energy level positions calculated in the IBA-1 are shown in Fig. 6-12.

The underlying group structure of IBA-1 lends itself also to three limiting cases that are of interest in nuclear structure. The U_6 group may be decomposed into a variety of subgroups. Among these decompositions, we shall limit ourselves to cases where the chain of reduction of U_6 contains the three-dimensional rotational group as one of the subgroups. This is necessary if angular momentum is to be retained as a constant of motion.

If the d-bosons are completely decoupled from the system, the Hamiltonian can be written in terms of s-boson operators alone,

$$H_{\text{seniority}} = \epsilon_s \boldsymbol{s}^\dagger \boldsymbol{s} + u_0 \boldsymbol{s}^\dagger \boldsymbol{s}^\dagger \boldsymbol{s}\boldsymbol{s} \tag{6-49}$$

This is the *seniority* scheme (A. de-Shalit and I. Talmi, *Nuclear Shell Theory,* Academic Press, New York, 1963), known to be useful in classifying many-nucleons states in the jj-coupling scheme (see §6-6) in which pairs of nucleons with their angular momentum coupled to zero are treated differently from those which are not coupled to $J = 0$. From this property, we see also that the IBA-1 has a pairing structure built into the Hamiltonian and can therefore account for many of the observed nuclear properties in which the pairing interaction dominates.

On the other hand, if all the terms related to the s-boson operator are ignored, we obtain a system dominated by quadrupole excitations induced by the d-bosons,

$$H_{\text{vib. limit}} = \epsilon_d \boldsymbol{d}^\dagger \cdot \boldsymbol{d} + \sum_J u_J (\boldsymbol{d}^\dagger \times \boldsymbol{d}^\dagger)_J \cdot (\overline{\boldsymbol{d}} \times \overline{\boldsymbol{d}})_J \tag{6-50}$$

In this limit, the interaction between d-bosons produces quadrupole vibrational motion in nuclei similar to that described in §6-1.

If we put all the parameters in (6-47) to zero except a_1 and a_2, we obtain the SU_3 limit,

$$H_{SU_3} = a_1 \boldsymbol{L} \cdot \boldsymbol{L} + a_2 \boldsymbol{Q} \cdot \boldsymbol{Q} \qquad (6\text{-}51)$$

which has been used with success in understanding rotation-like structure in *ds*-shell nuclei from oxygen to potassium. Because of the $\boldsymbol{L}^2$ term, the Hamiltonian gives a spectrum that has an $L(L+1)$-dependence. If the nucleon intrinsic spins are coupled together to $S = 0$, we have $\boldsymbol{J} = \boldsymbol{L}$, and an $L(L+1)$ spectrum is the same as a $J(J+1)$ spectrum we have seen earlier in rotational nuclei. The $\boldsymbol{Q} \cdot \boldsymbol{Q}$ term provides a constant in the energy of all the levels in a "band" and can therefore be interpreted as the dependence on the intrinsic structure of the rotating nucleus. In this way, we expect that the IBA-1 can explain rotational structure in nuclei as well.

The full model. In practice, the IBA-1 is found to be limited by the fact that only excitations of either neutrons or protons are allowed. To overcome this limitation, the Hamiltonian given in (6-47) must be expanded to include both neutron and proton bosons as well as interactions between them,

$$H_{\text{IBA-2}} = H_{nn} + H_{pp} + V_{np} \qquad (6\text{-}52)$$

where H_{nn} and H_{pp} are the neutron- and proton-boson Hamiltonians, respectively. The interaction between these two types of bosons is provided by V_{np}. The most general form, known as IBA-2, contains a maximum of 29 parameters, nine for H_{nn}, nine for H_{pp}, and eleven for V_{np}. This is too complicated, and a simplified version is found to be adequate for most applications.

The IBA-2 also has the advantage of allowing a connection to be made to the underlying single-particle basis of nuclear structure. All the nucleons in a nucleus can be divided into two groups, those in the inert core and those in the active or *valence* space. The core may be taken to be one of the closed shell nuclei (to be discussed in the next section) and, since the nucleons in the core are assumed to be inactive except in providing a binding energy to the valence nucleons, the core may be treated as the "vacuum" state for the problem. Active neutron pairs and proton pairs are added to the core by boson creation operators acting on the vacuum. The IBA-2 therefore provides a basis to study an even wider variety of nuclear structure phenomena, from single-particle to collective degrees of freedom (for more details, see A. Arima and F. Iachello, *Adv. Nucl. Phys.* **13** [1984] 139).

Interacting boson models belong to a more general type of approach to nuclear structure studies sometimes known as *algebraic* models. We have seen evidence that symmetries play an important role in nuclear structure. For each type of symmetry, there is usually an underlying mathematical group associated with it. Although there are very few exact symmetries, such as angular momentum or rotational symmetry in three dimensions, there is a large number of approximate or "broken" symmetries that are of physical interest and can be exploited. One good example of the latter category is isospin, or SU_2 symmetry, the symmetry in interchanges between protons and neutrons, or between u- and d-quarks. Although isospin invariance in nuclei is broken by the Coulomb interaction, it is nevertheless a useful concept, as we have seen earlier on many occasions. One of the aims of group theoretical approaches to nuclear structure problems is to make use of these symmetries to classify nuclear states according to the irreducible representations of the mathematical groups involved, as we have done for isospin. We have seen some features of such an approach in IBA-1; a few other elementary applications will also be used later to classify single-particle states in the nuclear shell model. A general discussion of algebraic models is, however, inappropriate here, in part because of the amount of preparation in group theory required.

§6-5 MAGIC NUMBER AND SINGLE-PARTICLE ENERGY

In §4-11, we have treated nuclear binding energy as a smooth function of A, N, and Z. Superimposed on this global picture, there are local variations that are useful in understanding nuclear physics. For example, nuclei with proton number $Z = 2, 8, 20, 40$, or 82, or neutron number $N = 2, 8, 20, 50, 82$, or 126 are known empirically to be special in that:

- Energies of the first few excited states are higher than those in nearby nuclei, as shown in Fig. 6-13;
- Single-neutron and single-proton removal energies, S_n and S_p, defined earlier in (4-78,79) are much larger than those in the neighboring even-even nuclei, as shown in Fig. 6-14; and
- The intrinsic shape of the ground states is spherical, as can be seen from observations such as electromagnetic transitions.

These properties are sufficiently prominent that the sequence of numbers, 2, 8, 20, 40(50), 82, and 126 are known as the magic numbers, and nuclei ^{4}He, ^{16}O, ^{40}Ca, ^{90}Zr, and ^{208}Pb, having both neutron and proton numbers in this sequence, are known as (doubly) magic nuclei. One of the early achievements of nuclear physics was explaining the cause of these magic numbers.

The existence of magic numbers may be explained by an independent-particle model. In this model, nucleons do not interact with each other; the effect of all other nucleons in the nucleus is replaced by an average or mean field which binds individual nucleons to the nucleus (see §6-7 for more detail). The nuclear Hamiltonian is then a sum of single-particle terms,

$$H = \sum_i \epsilon(i)\, \boldsymbol{n}_i \tag{6-53}$$

where the summation is over all single-particle states. The energy of each single-particle state is represented by $\epsilon(i)$ and $\boldsymbol{n}_i$ is the number operator which measures the occupancy (0 or 1) of single-particle state i.

The basic reason for the existence of magic numbers comes from the fact that the nuclear single-particle spectrum is not a smooth one; there are relatively large energy gaps between groups of single-particle states. When each group of states is completely filled, the Fermi energy of the nucleus is just below one of these large energy gaps. As a result, more energy than usual is required to excite the nucleus, for instance, by promoting one of the nucleons below the Fermi energy to an unoccupied state above. For this reason, nuclei fulfilling this condition for both neutrons and protons are also called closed shell nuclei. With all the orbits filled, the ground state of the nucleus is tightly bound and spherical in shape, as explained at the beginning of §6-3.

Harmonic oscillator single-particle spectrum. We need a simple model to see why energy gaps appear in the single-particle spectrum. A one-body Hamiltonian may be written in the form,

$$h(\boldsymbol{r}_i) = -\frac{\hbar^2}{2\mu_i}\nabla_i^2 + V(\boldsymbol{r}_i) \tag{6-54}$$

where r_i is the coordinate of nucleon i and μ_i is its reduced mass. The justification for using a one-body potential $V(\boldsymbol{r}_i)$ to represent the average effect of all the nucleon-nucleon interaction on nucleon i is based on Hartree-Fock calculations described in §6-7. Residual two-body effects are important to understand many of the detailed nuclear properties, but we do not need to be concerned with them here. Furthermore, we shall assume for mathematical convenience that the central binding potential may be approximated by a harmonic oscillator well,

$$V(r_i) = \tfrac{1}{2}\mu_i\omega_0^2 r_i^2 \tag{6-55}$$

where ω_0 is the frequency. This is a reasonable assumption, at least for all the bound nucleons. In order to provide binding to a nucleon, the potential must have a minimum and, near this minimum, it must have a quadratic dependence on the spatial coordinates, the form given by (6-55). Examples of single-particle radial wave functions generated by such a potential are shown in Table 6-2. The forms may not be realistic near the nuclear surface, especially for single-particle states around the Fermi energy. However, this is not a problem for us here.

For an isotropic, three-dimensional harmonic oscillator potential, each (major) *shell* is characterized by N, the number of oscillator quantum. All states belonging to a given shell are degenerate with energy

$$\epsilon_N = \left(N + \frac{3}{2}\right)\hbar\omega_0 \tag{6-56}$$

For each shell, the allowed orbital angular momenta are

$$\ell = N,\ N-2, \ldots,\ 1 \text{ or } 0 \tag{6-57}$$

(See, *e.g.,* C. Cohen-Tannoudji, B. Diu, and F. Laloë, *Quantum Mechanics*, English translation by S.R. Hemley, N. Ostrowsky, and D. Ostrowsky, Wiley, 1977,

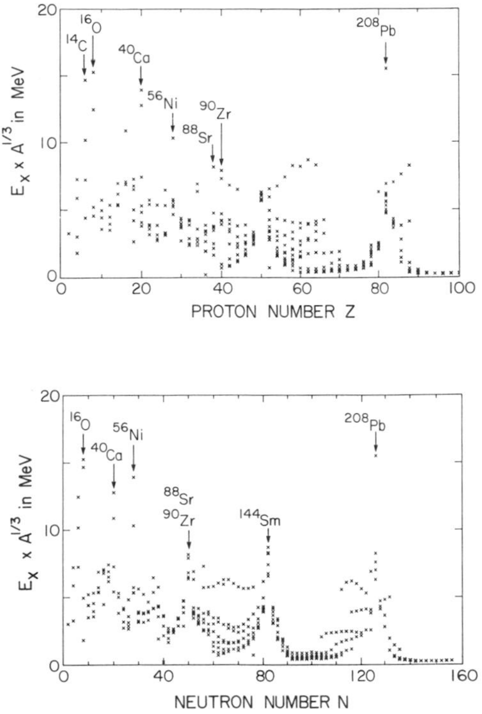

Fig. 6-13 Energy of the first excited state of even-even nuclei as a function of proton number (upper) and neutron number (lower). All energies are multiplied by a factor $A^{1/3}$ to account for the general decrease in excitation energy with increasing nucleon number A. A total of 389 nuclei, form ^{4}He to ^{256}Fm, have been surveyed. Among these, there are 372 with spin-parity 2^+ for the first excited state, and only seven with 0^+, three with 3^-, two with 1^- and five with unknown spin. The highest excitation energy is found in ^{4}He (not shown) with $E_x = 20.1$ MeV or $E_x \times A^{1/3} = 31$ MeV. The data are taken from *Tables of Isotopes*, 7th ed., edited by C.M. Lederer and V.S. Shirley, Wiley, New York, 1978.

p. 818). Since each nucleon also has an intrinsic spin $s = \frac{1}{2}$, the number of states, D_N, *i.e.*, the maximum number of neutrons or protons a harmonic oscillator shell can accommodate, is given by

$$D_N = 2 \sum_{\text{allowed } \ell} (2\ell + 1) = 2 \sum_{k=1}^{N+1} k = (N+1)(N+2) \tag{6-58}$$

where the factor of 2 in front of the summations is to account for the two possible orientations of nucleon intrinsic spin. The total number of states, $D_{\max}$, up to some maximum number of harmonic oscillator quantum $N_{\max}$, is given by a sum overall N-value to $N_{\max}$,

$$D_{\max} = \sum_{N=0}^{N_{\max}} D_N = \tfrac{1}{3}(N_{\max}+1)(N_{\max}+2)(N_{\max}+3)$$

$$\xrightarrow[N_{\max} \gg 1]{} \tfrac{1}{3}\left(N_{\max}+2\right)^3 \tag{6-59}$$

In arriving at the final result, we have made use of the identity

$$\sum_{k=1}^{n} k^2 = \tfrac{1}{6}n(n+1)(2n+1)$$

From (6-59) we obtain the values $D_{\text{max}} = 2, 8, 20, 40, 70, 112, 168, \ldots$, for $N_{\text{max}} = 0, 1, 2, \ldots$.

The harmonic oscillator frequency ω_0 may be related to the size of the nucleus and, hence, to the nucleon number A of a nucleus. The expectation value of r^2 in a state of $N\hbar\omega_0$ can be obtained from the expectation value of the harmonic oscillator potential energy,

$$\langle \tfrac{1}{2}\mu\omega_0^2 r^2 \rangle_N = \frac{1}{2}\left(N + \frac{3}{2}\right)\hbar\omega_0$$

where the factor $\frac{1}{2}$ on the right hand side comes from the fact that, for a particle in a three-dimensional harmonic oscillator potential, the average of potential energy is half of the total energy. Using this relation, we obtain the expectation value of r^2 in a state with N harmonic oscillator quanta,

$$\langle r^2 \rangle_N = \frac{\hbar}{\mu\omega_0}\left(N + \frac{3}{2}\right) \tag{6-60}$$

The mean-square radius of a nucleus made of A nucleons is given by the average over all occupied harmonic oscillator states for both neutrons and protons,

$$\begin{aligned} \langle R^2 \rangle &= \frac{2}{A}\sum_{N=0}^{N_{\text{max}}} D_N \langle r^2 \rangle_N \\ &= \frac{2}{A}\sum_{N=0}^{N_{\text{max}}} (N+1)(N+2)(N+\tfrac{3}{2})\frac{\hbar}{\mu\omega_0} \end{aligned} \tag{6-61}$$

where the factor 2 in front of the summations arises from the fact we must consider both neutrons and protons, assuming that neutron and proton numbers are equal for simplicity. The final result is obtained by substituting the explicit values of D_N given in (6-59) and $\langle r^2 \rangle_N$ in (6-60).

The summation over N in the final form of (6-61) may be carried out with the help of the mathematical identity

$$\sum_{k=1}^{n} k^3 = \left(\frac{n(n+1)}{2}\right)^2$$

and those for $\sum k^2$ and $\sum k$ given earlier. The result is

$$\sum_{N=0}^{N_{\text{max}}} (N+1)(N+2)(N+\tfrac{3}{2}) = \tfrac{1}{4}(N_{\text{max}}+1)(N_{\text{max}}+2)^2(N_{\text{max}}+3)$$

$$\xrightarrow[N_{\text{max}} \gg 1]{} \tfrac{1}{4}(N_{\text{max}}+2)^4$$

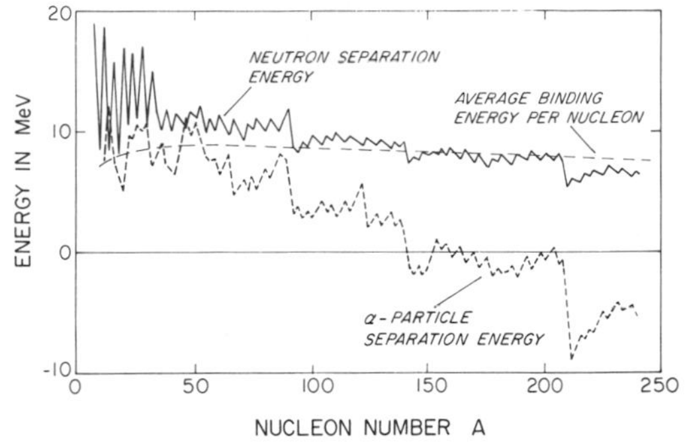

Fig. 6-14 Neutron and α-particle separation energies for stable nuclei as a function of nucleon number calculated from a table of binding energies.

In the limit of large $N_{\max}$, we obtain the result

$$\langle R^2\rangle = \frac{2}{A}\frac{\hbar}{\mu\omega_0}\frac{1}{4}(N_{\max}+2)^4$$

which relates the square of the nuclear radius to the value of $N_{\max}$.

Alternatively, we can use this relation to express $\hbar\omega_0$, a quantum of harmonic oscillator energy, in terms of $N_{\max}$,

$$\hbar\omega_0 = \frac{1}{A}\frac{\hbar^2}{\mu\langle R^2\rangle}\frac{1}{2}(N_{\max}+2)^4 \tag{6-62}$$

The number of nucleons A can be also expressed in terms of $N_{\max}$ using (6-59),

$$A = 2\sum_{N=0}^{N_{\max}} D_N = \tfrac{2}{3}(N_{\max}+2)^3$$

where the factor of 2 is used to account for the fact that each harmonic oscillator state can take a neutron as well as a proton. On inverting the relation, we obtain the expression

$$(N_{\max}+2) = \left(\tfrac{3}{2}A\right)^{1/3} \tag{6-63}$$

Combining the results of (6-61) to (6-63), we obtain

$$\begin{aligned}\hbar\omega_0 &= \frac{1}{A}\frac{\hbar^2}{\mu\langle r^2\rangle}\frac{1}{2}\left(\frac{3}{2}A\right)^{4/3}\\ &= \frac{\hbar^2}{\mu\frac{3}{5}\left(r_0A^{1/3}\right)^2}\frac{3}{4}\left(\frac{3}{2}A\right)^{1/3}\\ &= \frac{5}{4}\left(\frac{3}{2}\right)^{1/3}\frac{\hbar^2}{\mu r_0^2}A^{-1/3} \approx 41A^{-1/3}\ \text{MeV}\end{aligned} \tag{6-64}$$

where we have adopted a constant density sphere model to convert $\langle r^2\rangle$ into $\frac{3}{5}\left(r_0A^{1/3}\right)^2$ as done in (4-19) and used $r_0 = 1.2$ fm to arrive at the final result invoked earlier to characterize the energy required to excite a nucleon up one major shell.

Spin-orbit energy. Let us return to the question of magic numbers. From (6-59), we find that the first part of the sequence 2, 8, 20, and 40 is accounted for by filling up harmonic oscillator shells with either neutrons or protons up to $N_{\max} = 0$, 1, 2, and 3. This is an indication that the harmonic oscillator potential is a reasonable starting point for understanding the structure of single-particle states in nuclei. However, deviations are found beyond $N_{\max} = 3$. To correct for this, additional terms must be introduced to the single-particle Hamiltonian beyond the harmonic oscillator potential of (6-55).

The departure of the sequence of magic numbers from the values given by $D_{\max}$ in (6-59) is explained by single-particle spin-orbit energy, suggested by Mayer and Haxel, Jensen, and Suess in 1949. If the potential that binds a nucleon to the central well has a term that depends on the coupling between $\boldsymbol{s}$, the intrinsic spin of a nucleon, and $\boldsymbol{\ell}$, its orbital angular momentum, the single-particle energies will be a function of the j values of a state as well. Since $\boldsymbol{j} = \boldsymbol{s} + \boldsymbol{\ell}$, two possible states can be formed from a given ℓ and the energies of the two states are different, depending on whether $\boldsymbol{s}$ is parallel to $\boldsymbol{\ell}$ ($j = \ell + \frac{1}{2}$) or anti-parallel to $\boldsymbol{\ell}$ ($j = \ell - \frac{1}{2}$). The source of this single-particle spin-orbit term may be traced back to the spin dependence in the nucleon-nucleon interaction. For our purpose here, we shall, for simplicity, take a semi-empirical approach without being concerned about the origin.

Let a be the strength of the spin-orbit term. The single-particle Hamiltonian given in (6-54) now takes on the form

$$h(\boldsymbol{r}_i) = -\frac{\hbar^2}{2\mu}\nabla_i^2 + \tfrac{1}{2}\mu_i\omega_0^2 r_i^2 + a\,\boldsymbol{s}\cdot\boldsymbol{\ell} \tag{6-65}$$

where the parameter a may depend on the nucleon number A and can be determined, for example, by fitting the observed single-particle energies. When the spin-orbit term is included, the single-particle energy of (6-56) now becomes

$$\epsilon_{N\ell j} = \left(N + \frac{3}{2}\right)\hbar\omega_0 \begin{cases} +\frac{a}{2}\ell & \text{for } j = \ell + \frac{1}{2} \\ -\frac{a}{2}(\ell+1) & \text{for } j = \ell - \frac{1}{2} \end{cases} \tag{6-66}$$

The splitting in the energy between $j_> \equiv \ell + \frac{1}{2}$ single-particle level and $j_< \equiv \ell - \frac{1}{2}$ level is $a(2\ell+1)/2$, but the centroid energy of the two groups is not affected by the spin-orbit splitting.

For $a < 0$, a single-particle state with $j = \ell + \frac{1}{2}$ is lowered in energy. Since the amount of depression increases with increasing ℓ value, a $j_>$ state for large ℓ may be pushed down in energy by an amount comparable to $\hbar\omega_0$, the energy gap between two adjacent harmonic oscillator major shells. As a result, the $j_>$ states of the largest ℓ in a shell with N oscillator quanta may be moved closer to the group of states belonging to the $(N-1)$ shell below. (In practice one introduces also an $\boldsymbol{\ell}^2$-dependent term to the single-particle Hamiltonian to lower the centroid energy of states with large ℓ values; in this way, the $j_<$ states are prevented from

Table 6-2 Harmonic oscillator radial wave function. The oscillator parameter, $\nu = \frac{m\omega_0}{\hbar}$, is given approximately by $A^{-1/3}$ fm^{-2}.

$R_{1s}(r) = 2\left(\frac{\nu^3}{\pi}\right)^{1/4} e^{-\frac{1}{2}\nu r^2}$	$R_{1p}(r) = \sqrt{\frac{2^3}{3}}\left(\frac{\nu^5}{\pi}\right)^{1/4} r e^{-\frac{1}{2}\nu r^2}$
$R_{1d}(r) = \sqrt{\frac{2^4}{15}}\left(\frac{\nu^7}{\pi}\right)^{1/4} r^2 e^{-\frac{1}{2}\nu r^2}$	$R_{2s}(r) = \sqrt{\frac{2^3}{3}}\left(\frac{\nu^3}{\pi}\right)^{1/4} \left(\frac{3}{2} - \nu r^2\right) e^{-\frac{1}{2}\nu r^2}$
$R_{1f}(r) = \sqrt{\frac{2^5}{105}}\left(\frac{\nu^9}{\pi}\right)^{1/4} r^3 e^{-\frac{1}{2}\nu r^2}$	$R_{2p}(r) = \sqrt{\frac{2^4}{15}}\left(\frac{\nu^5}{\pi}\right)^{1/4} \left(\frac{5}{2} - \nu r^2\right) r e^{-\frac{1}{2}\nu r^2}$
$R_{1g}(r) = \sqrt{\frac{2^6}{945}}\left(\frac{\nu^{11}}{\pi}\right)^{1/4} r^4 e^{-\frac{1}{2}\nu r^2}$	$R_{2d}(r) = \sqrt{\frac{2^5}{105}}\left(\frac{\nu^7}{\pi}\right)^{1/4} \left(\frac{7}{2} - \nu r^2\right) r^2 e^{-\frac{1}{2}\nu r^2}$
$R_{3s}(r) = \sqrt{\frac{2^3}{15}}\left(\frac{\nu^3}{\pi}\right)^{1/4} \left(\frac{15}{4} - 5\nu r^2 + \nu^2 r^4\right) e^{-\frac{1}{2}\nu r^2}$	

moving up to join the states in the next major shell higher up.) Because of spin-orbit splitting, we find that $j = \frac{9}{2}$ states for $\ell = 4$ in the $N = 4$ shell are depressed sufficiently to be closer in energy to the $N = 3$ group. As a result, the $j = \frac{9}{2}$-states join the $N = 3$ states to form a major shell of 30 single-particle states instead of 20. For this reason, we have 50 instead of 40 as the magic number for neutrons. Similarly, the magic number 82 instead of 70 is obtained if $j = \frac{11}{2}$-states of the $\ell = 5$ group in the $N = 5$ shell are lowered to join the $N = 4$ group. The magic number 126 is formed by summing all the particles in the $N \leq 5$ shells (totalling 112) together with those filling the $j = \frac{13}{2}$-orbit (which accommodates $(2j + 1) = 14$ identical nucleons) from the major shell above. Following this line of reasoning, the first magic number beyond the known ones is 184.

A point to be noted here is the absence of doubly-magic nucleus with $Z = 50$. Because of Coulomb repulsion, nuclei beyond ^{40}Ca must have an excess of neutrons over protons to be stable and the amount of neutron excess required increases with Z. For ^{90}Zr ($Z = 40$), we find that the neutron excess is $(N - Z) = 10$ and for ^{208}Pb ($Z = 82$), the excess increases to $(N - Z) = 44$. To form a stable nucleus with $Z = 50$, we expect a neutron excess somewhere between 10 to 20. The next higher magic number after 50 is 82. Since $N = 82$ gives too large a neutron excess for $Z = 50$, a doubly-magic nucleus with $Z = 50$ cannot be formed. In spite of this, we do find that the element Sn ($Z = 50$) has more stable isotopes than elements nearby. Other properties of the stable tin isotopes also support the fact that empirically $Z = 50$ is one of the magic numbers, producing nuclei that are more tightly bound than their neighbors.

Superheavy nuclei. The heaviest closed-shell nucleus known is ^{208}Pb with $Z = 82$ and $N = 126$. Calculations indicate that the next stable proton number may be 114 because of the large separation in single-particle energy between proton orbits $1h_{9/2}$, $1i_{13/2}$, and $2f_{7/2}$, and proton orbits $3p_{3/2}$ and $2f_{7/2}$. There is a similar separation for the neutron orbits but the energy gap is smaller and no clear indication for the formation of a neutron subshell at $N = 114$ is found among

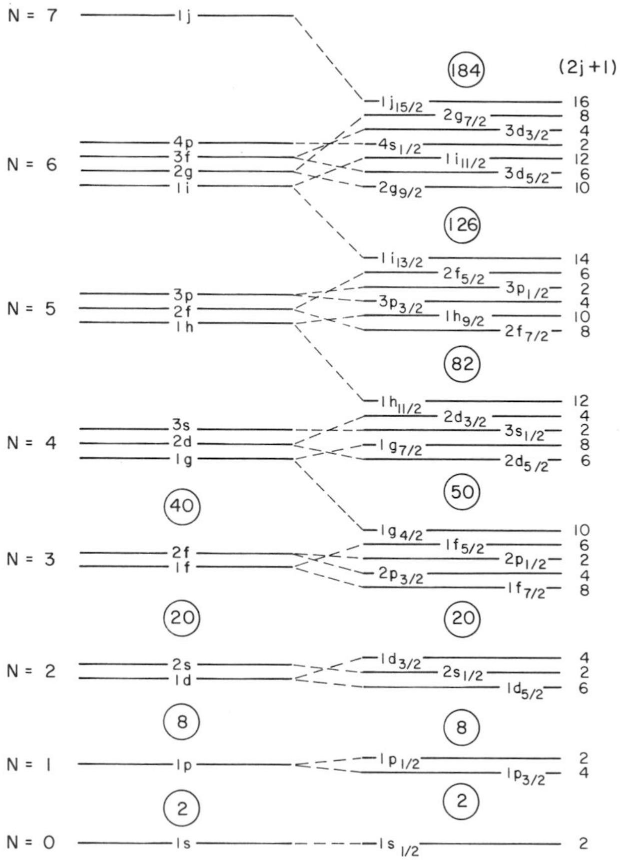

Fig. 6-15 Schematic diagram of single-particle energy level positions for the spherical shell model. The lowest three major shells, $1s$, $1p$ and $2s1d$, are the same as those produced by a three-dimensional, isotropic harmonic oscillator well. The higher major shells also include the orbit with the largest j-values lowered in energy from the major shell above by the spin-orbit term.

empirical evidence. Since $Z = 114$ is not too far from the end of the actinide series at $Z = 103$, there is some possibility that a "superheavy" element with $A = 298$ ($Z = 114$ and $N = 184$) can be made in the laboratory. Alternatively, we may use the known magic number of 126 as the proton number and end up with $A = 310$ as the possible candidate for a superheavy nucleus. Many experimental attempts have been made to find these nuclei and to discover a new group or "island" of stable nuclei around the next set of magic numbers. However, no firm evidence for the existence of these superheavy nuclei has yet been demonstrated. The discovery of one or more superheavy elements will give us new information to test our knowledge of the atomic nucleus. Furthermore, the creation of new and exotic elements is one of those challenges that excites human interest.

Spectroscopic notation. We shall end this section with a description of the standard nomenclature for single-particle orbits in nuclear physics. Each orbit is specified by three labels, N, ℓ, and j. The convention is to use spectroscopic notation with a single letter s, p, d, f, g, h, i, j and so on to stand for $\ell = 0$, 1, 2, 3, 4, 5, 6, 7, ... respectively. The j-value is indicated as a subscript following the letter and the label for major shell is given as a prefix. It is customary to replace N, the number of harmonic oscillator quanta in a major shell, by n the number of nodes in the (modified) radial wave function. There are at least two conventions to number n, depending whether the node at the origin is counted. We shall follow the one with $n = 1, 2, \ldots$ for the first time (one node), the second time (two nodes) and so on for a particular ℓ value to appear in the sequence of single-particle orbits arranged in ascending order according to energy. For example, the single-particle orbit with the lowest energy is $\ell = 0$ without a node in the wave function. It is called the $1s_{1/2}$-orbit. A higher $\ell = 0$ orbit found in the $N = 2$ shell is called the $2s_{1/2}$. The next higher $\ell = 0$ orbit occurs for $N = 4$ and is called the $3s_{1/2}$. Fig. 6-15 gives a more complete illustration of this way of numbering the single-particle orbits. The alternate convention is basically the same except that instead of starting with 1 for the first occurrence of a particular ℓ value, it starts with 0. Both conventions are equally well used, and some confusion may arise on occasion. Furthermore, both conventions are different from that used in atomic physics and in some quantum mechanic textbooks.

§6-6 MANY-BODY BASIS STATES

In a microscopic calculation the main aim is to solve the nuclear many-body eigenvalue problem using the location, spin, isospin, and other degrees of freedom associated with individual nucleons as the independent variables. In this space, the Hamiltonian H is made of a sum of the kinetic energy of each nucleon and the interaction between pairs of nucleons. The behavior of a nucleus is discussed in terms of those of nucleons and the correlations between them. The energy levels of the A-nucleon system is given by the eigenvalues E_α of the Schrödinger equation

$$H\Psi_\alpha(\boldsymbol{r}_1, \boldsymbol{r}_2, \ldots, \boldsymbol{r}_A) = E_\alpha \Psi_\alpha(\boldsymbol{r}_1, \boldsymbol{r}_2, \ldots, \boldsymbol{r}_A) \tag{6-67}$$

Other properties of the system are obtained from the matrix elements of the eigenvectors $\Psi_\alpha(\boldsymbol{r}_1, \boldsymbol{r}_2, \ldots, \boldsymbol{r}_A)$ for the corresponding operators. A microscopic description of the nuclear many-body problem is, therefore, centered around the solution to (6-67). In order to simplify the notation we shall not make any explicit reference here to the intrinsic spin and other degrees of freedom of the nucleons, and use $\boldsymbol{r}_i$ to represent all the independent variables of the system pertaining to nucleon i.

Matrix method to solve the eigenvalue problem. For many purposes, it is more convenient to solve the eigenvalue problem posted by (6-67) using a matrix method. In this approach, we need a complete set of basis states for the A-particle system, $\{\Phi_k(\boldsymbol{r}_1, \boldsymbol{r}_2, \ldots, \boldsymbol{r}_A)\}$ for $k = 1, 2, \ldots, D$, where D is the number of linearly independent states in the Hilbert space. Mathematical convenience dictates that the basis is an orthogonal and normalized one. Any eigenvector $\Psi_\alpha(\boldsymbol{r}_1, \boldsymbol{r}_2, \ldots, \boldsymbol{r}_A)$ may be expressed as a linear combination of these D basis states,

$$\Psi_\alpha(\boldsymbol{r}_1, \boldsymbol{r}_2, \ldots, \boldsymbol{r}_A) = \sum_{k=1}^{D} C_k^\alpha \, \Phi_k(\boldsymbol{r}_1, \boldsymbol{r}_2, \ldots, \boldsymbol{r}_A) \tag{6-68}$$

where C_k^α are the expansion coefficients for the α-th eigenfunction. In principle, the solution to (6-67) is independent of the basis states chosen; in practice, the ease of solving the problem depends critically on the choice. We shall return to this question in the next section.

The starting point of a matrix approach is to select a set of basis states. Once the basis is fixed, the unknown expansion coefficients C_j^α in (6-68) may be found by proceeding in the following way. First, we multiply both sides of (6-67) from the left with $\Phi_j^*(\boldsymbol{r}_1, \boldsymbol{r}_2, \ldots, \boldsymbol{r}_A)$ and integrate over all the independent variables. In terms of Dirac bra-ket notation, the result may be expressed in the form

$$\begin{aligned}\big\langle \Phi_j(\boldsymbol{r}_1, \boldsymbol{r}_2, \ldots, \boldsymbol{r}_A)\big|H\big|\Psi_\alpha(\boldsymbol{r}_1, \boldsymbol{r}_2, \ldots, \boldsymbol{r}_A)\big\rangle \\ = E_\alpha \big\langle \Phi_j(\boldsymbol{r}_1, \boldsymbol{r}_2, \ldots, \boldsymbol{r}_A)\big|\Psi_\alpha(\boldsymbol{r}_1, \boldsymbol{r}_2, \ldots, \boldsymbol{r}_A)\big\rangle\end{aligned}$$

Using the expansion given in (6-68) and the orthonormal property of the basis wave functions, the expression can be reduced into the form

$$\sum_{k=1}^{D} H_{jk} C_k^\alpha = E_\alpha C_j^\alpha \tag{6-69}$$

where H_{jk} is the matrix element of the Hamiltonian between basis states Φ_j and Φ_k,

$$H_{jk} \equiv \langle \Phi_j(\boldsymbol{r}_1, \boldsymbol{r}_2, \ldots, \boldsymbol{r}_A)\big|H\big|\Phi_k(\boldsymbol{r}_1, \boldsymbol{r}_2, \ldots, \boldsymbol{r}_A)\rangle$$

Eq. (6-69) may be written in a matrix form,

$$\begin{pmatrix} H_{11} & H_{12} & \cdots & H_{1D} \\ H_{21} & H_{22} & \cdots & H_{2D} \\ \vdots & \vdots & \ddots & \vdots \\ H_{D1} & H_{D2} & \cdots & H_{DD} \end{pmatrix} \begin{pmatrix} C_1^\alpha \\ C_2^\alpha \\ \vdots \\ C_D^\alpha \end{pmatrix} = E_\alpha \begin{pmatrix} C_1^\alpha \\ C_2^\alpha \\ \vdots \\ C_n^\alpha \end{pmatrix} \tag{6-70}$$

The eigenvalues E_α are the roots of the secular equation,

$$\det \begin{vmatrix} (H_{11} - E_\alpha) & H_{12} & \cdots & H_{1D} \\ H_{21} & (H_{22} - E_\alpha) & \cdots & H_{2D} \\ \vdots & \vdots & \ddots & \vdots \\ H_{D1} & H_{D2} & \cdots & (H_{DD} - E_\alpha) \end{vmatrix} = 0 \tag{6-71}$$

Once an eigenvalue E_α is found, the expansion coefficients C_i^α, $i = 1, 2, \ldots, D$ may be obtained by solving (6-70) as a set of D simultaneous algebraic equation. This gives the eigenvector corresponding to E_α. The complete set of eigenvectors for $\alpha = 1, 2, \ldots, D$ may be viewed as a matrix $\{C_i^\alpha\}$ that transforms the Hamiltonian from the basis representation into a diagonal form. In this way, the eigenvalue problem posted by (6-67) is solved by diagonalizing the Hamiltonian matrix $\{H_{jk}\}$.

Single-particle basis states. In microscopic nuclear structure calculations, the many-body basis wave functions are usually constructed out of products of single-particle wave functions $\phi_i(\boldsymbol{r}_j)$. In order to ensure proper antisymmetrization among the fermions, a many-body state is often written in the form of a Slater determinant,

$$\Phi_k(\boldsymbol{r}_1, \boldsymbol{r}_2, \ldots, \boldsymbol{r}_A) = \frac{1}{\sqrt{A!}} \det \begin{vmatrix} \phi_1(\boldsymbol{r}_1) & \phi_1(\boldsymbol{r}_2) & \cdots & \phi_1(\boldsymbol{r}_A) \\ \phi_2(\boldsymbol{r}_1) & \phi_2(\boldsymbol{r}_2) & \cdots & \phi_2(\boldsymbol{r}_A) \\ \vdots & \vdots & \ddots & \vdots \\ \phi_A(\boldsymbol{r}_1) & \phi_A(\boldsymbol{r}_2) & \cdots & \phi_A(\boldsymbol{r}_A) \end{vmatrix} \tag{6-72}$$

where the factor $(A!)^{-1/2}$ is required for normalization. If a different set of A single-particle states is taken, a different many-body basis state can be constructed. Choice of single-particle wave functions is therefore an essential step that precedes the construction of many-body basis states.

In principle, the single-particle spectrum is an infinite one, extending in energy without an upper bound. As a result, the number of single-particle states, and consequently D, the number of many-body basis states that can be constructed out of these single-particle states, is infinite. In such an infinite dimensional Hilbert space, the eigenvalue problem is impractical, if not impossible, to solve. One of the main goals of microscopic nuclear structure models is to find reasonable and realistic approximation schemes such that only a small portion of the Hilbert space is adequate to give a reasonable description of the nuclear states we wish to study.

There are many possible ways to specify the single-particle wave functions $\phi_i(\boldsymbol{r}_i)$ for our needs. In general, it is more convenient to take them as eigenfunctions of a single-particle Hamiltonian, such as the one given by (6-54),

$$h(\boldsymbol{r}_i)\phi_k(\boldsymbol{r}_i) = \epsilon_k \phi_k(\boldsymbol{r}_i) \tag{6-73}$$

Here ϵ_k is the single-particle energy. In terms of such a single-particle Hamiltonian, the many-body Hamiltonian in (6-67) may be expressed in the form

$$H = \sum_{i=1}^{A} h(\boldsymbol{r}_i) + \sum_{i \neq j=1}^{A} V(\boldsymbol{r}_i, \boldsymbol{r}_j) \tag{6-74}$$

where $V(\boldsymbol{r}_i, \boldsymbol{r}_j)$ is the *residual* two-body interaction, the original nucleon-nucleon interaction minus contributions already included in the first term.

The matrix elements of H in the many-body basis states can now be written as a sum of two terms,

$$H_{jk} = \delta_{jk} \sum_{n=1}^{A} \epsilon_n + V_{jk} \tag{6-75}$$

where ϵ_n is the single-particle energy defined in (6-73). The first term of (6-75) is diagonal in a many-body basis constructed out of products of the eigenfunctions of $h(\boldsymbol{r}_i)$. The second term,

$$V_{jk} \equiv \left\langle \Phi_j(\boldsymbol{r}_1, \boldsymbol{r}_2, \ldots, \boldsymbol{r}_A) \middle| \sum_{p \neq q=1}^{A} V(\boldsymbol{r}_p, \boldsymbol{r}_q) \middle| \Phi_k(\boldsymbol{r}_1, \boldsymbol{r}_2, \ldots, \boldsymbol{r}_A) \right\rangle \tag{6-76}$$

is the contribution to the Hamiltonian matrix element from the residual interaction.

Independent-particle model. In the basis we have adopted, the only contributions of the many-body Hamiltonian to the off-diagonal matrix elements come from V_{jk}, the two-body part of the Hamiltonian. If the residual interaction is sufficiently weak, the contributions will be small compared with those from the single-particle Hamiltonian. Under such conditions it may be possible to ignore the two-body part in (6-74), and the Hamiltonian will then be diagonal in the basis we have chosen. In this limit, the eigenvalues are simply given by sums of single-particle energies alone, and the many-body eigenvalue problem is reduced to a much simpler one-body eigenvalue problem. This is the idea behind the *independent-particle model.* It is a good approximation of a microscopic solution of the nuclear structure problem if a realistic single-particle Hamiltonian is available and, consequently, a physically reasonable set of single-particle states can be constructed. We shall see that this is the goal of the Hartree-Fock approach described in the next section.

In reality, many of the observed nuclear properties are the results of "correlations" between nucleons. In such cases, linear combinations of products of single-particle states are required to describe them. The independent-particle model no longer applies, and the effect of the residual interaction must be incorporated into the calculation of many-body wave functions. On the other hand, it is often the case that a wave function is dominated by a few many-body basis states. In particular this is true for the low-lying states of interest in many nuclei. For these states, we expect that the many-body eigenvalue problem may be approximated by diagonalizing the Hamiltonian matrix in a small, well-chosen subspace of the complete Hilbert space. The nuclear shell model, described in §6-8, was developed in the belief that this is possible, at least for a few low-lying states in nuclei near closed-shells.

§6-7 HARTREE-FOCK SINGLE-PARTICLE HAMILTONIAN

The complete nuclear Hamiltonian is a sum of one- and two-body parts, as given earlier in (6-74). In its most elementary form, the one-body part $h(\boldsymbol{r}_i)$ is made of the kinetic energy of individual nucleons and the two-body potential $V(\boldsymbol{r}_i, \boldsymbol{r}_j)$ expresses the interaction between nucleons. We have seen that it may be possible to represent a large part of the nucleon-nucleon interaction on nucleon i by an average field which we may include as a part of $h(\boldsymbol{r}_i)$. In this case, the residual two-body interaction is given by an "effective potential" between two nucleons and may therefore be much weaker than the original nucleon-nucleon interaction. One of the aims of the Hartree-Fock approach to the nuclear many-body problem is to find a single-particle representation such that the residual interaction is a minimum.

In principle, a complete set of single-particle states is needed before one can carry out any calculations. For the purpose of discussion, we shall start with an arbitrary set of single-particle states as the trial functions in the rest of this section. The final result is, in principle, independent of this choice; in practice, it is advantageous to take a set that is convenient from a mathematical point of view, such as the harmonic oscillator wave functions.

Variation of the trial wave function. In the absence of a two-body residual interaction, the ground state is given by the configuration with the lowest A single-particle states occupied. The wave function $|\,\Phi_0\,\rangle$ for the A-nucleon system is a Slater determinant having the form given by (6-72). In order to display that $|\,\Phi_0\,\rangle$ is made of a product of single-particle states without having to write the Slater determinant out explicitly, we shall adopt a shorthand notation,

$$|\Phi_0\rangle = |\phi_1\phi_2\cdots\phi_A\rangle \tag{6-77}$$

to represent the normalized and antisymmetrized product wave function of (6-72). If $|\,\Phi_0\,\rangle$ is the true ground state wave function of the system, it must satisfy the variational condition

$$\delta\langle\phi_1\phi_2\cdots\phi_A\big|H\big|\phi_1\phi_2\cdots\phi_A\rangle = 0 \tag{6-78}$$

that is, the wave function $|\,\Phi_0\,\rangle$ is one that gives a minimum in the energy. The aim of a Hartree-Fock calculation is to find a set of single-particle states that fulfills this condition.

Since variations on the bra $\langle\Phi_0|$ are not independent of the variations on the ket $|\Phi_0\rangle$, eq. (6-78) is equivalent to the condition

$$\big\langle\delta\Phi\big|H\big|\Phi_0\big\rangle = 0 \tag{6-79}$$

In other words, the variation in (6-78) can be accomplished by a variation of the bra (or the ket) alone. There are two possible ways to modify the many-body wave function. We can modify the single-particle wave functions themselves such that

(6-79) is satisfied or, alternatively, we can keep the single-particle basis fixed and alter Φ_0 by adding to it small amounts of Slater determinants made of products of different A single-particle states. As long as there is a complete set of states and all possible variations are applied, these two methods are equivalent.

We shall take the latter approach for our derivation here. With a fixed single-particle basis, each many-particle basis state is specified by the single-particle states that are occupied. For example, the trial ground state wave function Φ_0 in (6-77) is one with the occupancies of the lowest A single-particle states equal to unity and the occupancies of the rest of the single-particle states zero. A different many-body wave function will have a different set of single-particle states occupied. A linear combination of two or more such many-body basis states then means that some of the single-particle states are partially occupied; the occupancies of these states take on fractional value between 0 and 1 as a result.

Let us label the basis by numbering all the single-particle states with an index $r = 1, 2, \cdots, d$, for instance, according to ascending order in the single-particle energy ϵ_r. A many-body basis state can then be specified by giving the indices of the occupied single particle states. Thus, the lowest many-body state in terms of the sum of single-particle energies may be represented in the form

$$|\Phi_0\rangle = |1, 2, \ldots, A\rangle$$

sometimes referred to as the *occupancy representation*. In the limit that two-body interactions can be ignored, $|\Phi_0\rangle$ is the ground state of the A-nucleon system and the Fermi energy ϵ_F is given by the energy of the highest occupied single-particle state.

A variation on $|\Phi_0\rangle$ may be achieved by mixing a small amount of a state $|\Phi_{kt}\rangle$ made by promoting a particle below the Fermi energy, for example, single-particle state t, to a state above, for example, single-particle state k. Such a state may be represented in the form

$$|\Phi_{kt}\rangle = |1, 2, \ldots, t-1, t+1, \cdots, A, k\rangle$$

Many-body states made in this way by promoting a nucleon from an occupied single-particle state below the Fermi energy of the A-nucleon system to an unoccupied state above are called one-particle-one-hole states, or $1p1h$-states for short. Other variations, involving two-particle-two-hole ($2p2h$) and more complicated types of excitation, can also be constructed, but we shall ignore them here.

An arbitrary variation consisting of all possible $1p1h$-excitations may be written in the form

$$|\delta\Phi\rangle = \sum_{kt} \eta_{kt} |\Phi_{kt}\rangle$$

where η_{kt} gives the amount of each component in the variation. In order to ensure that the variations are carried out in small steps, $|\eta_{kt}|$ must be kept much smaller than unity. Using this form, (6-79) may be expressed in the following way:

$$\sum_{kt} \eta_{kt} \langle \Phi_{kt} | H | \Phi_0 \rangle = 0$$

Since different variations, represented by different occupied single-particle state k and empty single-particle state t, are independent of each other, it is necessary that each term in the sum vanishes, and we obtain the condition

$$\langle \Phi_{kt} | H | \Phi_0 \rangle = \langle \Phi_{kt} | \sum_{i=1}^{A} h(\boldsymbol{r}_i) + \sum_{i \neq j=1}^{A} V(\boldsymbol{r}_i, \boldsymbol{r}_j) | \Phi_0 \rangle = 0 \tag{6-80}$$

For the one-body part of the Hamiltonian, the only possible nonvanishing matrix elements are those with left and right hand sides differing by no more than the single-particle state of a nucleon. Similarly, for the two-body part of the Hamiltonian, only matrix element, with occupied single-particle states on the left and right hand side of $V(\boldsymbol{r}_i, \boldsymbol{r}_j)$ differing by no more than two can be different from zero. Upon integration over all the single-particle coordinates other than those acted upon by the operators, the condition expressed in (6-80) for a one- plus two-body Hamiltonian reduces to the form

$$\langle k | h | t \rangle + \sum_{r} \langle kr | V(\boldsymbol{r}_1, \boldsymbol{r}_2) | tr \rangle = 0 \tag{6-81}$$

where the summation is over all the occupied states.

The first term in (6-81), $\langle k|h|t \rangle$, is the matrix element of the one-body part of the Hamiltonian between single-particle states $|\, t\, \rangle$ and $|\, k\, \rangle$. Similarly, the matrix element of the two-body part, $\langle kr|V(\boldsymbol{r}_1, \boldsymbol{r}_2)|tr\rangle$, is between antisymmetrized two-particle states,

$$|kr\rangle \equiv \frac{1}{\sqrt{2}} \begin{vmatrix} \phi_k(\boldsymbol{r}_1) & \phi_k(\boldsymbol{r}_2) \\ \phi_r(\boldsymbol{r}_1) & \phi_r(\boldsymbol{r}_2) \end{vmatrix} = \frac{1}{\sqrt{2}} \Big\{ |\, \phi_k(\boldsymbol{r}_1)\phi_r(\boldsymbol{r}_2)\, \rangle - |\, \phi_r(\boldsymbol{r}_1)\phi_k(\boldsymbol{r}_2)\, \rangle \Big\}$$

$$|tr\rangle \equiv \frac{1}{\sqrt{2}} \begin{vmatrix} \phi_t(\boldsymbol{r}_1) & \phi_t(\boldsymbol{r}_2) \\ \phi_r(\boldsymbol{r}_1) & \phi_r(\boldsymbol{r}_2) \end{vmatrix} = \frac{1}{\sqrt{2}} \Big\{ |\, \phi_t(\boldsymbol{r}_1)\phi_r(\boldsymbol{r}_2)\, \rangle - |\, \phi_r(\boldsymbol{r}_1)\phi_t(\boldsymbol{r}_2)\, \rangle \Big\} \tag{6-82}$$

Using these results, the two-body matrix element in (6-81), *i.e.*, the second term on the left hand side, may be expressed explicitly in terms of single-particle wave functions, in the following form:

$$\begin{aligned} &\langle kr | V(\boldsymbol{r}_1, \boldsymbol{r}_2) | tr \rangle \\ &\quad = \tfrac{1}{2} \langle \phi_k(\boldsymbol{r}_1)\phi_t(\boldsymbol{r}_2) - \phi_t(\boldsymbol{r}_1)\phi_k(\boldsymbol{r}_2) | V(\boldsymbol{r}_1, \boldsymbol{r}_2) | \phi_t(\boldsymbol{r}_1)\phi_r(\boldsymbol{r}_2) - \phi_r(\boldsymbol{r}_1)\phi_t(\boldsymbol{r}_2) \rangle \\ &\quad = \langle \phi_k(\boldsymbol{r}_1)\phi_t(\boldsymbol{r}_2) | V(\boldsymbol{r}_1, \boldsymbol{r}_2) | \phi_t(\boldsymbol{r}_1)\phi_r(\boldsymbol{r}_2) \rangle \\ &\qquad\qquad - \langle \phi_k(\boldsymbol{r}_1)\phi_t(\boldsymbol{r}_2) | V(\boldsymbol{r}_1, \boldsymbol{r}_2) | \phi_r(\boldsymbol{r}_1)\phi_t(\boldsymbol{r}_2) \rangle \end{aligned} \tag{6-83}$$

where we have made use of the symmetry relation

$$\langle \phi_k(\boldsymbol{r}_1)\phi_r(\boldsymbol{r}_2)|V(\boldsymbol{r}_1,\boldsymbol{r}_2)|\phi_t(\boldsymbol{r}_1)\phi_r(\boldsymbol{r}_2)\rangle = \langle \phi_r(\boldsymbol{r}_1)\phi_k(\boldsymbol{r}_2)|V(\boldsymbol{r}_1,\boldsymbol{r}_2)|\phi_r(\boldsymbol{r}_1)\phi_t(\boldsymbol{r}_2)\rangle$$

The derivation from (6-80) to (6-81) may be carried out in a more elegant way using second-quantized notations where, as shown in Appendix F, antisymmetrization between single-particle operators is built into the commutation relations.

Hartree-Fock Hamiltonian. It is more instructive to write the relation expressed by (6-81) in an operator form. For this purpose, the left-hand side of the equation may be regarded as the matrix element of a one-body operator, since it operates only on the single-particle state k on the right (or the single-particle state t on the left). However, $V(\boldsymbol{r}_1,\boldsymbol{r}_2)$ is a two-body operator, and we shall see how to "reduce" it to a one-body operator.

Let us distinguish between two sets of single-particle states here by using Greek alphabets, α, β, ... to indicate the original, or trial, single-particle states and Roman alphabets r, s, ..., for the Hartree-Fock single-particle states that satisfy (6-81). A two-body interaction potential may be expressed in the original basis in the form of an operator in the following way:

$$\sum_{ij} V(\boldsymbol{r}_i,\boldsymbol{r}_j) = \sum_{\alpha\beta\gamma\delta} |\alpha\beta\rangle V_{\alpha\beta\gamma\delta}\langle\gamma\delta| \tag{6-84}$$

where

$$V_{\alpha\beta\gamma\delta} \equiv \langle\alpha\beta|V(\boldsymbol{r}_1,\boldsymbol{r}_2)|\gamma\delta\rangle$$

is the matrix element of $V(\boldsymbol{r}_1,\boldsymbol{r}_2)$ between antisymmetrized and normalized two-body wave functions $|\alpha\beta\rangle$ and $|\gamma\delta\rangle$, such as those given in (6-82). Using this form of the two-body potential, we can write the left hand side (6-81) as the one-body matrix element of the following operator,

$$h_{\rm HF} = h + \sum_r \sum_{\alpha\beta\gamma\delta} \langle r|\alpha\beta\rangle V_{\alpha\beta\gamma\delta}\langle\gamma\delta|r\rangle \tag{6-85}$$

This is the Hartree-Fock single-particle Hamiltonian operator. The left hand side of (6-81) is the matrix element of $h_{\rm HF}$ between single-particle bra $|\,k\,\rangle$ and single-particle ket $|\,t\,\rangle$. The quantity $\langle t|\alpha\beta\rangle$ in (6-85) is a one-body operator since it is the overlap of a two-body ket with a one-body bra. Except for the implied antisymmetrization in the two-body wave function, we may take the quantity as

$$\langle r|\alpha\beta\rangle \sim \langle r|\alpha\rangle|\,\beta\,\rangle$$

The second term of $h_{\rm HF}$ in (6-85) may be interpreted as the average one-body potential, or the *mean field*, experienced by a nucleon as the result of (two-body) interactions with all the other nucleons in the nucleus.

Eq. (6-81) can also be expressed as an eigenvalue equation using the operator form given in (6-85),

$$h|t\rangle + \sum_{r} \sum_{\alpha\beta\gamma\delta} \langle r|\alpha\beta\rangle V_{\alpha\beta\gamma\delta} \langle\gamma\delta|rt\rangle = \epsilon_t |t\rangle \tag{6-86}$$

where ϵ_t is the Hartree-Fock single-particle energy. The solution of (6-86) provides us with a transformation from the set of basis states $|\alpha\rangle$, $|\beta\rangle, \cdots$, used as the trial wave functions to the eigenstates of the Hartree-Fock single-particle Hamiltonian $|k\rangle$, $|t\rangle, \cdots$, defined by (6-86).

However, the calculation is not as straightforward as it may seem on the surface. To evaluate both $\langle r|\alpha\beta\rangle$ and $\langle\gamma\delta|rt\rangle$ in (6-86) or the equivalent quantity of $\langle kr|V|tr\rangle$ in (6-81), we need an *a priori* knowledge of the solution, as these matrix elements are evaluated in the Hartree-Fock basis that is a part of the end result of the calculation. This means that the calculation must be carried iteratively starting with an arbitrary set of single-particle wave functions, such as harmonic oscillator wave functions. Using these trial functions, we can evaluate all the necessary matrix elements and solve (6-86). The solutions obtained are only a first approximation, since we have not used the proper wave functions to evaluate the matrix elements to start with. On the other hand, the solutions represent a better approximation of the "true" Hartree-Fock wave functions than the trial wave functions which we have used as the input. Now we can improve the solutions by using the first approximation results as the input and carrying out the calculation again. The process can be repeated until self-consistency in the solution is achieved; that is, the solutions obtained are identical to the wave functions used to evaluate the matrix elements.

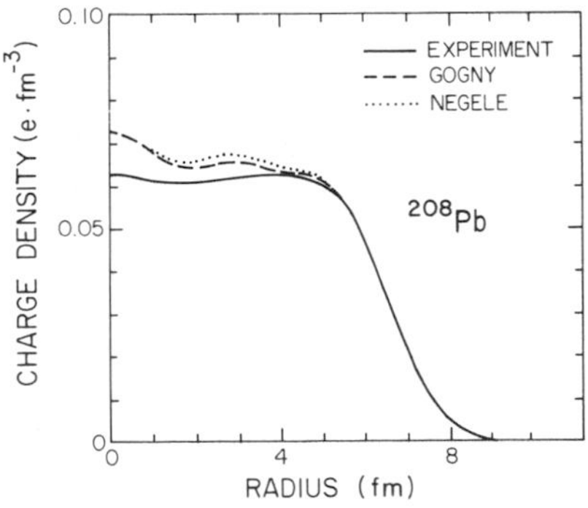

Fig. 6-16 Charge density of ^{208}Pb measured by Frois et al. (*Phys. Rev. Lett.* **38** [1977] 152) using elastic electron scattering. The experimental data (solid line) in the surface region of the nucleus are well described by the calculated Hartree-Fock results of Gogny (in *Nuclear Self-Consistent Fields* ed. by G. Ripka and M. Porneuf, North-Holland, Amsterdam, 1975; and Negele (*Phys. Rev. C* **1** [1970] 1260). The main discrepancies are in the interior region.

Hartree-Fock calculations have been used extensively to study low-lying states in nuclei. However, since each nuclear wave function is made of a single Slater determinant, it does not correspond to a state with definite spin and isospin.

In order to use them to calculate quantities that can be compared with experimental data, states of define J and T must be projected out of the Hartree-Fock wave function. We shall not go into the technical detail of how to carry out the projection or extensions of the topic to projected Hartree-Fock where the variational calculation is carried out after spin and isospin projections. The self-consistent single-particle basis obtained here is, however, important for understanding nuclear ground states as well as the nuclear shell model to be discussed next.

§6-8 SPHERICAL SHELL MODEL

The aim of the nuclear shell model is to describe nuclear states in terms of nucleon degrees of freedom. We have seen that for phenomena where $1p1h$ type of correlation is important, it is adequate to use an independent-particle model with a Hartree-Fock single-particle basis. An example of such a study is shown in Fig. 6-16 for the charge density distribution in a nucleus. For many other observables, higher-order correlations induced by the residual interaction play an important role. A study of these properties requires the presence of the two-body residual interaction and the shell model is designed to meet this need.

A shell model calculation starts with a set of single-particle states. The interaction between nucleons is, in general, sufficiently strong that calculations must be carried out in a Hilbert space far larger than what is practical. On the other hand, we have seen in the previous section that, by transforming the single-particle states to a Hartree-Fock basis, a large part of the nucleon-nucleon interaction may be included in the average one-body field for a nucleon. The residual interaction remaining may be weak enough that the nuclear many-body eigenvalue problem may be carried out in a small subset of the complete space. In other words, if a Hartree-Fock type of single-particle basis is used, the residual interaction may be sufficiently weak that most of the many-body basis states are not involved in any significant way in the low-lying states of interest to us. As a result, we can ignore these basis states without having to pay a large penalty in accuracy. We shall assume from now on that the single-particle basis for shell model investigations always satisfies this criterion.

In Chapter 3 we have seen that the nucleon-nucleon potential is usually in a form convenient for describing the scattering of one nucleon off another. In order to obtain the proper residual interaction for a shell model calculation, we must apply a "transformation" to such a nucleon-nucleon potential so that it is appropriate for the single-particle representation adopted. In addition, two further considerations must also be included. The first is, as mentioned earlier, that the interaction between nucleons bound in a nucleus differs from that between free nucleons. To simplify the discussion, we shall assume that an interaction suitable for bound nucleons is available. The second is related to the truncation of the shell model space. Since it is desirable from a practical point of view to reduce the size of the active space as much as possible, it may become necessary to leave out a large number of basis states which individually contribute only a very small amount to the nuclear states of interest. To account for the small contributions of these states, the residual interaction should be "renormalized" so that the effect of the truncated space may be taken care of in an average way.

So far we have outlined three steps that must be carried out before a shell model calculation can be performed: the choice of a single-particle basis, the selection of an active space, and the derivation of an effective interaction. It is easy to see that all three steps are intimately related to each other and all of them hinge on the single-particle basis chosen for the calculation. In this section, our main concern will be with calculations in the spherical bases. An alternative approach is to use a physically more realistic single-particle basis which includes deformation of the nuclear shape. This will be done in the next section.

Selection of the shell model space. We shall begin with the selection of a shell model space and effective interaction before going into any examples. Since many-body basis states are made of products of single-particle wave functions, a selection of the shell-model space is usually achieved (but not limited to) by restricting the number of active single-particle states. In the spherical shell model, each nucleon has an intrinsic spin $\boldsymbol{s}$ and occupies a state of definite orbital angular momentum $\boldsymbol{\ell}$. The many-body basis states consisting of A such nucleons may be coupled together to form states with definite total angular momentum $\boldsymbol{J}$. In addition, the wave function is antisymmetrized; for example, as done in (6-72).

There are two different ways to carry out the angular momentum coupling. In the LS-coupling scheme, the orbital angular momentum $\boldsymbol{\ell}_i$ and the intrinsic spin $\boldsymbol{s}_i$ of all the nucleons are first coupled separately to form total orbital angular momentum $\boldsymbol{L}$ and total intrinsic spin $\boldsymbol{S}$ for the A-nucleon state,

$$\boldsymbol{L} = \sum_{i=1}^{A} \boldsymbol{\ell}_i \qquad\qquad \boldsymbol{S} = \sum_{i=1}^{A} \boldsymbol{s}_i \tag{6-87}$$

The total angular momentum $\boldsymbol{J}$, more commonly referred to as the spin of the state, is then given by coupling together $\boldsymbol{L}$ and $\boldsymbol{S}$,

$$\boldsymbol{J} = \boldsymbol{L} + \boldsymbol{S} \tag{6-88}$$

Alternatively, in the jj-coupling scheme, the orbital angular momentum and the intrinsic spin of each nucleon is coupled together first to form the nucleon spin $\boldsymbol{j}_i$,

$$\boldsymbol{j}_i = \boldsymbol{\ell}_i + \boldsymbol{s}_i \tag{6-89}$$

and the nuclear spin $\boldsymbol{J}$ is given by coupling together the spins $\boldsymbol{j}_i$ of individual nucleons,

$$\boldsymbol{J} = \sum_{i=1}^{A} \boldsymbol{j}_i \tag{6-90}$$

Since the Hamiltonian is invariant under a rotation of the coordinate system, J is a good quantum number. Furthermore, isospin T is also a constant of motion if we ignore the symmetry breaking effects of electromagnetic interactions. For these reasons, Hamiltonian matrix elements between states of different J and T values vanish. If the many-body basis states are grouped together according to their (J, T)-values, the Hamiltonian matrix in the complete shell model space will appear in a block-diagonal form; that is, only square blocks of matrix elements along the diagonal corresponding to a given set of (J, T)-value are different from zero. The calculation can therefore be carried out separately within the subspace of a specific (J, T) of the full shell model space. In this way, angular momentum coupling greatly reduces the size of the active space in which a calculation has to be carried out.

On the other hand, unless the number of active single-particle states is finite, the number of A-nucleon states for given J and T is still infinite. To construct a finite shell model space, the active single-particle states must be restricted to a small number. This can be achieved from two different directions. Since we are primarily interested in a few nuclear states near the ground state region, we need only to excite particles occupying single-particle states just below the Fermi energy, the energy of the highest occupied single-particle state in the lowest configuration. These are the *active* or *valence* nucleons.

If nucleons in some of the low-lying single-particle states are never excited, they form an inert *core* and the single-particle states they occupy may be left out of the active space. Their contributions to the Hamiltonian may be separated into two parts. The first is a constant term in energy, made of the contributions from single-particle energies and mutual interaction between nucleons in the core. Such a constant can usually be absorbed into the definition of the zero point of the energy scale for the A-nucleon system and may therefore be ignored except in calculations involving the total binding energy of the nucleus. The second is the contributions of the core nucleons to the binding energy of the active nucleons. The single-particle energy of an active nucleon is the average interaction energy with all the other nucleons, including those in the inert core. Contributions from the core nucleons cannot be ignored but may be easily accounted for in the definition of the single-particle energies of the active nucleons. The net result is that contributions of the core nucleons in a shell model calculation are limited to the single-particle energy of the active nucleons and may be included without explicitly considering the core nucleons, as we shall see later.

By the same token, there are also single-particle states so high above the Fermi energy that any many-body basis states having nucleons occupying these states will lie very high in energy. If our interests are confined to the low-lying region of a nucleus, it is unlikely that there can be any significant contributions from these basis states. Such a set of single-particle states is therefore essentially always empty for our purposes and, as a result, may also be ignored.

The only remaining single-particle states are the few near the Fermi energy and they form the *active* or *valence* space from which reasonable approximate wave functions of the nuclear states of interest can be constructed. In summary, the complete single-particle space is now divided into three parts, the core state, the empty states, and the active states. The aim of the nuclear shell model is to solve the eigenvalue problem with many-body basis states consisting of single-particle states in the active space alone.

Effective Hamiltonian. What is the appropriate Hamiltonian to be used in a shell model space? From the beginning of this section we have assumed that we are in a Hartree-Fock single-particle basis and the Hamiltonian is a sum of one- and two-body parts. The one-body part is made of nucleon kinetic energy and a mean field acting on the individual nucleon due to the average interaction with all other nucleons in the nucleus. The two-body part consists of a residual interaction, the part of the nucleon-nucleon interaction not accounted for by the mean field. Here we wish to make a further transformation and use, instead, an *effective* Hamiltonian such that when the active shell model space is restricted to a manageable size, the effect of the states ignored in the calculation may be accounted for in an average way.

A formal definition of the effective Hamiltonian H_{eff} may be carried out in the following way. Let P be an operator that projects out a finite shell model space of dimension d in which we wish to carry out the calculations. If the "true" Hamiltonian is H, the eigenvalue problem in the complete Hilbert space may be expressed in the form

$$H\Psi_i = E_i\Psi_i \tag{6-91}$$

where E_i is an eigenvalue and Ψ_i is an eigenvector of H. An ideal effective Hamiltonian is one that satisfies the following condition,

$$H_{\text{eff}}P\Psi_i = E_iP\Psi_i \tag{6-92}$$

In other words, an effective Hamiltonian is one which produces the same eigenvalues in the shell model space as those obtained by solving the problem in the complete space using the true Hamiltonian. In general, it is impossible to satisfy this condition for all d eigenvalues in the truncated shell model space. This is, however, not a problem, since we are interested only in a small number of low-lying states that are much less than d.

The effective Hamiltonian may also be written as a sum of two terms,

$$H_{\text{eff}} = H_0 + V_{\text{eff}} \tag{6-93}$$

where the one-body part, H_0, may be taken to be a sum of Hartree-Fock single-particle Hamiltonian,

$$H_0 = \sum_i h(\boldsymbol{r}_i) = \sum_i \epsilon_i \boldsymbol{n}_i \tag{6-94}$$

Here ϵ_i is the energy and $\boldsymbol{n}_i$ is the number operator for single-particle state i.

It is understood that single-particle energy ϵ_i, defined in (6-86), also includes contributions from the core nucleons. In practice, it is common to replace ϵ_i by the observed energy level positions of single-particle states in the region. The measured values of ϵ_i may be found in nuclei with one nucleon away from closed shells where some of the low-lying states are predominantly formed by the coupling of one nucleon or one hole to the ground states of the closed shell nuclei. Such states are exactly the ones described by Hartree-Fock eigenvectors. If a realistic Hamiltonian were used in the Hartree-Fock calculation, one would have obtained the same eigenvalues as the experimental energies. For nuclei away from closed shells, correlations other than $1p1h$ play a role, and the Hartree-Fock states are no longer good approximations of the eigenstates of the complete Hamiltonian. This results in a situation where the strength of each single-particle state is shared by several nuclear states. If the fractional strength to each state is known, for example, from the spectroscopic factors of one-nucleon transfer reactions (see §7-2), the location of the single-particle state is then given by the centroid of the strength distribution. As long as such "experimental" single-particle energies are available, it is a good approximation to use these observed values in place of calculated Hartree-Fock values.

We shall assume that the effective interaction V_{eff} remains two-body in character, although there is no reason to rule out three-body and higher order terms caused by excitations to basis states outside the shell model space. Such terms are believed to be small in general and may be ignored.

A formal solution of the effective potential problem may be obtained in the following way. In addition to the operator P, which projects out the active part of the space from the entire many-body Hilbert space, we shall also define an operator Q which projects out the rest of the Hilbert space, such that

$$P + Q = 1 \tag{6-95}$$

By virtue of the fact that they are projection operators, they have the properties

$$P^2 = P \qquad\qquad Q^2 = Q \tag{6-96}$$

To economize on the notation, let us write the eigenvalue equation in the complete space in the form

$$H\Psi = E\Psi \tag{6-97}$$

and

$$H = H_0 + V \tag{6-98}$$

On applying the operator P on the left to both sides of (6-97), we obtain

$$PH\Psi = PE\Psi \tag{6-99}$$

Using (6-95), and the fact that E is a number and therefore commutes with operator P, we may express (6-99) in the form

$$PH(P+Q)\Psi = EP\Psi$$

or

$$PHP\Psi + PHQ\Psi = EP\Psi \tag{6-100}$$

Similarly, instead of P, we can apply Q on the left to both sides of (6-97) and obtain an equation similar to (6-100) except with the roles of P and Q interchanged,

$$QHP\Psi + QHQ\Psi = EQ\Psi \tag{6-101}$$

Eqs. (6-100) and (6-101) may be regarded as a set of two coupled equations for $P\Psi$ and $Q\Psi$.

We can now proceed to solve these equations by expressing $Q\Psi$ in terms of $P\Psi$. Eq. (6-101) may be written in the form,

$$EQ\Psi - QHQ\Psi = QHP\Psi$$

This can be rewritten in the following way:

$$(E - QH)Q\Psi = QHP\Psi$$

or

$$Q\Psi = \frac{1}{E - QH} QHP\Psi \tag{6-102}$$

We must be a little careful with the order of operations in this "formal" solution, since not all the operators commute with each other. Furthermore, the meaning of having operators in the denominator, $(E - QH)^{-1}$, has to be clarified and we shall do this later.

Substituting the "value" of $Q\Psi$ given by (6-102) into (6-100), we obtain the result

$$PHP\Psi + PH\frac{1}{E - QH}QH\ P\Psi = EP\Psi$$

or

$$P\Big\{H + H\frac{1}{E - QH}QH\Big\}P\Psi = E\ P\Psi \tag{6-103}$$

It is useful to recall that the Hamiltonian has the form $H = H_0 + V$, where H_0 is a one-body Hamiltonian and V is a two-body potential. In terms of H_0 and V, eq. (6-103) takes on the form

$$P\Big\{H_0 + V + (H_0 + V)\frac{1}{E - QH}Q(H_0 + V)\Big\}P\Psi = E\ P\Psi \tag{6-104}$$

We are now in a position to simplify this equation and put it in a form that can be compared with (6-92).

If all the single-particle states are chosen to be eigenfunctions of $h(\boldsymbol{r})$ and our truncation of the many-body Hilbert space is carried out by restricting the number of active single-particle states, we have the commutation relation

$$PH_0 = H_0P$$

Since P and Q are mutually exclusive, *i.e.*, $PQ = QP = 0$, we can eliminate the last two H_0s on the left hand side of (6-104). Among the three H_0s in the expression it is easy to see that the one furthest to the right does not make any contribution to the equation, since $QH_0P\Psi = 0$. The H_0 to its left occurs in the product $P\{H_0(E - QH)^{-1}QV\}P\Psi$. Since everything to the right of this H_0 acts on states in the space projected out by Q, the term is equivalent to $PH_0Q(E - QH)^{-1}QVP\Psi$ and vanishes because $PH_0Q = H_0PQ = 0$. Upon eliminating these two H_0s, (6-103) may now be written in a form that can be compared with (6-92) and (6-93),

$$P\Big\{H_0 + V + V\frac{1}{E - QH}QV\Big\}P\Psi = E\,P\Psi \tag{6-105}$$

From the comparison we identify that

$$V_{\text{eff}} = V + V\frac{1}{E - QH}QV \tag{6-106}$$

This is still a formal solution and its usefulness lies mainly in the possibility of making an infinite series expansion for $(E - QH)^{-1}$. This is an advantage, since an order by order calculation of the effective potential can be carried out.

An equivalent way of writing the second term on the right hand side of (6-106) is the following:

$$V\frac{1}{E - QH}QV = VQ\frac{1}{E - H_0 - QV}QV$$

In a Hartree-Fock single-particle basis, it is likely that the expectation values of the residual interaction V are smaller than those for H_0 and perhaps smaller than those for $(E - H_0)$ as well. As a result, it is possible that the following condition is true:

$$\frac{1}{E - H_0}QV < 1$$

This makes it possible for us to take the operator $(E - H_0 + V)^{-1}$ as an infinite series expansion in powers of $(E - H_0)^{-1}QV$,

$$\begin{aligned} VQ\frac{1}{E - H_0 - QV}QV &= VQ\frac{1}{E - H_0}QV + VQ\frac{1}{E - H_0}QV\frac{1}{E - H_0}QV \\ &\quad + VQ\frac{1}{E - H_0}QV\frac{1}{E - H_0}QV\frac{1}{E - H_0}QV + \cdots \\ &= VQ\sum_{n=1}^{\infty}\Big(\frac{1}{E - H_0}QV\Big)^n \end{aligned}$$

The effective interaction of (6-106) can now be expressed in the following form,

$$V_{\text{eff}} = V + VQ\sum_{n=1}^{\infty}\Big(\frac{1}{E-H_0}QV\Big)^n \tag{6-107}$$

If the series converges, we have a method to evaluate the effective interaction to any order of accuracy desired. Furthermore, since H_0 is diagonal in the basis states we have chosen,

$$H_0|\Psi_i\rangle = \sum_r \epsilon_r |\Psi_i\rangle$$

where ϵ_r are the single-particle energies and the summation is over all the occupied single-particle states in $|\,\Phi\,\rangle$, operator H_0 in the denominator may be replaced by sums of single-particle energies.

There is, however, no known proof that the series is actually convergent. Furthermore, in practice it is not easy to carry the calculation beyond the third order or so in a nontrivial P-space. In spite of these difficulties, the effective interaction obtained by using (6-107) to roughly the second order, starting with a realistic potential that fits the free nucleon-nucleon scattering data, has been shown to give shell model results that are in agreement with a variety of experimental data.

The procedure outlined above to find the effective interaction in a shell model space is also known as a *renormalization* procedure, since it "renormalizes" an interaction for the complete Hilbert space to one suitable for the truncated space. It is worthwhile to review the connection of an effective interaction with the interaction between nucleons. In Chapter 2, we started the discussion by saying that nucleon-nucleon interaction is an aspect of the fundamental strong interaction between quarks. However, at the moment our knowledge of QCD is inadequate to derive the nuclear interaction from the fundamental strong interaction; as a result, a more empirical approach of a meson-exchange picture plus an empirical hard core was adopted. A realistic interaction in nuclear physics is usually taken to be one that fits data on nucleon-nucleon scattering and other two-nucleon systems. In order to adapt such an interaction between free nucleons to nuclear structure problems, we must resolve the difference between free and bound nucleons and the problem of off-shell effects (see §3-9), for example, by constructing a realistic potential. With such a potential, a Hartree-Fock type of calculation can be used to reduce the interaction to a residual interaction and thence to an effective interaction for calculations in a manageable shell model space using, for example, the method outlined above. The complete process from nucleon-nucleon potential to an effective potential is a very involved one and is carried out in practice on only a few occasions. The important point to realize here is that a clear connection of the "effective" shell model potential and the "fundamental" nucleon-nucleon can be made, at least in principle, through a set of well-defined procedures. In spite of the fact that some of the intervening steps, such as low-energy QCD calculation and off-shell effects, are yet to be resolved, the connection exists. For practical applications, simplifications are needed and several semi-empirical approaches have been developed to obtain the shell model effective interaction.

Two-body matrix elements. In a finite active space, a two-body operator is completely specified if all the independent two-body matrix elements of the operator in the space are given. Thus an effective interaction V_{eff} may also be defined in terms of all the independent two-body matrix elements it has in the active space. We have already seen an example of this in (6-84), where we defined the two-body part of the Hamiltonian for a Hartree-Fock calculation in terms of $V_{\alpha\beta\gamma\delta}$.

The two-body matrix elements required here are slightly different from $V_{\alpha\beta\gamma\delta}$, as we wish to work in a subspace of the shell model space with definite spin J and isospin T. For this purpose, it is convenient to have the defining two-body matrix elements given in terms of two-particle states with definite J and T. In this scheme, an antisymmetrized and normalized two-body matrix element may be written in the form

$$W^{JT}_{rstu} \equiv \langle rsJT|V|tuJT\rangle \tag{6-108}$$

where $|rsJT\rangle$ is an antisymmetrized and normalized wave function for two particles, one in single-particle state r and the other in state s, similar to that given in (6-82). The additional feature here is that the two single-particle wave functions are angular momentum coupled together to final spin J and isospin T. The two-particle wave function $|tuJT\rangle$ is defined in the same way except that the single-particle states involved are t and u instead of r and s. In terms of two-body matrix elements, the effective interaction can be written in the form

$$V_{\text{eff}} = \sum_{\substack{rstu \\ JT}} |rsJT\rangle W^{JT}_{rstu} \langle tuJT| \tag{6-109}$$

in analogy with (6-84). Since the nuclear Hamiltonian is a scalar in spin and isospin, only two-body matrix elements diagonal in J and T are nonvanishing as a result of angular momentum and isospin selection rules.

Other symmetries of the nuclear Hamiltonian can also help to reduce the number of independent two-body matrix elements required to define an effective interaction. Because of time reversal invariance, the matrix elements may be taken to be real and symmetric, *i.e.,*

$$W^{JT}_{rstu} = W^{JT}_{turs} \tag{6-110}$$

Furthermore, since the wave functions are antisymmetrized, two functions differing only by a permutation of the two single-particle wave functions involved are related by a phase factor,

$$|rsJT\rangle = (-1)^{j_r+j_s-J-T}|srJT\rangle$$

This is made of a combination of three separate factors: a minus sign due to the permutation of two fermion states, a factor $(-1)^{\frac{1}{2}+\frac{1}{2}-T}$ due the recoupling of the isospins of the two single-particle states as given by (B-34), and a similar factor

$(-1)^{j_r+j_s-J}$ for recoupling the spins. Because of these relations, the following symmetries exist among the two-body matrix elements:

$$W^{JT}_{rstu} = (-1)^{j_r+j_s-J-T} W^{JT}_{srtu} = (-1)^{j_t+j_u-J-T} W^{JT}_{rsut}$$
$$= (-1)^{j_r+j_s-j_t-j_u} W^{JT}_{srut} \tag{6-111}$$

The relations given by (6-110,11) reduce the number of independent two-body matrix elements required to define an effective interaction in a finite shell model space. In fact, the number for a modest active space may be sufficiently small that one may be able to determine the effective interaction empirically by fitting all the required two-body matrix elements to the available data in the space.

Semi-empirical effective interaction. A simple example may be used to illustrate the semi-empirical approach to effective interaction. Some of the low-lying levels in the calcium isotopes, ^{41}Ca to ^{48}Ca, may be approximated by a shell model space made of the $1f_{7/2}$-orbit only. The inert core is taken to be the ^{40}Ca nucleus and the 40 nucleons filling the $1s$, $1p$, $1d$, and $2s$-orbits are not to be excited out of these orbits. All the active nucleons in this case are neutrons. Since a $f_{7/2}$-orbit can take a maximum of $(2j+1) = 8$ neutrons, the active space is completely filled by the time we come to ^{48}Ca.

The binding energy difference between ^{41}Ca and ^{40}Ca provides us with the single-particle energy of the $1f_{7/2}$-orbit,

$$\epsilon_{1f_{7/2}} = -8.36 \text{ MeV}$$

obtained from a table of mass excess. In a similar way we can calculate the binding energy of the two neutrons in ^{42}Ca with respect to the ^{40}Ca core. The result, -19.84 MeV, is different from twice the value of $\epsilon_{1f_{7/2}} = -8.36$ MeV obtained from ^{41}Ca because of the residual interaction between two neutrons. This provides us with one piece of experimental information required to specify the effective interaction. Since the ground state of ^{42}Ca has $J = 0$ and $T = 1$, we obtain the two-body matrix element for $(J,T) = (0,1)$ from the binding energy of ^{42}Ca with respect to ^{40}Ca after removing the contributions due to the single-particle energies of the two neutrons,

$$W^{01}_{1f_{7/2}1f_{7/2}1f_{7/2}1f_{7/2}} = -19.84 - (2\times -8.36) = -3.12 \text{ MeV}$$

Because of antisymmetrization requirements, two neutrons in $f_{7/2}$-orbit can only be coupled to $J =$ 0, 2, 4, and 6. Three additional two-body matrix elements are therefore needed to complete the definition of the effective interaction in this simple shell model space. These can be found from the energy level positions of the $J = 2$, 4, and 6 excited states in ^{42}Ca at 1.53, 2.76, and 3.19 MeV, respectively.

To simplify the notation, we shall drop the superscript for isospin (since we are dealing with neutrons only) and the subscript $1f_{7/2}$, since this is the only orbit with which we are concerned in this example. The complete effective Hamiltonian for the $1f_{7/2}$-neutron shell model is therefore given by five pieces of information, one single-particle energy and four two-body matrix elements,

$$\epsilon = -8.36 \text{ MeV}, \qquad W^0 = -3.12 \text{ MeV}, \qquad W^2 = -1.59 \text{ MeV}$$
$$W^4 = -0.364 \text{ MeV}, \qquad W^6 = 0.0728 \text{ MeV}$$

With these five pieces of input obtained from the energies of ^{41}Ca and ^{42}Ca, we are now in a position to calculate all the energy levels in the $1f_{7/2}$-shell model space from ^{43}Ca to ^{48}Ca.

Table 6-3 A shell model calculation in the $1f_{7/2}$-space for the energy levels of calcium isotopes.

^{43}Ca			^{44}Ca			^{45}Ca			^{46}Ca		
J	Exp.	Calc.	J	Exp.	Calc.	J	Exp.	Calc.	J	Exp.	Calc.
E_B	27.78	28.27	E_B	38.91	39.83	E_B	46.32	48.35	E_B	56.72	59.98
5/2	0.37	0.28	2	1.16	1.52	5/2	0.17	0.28	2	1.35	1.52
3/2	0.59	1.31	4	2.28	2.51	3/2	1.43	1.31	4	2.58	2.75
11/2	1.68	1.77	4	3.05	2.75	11/2		1.77	6	2.97	3.19
9/2	2.09	2.05	6	3.28	3.19	9/2		2.05			
15/2	2.75	3.12	8	5.09	5.30	15/2		3.12			

The calculated results for ^{43}Ca to ^{46}Ca are listed in Table 6-3. The binding energies with respect to the ^{40}Ca core are given in the first row. In addition to the entries in the table, the calculated binding energies for ^{47}Ca and ^{48}Ca are found to be, respectively, 68.58 and 80.29 MeV, compared with the measured values of 63.99 and 73.94 MeV. When we examine the six binding energies in more detail, we find the difference between the calculated and observed values gets progressively worse as the number of active neutrons increases from three ($A = 43$) to eight ($A = 48$). In fact, a little calculation will show that the difference is roughly proportional to the factor $n(n-1)/2$, the number of neutron pairs in the $1f_{7/2}$-shell model space. This means that the effective interaction deduced from the difference between ^{42}Ca and ^{40}Ca turns out to be a little too strong. If we reduce the contribution due to binding energy of ^{42}Ca in the effective Hamiltonian by 0.21 MeV, *i.e.*, increasing each one of the five two-body matrix by 0.21 MeV, the calculated binding energies change to 27.64, 38.58, 46.26, 56.83, 64.20, and 74.44 MeV for ^{43}Ca to ^{48}Ca, respectively, in much better overall agreement with the observed values. (On the other hand, if the difference were linearly proportional to the number of active neutrons, the cause would have to be attributed to the single-particle energy instead.) Such overall changes in the defining matrix elements in general do not affect the excitation energies given in the table.

The calculated energy level positions of the excited states in ^{43}Ca to ^{46}Ca are also compared with experimental values in Table 6-3. There are obviously more excited states in these nuclei than those listed. Since our model space is restricted to the $1f_{7/2}$-orbit alone, only those observed levels that belong to this space may be used in the comparison. Experimentally, in principle, one can identify a $1f_{7/2}$-level by one-nucleon transfer reactions (see §7-2). In practice, the identification is not always simple, since substantial admixture of the contributions from other single-particle states, such as $2p_{3/2}$ and $2p_{1/2}$, can be expected. The comparison between calculated and observed values in the table must therefore be viewed with the simplicity of the model space used for this illustrative example in mind. In fact, the agreement is better than we could have expected.

Table 6-4 Allowed two-particle states in $1p$-shell.

	$T=0$			$T=1$		
$J=$	1	2	3	0	1	2
$1p_{3/2}^2$	1		1	1		1
$1p_{1/2}^2$	1			1		
$1p_{3/2}1p_{1/2}$	1	1		1	1	
Total	3	1	1	2	1	2
d_{JT}	6	1	1	3	1	3

A nontrivial but still relatively simple example is the $1p$-shell, composed of nuclei from ^{5}He to ^{16}O. The inert core here is the ^{4}He nucleus with two protons and two neutrons completely filling up the $1s_{1/2}$-orbit. All single-particle states above the $1p$-shell, starting from the ds-shell, are empty. There are two valence orbits in this space in the jj-coupling scheme, $1p_{3/2}$ and $1p_{1/2}$, and the one-body part of the effective Hamiltonian is therefore defined by two single-particle energies, $\epsilon_{1p_{3/2}}$ and $\epsilon_{1p_{1/2}}$. The number of two-particle states for each allowed set of J and T values in the $1p$-shell is given in Table 6-4. Because of the symmetries given in (6-110,11), the number of independent two-body matrix elements required to specify the two-body residual interaction for a given J and T is $d_{JT} = n(n+1)/2$, where n is the number of two-particle states with spin-isospin (J,T). The total number of two-body matrix elements in this shell model space is then $\sum_{JT} d_{JT} = 15$. The complete $1p$-shell effective interaction comprises a total of seventeen parameters, two single-particle energies and fifteen two-body matrix elements. It is not possible to find seventeen energy levels in ^{5}He and mass 6 nuclei in a similar way as we have done earlier to specify the effective Hamiltonian for the simpler case of $1f_{7/2}$ shell model space for calcium isotopes. On the other hand, since all the $1p$-shell states in nuclei from $A=5$ to 16 can be calculated from these 17 parameters, we can use any $1p$-shell data in these nuclei to determine the effective interaction matrix elements. In fact, more than 17 pieces of experimental information can be found in this mass region and a least-square procedure may be used to deduce the values of the parameters that best fit the data.

This was done by Cohen and Kurath (*Nucl. Phys.* **73** [1965] 1). The calculation serves two useful purposes. The first is to demonstrate that the idea behind a semi-empirical effective Hamiltonian is a sound one. Once the seventeen parameters are determined from fitting a given number of pieces of data, the effective Hamiltonian obtained may be used to calculated the shell model values corresponding to the data used. Since a fitting procedure was used, the calculated results will not necessarily be identical to the observed values used as input. Normally the success of a least-square fit to some functional form is given by the value of the χ^2 for the overall fit and the standard deviations in the values obtained for each individual pieces of independent parameter. A small χ^2, among other things, indicates that the functional form used to fit it with is a reasonable one. Here we are dealing with a highly nonlinear least-square fitting procedure involving matrix calculations. It is therefore not easy to give a figure of merit analogous to the role of a χ^2-value for the functional form used. The agreement obtained when the calculated results are compared with the original input is a good indication of the validity of the shell model and the effective Hamiltonian approach used in the calculation. A second use of the seventeen parameters is that we now have an effective $1p$-shell Hamiltonian that can be employed for investigating other nuclear properties in the same space. This has been used extensively with success.

A similar project for the ds-shell composed of nuclei from ^{17}O to ^{40}Ca has also been carried out (B.H. Wildenthal and W. Chung, in ***Mesons in Nuclei,*** edited by M. Rho and D.H. Wilkinson, North-Holland, Amsterdam, 1979). Here, the valence orbits are $1d_{5/2}$, $1d_{3/2}$, and $2s_{1/2}$. The effective Hamiltonian is given by 3 single-particle energies and 63 two-body matrix elements (see Problem 6-10). The calculated results represent some of the best description of the low-lying states in nuclei from mass 17 to 40.

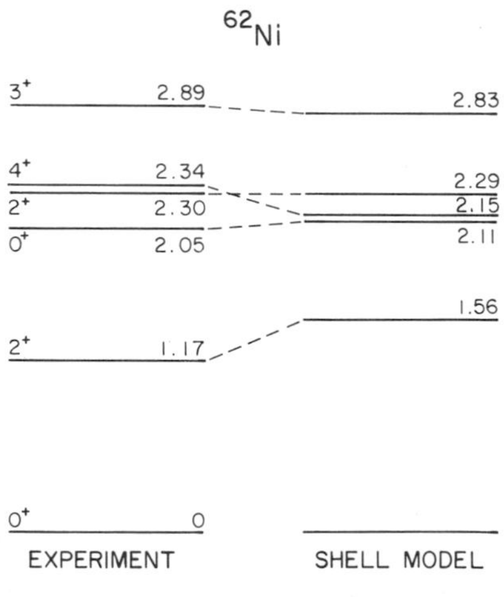

Fig. 6-17 Comparison of the observed energy levels of ^{62}Ni with the results of a shell model calculation in the $1f_{5/2}$, $2p_{3/2}$, and $2p_{1/2}$ space using a renormalized effective interaction. Note that the observed quadrupole vibrational features of the nucleus is well reproduced by the microscopic calculation.

Examples of shell model results. It is useful here to give a few examples of what a microscopic shell model calculation can produce. Instead of introducing new physical phenomena, we shall make use of observations that are already familiar. In §6-1 we have seen that some of the excited states in certain nuclei can be understood as collective vibrations of nucleons. There, the observed properties were described in terms of harmonic vibrations of the collective coordinates $\alpha_{\lambda\mu}$, the shape parameters. One of the nuclei exhibiting such properties is ^{62}Ni. The ground state spin-parity of ^{62}Ni is 0^+, typical of an even-even nucleus, and the first excited state is 2^+ at 1.17 MeV. A triplet of states, 0^+, 2^+, and 4^+, is observed at slightly less than twice this energy at 2.05 to 2.34 MeV. These three groups of states are interpreted in the vibrational model as quadrupole vibrations built upon a spherical nucleus with the ground state as the zero-phonon state, the first excited state as the one-phonon state, and the triplet of 0^+, 2^+, and 4^+ states as the two-phonon states. The $E2$-transition rates, given earlier in Table 6-1, also confirm this interpretation. Here, we shall treat ^{62}Ni using the shell model. Six neutrons outside a ^{56}Ni core are considered to be the active particles in the shell model space made of single-particle orbits $1f_{5/2}$, $2p_{3/2}$, and $2p_{1/2}$. The three single-particle energies are taken from ^{57}Ni. The low-lying energy level positions calculated with an effective interaction, obtained using a renormalization procedure, are compared with the observed values in Fig. 6-17. It is seen that the energy level positions obtained indeed display the typical structure of a vibrational nucleus in agreement with the data.

Table 6-5 Values of $B(E2; J \to J-2)$ between $K = 0^+$ band members in ^{20}Ne.

J_i	J_f	Experimental e^2fm^4	W.u.	Shell Model* e^2fm^4	Rotation Model e^2fm^4
2	0	57 ± 8	17.8 ± 2.5	48	57
4	2	71 ± 7	21.9 ± 2.1	58	83
6	4	66 ± 8	20.4 ± 2.4	43	91
8	6	24 ± 8	7.5 ± 2.5	28	96

* An effective charge of $e_p = 1.5e$ and $e_n = 0.5e$ is used.
1 W.u.(Weisskopf unit) = 3.2 e^2fm^4 for ^{20}Ne.

Using the eigenvectors obtained we can calculate the electromagnetic properties of the states involved. Here we encounter the question of effective operator in a truncated shell model space. For a vibrational nucleus, the dominant electromagnetic transitions are $E2$, induced by charged currents. Since electric currents in a nucleus are usually associated with the motion of protons, our calculated results using the *bare* charge of active nucleons in the space, *i.e.*, the free nucleon values used to define electromagnetic operators in §5-2, will be zero as we have only active neutrons. This is clearly a problem caused by the truncation procedure. The $E2$-transition operator must also be *renormalized* in the shell model space adopted in order to describe what is going on inside a nucleus. The usual practice is to give an *effective* charge to a neutron and adjust its size to fit the observed $E2$-transition rates and quadrupole moment.

A second example is the low-lying positive parity states of ^{20}Ne shown in Fig. 6-18. Here we see that the energy level positions display a rotational structure with E_J essentially proportional to $J(J+1)$ up to $J = 8$. For a shell model calculation, we can take ^{16}O as the inert core and the four active nucleons, two protons and two neutrons, are in single-particle orbits $1d_{5/2}$, $1d_{3/2}$, and $2s_{1/2}$. The results also give a reasonable description of the energy level positions, as can be seen in the figure. For the calculated $E2$-transitions rates listed in Table 6-5, an effective charge of $0.5e$ is used; that is, the charge of an active proton is $1.5e$ and that of an active neutron is $0.5e$. Here we see again that a microscopic interpretation of a collective phenomenon is given in terms of a very small number of active nucleons in a highly truncated active space.

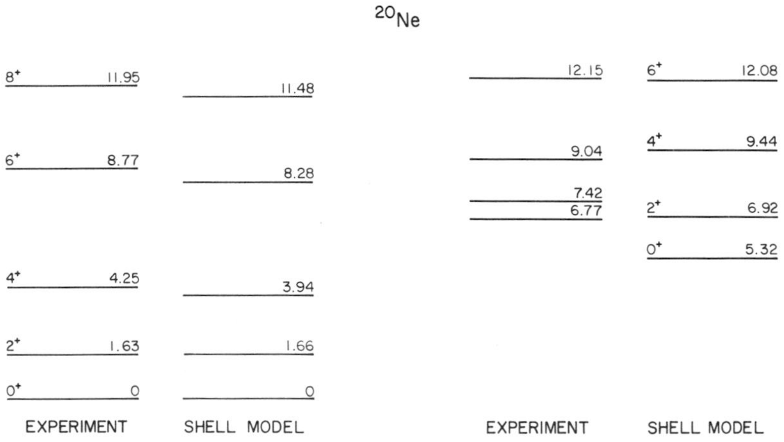

Fig. 6-18 Comparison of the observed energy levels of the two lowest $K = 0^+$ bands in ^{20}Ne with the results of a shell model calculation in the $1d_{5/2}$, $1d_{3/2}$ and $2s_{1/2}$ space using a renormalized effective interaction.

Effective operator. The reasons behind the large effective charge for $E2$-transition operators may be traced to the relatively small number of active nucleons used in the shell model calculations. Although the energy level positions are well accounted for, the small number is inadequate to produce the large enhancements seen in the $E2$-transition rates for collective states. Since deformations, whether in the equilibrium shape as in the case of ^{20}Ne or in the form of shape vibration as in the case of ^{62}Ni, are predominantly quadrupole in nature in these examples, it is not surprising that we find the difference between the effective and real charge for $E2$-transitions to be the most pronounced ones. Since collective motion involves the action of a large number of nucleons, including some of those considered to be a part of the "inert" core in the shell model calculation, a large effective charge is required. It is also interesting to note that such large enhancements involving the core nucleons can be accounted for by a simple factor in the form of an effective charge. The possibility of making the corrections in a simple way is the basic idea behind renormalizing the operators.

Another simple demonstration of effective operators can be found in the square of the charge form factor $F^2(q)$ obtained, for example, from electron scattering off nuclei. As we have seen earlier in §4-2, the charge form factor is the Fourier transfer of the charge distribution in a nucleus. The measured values for ^{12}C and ^{16}O are shown in Fig. 6-19. For ^{12}C we see that there is only one minimum in $F^2(q)$ at around $q = 1.5\ \text{fm}^{-1}$. This is exactly what is expected from the Fourier transform of the density of a nucleon in the $1p$-shell (see Problem 6-11). On the other hand, two minima are observed in the value of $F^2(q)$ for ^{16}O. A simple shell model puts the active nucleons for the ground state of ^{16}O also in the $1p$-shell. The appearance of the second minimum implies that there is a substantial admixture of configurations having nucleons excited into the ds-shell. If we insist on carrying out the form factor calculation for ^{16}O using only active nucleons in the $1p$-shell space, a correction factor, for example in the form

$$f(r) = 1 - e^{-\alpha r^2}$$

must be introduced. This will give the observed second minimum without having to invoke configurations involving particles in the ds-shell. Such a correction factor may be regarded as a renormalization of the operator for charge form factor. The effective operator produced as a result simulates the shapes of form factor for nucleons in the ds-shell for active nucleons in the $1p$-shell.

Not every operator requires a large renormalization, as we have seen for electric quadrupole transitions. For example, Gamow-Teller transitions throughout the ds-shell have been found to be given by the bare operator without noticeable modifications. This may be related to the fact that β-decay is not a collective phenomenon like, for instance, $E2$-transitions.

Besides effects due to truncation of the shell model space, renormalization of the excitation operators from their bare nucleon values may also be required because of mesonic and other degrees of freedom in nuclei. When a nucleon is imbedded inside a nucleus, we expect such processes as the exchange of virtual mesons to be different from the situation when the nucleon is a free, isolated particle. In recent years one of the interesting developments in the nuclear shell model has been to obtain such renormalization effects from field theoretical approaches. In this way, a better and more fundamental understanding of the behavior of nucleons inside a nucleus may be reached.

Because of its direct connection with the individual nucleon degrees of freedom in a finite nucleus, the shell model can also be used to "simulate" data for

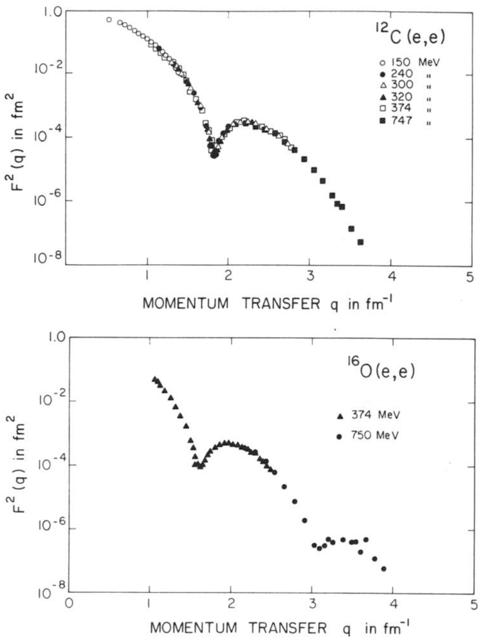

Fig. 6-19 The square of charge form factors for ^{12}C and ^{16}O obtained from elastic electron scattering. The existence of a single minimum in $F^2(q)$ in ^{12}C is expected from the fact that the active nucleons are in the $1p$-shell. The second minimum in ^{16}O shows the presence of active particles in higher single-particle orbits.

testing other models. This is similar to numerical simulations used in many other fields to supplement data and to help us to probe aspects of nature that are difficult to examine experimentally. Such numerical "experiments" do not substitute for actual observation involving real nuclei, but they provide a convenient avenue to test our models and our understanding of the physical situation and therefore are useful as a tool for furthering our knowledge of nuclei.

§6-9 DEFORMED SHELL MODEL

Deformed single-particle states. So far we have assumed that single-particle states are generated by a spherical potential well made of the average interaction of a nucleon with all the other nucleons in the nucleus. This is a reasonable assumption as long as the equilibrium shape of the nucleus itself is spherical. We have seen earlier that, for many nuclei, a deformed shape is more stable and, for such nuclei, a deformed average potential well is a more reasonable one with which to generate the basis single-particle states.

Let us again assume axial symmetry for simplicity. Instead of (6-65), the single-particle Hamiltonian now takes on the form

$$h = h_0 + h_\delta + a\boldsymbol{\ell}\cdot\boldsymbol{s} + b\boldsymbol{\ell}^2 \tag{6-112}$$

where h_0 is the spherical part, generally taken to be the Hamiltonian of an isotropic three-dimensional harmonic oscillator similar to that given in (6-54),

$$h_0 = -\frac{\hbar^2}{2\mu}\nabla^2 + \frac{1}{2}\mu\omega_0^2 r^2 \tag{6-113}$$

The deformation is produced by the potential h_δ and it is usually taken to be that due to a quadrupole field,

$$h_\delta \equiv -\delta_{\text{osc}}\frac{1}{3}\mu\omega_0^2 r^2\sqrt{\frac{16\pi}{5}}\,Y_{20} \tag{6-114}$$

Here δ_{osc}, to be defined later in (6-117), provides a measure of the departure from a spherical shape. We have already encountered the other two terms in the single-particle Hamiltonian. The spin-orbit term $a\,\boldsymbol{\ell}\cdot\boldsymbol{s}$, given in §6-6, is required to account for the magic numbers, and the term $b\,\boldsymbol{\ell}^2$ is used to give the proper ordering of single-particle states in the spherical limit. Deformed single-particle states produced as the eigenvectors of the Hamiltonian given in (6-112) are usually referred to as the Nilsson states or Nilsson orbitals.

Labels for deformed single-particle states. With a deformed Hamiltonian, the spin j is no longer a constant of motion and a new set of labels must be found to specify a single-particle state. As we have seen earlier in the discussion of the rotational model, the third component of $\boldsymbol{j}$ remains a good quantum number for axially symmetric nuclei. This gives us Ω, the projection of $\boldsymbol{j}$ on the body-fixed quantization axis (the 3-axis), as one of the labels.

The wave function of a Nilsson state may be expanded in terms of spherical harmonic oscillator states $|N\ell j\Omega\rangle$,

$$|N\Omega\rangle = \sum_{\ell j} C_{N\ell j}|N\ell j\Omega\rangle \tag{6-115}$$

where the expansion coefficients $C_{N\ell j}$ depend on the value of the deformation parameter δ_{osc}, as different deformation causes different amounts of mixing between spherical orbits. In the limit of zero deformation, the basis states coincide with those in the spherical case used in the previous sections. If admixtures between spherical states belonging to different major harmonic oscillator shells are forbidden in constructing the deformed single-particle basis, N, the number of harmonic oscillator quanta remains a constant and may be used as one of the labels to specify a state.

Labels N and Ω alone cannot uniquely specify a deformed state; two additional quantities, n_3 and λ, are commonly used. The origin of these labels may be summarized in the following way. For large deformations, we can ignore the effects of the $\boldsymbol{\ell}\cdot\boldsymbol{s}$ and $\boldsymbol{\ell}^2$ terms and treat $h_0 + h_\delta$ in a cylindrical coordinate system. The Hamiltonian of (6-112) can now be written in the form

$$h = -\frac{\hbar^2}{2\mu}\nabla^2 + \frac{1}{2}\mu\{\omega_3^2 x_3^2 + \omega_\perp^2(x_1^2 + x_2^2)\} \tag{6-116}$$

where the oscillator frequency along the symmetric axis is

$$\omega_3 = \omega_0\sqrt{1-\tfrac{4}{3}\delta_{\rm osc}} \approx \omega_0(1-\tfrac{2}{3}\delta_{\rm osc})$$

and in the directions perpendicular to it,

$$\omega_\perp = \omega_0\sqrt{1+\tfrac{2}{3}\delta_{\rm osc}} \approx \omega_0(1+\tfrac{1}{3}\delta_{\rm osc})$$

In terms of $\omega_\perp$ and ω_3, the deformation parameter $\delta_{\rm osc}$ is given by the relation

$$\delta_{\rm osc} = \frac{\omega_\perp - \omega_3}{\omega_0} \qquad (6\text{-}117)$$

where the average frequency

$$\omega_0 = \tfrac{1}{3}(\omega_1+\omega_2+\omega_3) = \tfrac{1}{3}(2\omega_\perp+\omega_3) \qquad (6\text{-}118)$$

may be taken to be the same as the harmonic oscillator frequency in the spherical limit. The difference between the parameter $\delta_{\rm osc}$ defined here and δ given in (6-32) is that the former is given in terms of the harmonic oscillator frequencies for different directions, whereas the latter is given in terms of the lengths of various radii. These two quantities are closely related but not identical to each other; however, we shall not go into the details of their differences here.

In the limit of large deformation, the single-particle energy of a deformed state may be expressed in terms of the number of oscillator quantum n_i along each of the three principal (body-fixed) axes,

$$\begin{aligned}\epsilon_{Nn_3\Omega} &= (n_3+\tfrac{1}{2})\hbar\omega_3 + (n_2+\tfrac{1}{2})\hbar\omega_2 + (n_1+\tfrac{1}{2})\hbar\omega_1\\ &= (n_3+\tfrac{1}{2})\hbar\omega_3 + (n_\perp+1)\hbar\omega_\perp\end{aligned}$$

where, for the axially symmetric case under discussion here, $n_\perp = n_1+n_2$ is the number of quanta in the direction perpendicular to the symmetry axis. Since the total number of harmonic oscillator quanta is fixed, we have

$$N = n_3 + n_\perp$$

The label n_3 (or $n_\perp$) is a good quantum number in the limit of large deformation and can therefore be used as a label for Nilsson states.

When terms $a\,\boldsymbol{\ell}\cdot\boldsymbol{s}$ and $b\,\boldsymbol{\ell}^2$ are included, states with different projections of orbital angular momentum on the symmetry axis are no longer degenerate. Hence, in the limit of large deformation, the projection of the orbital angular momentum along the symmetry axis,

$$\lambda = \pm n_\perp, \pm(n_\perp-2), \cdots, \pm 1 \text{ or } 0 \qquad (6\text{-}119)$$

may be used as the fourth label. The set of four labels $\{Nn_3\lambda\Omega\}$ completely specifies a deformed state within one major shell. Examples of the variation of energy with deformation for the low-lying Nilsson states are shown in Fig. 6-20. More complete results can be found, for example, in *Table of Isotopes* edited by C.M. Lederer and V. Shirley (Wiley, New York, 1978).

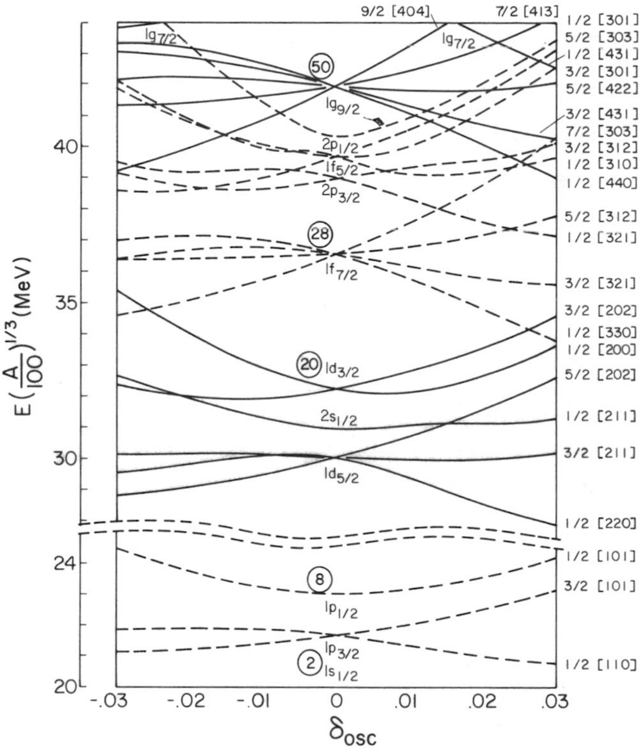

Fig. 6-20 Low-lying Nilsson single-particle energies as a function of deformation parameter $\delta_{\rm osc}$, obtained by solving (6-112).

Many-body states in the Nilsson scheme. The Nilsson orbitals for a deformed nucleus may be thought of as the equivalent of Hartree-Fock single-particle states in the spherical shell model. Since a deformed single-particle potential may be the more realistic average potential experienced by a nucleon in many nuclei, the residual interaction is likely to be even less important here than in the corresponding spherical case. Consequently, an observed state is likely to be dominated by even fewer components in the deformed basis. On the other hand, any calculation involving a two-body potential here is more difficult to carry out, since we no longer have angular momentum algebra to help us to reduce the mathematical complexity.

For an independent-particle approximation in the deformed basis, we can ignore the residual interaction. The nuclear Hamiltonian is then a sum of the single-particle Hamiltonian given in (6-112) over all the active nucleons. Like the case of the spherical shell model, we can separate all the single-particle states into three groups: the core states, the valence states, and the empty states. Since the single-particle Hamiltonian $h(\boldsymbol{r}_i)$ is, in part, a result of the interaction of the

Table 6-6 Examples of using Nilsson orbitals.

Nucleus	Proton Conf.	Neutron Conf.	$K = \sum \Omega_i$
^{19}F	$\frac{1}{2}[220]^1$	$\frac{1}{2}[220]^2$	$\frac{1}{2}$
^{19}Ne	$\frac{1}{2}[220]^2$	$\frac{1}{2}[220]^1$	$\frac{1}{2}$
^{21}Ne	$\frac{1}{2}[220]^2$	$\frac{1}{2}[220]^2\frac{3}{2}[211]^1$	$\frac{3}{2}$
^{23}Ne	$\frac{1}{2}[220]^2\frac{3}{2}[211]^2$	$\frac{1}{2}[220]^2\frac{3}{2}[211]^1$	$\frac{3}{2}$

valence nucleons with the core, the value of the deformation parameter depends on the equilibrium shape of the core. However, the core is not a part of the active space, and the determination of $\delta_{\rm osc}$ is, therefore, outside the scope of a simple Nilsson model in the same manner as the determination of the value of the moment of inertia $\mathcal{I}$ is outside a simple rotational model.

For a given value of $\delta_{\rm osc}$, the single-particle energies are given by the eigenvalues of the deformed Hamiltonian (6-112) and some of the low-lying ones are shown as a function of $\delta_{\rm osc}$ in Fig. 6-20. Nilsson orbits are degenerate in energy with respect to the sign of K and, as a result, each orbit can accommodate two identical nucleons, one with $K = +\Omega$ and the other with $K = -\Omega$. To construct the ground state configuration of n active nucleons, we follow the same procedure as in the spherical case by filling up the lowest available single-particle states first. The order of deformed orbits according to single-particle energy $\epsilon_{Nn_3\lambda\Omega}$ changes with deformation, as can be seen in Fig. 6-20. For this reason, the order of filling the single-particle orbits depends also on the value of $\delta_{\rm osc}$. This is best illustrated by specific examples.

For simplicity we shall use nuclei in the beginning of the *ds* shell. Again, ^{16}O is taken as the inert core and the *ds* shell as the active space. In a spherical basis, the $1d_{5/2}$ single-particle energy is the lowest one among the *ds*-shell orbits. As a result, we expect that the lowest energy configuration of nuclei in the beginning of the *ds* shell, from ^{17}O to ^{28}Si, is made by filling the $1d_{5/2}$ orbit with nucleons. The even-even nuclei are not of interest here, since dominance of the pairing interaction dictates that the ground states have $J^\pi = 0^+$. For odd-mass nuclei, the ground state spin is expected to be given by the single-particle state occupied by the unpaired nucleon. Consequently, nuclei ^{19}F, ^{19}Ne, ^{21}Ne, and ^{23}Na are expected to have $J^\pi = \frac{5}{2}^+$ for their ground state spins. Experimentally, they turn out to be different and have, instead, values $\frac{1}{2}^+$, $\frac{1}{2}^+$, $\frac{3}{2}^+$, and $\frac{3}{2}^+$, respectively.

The deformed shell model explains the observed data in the following way. From a variety of evidence it is well known that the nuclei in the lower half of the *ds* shell are deformed and have predominantly prolate spheroidal shapes (axially symmetric with $\delta_{\rm osc} > 0$). From Fig. 6-20, we see that the positions of the deformed orbitals above the ^{16}O core in ascending order according to energy are $\Omega[Nn_3\lambda] = \frac{1}{2}[220]$, $\frac{3}{2}[211]$, and $\frac{5}{2}[211]$ for positive deformation in this mass region. Since each Nilsson orbit can accommodate two identical nucleons, the ground state configurations of the four nuclei in question are those shown in Table 6-6.

With deformed single-particle states, we do not have a definite spin $\boldsymbol{j}$ for each nucleon. As a result, it is not possible to couple the angular momenta of all the active nucleons to form the nuclear ground state spin $\boldsymbol{J}$, as we have done in the spherical case. Instead, we shall proceed in the same manner as we have done in the case of the rotational model. Since the projection of $\boldsymbol{j}$ on the body-fixed 3-axis is a constant of motion, the sum

$$K = \sum_i \Omega_i \tag{6-120}$$

for all the active nucleons is also a constant of motion. Using the fact that K is the projection of $\boldsymbol{J}$ on the 3-axis, the ground state spin of a deformed nucleus must have spin $J = K$. This is the same argument as was used earlier to deduce spin in the rotational model. From the last column of Table 6-6, we find that $K = \frac{1}{2}$ for ^{19}F and ^{19}Ne, and $\frac{3}{2}$ for ^{21}Ne and ^{23}Ne. Hence $J = \frac{1}{2}$ for the first pair of nuclei and $J = \frac{3}{2}$ for the second pair, in agreement with observation. If the deformation were oblate ($\delta_{\text{osc}} < 0$), the unpaired nucleon would have been in orbitals $\frac{5}{2}[202]$, $\frac{5}{2}[202]$, $\frac{1}{2}[220]$ and $\frac{1}{2}[220]$ instead, and the ground spin of the nuclei would have been $\frac{5}{2}$, $\frac{5}{2}$, $\frac{1}{2}$, and $\frac{1}{2}$, in contradiction to the observed values.

The residual interaction left over from the deformed average single-particle potential given in (6-112) is, in principle, smaller than that given in (6-74) for the spherical case. For a discussion of detailed properties of nuclei, it is essential to include the residual interaction in the deformed shell model as well. The mathematics involved here is, however, far more complicated than in the spherical shell model, since we no longer have the simplicity of rotational symmetry, and hence spherical tensors and angular momentum algebra, to help us. For this reason, only limited attempts have been made in this direction.

§6-10 OTHER MODELS

In previous sections we have seen that some nuclear properties can be understood from a macroscopic point of view in terms of the collective degrees of freedom involving vibrational and rotational motion of the nucleons. Alternatively one can start from the individual nucleon degrees of freedom and try to understand the observed nuclear phenomena from a microscopic, shell model point of view. The latter approach is attractive, since we can make connections with the interaction between nucleons. Such an interaction can, in turn, be traced to the fundamental strong interaction between quarks in the underlying sub-structure of nucleons. However, the nuclear many-body problem, similar to many-body problems in other branches of physics, is not a simple one to solve. For this reason, many techniques in addition to the ones described above have been developed so that we may be able to examine specific aspects of a problem in a more convenient way. It is perhaps useful to mention some of these techniques very briefly here, even though both the scope of this book and the background knowledge required preclude any detailed discussions.

If we are concerned only with a limited range of behavior of a few special states, it seems superfluous to invoke the nuclear shell model and solve the complete eigenvalue problem. As we have seen earlier in the discussion on Hartree-Fock techniques, $1p1h$ (one-particle-one-hole) excitations in a many-body system can be handled with relative ease. This is a blessing, since $1p1h$ excitations are the dominant components in a variety of processes. For example, nuclear states that have large $1p1h$ components are strongly excited by electromagnetic processes, such as inelastic electron scattering, and by strong interaction probes in the form of intermediate energy nucleon scattering. We shall see more details of this type of excitation process later in §7-4. In a study of the properties of such states, one of our primary concerns is to establish the correct correlations between different $1p1h$ components so as to be able to produce, for example, the observed strong enhancement in strengths. One of the difficulties here is in the ground state wave function, upon which the excitations are built. A simple independent-particle model ground state will not be adequate for our purpose, since correlations caused by the residual two-body interaction are largely absent. The random phase approximation (RPA) solves this problem by allowing only certain types of correlations and is thus able to account for the strong $1p1h$ excitations observed in many nuclei with relatively a simple calculation. A more detailed discussion of RPA can be found in A.L. Fetter and J.D. Walecka (*Quantum Theory of Many-Particle Systems*, McGraw-Hill, New York, 1971).

The idea behind independent-particle approximation can also be generalized to include other types of correlation. Since one-body terms are so much easier to handle than two-body residual interactions, it is preferable to include, as much as possible, the effect of nucleon-nucleon interaction in the mean field experienced by individual nucleons. There is a wide variety of mean field theories. One of the attractive features here is that the approach can also be generalized relatively easily to a relativistic theory (see, *e.g.,* L.S. Celenza and C.M. Shakin, ***Relativistic Nuclear Physics***, World Scientific, Singapore, 1986). Although we have limited ourselves to nonrelativistic quantum mechanics here, many aspects of nuclear structure and nuclear scattering, especially those involving intermediate and high-energy probes, require a relativistic treatment. For this reason, mean field theories are becoming an increasingly more important tool for nuclear studies.

Besides vibrations and rotations, nuclei also display clustering behavior. The simplest example is the decay of the ground state of ^{8}Be to two α-particles. One explanation of this phenomenon is that nucleons prefer to form α-particle clusters in nuclei. Since the binding energy per nucleon of an α-particle is approaching the maximum value that can be attained inside a nucleus, there is relatively little force of attraction left between different α-particle clusters. In the case of ^{8}Be, binding energy actually favors the formation of two separate α-particles and, as a result, the ground state of ^{8}Be is unstable toward α-particle emission even though it is stable against β-decay and nucleon emission. Another example is the observation of nuclear molecular states such as in the separation of an excited ^{24}Mg nucleus into two ^{12}C clusters (D.A. Bromley, *Scientific American* **76** [1978] 277). In order for a shell model to account for the splitting of a group of nucleons into two or more separate clusters, a single-particle basis, far larger than anything one can contemplate in practice, is required. Special techniques such as the generator coordinate method (K. Wildermuth and Y.C. Tang, *A Unified Theory of the Nucleus,* Vieweg, Braunschweig, Germany, 1977) and two-centered shell model have been developed for studying such phenomena.

We have seen from shell model studies that the presence of energy gaps in a single-particle level spectrum is an important point in understanding nuclear structure. The most naive collective models, however, ignore this feature, since only smooth variations with nucleon number and other macroscopic properties of nuclei are incorporated into the picture. Improvements to the collective models can be achieved if local variations due to shell closures can be included in the description in terms of collective degrees of freedom. Such "shell" corrections are essential since, as deformation grows, the energies of individual single-particle states are modified in such a way that the energy gaps observed for spherical nuclei disappear and new ones at different neutron and proton numbers appear. For example, such shell corrections have been found to be important in improving the vibrational model description of many of the bulk properties of nuclei (see, *e.g.,* W.D. Myers and W.J. Swiatecki, *Ann. Rev. Nucl. Part. Sci.* **32** [1982] 309).

For illustrative purposes, we have separated nuclear properties into collective and single-particle behaviors. In practice, both types of phenomena are present in a given nucleus. Furthermore, they can couple with each other to form states with both types of behavior coexisting with each other. What we have mainly done so far is to examine both extremes to illustrate the underlying physics. Specific states are identified as being either single-particle or collective depending on which one of these two extremes dominates the property of the state. In fact, states which can be identified in such a simple manner constitute the minority among the multitude of nuclear states known. For the bulk of states, all physical principles are at play. Although a thorough understanding of nuclear structure will require us to examine these states as well, we shall ignore them here.

Problems

6-1. When two identical phonons, each carrying angular momentum $\boldsymbol{\lambda}$, are coupled together, only states with even J-values ($\boldsymbol{J} = \boldsymbol{\lambda} + \boldsymbol{\lambda}$) are allowed. Show that this is true by counting the number of states for a given total M, the projection of angular momentum on the quantization axis. Use the same method to show that when three quadrupole phonons are coupled together, only states with $J^\pi = 0^+$, 2^+, 3^+, 4^+ and 6^+ are allowed.

6-2. Three rotational bands have been identified in ^{25}Mg: a $K^\pi = 5/2^+$ band starts from the ground state ($J^\pi = 5/2^+$) and has three other members, $7/2^+$ at 1.614 MeV, $9/2^+$ at 3.405 MeV, and $11/2^+$ at 5.45 MeV; a $K = 1/2^+$ band with six members, $1/2^+$ at 0.585 MeV, $3/2^+$ at 0.975 MeV, $5/2^+$ at 1.960 MeV, $7/2^+$ at 2.738 MeV, $9/2^+$ at 4.704 MeV, and $11/2^+$ at 5.74 MeV; and a second $K = 1/2^+$ band with four members, $1/2^+$ at 2.562 MeV, $3/2^+$ at 2.801 MeV, $5/2^+$ at 3.905 MeV, and $7/2^+$ at 5.005 MeV. Calculate the moment of inertia and the decoupling parameter, where applicable, for each band.

6-3. The following γ-ray transition energies in keV are known among the yrast levels in ^{154}Dy (Azgui et al., *Nucl. Phys.* **A439** [1985] 537): $2^+ \to 0^+$ 334.5, $4^+ \to 2^+$ 412.2, $6^+ \to 4^+$ 477.0, $8^+ \to 6^+$ 523.6, $10^+ \to 8^+$ 557.0, $12^+ \to 10^+$ 588.7, $14^+ \to 12^+$ 616.2, $16^+ \to 14^+$ 581.8, $18^+ \to 16^+$ 546.6, $20^+ \to 18^+$ 612.3, $22^+ \to 20^+$ 685.2, $24^+ \to 22^+$ 755.9, $26^+ \to 24^+$ 823.0, $28^+ \to 26^+$ 887.3, and $30^+ \to 28^+$ 735.7. Plot the excitation energy as a function of $J(J+1)$ and calculate the moment of inertia for each state. Use this result to plot $2\mathcal{I}/\hbar^2$ as a function of $\hbar^2\omega^2$ and see if there is any sudden change in the moment of inertia generally known as "backbending."

6-4. The following energy level positions in MeV are known for two rotational bands in ^{19}F: $1/2^+$ 0.000, $1/2^-$ 0.110, $5/2^+$ 0.197, $5/2^-$ 1.346, $3/2^-$ 1.459, $3/2^+$ 1.554, $9/2^+$ 2.780, $7/2^-$ 3.999, $9/2^-$ 4.033, $13/2^+$ 4.648, and $7/2^+$ 5.465. Calculate the moment of inertia and the decoupling parameter for each band. Comment on the likelihood of a $11/2^+$ level at 6.5 MeV to be a member of the $1/2^+$-band.

6-5. The ground state of $^{152}_{63}$Eu is known to be 3^- with an electric quadrupole moment of $+3.16 \times 10^2$ $e$fm^2. Find the intrinsic quadrupole moment of the nucleus at its ground state and deduce the value of δ defined in (6-32), the difference between R_3 and $R_\perp$. What is the shape of the nucleus?

6-6. The following $E2$-transition rates appear in a table of nuclei in terms of natural line width Γ for the $K = 0^+$ band in ^{20}Ne: 2^+ (1.63 MeV) $\to 0^+$ (ground) 6.3×10^{-4} eV, 4^+ (4.25 MeV) $\to 2^+$ (1.63 MeV) 7.1×10^{-3} eV, 6^+ (8.78 MeV) $\to 4^+$ (4.25 MeV) 0.100 eV, and 8^+ (11.95 MeV) $\to 6^+$ (8.78 MeV) 1.2×10^{-3} eV. From the information provided, find

(a) the moment of inertia of the band

(b) the intrinsic quadrupole moment of the band

(c) predict the quadruple moment of the 2^+ member

6-7. If $^{25}_{12}$Mg is a spherical nucleus, what is the most likely spin, parity, and isospin of its ground state? If, instead, it is a deformed nucleus with prolate deformation, what is the most likely spin and parity?

6-8. The energies of ^{17}O with respect to the ^{16}O core are -4.15 MeV for the $5/2^+$ state, -3.28 MeV for the $1/2^+$ state, and $+0.93$ MeV for the $3/2^+$ state. Assuming these values are the single-particle energies of the *ds*-orbits, use an independent-particle model to find the energies of the lowest $1/2^+$, $3/2^+$, and $5/2^+$ states in ^{39}Ca with respect to the ^{40}Ca core.

6-9. The *ds*-shell single-particle energies with respect to ^{16}O core are: $\epsilon_{1d5/2} = -4.15$ MeV, $\epsilon_{2s1/2} = -3.28$ MeV, and $\epsilon_{1d3/2} = +0.93$ MeV. A particular effective interaction has the following set of two-body matrix elements:

$$\begin{aligned}
\langle 1d_{5/2}, 1d_{5/2}\ ; J=0, T=1|V|1d_{5/2}, 1d_{5/2}\ ; J=0, T=1\rangle &= -2.0094 \text{ MeV}\\
\langle 1d_{5/2}, 1d_{5/2}\ ; J=0, T=1|V|1d_{3/2}, 1d_{3/2}\ ; J=0, T=1\rangle &= -3.8935 \text{ MeV}\\
\langle 1d_{5/2}, 1d_{5/2}\ ; J=0, T=1|V|2s_{1/2}, 2s_{1/2}\ ; J=0, T=1\rangle &= -1.3225 \text{ MeV}\\
\langle 1d_{3/2}, 1d_{3/2}\ ; J=0, T=1|V|1d_{3/2}, 1d_{3/2}\ ; J=0, T=1\rangle &= -0.8119 \text{ MeV}\\
\langle 1d_{3/2}, 1d_{3/2}\ ; J=0, T=1|V|2s_{1/2}, 2s_{1/2}\ ; J=0, T=1\rangle &= -0.8385 \text{ MeV}\\
\langle 2s_{1/2}, 2s_{1/2}\ ; J=0, T=1|V|2s_{1/2}, 2s_{1/2}\ ; J=0, T=1\rangle &= -2.3068 \text{ MeV}
\end{aligned}$$

(a) Calculate the ground state binding energy of ^{18}O with respect to ^{16}O and compare the result obtained from a table of mass excess. What are the excitation energies of the two other 0^+ states in this space?

(b) Obtain the ground state wave function of ^{18}O and use it to calculate the relative probability for finding a neutron in the $1d_{5/2}$, $2s_{1/2}$, and $1d_{3/2}$ single-particle states in ^{18}O. The results are essentially the spectroscopic factors for one-neutron pickup reactions.

(c) If the wave function of the lowest 1^+ state in ^{18}O is

$$|J^\pi = 1^+, T=1\rangle = |1d_{3/2}2s_{1/2}; J^\pi = 1^+, T=1\rangle$$

find the magnetic dipole moment of this state.

6-10. Use the same method as outlined in Problem 6-1 to show that when two nucleons are in an orbit with spin j, the allowed J-values for two-particles states are 0, 2, 4, ..., $2j-1$ for $T=1$, and $J = 1, 3, 5, \ldots, 2j$ for $T=0$. Construct a table similar to Table 6-4 to give the number of states of each J and T for two nucleons in the three *ds*-shell orbits, $1d_{5/2}$, $1d_{3/2}$, and $2s_{1/2}$. Use this information to show that the total number of two-body matrix elements required to define a two-body potential in the *ds*-shell is 63.

6-11. The explicit forms of the radial wave functions for a spherical harmonic oscillator potential well are given in Table 6-2. Use these to show that the form factor, the Fourier transform of the radial density $\rho_{n\ell}(r) = |R_{n\ell}(r)|^2$, is positive for the 1*s*-orbit in the region $q = 0$ to infinity [*i.e.*, no node in $F^2(q)$] and changes sign once at $q^2 = 6\nu$ for the 1*p*-orbit [one node in $F^2(q)$]. For the 1*d*-orbit, the sign changes twice, *i.e.*, there are two nodes in $F^2(q)$. This may be shown, for example, by carrying out the radial integral involved numerically.

6-12. The following nuclei ${}^{5}_{2}\mathrm{He}$, ${}^{5}_{3}\mathrm{Li}$, ${}^{17}_{8}\mathrm{O}$, ${}^{17}_{9}\mathrm{F}$, ${}^{41}_{20}\mathrm{Ca}$, ${}^{41}_{21}\mathrm{Sc}$, ${}^{209}_{82}\mathrm{Pb}$, and ${}^{209}_{83}\mathrm{Bi}$ may be considered to be made of a neutron or a proton outside a closed shell core. Use an independent-particle model to deduce their ground state spin and parity. From this information calculate the ground state magnetic dipole moment of each nucleus. Do the same for the one-hole nuclei of ${}^{3}_{1}\mathrm{H}$, ${}^{3}_{2}\mathrm{He}$, ${}^{15}_{7}\mathrm{N}$, ${}^{15}_{8}\mathrm{O}$, ${}^{39}_{19}\mathrm{K}$, ${}^{39}_{20}\mathrm{Ca}$, ${}^{207}_{81}\mathrm{Tl}$, and ${}^{207}_{82}\mathrm{Pb}$.

Chapter 7

NUCLEAR REACTIONS

We have seen in previous chapters that a large percentage of our knowledge of the properties of nuclei is derived from nuclear reactions. When an incoming particle is scattered off a target nucleus, the outcome depends on a combination of three factors: reaction mechanism, interaction between the projectile and the target and internal structure of the nuclei involved. In the case of electromagnetic probes, the interaction and the reaction mechanism are relatively well known, and one may be able to concentrate more on the nuclear structure question as a result. On the other hand, for strong interaction processes, both the reaction mechanism and the interaction can be studied. Different probes therefore complement each other in what we can learn from an investigation. Furthermore, it is often possible to select the bombarding energy and the reaction in such a way that we can focus on a particular aspect of the problem, as we shall see later in this chapter.

Nuclear reaction is a large subject by itself. In this chapter we can give only an overview of some of the more important topics. In order to highlight the basic points, it may be necessary to sacrifice some of the details in our discussion. For some of the more established topics, such as Coulomb scattering, excellent review articles are available. These will be referred to at the appropriate places. For some of the fast developing topics, such as heavy-ion reactions, only current papers and recent conference proceedings will provide the latest information.

§7-1 COULOMB EXCITATION

When a projectile carrying a charge ze approaches a target consisting of Z protons, the strength of the Coulomb field between these two particles may be characterized by the Sommerfeld number,

$$\eta = \left[\frac{1}{4\pi\epsilon_0}\right]\frac{zZe^2}{\hbar v} = \alpha z Z\frac{c}{v} \tag{7-1}$$

where v is the velocity with which one particle approaches the other when they are separated by large distances and α is the fine structure constant. Classically, the distance of closest approach R_s is given by the condition,

$$\left[\frac{1}{4\pi\epsilon_0}\right]\frac{zZe^2}{R_s} = \tfrac{1}{2}\mu v^2 \tag{7-2}$$

where μ is the reduced mass of the projectile in the center of mass. In terms of the Sommerfeld number, we have the result,

$$\tfrac{1}{2}R_s = \frac{\alpha\hbar czZ}{\mu v^2} = \eta\frac{\hbar}{\mu v}$$

From this expression, we see that the Sommerfeld number may be viewed as the ratio between $\frac{1}{2}R_s$ and the reduced de Broglie wave length,

$$\frac{1}{2\pi}\lambda = \frac{1}{2\pi}\frac{h}{p} = \frac{\hbar}{\mu v}$$

A small η means that the Coulomb field is weak compared with the available kinetic energy in the scattering. Under such conditions, the wave function of the incident particle is not greatly modified by the Coulomb field and the Born approximation applies. On the other hand, in Coulomb excitations we are primarily interested in the opposite limit with $\eta \gg 1$. In this case, the two particles are never close enough to each other for the nuclear force to play a role, and the excitation of the target or the projectile nucleus is accomplished through the Coulomb field. Such a process is known as *Coulomb excitation*. Because of its intrinsic interest, it is a subject also treated in many standard quantum mechanics texts such as Merzbacher (*Quantum Mechanics*, 2nd ed., Wiley, New York, 1961) and Messiah (*Quantum Mechanics*, North-Holland, Amsterdam, 1966).

There are several reasons why Coulomb excitation is of interest in nuclear physics. In the first place, the reaction mechanism is well known and may be regarded essentially as the inverse of electromagnetic decay, discussed in §5-2. Experimentally very strong Coulomb fields can be created by bombarding nuclei with a beam of heavy ions. When this advantage is coupled with the precision that can be achieved in charged particle experiments, we have a powerful tool for investigating some of the detailed properties of nuclei.

Multipole expansion. We shall follow the same approach used in §5-2 and treat electromagnetic interaction between the projectile and the target as a perturbation to the nuclear wave functions. The perturbing Hamiltonian due to Coulomb excitation may be written in the form

$$H'(t) = \left[\frac{1}{4\pi\epsilon_0}\right]\frac{zZe^2}{|\boldsymbol{r}_p(t) - \boldsymbol{r}|} - \text{monopole term} \tag{7-3}$$

where $\boldsymbol{r}_p(t)$ is the location of the projectile and $\boldsymbol{r}$ that of the target nucleus. It is necessary to eliminate contributions from the "monopole term" here, since such a term can only deflect the projectile without exciting any of the internal degrees of

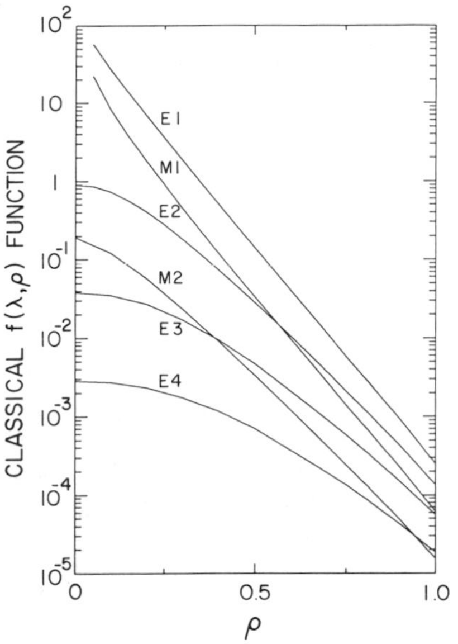

Fig. 7-1 Classical Coulomb excitation function $f(\lambda, \varrho)$ for electric ($E\lambda$) and magnetic ($M\lambda$) transition of different multipoles. The variable ϱ is related to the excitation energy, as can be seen from (7-10). The function $f(\lambda, \varrho)$ gives the total excitation cross section in the classical limit $\eta \rightarrow \infty$, corresponding to the situation of large Z and very low-energy incident particles. (Plotted from values given in Alder et al., *Rev. Mod. Phys.* **28** [1956] 432.)

freedom of the particles involved. For simplicity, we shall not mention this term explicitly from now on.

The perturbation may be expanded in the region $r_p > r$ in terms of spherical harmonics

$$\frac{1}{|\boldsymbol{r}_p(t) - \boldsymbol{r}|} = \sum_{\lambda=0}^{\infty} \sum_{\mu=-\lambda}^{\lambda} \frac{4\pi}{2\lambda+1} \frac{r^\lambda}{r_p^{\lambda+1}(t)} Y_{\lambda\mu}(\theta, \phi) Y^*_{\lambda\mu}(\theta_p, \phi_p) \tag{7-4}$$

as we have done earlier in §4-7. In the above expression, quantities pertaining to the target state are of the form $r^\lambda Y_{\lambda\mu}(\theta, \phi)$, the same form as $\boldsymbol{O}_{\lambda\mu}(E\lambda)$, the operator for $E\lambda$ excitation given in (5-35). We can make use of this similarity to make a connection between Coulomb excitation and electromagnetic decay.

Furthermore, the scattering is reminiscent of electron scattering off nuclei discussed in §4-2. There we found that, since the scattering is due primarily to electromagnetic interactions, the cross section is proportional to the Rutherford cross section. Nuclear effects enter as form factors modifying the purely electromagnetic results. Analogous to (4-5), the Coulomb excitation cross section from an initial state i to a final state f may be expressed in the following form:

$$\left(\frac{d\sigma}{d\Omega}\right)_{fi} = \frac{1}{2J_i + 1} \sum_{M_f M_i} \left|P_{M_f M_i}\right|^2 \left(\frac{d\sigma}{d\Omega}\right)_{\text{Rutherford}} \tag{7-5}$$

where J_i is the initial spin of the target nucleus, and M_i and M_f are, respectively, the projections of the initial and final spin on the quantization axis. The square root of the constant of proportionality is given by the expression

$$P_{M_f M_i} = \frac{1}{i\hbar}\int_{-\infty}^{+\infty} \langle J_f M_f \xi | H'(t) | J_i M_i \zeta \rangle e^{i(E_f - E_i)t/\hbar} dt \tag{7-6}$$

The integral may be written as a product of two parts, a nuclear transition matrix element and an integral over time, independent of the nuclear states involved.

On substituting the result of (7-4) into $H'(t)$ given in (7-3), $P_{M_f M_i}$ may be reduced to a sum over products in the following form:

$$P_{M_f M_i} = \frac{4\pi ze}{i\hbar}\sum_{\lambda\mu}\frac{1}{2\lambda+1}\langle J_f M_f \xi | \boldsymbol{O}_{\lambda\mu}(E\lambda) | J_i M_i \zeta \rangle S_{\lambda\mu}(E\lambda). \tag{7-7}$$

The matrix element $\langle J_f M_f \xi | \boldsymbol{O}_{\lambda\mu}(E\lambda) | J_i M_i \zeta \rangle$ gives the dependence of the cross section on nuclear states and may be related to the reduced transition probability $B(E\lambda; J_i\zeta \to J_f\xi)$ for $E\lambda$-transition given in (5-38). The integral over time is contained in the factor $S_{\lambda\mu}(E\lambda)$ and has the form

$$S_{\lambda\mu}(E\lambda) = \int_{-\infty}^{+\infty} e^{i(E_f - E_i)t/\hbar}\frac{1}{r_p^{\lambda+1}(t)} Y_{\lambda\mu}(\theta_p, \phi_p)\, dt \tag{7-8}$$

Because of the spherical harmonics in the integrand, the integral is a function of the scattering angles; however, it is independent of the nuclear states involved. The derivation of $S_{\lambda\mu}(E\lambda)$ is basically simple, even though the actual steps are complicated by angular momentum couplings. The final form, given by Alder et al. (*Rev. Mod. Phys.* **28** [1956] 432; **30** [1958] 353) may be expressed in the following manner:

$$S_{\lambda\mu}(E\lambda) = \frac{1}{va^{\lambda}} Y_{\lambda\mu}(\tfrac{1}{2}\pi, 0)\mathcal{F}_{\lambda\mu}(\theta, \varrho) \tag{7-9}$$

where ϱ, the "adiabaticity parameter," is related to the energy required to excite the target nucleus from an initial state at energy E_i to a final state at E_f,

$$\varrho = \eta\frac{E_f - E_i}{2E_i} \tag{7-10}$$

The quantity $\mathcal{F}_{\lambda\mu}(\theta, \varrho)$ is an integral having the form

$$\mathcal{F}_{\lambda\mu}(\theta, \varrho) = \int_{-\infty}^{+\infty} e^{i\varrho(\varepsilon\sinh x + x)}\frac{(\cosh x + \varepsilon + i\sqrt{\varepsilon^2 - 1}\sinh x)^{\mu}}{(\varepsilon\cosh x + 1)^{\lambda+\mu}}\, dx \tag{7-11}$$

where $\varepsilon = (\sin\frac{1}{2}\theta)^{-1}$

The spherical harmonics in (7-9), with $\theta = \frac{1}{2}\pi$ and $\phi = 0$, may be written explicitly in terms of λ and μ,

$$Y_{\lambda\mu}(\tfrac{1}{2}\pi, 0) = \begin{cases} (-1)^{(\lambda+\mu)/2}\sqrt{\frac{2\lambda+1}{4\pi}}\frac{\sqrt{(\lambda-\mu)!(\lambda+\mu)!}}{(\lambda-\mu)!!(\lambda+\mu)!!} & \text{for } \lambda+\mu = \text{even}, \\ 0 & \text{otherwise.} \end{cases}$$

The differential scattering cross section for Coulomb excitation in (7-5) now reduces to the form

$$\left(\frac{d\sigma}{d\Omega}\right)_{fi} = \sum_{\lambda=1}^{\infty}\left(\frac{2ze}{a^{\lambda}\hbar v}\right)^2 B(E\lambda; J_i\zeta \rightarrow J_f\xi)\,\frac{df(E\lambda,\varrho)}{d\Omega} \tag{7-12}$$

Here $a = \frac{1}{2}R_s$, half the distance of closest approach given in (7-2). The angular dependence is contained in the differential,

$$\frac{df(E\lambda,\varrho)}{d\Omega} = \frac{4\pi^2}{(2\lambda+1)^3}\sum_{\mu}\left|Y_{\lambda\mu}(\tfrac{1}{2}\pi,0)\mathcal{F}_{\lambda\mu}(\theta,\varrho)\right|^2\left(\frac{d\sigma}{d\Omega}\right)_{\text{Rutherford}} \tag{7-13}$$

The integral,

$$f(E\lambda,\varrho) = \int \frac{df(E\lambda,\varrho)}{d\Omega}\,d\Omega$$

is the total excitation cross section in the classical limit. Explicit values for different multipole λ are shown in Fig. 7-1 for the limiting case of $\eta \rightarrow \infty$.

So far we have discussed only electric multipole excitations. It is clear that magnetic multipole excitations are also present in a Coulomb excitation. The form of the differential cross section for magnetic multipole scattering is similar to that for electric multipoles given by (7-12) except that $B(E\lambda)$ is now replaced by the corresponding $B(M\lambda)$, reduced transition probability for magnetic multipole λ. Differences in the angular distribution between $M\lambda$ and $E\lambda$ transitions are contained in the differences between $\frac{df(M\lambda)}{d\Omega}$ and $\frac{df(E\lambda)}{d\Omega}$. These are also given in Alder et al.; the values for low-order magnetic multipoles are given in Fig. 7-1 as well.

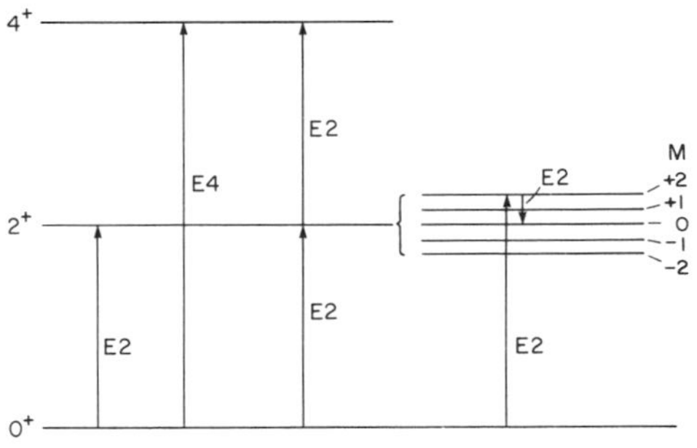

Fig. 7-2 Low-lying energy level spectrum of a hypothetical even-even nucleus showing first order and second order Coulomb excitation processes.

Multiple scattering. Coulomb excitation is useful in creating excited states in the target or projectile nucleus. Because of the intense electromagnetic fields brought along by heavy-ion scattering, very high multipolarity excitations may be attained. On the other hand, an examination of Fig. 7-1 will show that, for example, the probability of $E4$ excitation is reduced by about two orders of magnitude compared with $E2$ processes. Furthermore, in general, the strengths for high multipole transitions are weaker, since the nuclear matrix elements involved are smaller. This can be seen, for example, from the values of single-particle estimates given in §5-3. As a result, multiple low-order excitations may become competitive with a single higher multipolarity transition in exciting high-spin states.

Consider the hypothetical even-even nucleus with low-lying level scheme shown in Fig. 7-2 as an example. On the left we have first order processes exciting the nucleus from the 0^+ ground state to 2^+ and 4^+ excited states. Since $E2$-Coulomb excitations are so much stronger than $E4$ excitations, a succession of two $E2$-excitations may be comparable or even stronger in strength than a single $E4$ process in reaching the 4^+ excited state. In fact, because of the large reduction in the size of $f(\lambda, \varrho)$ with increasing λ, multiple processes through successive low multipolarity Coulomb excitations may become the preferred path for a nucleus to reach final states of relatively high angular momenta.

There is also another class of second order processes shown on the right of Fig. 7-2. In this example, the first excitation brings the nucleus to a particular magnetic substate M of a $J^\pi = 2^+$ level. Instead of proceeding to a higher state, the second "scattering" may take the nucleus to a state having a different M-value of the same 2^+ level. This is known as a *reorientation* effect. In addition to the value of $B(E2)$, such a process is also sensitive to the matrix element of $O(E2)$ between wave functions of the same level, the 2^+-state in the example here. Such a matrix element is related to the quadrupole moment of the excited state and, in this way, the quadrupole moment of excited states may be deduced from the cross section of Coulomb excitation processes. The value of the quadrupole moment extracted depends somewhat on the nuclear model used: however, this is not a serious problem in general, since reliability of the model may be checked against several other properties of the nucleus at the same time. Besides electric quadrupole moments, magnetic dipole and electric hexadecapole (2^4) moments can also be deduced through second order Coulomb excitation processes. In this way the static moments of excited states are determined for a large number of nuclei.

§7-2 COMPOUND NUCLEUS FORMATION

We have seen in the previous chapter that single-particle and collective degrees of freedom form the two extreme points of view of nuclear structure. A parallel situation exists in nuclear reaction studies in terms of the two limiting situations of direct reaction and compound nucleus formation. In the former case, one assumes that only one nucleon in the projectile interacts with one of the nucleons in the target while the rest of the nucleons in both the target and the projectile remain unaffected by the reaction. The basis for taking this point of view is the short time, of the order of 10^{-22} s, available for the projectile and the target to interact with each other. Since this time is comparable to the transit time for an incident particle with kinetic energy greater than the order of 1 MeV per nucleon to travel over a distance comparable to the nuclear diameter, the probability for interacting more than once is small. On the other hand, if the incident energy is much lower, the projectile and the target may "fuse" together for a much longer time, say, of the order of 10^{-14} s, and a *compound nucleus* is formed as the intermediate state. In this section we shall be mainly concerned with reactions involving compound nucleus formation. We shall return to a discussion of direct reactions in the next section.

Notation. Before we begin, it is useful to define some of the notations used to represent a reaction. Among the possible reactions induced by bombarding a target nucleus A with projectile nucleus a, we choose to observe a particular exit channel with particle b emerging and nucleus B remaining behind. The reaction is usually represented in either one of the following two ways:

$$A(a,b)B \qquad \text{or} \qquad a + A \longrightarrow b + B$$

For example, if a proton incidents on a ^{48}Ca target and a neutron is observed to emerge from the reaction, the residual nucleus is ^{48}Sc. The reaction may be represented in the form

$$^{48}\text{Ca}(p,n)^{48}\text{Sc} \qquad \text{or} \qquad p + {}^{48}\text{Ca} \rightarrow n + {}^{48}\text{Sc}$$

Other reactions may also take place as a result of proton incident on ^{48}Ca; for example, a proton may come out, leaving the ^{48}Ca nucleus in an excited state. Each one of these final states of the emerging particle and residual nucleus is called an *exit channel*, and the possible or "open" exit channels are governed by conservation laws and selection rules operating in the scattering. In general, the number of open channels increases very fast with increasing kinetic energy of the incident particle.

The allowed exit channel is not restricted to final states consisting of two particles in the examples cited above. For example, an experiment may be carried out using deuteron as the incident particle and ^{48}Ca as the target. A possible exit channel may involve a breakup of the deuteron into a proton and a neutron. The reaction is represented in the form

$$^{48}\text{Ca}(d,pn)^{48}\text{Ca} \qquad \text{or} \qquad d + {}^{48}\text{Ca} \rightarrow p + n + {}^{48}\text{Ca}$$

In order to simplify the discussion, we shall ignore reactions involving three or more particles in the final state. Furthermore, the separation between projectile and target nucleus and that between the scattered particle and residual nucleus is useful only in the laboratory, where there is often a natural way to make the distinction. For most of our discussions, we shall be working in the center of mass of the two-body system and the distinction reduces to a simple question of semantics.

It is often possible to arrive at the same exit channel using different combinations of incident particle and target nucleus. For example, the final state of neutron plus ^{48}Sc in the example above may also be obtained by scattering neutrons from a ^{48}Sc target. The $n + {}^{48}$Sc system here constitutes a different *incident channel* from the proton plus ^{48}Ca channel. Since we are dealing with microscopic objects, time reversal invariance holds. Thus the reaction $p+{}^{48}\text{Ca} \rightarrow n+{}^{48}\text{Sc}$ may also take place with time order going in the opposite direction, $n+{}^{48}\text{Sc} \rightarrow p+{}^{48}\text{Ca}$, and the roles of incident and exit channels are reversed. On occasion it may be more convenient to speak of a *reaction channel* without specifying whether it is an incident channel or an exit channel.

For a reaction channel involving two particles we need three sets of quantum numbers to label the channel in an unambiguous way. For instance, for an exit channel we need labels to specify the wave function of the emerging particle, labels to identify the wave function of the residual nucleus, and labels describing the relative motion between these two particles, as illustrated in §C-4. Returning again to the $^{48}\text{Ca}(p,n)^{48}\text{Sc}$ example, the reaction may leave the ^{48}Sc nucleus in an excited state. Since the wave function for an excited state of ^{48}Sc is different from that for the ground state, we have a different exit channel, distinguishable from the ground state channel by the wave function of the residual nucleus. Furthermore, the wave function for the relative motion of the two particles may be decomposed according to partial waves, each with a definite orbital angular momentum ℓ. In principle, each partial wave forms a different reaction channel. On the other hand, the orbital angular momentum between two particles is not usually observed in a reaction, and may sometimes wish to refer to a reaction channel as the sum of all the partial waves instead.

At low incident energies, where the kinetic energy is less than 1 MeV per nucleon, the de Broglie wave length is longer than the dimension of a typical nucleus. Under such situations, the scattering cannot be very sensitive to the details of the structure of nuclei involved. Once the two nuclei in the incident channel come into contact with each other, their nucleons have a chance of interacting with each other many times, resulting in the loss of identities of the two original nuclei. For a short time these two nuclei form a single entity, the *compound nucleus.* Once a compound nucleus is formed, the memory of the entrance channel is lost because of the numerous intervening interactions and the decay of the system is determined primarily by the amount of excitation energy available in the system. At low energies, the lifetimes of the excited states in the system are relatively long, since the number of open exit channels is small. As a result, the width of a compound nuclear state Γ is narrow. At the same time, the density of states at such low energies is small so that D, the mean spacing between neighboring states, is large. With $D \gg \Gamma$, isolated resonances dominate the reaction cross section.

Scattering cross section. One of the main characteristics of compound nucleus reaction is the absence of any dependence between formation and decay of the compound nucleus. This is to be expected, since excitation energy in the compound nucleus brought in through the incident channel is shared among many nucleons, arising as a result of the large number of collisions between individual nucleons in the relatively long lifetime of the compound nucleus state (compared with the typical time for a nuclear interaction). Let the cross section for forming a compound nucleus $\mathcal{N}$ through incident channel α be represented by σ_α. The decay of $\mathcal{N}$ to a particular exit channel β with final state consisting of particles b and B is characterized by transition probability $\mathcal{W}_\beta$ or partial width $\Gamma_\beta = \hbar\mathcal{W}_\beta$. There may be several such channels open, for example,

		decay product	exit channel
$\mathcal{N}$	$\longrightarrow$	$a + A$	α
	$\longrightarrow$	$b + B$	β
	$\longrightarrow$	$c + C$	γ
		$\cdots$	$\cdots$

The total width of the decay is given by the sum of all partial widths,

$$\Gamma = \Gamma_\alpha + \Gamma_\beta + \Gamma_\gamma + \cdots \tag{7-14}$$

and Γ_β/Γ is the probability for decaying through channel β. The reaction cross section from an entrance channel α to an exit channel β is then given by the product of the probabilities to form the compound nucleus $\mathcal{N}$ and for $\mathcal{N}$ to decay through β,

$$\sigma_{\beta\alpha} = \sigma_\alpha \frac{\Gamma_\beta}{\Gamma} \tag{7-15}$$

To make further progress we need some of the result of scattering theory given in §C-4.

We shall assume that in each reaction channel there is a radius R_c, the *channel radius*, outside which there is no interaction between the scattered particle and the residual nucleus (ignoring the long-range Coulomb interaction here for simplicity). Thus, in the outside region ($r > R_c$), the particles may be considered to be free and the wave functions are given by plane waves (or Coulomb wave functions in the more general case). In the inside region ($r < R_c$), the wave function may be complicated due to interaction between nucleons in the two components, and there is little hope of obtaining a reasonable solution. At the boundary $r = R_c$, the logarithmic derivative of the modified radial wave function u_c of each channel,

$$\rho_c = \left(\frac{r}{u_c}\frac{du_c}{dr}\right)_{r=R_c} \tag{7-16}$$

must be continuous from the inside to the outside region. The quantity ρ_c is complex in general and we shall make use of it to parametrize the relevant information of the interior region.

In scattering problems we are interested in the asymptotic behavior of the system. The only knowledge of the wave function for $r < R_c$ we need is the value of its logarithmic derivative at the boundary. In other words, the information of the inside region is completely contained in a set of logarithmic derivatives, and the values of these derivatives may be used as the "boundary conditions" to fix the asymptotic wave function of interest to a scattering problem. In the absence of better knowledge we can take the logarithmic derivatives as parameters describing the inside region. This method of treating the scattering problem is akin to that used in solving electrostatic problems, where we exclude regions containing sources and replace them by appropriate boundary conditions. In this way, the problem is reduced to a simpler one.

In order to simplify the discussion further, we shall assume that only s-wave scattering is different from zero and, as a result, a single logarithmic derivative ρ_0 is adequate to specify the problem completely. The reduced radial wave function of (C-12) has the asymptotic form

$$u_0(r) \sim \left\{e^{-ikr} - \eta_0 e^{ikr}\right\} \tag{7-17}$$

The quantity $\eta_0 = \exp 2i\delta_0$ is the inelasticity parameter and δ_0 is the complex phase shift for the $\ell = 0$-channel. By taking the logarithmic derivative of $u_0(r)$ and equating it to ρ_0 at $r = R_c$, we can relate η_0 to ρ_0,

$$\eta_0 = \frac{\rho_0 + ikR_c}{\rho_0 - ikR_c} e^{-2ikR_c} \tag{7-18}$$

Using (C-21), we obtain the elastic scattering cross section in the form

$$\sigma^{\text{el}} = \frac{\pi}{k^2}\left|1 - \eta_0\right|^2 = \frac{\pi}{k^2}\left|e^{2ikR_c} - 1 - \frac{2ikR_c}{\rho_0 - ikR_c}\right|^2 \tag{7-19}$$

and from (C-48), we obtain the expression for the reaction cross section

$$\sigma^{\text{re}} = \frac{\pi}{k^2}\left(1 - |\eta_0|^2\right) = \frac{\pi}{k^2}\frac{-4kR_c\text{Im}\,\rho_0}{(\text{Re}\,\rho_0)^2 + (\text{Im}\,\rho_0 - kR_c)^2} \tag{7-20}$$

If ρ_0 is real, corresponding to the case of scattering from a real potential, σ^{re} vanishes as expected. Furthermore, since the reaction cross section cannot be negative, the absolute value of η_0 must be equal or less than unity. This, in turn, implies that the imaginary part of ρ_0 must be negative.

Breit-Wigner formula for isolated resonances. For a reaction of this type, the cross section has a resonance structure similar to that of an alternating current electric circuit. The maximum of $\sigma^{\rm re}$ occurs at $\operatorname{Re}\rho_0 = 0$. Let E_c represent the energy where this takes place. The real part of ρ_0 may be expanded as a power series in E around E_c,

$$\operatorname{Re}\rho_0 = a(E - E_c) + \cdots$$

where a is a parameter characterizing the leading order term of the real part of ρ_0. Similarly, the leading order of the imaginary part of ρ_0 may be taken to be a (positive) parameter b

$$\operatorname{Im}\rho_0 = -b + \cdots$$

The two cross sections in (7-19) and (7-20) near the resonance energy E_c may now be expressed in the following forms:

$$\sigma^{\rm el} = \frac{\pi}{k^2}\left|e^{2ikR_c} - 1 - \frac{2ikR_c}{a(E - E_c) - i(b + kR_c)}\right|^2 \tag{7-21}$$

$$\sigma^{\rm re} = \frac{\pi}{k^2}\frac{4kR_c b}{\{a(E - E_c)\}^2 + (b + kR_c)^2} \tag{7-22}$$

in terms of a and b.

We now make the following identifications:

$$\Gamma = 2\frac{b + kR_c}{a} \qquad \Gamma_\alpha = \frac{2kR_c}{a} \qquad \Gamma^{\rm re} = 2\frac{b}{a}$$

where Γ is the total width and Γ_α is the partial width for the entrance channel. The total reaction width is then

$$\Gamma^{\rm re} = \sum_{i\neq\alpha} \Gamma_i, \tag{7-23}$$

and the total width is $\Gamma = \Gamma_\alpha + \Gamma^{\rm re}$. This allows us, in turn, to rewrite (7-19) and (7-20) in the following forms:

$$\sigma^{\rm el} = \frac{\pi}{k^2}\left|e^{2ikR_c} - 1 - \frac{i\Gamma_\alpha}{(E - E_c) - i\frac{1}{2}\Gamma}\right|^2 \tag{7-24}$$

$$\sigma^{\rm re} = \frac{\pi}{k^2}\frac{\Gamma^{\rm re}\Gamma_\alpha}{(E - E_c)^2 + (\frac{1}{2}\Gamma)^2} \tag{7-25}$$

The cross sections are expressed here in terms of the relevant widths instead of parameters a and b.

The elastic channel has two parts, a nonresonating part with amplitude proportional to $1 - \exp\{2ikR_c\}$ and a resonating part containing an energy-dependent factor $(E - E_c)$ in the denominator. The contribution of the nonresonating part corresponds to a smooth background in the cross section and is usually called *shape-elastic* or *potential* scattering. At $E \approx E_c$, the elastic cross section is dominated by the resonating part,

$$\sigma^{\text{cel}} \approx \frac{\pi}{k^2} \frac{\Gamma_\alpha^2}{(E - E_c)^2 + (\frac{1}{2}\Gamma)^2} \tag{7-26}$$

This is called *compound elastic* scattering cross section, since it differs from shape elastic scattering by the fact that a compound nucleus is formed before the system returns to the entrance channel.

We can now come back to (7-15) and calculate σ_α, the cross section for forming the compound nucleus through entrance channel α. Since shape-elastic scattering does not involve the formation of a compound nucleus, we can ignore it here. The cross section for compound nucleus formation is, then, a sum of compound-elastic and reaction contributions,

$$\sigma_\alpha = \sigma^{\text{cel}} + \sigma^{\text{re}} = \frac{\pi}{k^2} \frac{\Gamma^{\text{re}}\Gamma_\alpha + \Gamma_\alpha^2}{(E - E_c)^2 + (\frac{1}{2}\Gamma)^2} = \frac{\pi}{k^2} \frac{\Gamma\Gamma_\alpha}{(E - E_c)^2 + (\frac{1}{2}\Gamma)^2} \tag{7-27}$$

where we have made use of the fact that $\Gamma = \Gamma_\alpha + \Gamma^{\text{re}}$. The cross section for the reaction from an entrance channel α to an exit channel β given in (7-14) may now be written in the form

$$\sigma_{\beta\alpha} = \frac{\pi}{k^2} \frac{\Gamma_\beta\Gamma_\alpha}{(E - E_c)^2 + (\frac{1}{2}\Gamma)^2} \tag{7-28}$$

This is known as the Breit-Wigner one-level formula.

Overlapping resonances. So far the discussion has been confined to the low energy region where the density of states is low. The idea of compound nucleus formation applies also to higher energy regions where the individual level width is comparable or greater than the average level spacing ($\Gamma \gtrsim D$). Since resonances are now overlapping each other, it is more meaningful to examine the average values of the various cross sections that enter the scattering.

Assuming that the cross section to form a compound nucleus for a particular state is still given by the Breit-Wigner form of (7-27), we can define an average cross section in a small energy interval W in the following way:

$$\overline{\sigma}_\alpha = \frac{1}{W} \int_{E-W/2}^{E+W/2} \sum_i \frac{\pi}{k^2} \frac{\Gamma^i\Gamma_\alpha^i}{(E' - E_i)^2 + (\frac{1}{2}\Gamma^i)^2} dE' = \frac{\pi}{k^2} \frac{2\pi}{W} \sum_i \Gamma_\alpha^i \tag{7-29}$$

where Γ^i is the total width of the i-th resonance and Γ_α^i is the partial width for decaying into channel α. The summation is over all the resonances in the energy region W centered at E. The energy interval must be small enough that the underlying conditions for the resonances are not too different from each other and yet large enough so that $W \gg \Gamma^i$. Our discussion is still limited to s-wave scattering; a more general discussion will also involve angular momentum coupling factors.

Using the fact that the number of levels in the interval is given by W/D, where D is the average level spacing, we can define a mean width,

$$\overline{\Gamma}_\alpha = \frac{D}{W}\sum_i \Gamma_\alpha^i \tag{7-30}$$

The quantity $\overline{\Gamma}_\alpha/D$ is known as the (s-wave) *strength function* and the average compound nucleus formation cross section may be expressed in terms of it,

$$\overline{\sigma}_\alpha = \frac{\pi}{k^2}2\pi\frac{\overline{\Gamma}_\alpha}{D} \tag{7-31}$$

This quantity may be related to ρ_0, the logarithmic derivative of the wave function at $r = R_c$ given by (7-16). Since the density of states is now high, the probability for the compound nucleus to decay through the entrance channel is small. We can say that the nucleus appears to be "black" to the incident channel.

In the limit of a completely absorptive nucleus, the wave function of the interior region may be approximated by an incoming term, $u_0(r) \sim \exp\{i\kappa r\}$, alone. As a result, ρ_0 is imaginary and may be written in the form:

$$\rho_0 = -i\kappa R_c$$

where κ is the wave number for $r < R_c$. On substituting this value into (7-18) and (7-20), we obtain the average value of compound nucleus formation cross section for channel α in the energy region,

$$\sigma_\alpha = \frac{\pi}{k^2}\frac{4\kappa k}{(\kappa+k)^2} \approx \frac{4\pi}{\kappa k} \tag{7-32}$$

since $k \ll \kappa$ for low-energy scattering from an attractive potential well. Comparing this expression with (7-31), we obtain the result,

$$\frac{\overline{\Gamma}_\alpha}{D} = \frac{2k}{\pi\kappa} \tag{7-33}$$

Furthermore, no resonance can be expected from (7-32).

In practice, resonances are observed at high energies. These are due primarily to the coupling of a large number of small resonances, for example, to a state in the vicinity that is strongly excited due to some special reasons related to nuclear structure. Such a strongly excited state is often called a *doorway* state.

The decay of a compound nucleus in the high-energy region depends on the number of accessible final states and is therefore dominated by the density of final states and other statistical considerations. This is the subject of Hauser-Feshbach theory, for which we shall refer the reader to the original literature (W. Hauser and H. Feshbach, *Phys. Rev.* **87** [1952] 366).

§7-3 DIRECT REACTION

Stripping and pickup reactions. A good example of direct reaction is a (d,p) process in which a deuteron, with more than a few MeV of kinetic energy, incidents on a target nucleus and a proton is observed to emerge. Since the deuteron is a loosely bound system of a proton and a neutron, it is easy to envisage that the neutron is captured into a single-particle orbit of the target nucleus without disturbing the structure of the rest of the nucleons and the proton continues on to become the scattered particle. The process may be viewed as one in which a neutron is "stripped" from the projectile. For this reason, the reaction is known as a one-neutron *stripping* reaction. States in the final nucleus that are strongly excited by such a stripping reaction are those formed predominantly by a nucleon coupled to the ground state of the target nucleus. Other stripping reactions, such as (t,p), transfer two nucleons from the projectile to the target nucleus. Even more complicated reactions, such as those involving the transfer of a cluster of nucleons, are also possible. To qualify as a direct reaction, both the target nucleus and the internal structure of the cluster transferred must be undisturbed by the reaction; the residual nucleus is simply the coupling of the cluster and the ground state of the target nucleus. This condition is generally difficult to meet for transfer reactions involving a large number of nucleons.

The complement of a stripping reaction is a *pickup* reaction. In this case one or more nucleons are taken away from the target nucleus without changing the structure among the rest of the nucleons. A good example is the reaction $^{40}\text{Ca}(p,d)^{39}\text{Ca}$. The states in the residual nucleus, ^{39}Ca here, that are strongly excited by a pickup reaction are the one-hole states; *i.e.,* those formed by removing one of the particles in ^{40}Ca and leaving the remaining 39 nucleons unchanged in their relative motion.

The scattering cross section in a direct reaction, stripping as well as pickup, may be derived in a straightforward way using the first Born approximation. The reaction mechanism is relatively straightforward here because of the simple relation between initial and final nuclear states underlying the direction reaction assumption. The Schrödinger equation for the scattering may be written in the form of a standard second-order differential equation,

$$(\nabla^2 + k^2)\psi(\boldsymbol{r}) = \frac{2\mu}{\hbar^2}V(\boldsymbol{r})\,\psi(\boldsymbol{r}) \tag{7-34}$$

where $k^2 = 2\mu E/\hbar^2$. A formal solution of (7-34) for the outgoing wave function may be expressed in terms of a Green's function $G(\boldsymbol{r},\boldsymbol{r}')$ as done in §C-6,

$$\psi(\boldsymbol{r}) = e^{i\boldsymbol{k}_i\cdot\boldsymbol{r}} + \frac{2\mu}{\hbar^2}\int G(\boldsymbol{r},\boldsymbol{r}')V(\boldsymbol{r}')\,\psi(\boldsymbol{r}')d^3r' \tag{7-35}$$

where we have chosen $\boldsymbol{k}_i$ to be along the direction of the incident particle and the function $\exp i\boldsymbol{k}_i\cdot\boldsymbol{r}$ is the solution of the homogeneous part of (7-34); *i.e.,* for $V(\boldsymbol{r}) = 0$.

We shall take the Green's function here to have the explicit form,

$$G(\boldsymbol{r},\boldsymbol{r}') = -\frac{1}{4\pi}\frac{e^{ik|\boldsymbol{r}-\boldsymbol{r}'|}}{|\boldsymbol{r}-\boldsymbol{r}'|} \tag{7-36}$$

It is the solution of the following equation:

$$(\nabla^2 + k^2)G(\boldsymbol{r},\boldsymbol{r}') = \delta(\boldsymbol{r}-\boldsymbol{r}')$$

More generally, we can include within the defining equation for $G(\boldsymbol{r},\boldsymbol{r}')$ a part of $V(\boldsymbol{r})$, for example, the part representing the averaging effect of the target nucleons on the incident particle. This is similar in spirit to the mean field approach used in nuclear structure investigations to obtain the single-particle states for shell model calculations in §6-7. In order to keep the discussion simple here, we shall take the Green's function to have the elementary form given by (7-36).

Using (7-36), the formal solution for the scattering wave function in (7-35) may be written in the form:

$$\psi(\boldsymbol{r}) = e^{i\boldsymbol{k}_i\cdot\boldsymbol{r}} - \frac{\mu}{2\pi\hbar^2}\int \frac{e^{ik|\boldsymbol{r}-\boldsymbol{r}'|}}{|\boldsymbol{r}-\boldsymbol{r}'|}V(\boldsymbol{r}')\psi(\boldsymbol{r}')\,d^3r' \tag{7-37}$$

Except for contributions from the Coulomb interaction, which we shall discuss as a part of the optical model potential in the next section, the potential $V(\boldsymbol{r}')$ is short-ranged, and we may approximate the argument of the exponential function in the asymptotic region by the first two terms in the following expansion:

$$ik(|\boldsymbol{r}-\boldsymbol{r}'|) = k\sqrt{r^2 - 2\boldsymbol{r}\cdot\boldsymbol{r}' + r'^2} = kr - \boldsymbol{k}_f\cdot\boldsymbol{r}' + O(r'^2) \approx kr - \boldsymbol{k}_f\cdot\boldsymbol{r}'$$

where $\boldsymbol{k}_f = k\boldsymbol{r}/r$ is taken along the direction of the emerging particle. The formal solution of the scattering equation now takes on the form,

$$\psi(\boldsymbol{r}) \approx e^{i\boldsymbol{k}_i\cdot\boldsymbol{r}} - \frac{\mu}{2\pi\hbar^2}\frac{e^{ikr}}{r}\int e^{-i\boldsymbol{k}_f\cdot\boldsymbol{r}'}V(\boldsymbol{r}')\,\psi(\boldsymbol{r}')\,d^3r' \tag{7-38}$$

where we have taken $|\boldsymbol{r}-\boldsymbol{r}'| \simeq r$, correct in the asymptotic region where the scattered particle is observed.

Comparing (7-38) with the asymptotic form of the scattering wave function given in (C-5), we obtain the scattering amplitude

$$f(\theta) = -\frac{\mu}{2\pi\hbar^2}\int e^{-i\boldsymbol{k}_f\cdot\boldsymbol{r}'}V(\boldsymbol{r}')\,\psi(\boldsymbol{r}')d^3r' \tag{7-39}$$

This is only a formal or integral equation solution for the scattering amplitude, since the expression involves the unknown function $\psi(\boldsymbol{r}')$, the solution to the scattering equation (7-34). Eq. (7-39) is useful in that it provides us with a starting point to expand scattering cross section in terms of a Born series. In the (first) Born approximation, the unknown function $\psi(\boldsymbol{r}')$ in (7-39) is replaced by its first term in (7-35). In this way, we obtain an approximate form of the scattering amplitude,

$$f(\theta) \approx -\frac{\mu}{2\pi\hbar^2}\int e^{-i\boldsymbol{k}_f\cdot\boldsymbol{r}'}V(\boldsymbol{r}')\,e^{i\boldsymbol{k}_i\cdot\boldsymbol{r}'}\,d^3r' \tag{7-40}$$

We shall make use of this result to obtain the differential scattering cross section for stripping and pickup reactions.

The expression in (7-40) may be simplified further by expressing the results in terms of the momentum transfer vector,

$$\boldsymbol{q} = \boldsymbol{k}_i - \boldsymbol{k}_f \tag{7-41}$$

and by expanding the plane wave in terms of spherical harmonics given in (C-12),

$$e^{i\boldsymbol{q}\cdot\boldsymbol{r}'} = \sum_{\ell} i^{\ell}\sqrt{4\pi(2\ell+1)}\,j_{\ell}(qr')\,Y_{\ell 0}(\theta') \tag{7-42}$$

The angle θ' is between vectors $\boldsymbol{q}$ and $\boldsymbol{r}'$ and is therefore one of the variables of integration in (7-40); in contrast, the scattering angle θ is between vectors $\boldsymbol{k}_f$ and $\boldsymbol{k}_i$.

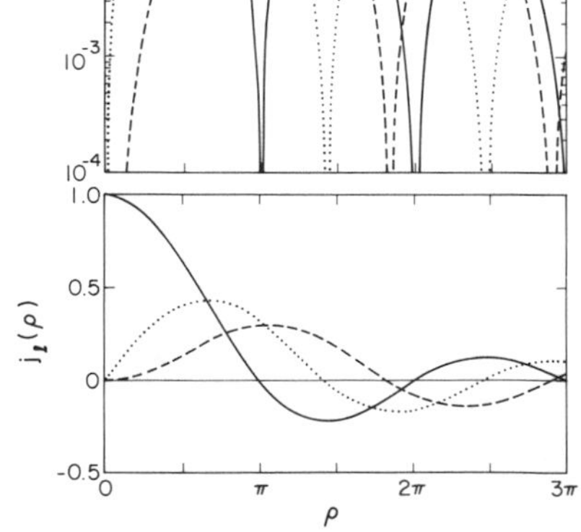

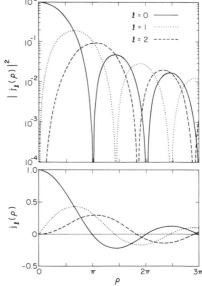

Fig. 7-3 Spherical Bessel functions $j_\ell(\rho)$ and characteristic angular distribution of stripping reaction given by $j_\ell^2(\rho)$. The plots are made as functions of ρ, related to the scattering angle θ and the momentum transfer q by (7-41).

Angular distribution. The discussion in the previous paragraph has ignored the internal structure of particles participating in the scattering. Since in stripping and pickup reactions we are dealing with a change in the structure of nuclei involved, the wave functions of both the initial nuclei and the final nuclei must enter into the expression for the scattering amplitude. Let us take the asymptotic forms of the initial and final wave functions of the scattering system to be

$$\Psi_i \rightarrow e^{i\boldsymbol{k}_i\cdot\boldsymbol{r}}\Phi_i \qquad\qquad \Psi_f \rightarrow e^{i\boldsymbol{k}_f\cdot\boldsymbol{r}}\Phi_f \tag{7-43}$$

where Φ_i and Φ_f are, respectively, the product of the internal wave functions of the incident particle and the target nucleus, and the product of the wave functions of the scattered particle and the residual nucleus. As a concrete example, let us take the case of $^{40}\text{Ca}(d,p)^{41}\text{Ca}$. For this reaction, we have

$$\Phi_i = \left\{\phi(d) \times \phi(^{40}\text{Ca})\right\} \qquad\qquad \Phi_f = \left\{\phi(p) \times \phi(^{41}\text{Ca})\right\}$$

where $\phi(d)$, $\phi(p)$, $\phi(^{40}\text{Ca})$ and $\phi(^{41}\text{Ca})$ are the wave functions describing the internal structure of deuteron, proton, ^{40}Ca, and ^{41}Ca, respectively. The multiplication symbols here imply that the wave functions are coupled together to some definite values in angular momentum and isospin.

In the spirit of direct reaction, the deuteron wave function may be taken as the (weakly) coupled state of a proton and a neutron,

$$\phi(d) = \big(\phi(p) \times \phi(n)\big) \tag{7-44}$$

In order to simplify the argument and to avoid complications due to angular momentum recoupling, we shall treat the proton purely as a spectator in the entire scattering process. If the neutron is captured into a single-particle state of the target nucleus with orbital angular momentum ℓ_t, the wave function of the residual nucleus may be expressed in the form

$$\phi_{\ell_t}(^{41}\text{Ca}) \sim \big(\phi(n)\phi(^{40}\text{Ca})Y_{\ell_t m_t}(\theta',\phi')\big) \tag{7-45}$$

Using these wave functions, the scattering amplitude for $^{40}\text{Ca}(d,p)^{41}\text{Ca}$ may be written in the form,

$$f(\theta) \approx -\frac{\mu}{2\pi\hbar^2}\int e^{-i\boldsymbol{q}\cdot\boldsymbol{r}'}\Big\langle\Big\{\phi(p)\times\big(\phi(n)\phi(^{40}\text{Ca})Y_{\ell_t m_t}(\theta',\phi')\big)\Big\}\Big| \\ V(\boldsymbol{r}')\Big|\Big\{\phi(^{40}\text{Ca})\times\big(\phi(p)\times\phi(n)\big)\Big\}\Big\rangle\, d^3r' \tag{7-46}$$

The role of the potential $V(\boldsymbol{r}')$ here is to strip the neutron from the deuteron and put it into the residual nucleus. For our purposes, it may be approximated by a delta function at the nuclear surface

$$V(\boldsymbol{r}') = V_0\,\delta(r' - R) \tag{7-47}$$

so as to simplify the derivation. Here R is the radius of the residual nucleus. The meaning of this approximation is that the neutron is stripped off the incident deuteron and captured by the ^{40}Ca on contact. The strength V_0 represents the probability for such a process to take place and may be treated as a parameter related to the absolute magnitude of the scattering cross section.

When we integrate (7-46) over the coordinates of both nucleons and ^{40}Ca, the nuclear wave functions drop out. The exponential factor may be expanded in terms of spherical harmonics using (7-42), and the scattering amplitude reduces to the form

$$\begin{aligned} f(\theta) &\approx -\frac{2\mu}{\hbar^2}V_0\sum_{\ell} i^{\ell}\sqrt{\frac{2\ell+1}{4\pi}}\, j_\ell(qR)\int Y_{\ell 0}(\theta')Y^*_{\ell_t m_t}(\theta',\phi')\sin\theta'\,d\theta'\,d\phi' \\ &= -\frac{2\mu}{\hbar^2}V_0\sum_{\ell} i^{\ell}\sqrt{\frac{2\ell+1}{4\pi}}\, j_\ell(qR)\delta_{\ell\ell_t}\delta_{m_t 0} \\ &= -\frac{2\mu}{\hbar^2}V_0 i^{\ell_t}\sqrt{\frac{2\ell_t+1}{4\pi}}\, j_{\ell_t}(qR) \end{aligned} \tag{7-48}$$

In integrating over the angular variables, we have made use of the orthonormal condition of spherical harmonics given in (B-10). The only angular dependence in (7-46) is in the spherical Bessel function through the relation between momentum transfer q and scattering angle θ. Since we have used a plane wave to approximate the solution to the scattering equation, the result is generally known as the plane wave Born approximation (PWBA) result. A more rigorous derivation can be found in standard references for direct reactions such as Tobocman (*Theory of Direct Nuclear Reaction*, Oxford University Press, London, 1961) and Satchler (*Direct Nuclear Reactions*, Oxford University Press, Oxford, 1983).

From (7-48), the differential cross section for direct reaction is given by

$$\frac{d\sigma}{d\Omega} \sim |j_{\ell_t}(qR)|^2 \tag{7-49}$$

The momentum transfer depends on the scattering angle θ as shown in (D-41),

$$q = \sqrt{k_i^2 + k_f^2 - 2k_ik_f\cos\theta} \approx 2k\sin(\tfrac{1}{2}\theta) \tag{7-50}$$

where we have taken $k \approx k_i \approx k_f$, valid if the incident energy is sufficiently high. The angular distribution is characterized by the angular momentum transferred and given by the factor $|j_{\ell_t}(2kR\sin(\frac{1}{2}\theta))|^2$, as shown in Fig. 7-3. For example, since $j_0(\rho) \sim \sin\rho/\rho$, we see that, for an $\ell_t = 0$ transfer, the angular distribution peaks at 0°. On the other hand, for higher ℓ_t-value transfers, there is no longer a maximum at 0°, as can be seen, for example, from $j_1(\rho) \sim \sin\rho/\rho^2 - \cos\rho/\rho$ for $\ell_t = 1$. As the value of ℓ_t is increased, the first maximum in the angular distribution moves out to successively larger angles, since the first peak of $|j_{\ell_t}(\rho)|^2$ appears at successively large values of ρ with increasing ℓ_t value. This is a feature

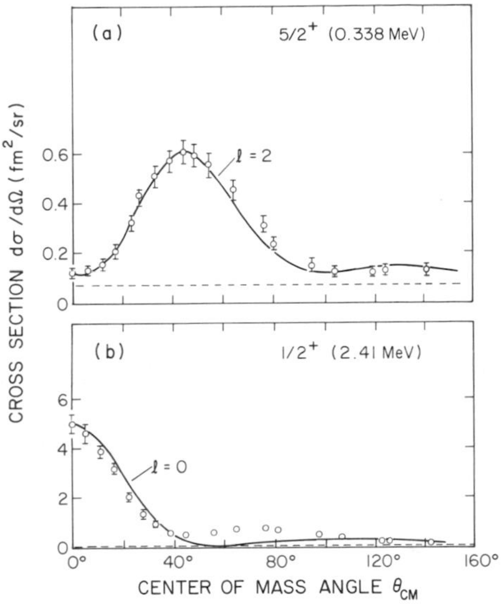

Fig. 7-4 Angular distribution of neutrons observed in the reaction $^{20}\text{Ne}(d,n)^{21}\text{Ne}$ leading to (a) the 0.338 MeV, $J^\pi = 5/2^+$ level in ^{21}Ne showing a typical $\ell = 2$ one-nucleon transfer direct reaction angular distribution, and (b) the 2.41 MeV, $J^\pi = 1/2^+$ level, an $\ell = 0$ transfer. Solid lines are calculated results using DWBA for direct reaction and dashed lines are the results of a Hauser-Feshbach calculation for compound nucleus reaction.

observed in direct reactions, as can be seen, for example, in the $^{20}\text{Ne}(d,n)^{21}\text{Ne}$ reaction shown in Fig. 7-4.

Although the PWBA results correctly give the essential features of the angular distribution for direction reactions, they lack predictive power. This is caused in part by the *distortion* of the incident and scattered wave due to the average or optical potential experienced by the incoming and scattered particle. This will be discussed in the next section. Furthermore, there does not seem to be an easy way to derive the interaction strength V_0 of (7-47) and, as a result, the absolute magnitude of the angular distributions cannot be deduced from PWBA. A more accurate picture of the scattering is given by the distorted wave Born approximation (DWBA) where, instead of plane waves, more realistic relative wave functions between the projectile and the target nucleus and between the scattered particle and the residual nucleus are used.

§7-4 THE OPTICAL MODEL

Besides compound nucleus formation and direct reaction, we may also be interested in the average result of a reaction at a given bombarding energy. For such purposes, we often invoke the analogy of an optical wave through a "cloudy" crystal ball. In a nuclear reaction, the scattered wave may be divided into two categories: elastic scattering in which only the direction of the wave propagation is changed and inelastic scattering in which the particle is scattered into an exit channel different from the incident one. The former corresponds to a refraction of the optical waves and the latter corresponds to an absorption due to the fact that the crystal ball is cloudy.

The aim of the optical model is to find a potential to describe smooth variations of the scattering cross section as a function of energy E and target nucleon number A. The general situation of scattering may be quite complex; however, great simplification may be obtained if we are interested only in the average properties, away from resonances and states strongly excited by direct reactions. The basic idea is therefore very similar to the mean field approach we have seen in the previous chapter for nuclear structure studies.

To simplify the discussion, we shall for the most part consider only elastic scattering. There are two main sources of contribution to the cross section. The first is potential scattering described earlier in (7-24). The second comes from multiple scattering with intermediate states involving the excited states of nuclei participating in the scattering. Not all such multiple scatterings return the system back to the incident channel and, as a result, some of the incident flux is lost to other reaction channels. Rather than trying to calculate the cross section of each one of the inelastic channels exactly, we shall attempt to represent the sum of their contributions by the imaginary part of an optical potential. The same idea may also be extended to scattering between hadrons in general, but we shall not do it here. Our primary concern will be with nucleon-nucleus scattering, and we shall return later for a brief discussion of applying the optical model for pion-nucleus scattering.

There are three aspects of the optical model potential we shall briefly describe in this section. First, we shall give a formal derivation to make a connection between optical model potential and averaging over contributions from a large number of intermediate states. Second, semi-empirical forms of the optical potential have been used over the years with great success for low-energy (< 200 MeV) scattering, and we shall give an example here to provide a feeling of the form of the optical potential and its dependence on incident energy and other variables. Third, we wish to make some contact to scattering at the nucleon-nucleon level to give a "microscopic" foundation to the optical model potential.

Formal derivation of the optical model potential. Consider the case of a free nucleon scattering off a nucleus made of A nucleons. Let $\boldsymbol{r}_0$ represent the coordinate of the projectile and $\boldsymbol{r}_i$, for $i = 1$ to A, represent the coordinates of the A nucleons in the target. In order to keep the notation simple, we shall suppress any reference to spin and other degrees of freedom. Our aim is to solve the Schrödinger equation,

$$H(\boldsymbol{r}_0; \boldsymbol{r}_1, \boldsymbol{r}_2, \ldots, \boldsymbol{r}_A)\Psi(\boldsymbol{r}_0; \boldsymbol{r}_1, \boldsymbol{r}_2, \ldots, \boldsymbol{r}_A) = E\Psi(\boldsymbol{r}_0; \boldsymbol{r}_1, \boldsymbol{r}_2, \ldots, \boldsymbol{r}_A) \tag{7-51}$$

under boundary conditions appropriate for scattering. As usual, it is impossible to solve exactly the many-body problem posted by (7-51) and we shall seek an approximate solution suitable for understanding the average results in a scattering.

For the time being we shall ignore the necessary antisymmetrization between the projectile and the nucleons in the target. The Hamiltonian for the complete system, consisting of the projectile and the target nucleus, may be separated into three parts,

$$H(\boldsymbol{r}_0;\boldsymbol{r}_1,\boldsymbol{r}_2,\ldots,\boldsymbol{r}_A)=T_0+\sum_{i=1}^{A}V(\boldsymbol{r}_{0i})+H_A(\boldsymbol{r}_1,\boldsymbol{r}_2,\ldots,\boldsymbol{r}_A) \tag{7-52}$$

where $\boldsymbol{r}_{0i}=\boldsymbol{r}_0-\boldsymbol{r}_i$ for $i\neq 0$ is the relative coordinate between the projectile and the i-th nucleon in the target. The operator T_0 describes the kinetic energy of the projectile and $H_A(\boldsymbol{r}_1,\boldsymbol{r}_2,\ldots,\boldsymbol{r}_A)$ is the Hamiltonian operating only among the nucleons in the target. The interaction between the projectile and the target nucleons is given by the potential $V_r(\boldsymbol{r}_{0i})$.

We shall assume that the eigenvalue problem for the target nucleus has already been solved and that a complete set of solutions $\{\Phi_i\}$ is available for the Schrödinger equation:

$$H_A(\boldsymbol{r}_1,\boldsymbol{r}_2,\ldots,\boldsymbol{r}_A)\Phi_i(\boldsymbol{r}_1,\boldsymbol{r}_2,\ldots,\boldsymbol{r}_A)=\epsilon_i\Phi_i(\boldsymbol{r}_1,\boldsymbol{r}_2,\ldots,\boldsymbol{r}_A) \tag{7-53}$$

Furthermore, we shall take that Φ_i is normalized to unity and is a part of an orthogonal set of eigenfunctions. The general solution for the complete system, including both the projectile particle and the target nucleus, may be expressed as a linear combination of products of $\chi_i(\boldsymbol{r}_0)$ and $\Phi_j(\boldsymbol{r}_1,\boldsymbol{r}_2,\ldots,\boldsymbol{r}_A)$,

$$\Psi(\boldsymbol{r}_0;\boldsymbol{r}_1,\boldsymbol{r}_2,\ldots,\boldsymbol{r}_A)=\sum_{ij}\chi_i(\boldsymbol{r}_0)\Phi_j(\boldsymbol{r}_1,\boldsymbol{r}_2,\ldots,\boldsymbol{r}_A) \tag{7-54}$$

where $\chi_i(\boldsymbol{r}_0)$ is the wave function of the projectile. Since our primary interest is in elastic scattering, the only part of Ψ that is of interest to us here is $\chi_0\Phi_0$, where both the target nucleus and the projectile are in their respective lowest energy states. Our problem here is to obtain χ_0 (Ψ_0 is assumed to be already known).

We shall first construct an equation for χ_0 using the method of projection operators. The approach is very similar to that used earlier in §6-8 to find a renormalized Hamiltonian in nuclear structure calculations, where the active space is reduced to a small subset of the complete shell model space. Let P be a projection operator for the ground state of the target. We may write P in the form

$$P=|\Phi_0\rangle\langle\Phi_0| \tag{7-55}$$

with the understanding that any integration to be carried out is over the coordinates of the target nucleons only. When P acts on the wave function of (7-54), we obtain the result,

$$P\Psi=\chi_0\Phi_0$$

We may also define an operator Q which projects out the rest of the states,

$$Q=1-P \tag{7-56}$$

It is easy to see that

$$P^2\Psi = P\Psi \qquad Q^2\Psi = Q\Psi \qquad PQ\Psi = QP\Psi = 0 \tag{7-57}$$

from the fact that P and Q are operators projecting out different parts of the complete space.

Since $P + Q = 1$, the Schrödinger equation (7-51) may be written in the form

$$(E - H)(P + Q)\Psi = 0 \tag{7-58}$$

On multiplying from the left by P and making use of the relations given in (7-57), we obtain $P\Psi$ in terms of $Q\Psi$,

$$(E - PHP)\,P\Psi = (PHQ)\,Q\Psi \tag{7-59}$$

Similarly, on multiplying (7-58) from the left by Q, we obtain

$$(E - QHQ)\,Q\Psi = (QHP)\,P\Psi \tag{7-60}$$

This result may be used to express $Q\Psi$ formally in terms of $P\Psi$,

$$Q\Psi = \frac{1}{E - QHQ}\,QHP\,P\Psi \tag{7-61}$$

When this form of $Q\Psi$ is substituted into the right hand side of (7-59), we obtain an expression for $P\Psi$,

$$\Big\{E - PHP - PHQ\frac{1}{E - QHQ}QHP\Big\}\,P\Psi = 0 \tag{7-62}$$

On multiplying this expression from the left by $\langle\Phi_0|$ and integrating over the coordinates of the target nucleons with the help of the explicit form of P given in (7-55), we arrive at an equation for χ_0,

$$\Big\{E - \langle\Phi_0|H|\Phi_0\rangle - \Big\langle\Phi_0\Big|HQ\frac{1}{E - QHQ}QH\Big|\Phi_0\Big\rangle\Big\}\,\chi_0 = 0 \tag{7-63}$$

This is the equation we must solve in order to obtain the unknown function χ_0.

The zero point of the energy scale is still arbitrary at this point and we may set it at the ground state of the target nucleus; that is, we let ϵ_0 in (7-53) equal 0, to simplify the form of (7-63). Using this definition, we obtain the result

$$H_A\Phi_0 = \epsilon_0\Phi_0 = 0$$

Eq. (7-63) can now be written in the form

$$\Big\{E - T_0 - \langle\Phi_0|V|\Phi_0\rangle - \Big\langle\Phi_0\Big|VQ\frac{1}{E - QHQ}QV\Big|\Phi_0\Big\rangle\Big\}\,\chi_0 = 0 \tag{7-64}$$

where V is the potential acting between the projectile and target nucleons. In arriving at the result we have made use of (7-52), and the fact that T_0 operates only on the projectile coordinates and therefore cannot act on Φ_0. This gives us the relation:

$$QT_0|\Phi_0\rangle = T_0Q|\Phi_0\rangle = 0$$

The operator $(E-QHQ)^{-1}$ in (7-64) is meaningful only in the sense of an infinite series expansion of the form:

$$\frac{1}{E-QHQ} = \frac{1}{E}\left\{1+\frac{1}{E}QHQ+\frac{1}{E}QHQ\frac{1}{E}QHQ+\cdots\right\} \tag{7-65}$$

The physical meaning of each term in this expansion may be interpreted in the following way. Each time the Hamiltonian acts between a pair of nucleons, there is an interaction or "scattering" between these two nucleons. The product (QHQ) implies that the interaction takes place with the target nucleus in one of its excited states and higher power terms such as $(E^{-1}QHQ)^n$ mean multiple interaction of order n. The last term of (7-64), therefore, contains multiple scattering to all orders weighted by energy factor E^{-1} to the appropriate powers.

Eq. (7-63) may be put in the familiar form of an eigenvalue equation,

$$\big(E - T_0 - \mathcal{V}(\boldsymbol{r}_0)\big)\chi_0 = 0 \tag{7-66}$$

where the equivalent potential has the form

$$\mathcal{V}(\boldsymbol{r}_0) = \langle\Phi_0|V|\Phi_0\rangle + \langle\Phi_0|VQ\frac{1}{E-QHQ}QV|\Phi_0\rangle \tag{7-67}$$

Since we have not yet made any approximation in arriving at (7-66), we do not have any better chance of solving it than the original equation given by (7-51). The aim of an optical model is to replace the equivalent potential $\mathcal{V}(\boldsymbol{r}_0)$ by an optical model potential U_{opt}, such that the equation

$$\big(E - T_0 - U_{\text{opt}}\big)\chi_0 = 0 \tag{7-68}$$

can be solved.

In general $\mathcal{V}(\boldsymbol{r}_0)$ is *nonlocal*; that is, the potential acting at one point of space may depend on the value of the wave function at a different point. The eigenvalue equation takes on the form

$$(E - T_0)\chi_0(\boldsymbol{r}_0) = \mathcal{V}(\boldsymbol{r}_0)\chi_0(\boldsymbol{r}_0) + \int f(\boldsymbol{r}_0,\boldsymbol{r}'_0)\chi_0(\boldsymbol{r}'_0)d\boldsymbol{r}'_0 \tag{7-69}$$

where $f(\boldsymbol{r}_0,\boldsymbol{r}'_0)$ is a function of both $\boldsymbol{r}_0$ and $\boldsymbol{r}'_0$. This greatly complicates the problem, and one may wish to approximate it with a local potential, often done in practice. Furthermore, the derivation here may have given the impression that all the scattering into the Q-space eventually returns the target to the ground state. This is certainly not true in general. In order to represent the loss of flux from the incident channel by scattering that ends up in other exit channels, the optical potential is usually complex.

Phenomenological optical model potential. The origin of optical model potential is the interaction between nucleons in the projectile with those in the target nucleus. It is, in principle, possible to calculate the potential from a nucleon-nucleon interaction. Before carrying out such a calculation, it is instructive to take a more phenomenological approach to the problem.

Based on the fact that the range of nuclear force is short, we expect the radial dependence of an optical model potential $U_{\text{opt}}(\boldsymbol{r})$ to follow closely the nucleon distribution in a nucleus. For simplicity, a two-parameter Fermi form given earlier in (7-39) may be used to describe such a distribution,

$$f(r, r_0, a) = \frac{1}{1 + \exp\{(r - r_0 A^{1/3})/a\}} \tag{7-70}$$

In optical model studies this is also known as the Woods-Saxon form. The optical model potential is a complex one in general,

$$U_{\text{vol}}(\boldsymbol{r}) = -\{V_0 f(r, r_v, a_v) + iW_0 f(r, r_w, a_w)\} \tag{7-71}$$

In a phenomenological approach V_0 and W_0, the depths of the real and imaginary parts of the potential well, may be taken as free parameters to be determined from experimental data. If we assume the radial dependence of both parts, $f(r, r_v, a_v)$ and $f(r, r_w, a_w)$, to follow the form given by (7-70), we may take the radii r_v and r_w and the surface diffuseness a_v and a_w also as adjustable parameters.

The potential given by (7-71) is usually referred to as the *volume term* of $U_{\text{opt}}(\boldsymbol{r})$, since the interaction follows the volume distribution of nucleons. In addition to this, optical potentials are also known to have a spin dependence. For example, when a nucleon is scattered from a nucleus, the result depends on the relative orientation of the nucleon spin before and after the scattering. One way to measure such a dependence is through the *analyzing power* parameter A_y. Let us define the transverse polarization of the incident and scattered nucleons as positive (or up) if they are oriented in the same direction as the unit vector

$$\hat{\boldsymbol{n}} = \frac{\boldsymbol{k} \times \boldsymbol{k}'}{|\boldsymbol{k}|\,|\boldsymbol{k}'|}$$

given in (C-6), and negative if they are aligned opposite to $\hat{\boldsymbol{n}}$. Writing the cross section for scattering from positive initial polarization to both positive and negative final polarization as σ_+ and from initial negative to both positive and negative final polarization as σ_-, the analyzing power is given by the ratio of their difference to their sum,

$$A_y = \frac{\sigma_+ - \sigma_-}{\sigma_+ + \sigma_-} \tag{7-72}$$

For elastic scattering, the same information can also be obtained by starting from an unpolarized incident beam and measuring the difference between the cross sections leading to positive and negative polarization for the scattered particle.

The result, normalized in the same way as (7-72) for the analyzing power, is called *polarization.*

The fact that A_y is nonzero in general in nucleon-nucleus scattering is strong evidence for the presence of spin dependence in the optical potential. A *spin-orbit* term may be used to represent such an effect,

$$U_{\text{s.o.}}(\boldsymbol{r}) = \boldsymbol{\sigma}\cdot\boldsymbol{\ell}\left(\frac{\hbar}{m_\pi c}\right)^2 \frac{1}{r}\left\{V_s\frac{d}{dr}f(r,r_{sv},a_{sv}) + iW_s\frac{d}{dr}f(r,r_{sw},a_{sw})\right\} \qquad (7\text{-}73)$$

Again there are six parameters, V_s, r_{sv}, a_{sv}, W_s, r_{sw}, and a_{sw} which may be adjusted to fit scattering data. Note that the value of $(\hbar/m_\pi c^2)^2$ is roughly 2; the approximate numerical value is often used in the place of the square of the pion Compton wave length.

The reason for the radial dependence of the spin-orbit term to be proportional to the derivative of the volume density distribution of nucleons comes from the analogy of the Thomas spin-orbit potential for an atomic electron in the Coulomb field of a nucleus. For an electron, the spin-orbit term comes from the interaction of its intrinsic magnetic dipole moment $\boldsymbol{\mu}_e$ and the magnetic field $\boldsymbol{B}(\boldsymbol{r})$ it feels due to its own orbital motion around the nucleus. The value of $\boldsymbol{B}(\boldsymbol{r})$ may be found by a Lorentz transformation of the electrostatic field $\boldsymbol{E}(\boldsymbol{r})$ provided by the nucleus, stationary in the laboratory, into a frame of reference at rest with the orbiting electron. The result is

$$\boldsymbol{B}(\boldsymbol{r}) = -\frac{1}{c[c]}\boldsymbol{v}\times\boldsymbol{E}(\boldsymbol{r}) = -\frac{\hbar}{m_e c r[c]}\frac{dV}{dr}\boldsymbol{\ell}$$

where the factor inside the square brackets is needed to convert the expression from cgs to SI units. The orbital angular momentum of the electron is given by

$$\boldsymbol{\ell}\hbar = \boldsymbol{r}\times\boldsymbol{p} = m_e(\boldsymbol{r}\times\boldsymbol{v})$$

Furthermore, in order to relate to an electrostatic potential $V(\boldsymbol{r})$, we have made use of the relation

$$\boldsymbol{E}(\boldsymbol{r}) = -\frac{dV}{dr}\frac{\boldsymbol{r}}{r}$$

The spin-orbit interaction energy for an atomic electron is then given by

$$W(\boldsymbol{r}) = -\boldsymbol{\mu}_e\cdot\boldsymbol{B}(\boldsymbol{r}) = \frac{e\hbar^2}{m_e^2c^2r}\frac{dV}{dr}\boldsymbol{\sigma}\cdot\boldsymbol{\ell}$$

using the relation $\boldsymbol{\mu}_e = (e\hbar[c]/m_e c)\,\boldsymbol{\sigma}$. Eq. (7-73) has the same form; the use of pion mass instead of that of electron may be regarded as a convention for the definition of the spin-obit potential well depths V_s and W_s. For the same reason,

Table 7-1 Proton-nucleus scattering optical potential parameters for $40 \leq A \leq 208$ and proton laboratory energy $80 \leq T_p \leq 180$ MeV.

$$
\begin{aligned}
V_0 = \quad & 105.5(1 - 0.1625 \ln T_p) + 16.5\frac{N-Z}{A} \\
& r_v = \begin{cases} 1.125 + 0.001T_p & \text{for } T_p \leq 130 \text{ MeV} \\ & (\text{except } T_p \leq 180 \text{ MeV for Ca}) \\ 1.255 & \text{for } T_p > 130 \text{ MeV (except Ca)} \end{cases} \\
& a_v = 0.675 + 3.1 \times 10^{-4} T_p \\
W_0 = \quad & 6.6 + 0.0273(T_p - 80) + 3.87 \times 10^{-6}(T_p - 80)^3 \\
& r_w = 1.65 - 0.0024T_p \\
& a_w = 0.32 + 0.0025T_p \\
V_{\text{s.o.}} = \quad & 19.0(1 - 0.166 \ln T_p) - 3.75\frac{N-Z}{A} \\
& r_{sv} = 0.920 + 0.0305A^{1/3} \\
& a_{sv} = \begin{cases} 0.768 - 0.0012T_p & \text{for } T_p \leq 140 \text{ MeV} \\ 0.60 & \text{for } T_p > 140 \text{ MeV} \end{cases} \\
W_{\text{s.o}} = \quad & 7.5(1 - 0.248 \ln T_p) \\
& r_{sw} = 0.877 + 0.0360A^{1/3} \\
& a_{sw} = 0.62
\end{aligned}
$$

V_0, W_0, $V_{\text{s.o.}}$, $W_{\text{s.o.}}$ and T_p are in MeV
r_v, a_v, r_w, a_w, r_{sv}, a_{sv}, r_{sw} and a_{sw} are in fm.

the electron charge is inappropriate and is also absorbed into the definition of the well depths.

For charged particle scattering, a Coulomb term may be included in the optical model potential. The form is usually obtained by approximating the target nucleus as a uniformly charged sphere,

$$
U_C(r) = \begin{cases} \left[\frac{1}{4\pi\epsilon_0}\right] \frac{zZe^2}{2R_c}\left(3 - \frac{r^2}{R_c^2}\right) & \text{for } r < R_c \\ \left[\frac{1}{4\pi\epsilon_0}\right] \frac{zZe^2}{r} & \text{for } r \geq R_c \end{cases} \tag{7-74}
$$

where R_c is the Coulomb radius. The quantities z and Z are, respectively, the charge number of the projectile and the target nucleus. It may be tempting to treat R_c as a free parameter also. In practice, the scattering results are not sensitive to the details of the Coulomb potential, and it is quite adequate to use the value $R_c = 1.2A^{1/3}$ fm.

The complete phenomenological optical model potential is then the sum of volume, spin-orbit, and Coulomb terms;

$$
U_{\text{opt}}(\boldsymbol{r}) = U_{\text{vol}}(\boldsymbol{r}) + U_{\text{s.o.}}(\boldsymbol{r}) + U_C(\boldsymbol{r}) \tag{7-75}
$$

The total number of adjustable parameters is twelve if we do not include R_c as one. In a typical scattering experiment, the angular distribution of the differential scattering cross section, as well as the analyzing power where possible, are observed. The number of independent pieces of data is usually greater than twelve, and there is no difficulty obtaining a complete set of the parameters by fitting the calculated results to the measured quantities. A large amount of information has been accumulated in this way on the phenomenological optical potential, and we have now a very clear picture of the energy and mass dependence of these parameters for proton scattering off nuclei up to a laboratory energy of 200 MeV. One of the possible sets of the parameters obtained by fitting proton scattering data on a variety of nuclei from $A = 40$ to $A = 208$ and laboratory proton energy from 80 to 180 MeV is given in Table 7-1 (P. Schwandt et al., *Phys. Rev. C* **26** [1982] 55).

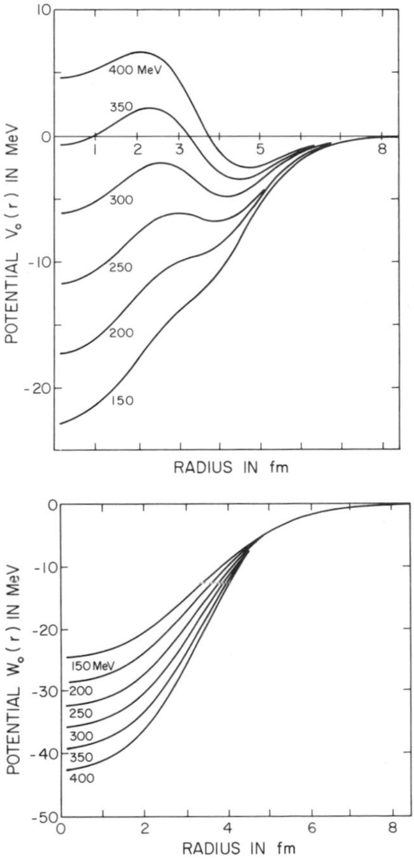

Fig. 7-5 Radial shapes of the volume term of a proton-nucleus optical potential at different bombarding energies. The upper curves are the real part and the lower one the imaginary part of the potential. The results are calculated using a folding procedure with a Paris potential as the interaction between nucleons. (Adapted from H.V. von Geramb, in *Studying Nuclei with Medium Energy Protons,* TRIUMF report [1983] TRI-83-3.)

There are, however, several problems associated with the phenomenological approach. The first is that, although we have a fairly detailed picture for proton scattering, this knowledge does not extend to other projectiles. For example, even the neutron optical potential is not as well known since there are far fewer experimental data available for neutron-nucleus scattering. Because of its phenomenological nature, the approach does not lend itself easily to extrapolation to regions where experimental data are scarce. The second is that the forms of radial dependence used in (7-71) and (7-73) are found to be inadequate as we go to higher bombarding energies. One remedy is to use more complicated expressions involving additional parameters; however on aesthetic grounds alone, this is not desirable. Finally, the parametrization is not unique; other ways of characterizing the potential can also give equally good fits to the data. Furthermore, the twelve parameters are interdependent in a complicated way and there are other sets of values which can often provide an equally good description of the data.

Microscopic optical model potential. Since an optical model potential for nucleon-nucleus scattering is the result of the average interaction between the incident nucleon and nucleons in the target nucleus, it is a function of the nucleon-nucleon interaction weighted by the nucleon distribution inside the target nucleus. A microscopic model of the potential may therefore be built on this idea of convoluting the fundamental nucleon-nucleon interaction with the nuclear density. Such a *folding* model has been known to be quite successful in describing nucleon scattering data as long as an appropriate nucleon-nucleon interaction is used as the starting point. For simplicity we shall consider the incident particle to be a nucleon and ignore any internal structure it may have. The first term of (7-67) suggests that we may be able to approximate the nucleon-nucleus optical model potential by the integral

$$U_{\rm opt}(\boldsymbol{r}_0) \approx \big\langle \Phi_0(\boldsymbol{r}_1, \boldsymbol{r}_2, \ldots, \boldsymbol{r}_A) \big| \sum_{i=1}^{A} V(\boldsymbol{r}_{0i}) \big| \Phi_0(\boldsymbol{r}_1, \boldsymbol{r}_2, \ldots, \boldsymbol{r}_A) \big\rangle \tag{7-76}$$

where the integration is taken only over the coordinates of the target nucleons.

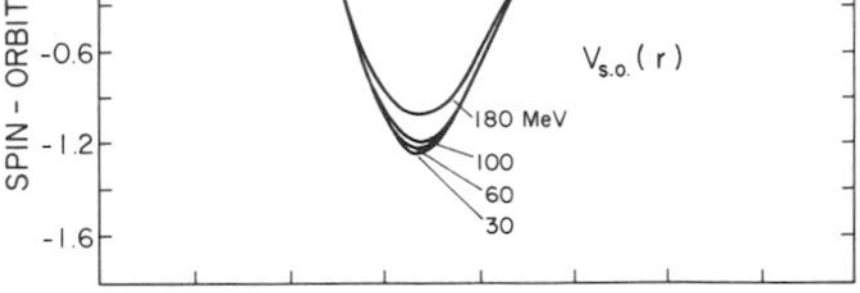

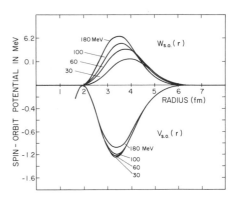

Fig. 7-6 Radial shapes of the real and imaginary parts of the spin-orbit term of an optical potential for proton-nucleus scattering at different bombarding energies. The calculation is based on a folding procedure using a Paris potential for the interaction between nucleons. (Adapted from W. Bauhoff, H.V. von Geramb and G. Palla, *Phys. Rev. C* **27** [1983] 2466.)

One must be careful here with antisymmetrization between the incident nucleon and the nucleon in the target nucleus with which it interacts. Let us consider the simplest case in which the incident nucleon interacts only once with one of the nucleons in the target. When a nucleon emerges from the scattering, there is no way to identify whether the observed particle is the same one as the incident nucleon or the nucleon in the target nucleus with which it interacted. Both of these possibilities must be included. For this reason, the matrix element on the right hand side of (7-76) should be a sum of two terms,

$$\langle\Phi_0|V|\Phi_0\rangle = \langle\Phi_0|t_D|\Phi_0\rangle + \langle\Phi_0|t_E|\Phi_0\rangle \tag{7-77}$$

where t_D is the operator for the direct part of the interaction in which the scattered nucleon is the same one as the incident particle, and t_E is the operator for the exchange part in which the incident nucleon is absorbed by the target nucleus and the scattered particle is one of the nucleons that was originally in the target. We shall see later that the difference between the contributions from these two terms is important in understanding the radial shape of the optical potential at high scattering energies.

For the nucleon-nucleon potential $V(\boldsymbol{r}_{0i})$, one could take a naive approach and replace it by a free nucleon-nucleon interaction obtained, for example, from nucleon-nucleon scattering. This is known as the *impulse approximation* (IA) and, in practice, is found to be too crude an approximation to fit experimental data on nucleon-nucleus scattering. Just as with nuclear shell model calculations, an effective nucleon-nucleon interaction is required here, since one of the two interacting nucleons is imbedded in the nucleus. The requirements on the effective interaction are, however, somewhat less stringent than in the shell model case. It is usually possible to approximate the nuclear medium here as an infinite nuclear matter; this greatly simplifies the calculation. A finite nucleus, however, has a large surface region where the density varies from very small values up to the saturation value in infinite nuclear matter. To account for this variation, a *density-dependent* effective potential is often used. In other words, the operators t_D and t_E in (7-77) are also functions of the nuclear density ρ. The effective potential in different density regions is obtained from a calculation using nuclear matter of the same density. Furthermore, the free nucleon-nucleon interaction itself is energy-dependent and, as a result, both t_D and t_E are functions of the bombarding energy as well.

In terms of single-particle wave functions $\phi_i(\boldsymbol{r}_i)$ of target nucleons, the density of target nucleus may be expressed in the following operator form,

$$\rho(\boldsymbol{r}) = |\Phi_0\rangle\langle\Phi_0| \approx \sum_{i=1}^{A} \phi_i^*(\boldsymbol{r})\phi_i(\boldsymbol{r}) \tag{7-78}$$

Using this, the optical model potential may be related to an integral over a function

$$f(\boldsymbol{r}_0, \boldsymbol{r}) = \sum_{i=1}^{A} \phi_i^*(\boldsymbol{r})t_D(\boldsymbol{r}_0, \boldsymbol{r}, \rho, E)\phi_i(\boldsymbol{r}) + \sum_{i=1}^{A} \phi_i^*(\boldsymbol{r})t_E(\boldsymbol{r}, \boldsymbol{r}_0, \rho, E)\phi_i(\boldsymbol{r}_0) \tag{7-79}$$

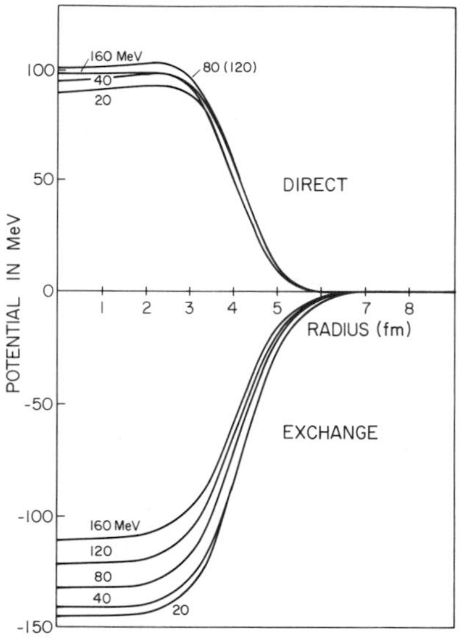

Fig. 7-7 Contributions from the direct and exchange terms to $V_0(r)$, the real part of the volume term of a proton-nucleus optical potential, at different bombarding energies. The upper curves are the results for the direct part of a microscopic potential given by (7-80) and the lower curves are the exchange part given by (7-81). The nucleon-nucleon interaction used in this microscopic calculation is taken from a Paris potential. (Taken from H.V. von Geramb, F.A. Brieva, and J.R. Rook, in *Microscopic Optical Potentials,* Lecture Notes in Phys. **89**, ed. by H.V. von Geramb, Springer-Verlag, Berlin, 1979.)

constructed by convoluting or "folding" the nucleon-nucleon interaction between the incident and the target nucleons with the nuclear density. In order to emphasize the dependence on nuclear density and bombarding energy, we have put ρ and E explicitly into the arguments of t_D and t_E.

The direct term of the optical model potential is relatively easy to evaluate, since the integral for the first term of (7-79) may be expressed in the form

$$U^D_{\text{opt}}(\boldsymbol{r}_0, E) = \int \rho(\boldsymbol{r}) t_D(\boldsymbol{r}_0, \boldsymbol{r}, \rho, E)\, d^3r \tag{7-80}$$

using the approximate form of $\rho(\boldsymbol{r})$ given in (7-78). On the other hand, the same transformation cannot be carried out for the exchange term, as the two single-particle wave functions in the second term of (7-79) are functions of two different coordinates, $\boldsymbol{r}$ and $\boldsymbol{r}_0$. As a result, the contribution of the exchange term is nonlocal in general. A "local momentum" approximation is usually used to reduce the exchange term to a simpler form,

$$U^E_{\text{opt}}(\boldsymbol{r}_0, E) \approx \int \rho(\boldsymbol{r}_0, \boldsymbol{r}) t_E(\boldsymbol{r}, \boldsymbol{r}_0, \rho, E) j_0(k|\boldsymbol{r}_0 - \boldsymbol{r}|) d^3r \tag{7-81}$$

where

$$\rho(\boldsymbol{r}_0, \boldsymbol{r}) = \sum_{i=1}^{A} \phi_i^*(\boldsymbol{r}) \phi_i(\boldsymbol{r}_0)$$

in analogy with (7-78). Here, $j_0(\xi)$ is the spherical Bessel function of order zero.

At laboratory energy below 200 MeV, the folding potential produces results very similar to those derived from phenomenological potentials. At higher energies, however, the Woods-Saxon radial shape used in the semi-empirical approach is known to be inadequate. From the folding potential calculation we find that, as the bombarding energy is increased, the radial shape of the volume term of the optical model potential changes to a "wine-bottle" shape shown in Fig. 7-5. [The radial shape for the spin-orbit potential, shown in Fig. 7-6, however, retains essentially the same form as given by (7-73)]. The shape change comes from the differences in the energy dependence of the direct and exchange parts of the folding potential. As shown in Fig. 7-7, the direct part is repulsive and the exchange part attractive. Since the exchange part has a slightly sharper energy dependence than the direct part, the cancellation between the repulsive and attractive potential, when we sum the two terms to produce $U_{\rm opt}$, gives an energy dependence that cannot be represented by a simple Woods-Saxon form. In Fig. 7-8 the results for the elastic scattering of 362 MeV protons from ^{40}Ca are given as an illustrative example to show that a microscopic optical potential is quite capable of describing intermediate energy proton-nucleus scattering to very large momentum transfers.

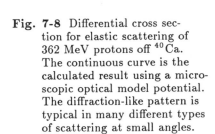

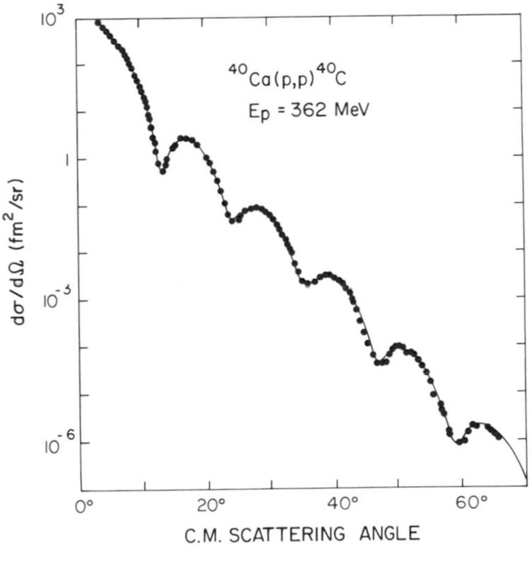

Fig. 7-8 Differential cross section for elastic scattering of 362 MeV protons off ^{40}Ca. The continuous curve is the calculated result using a microscopic optical model potential. The diffraction-like pattern is typical in many different types of scattering at small angles.

Besides elastic scattering, the optical potential is also important in understanding the cross section for other types of reactions. For example, in cases where the scattering is dominated by direct reactions, contributions from potential scattering and multipole scattering to the same final state form the "background" and are represented by the optical model potential. Furthermore, incident flux removed as a result of the imaginary part of the optical potential has an effect on the result of a direct reaction, since the net amount of available incident flux is reduced by other open reaction channels.

From a wave mechanics point of view, we can say that the optical potential is an average potential that "distorts" both the incident and scattered waves from their plane wave states. The effect of direct reactions to specific states may be regarded as terms in addition to scattering due to the optical potential. These contributions favor only specific final states. The essence of distorted wave Born approximation (DWBA) to direct reaction is to separate the scattering into two parts, a background given by the optical model potential and a direct reaction contribution because of the special structure of the final state involved. The calculated results obtained this way, as we shall see in the next section, have been found to give a fairly good description of the observed cross sections.

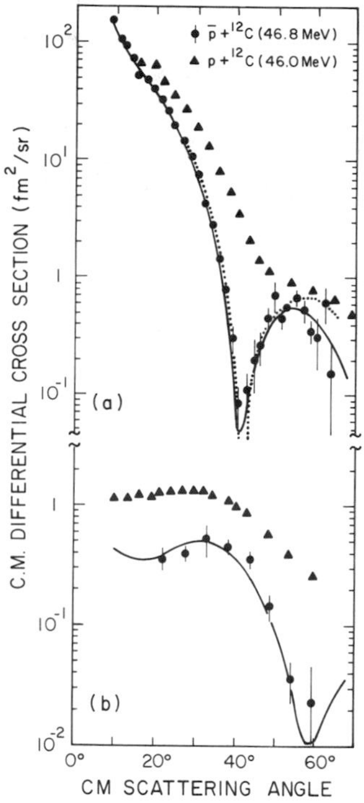

Fig. 7-9 Differential cross section for proton (triangles) and antiproton (circles) scattering off ^{12}C at 46 MeV incident energy. Figure (*a*) is for elastic scattering and (*b*) for inelastic scattering leading to the 2^+-state at 4.44 MeV excitation. The solid curves are obtained from a coupled channel calculation and the dotted curve is a theoretical calculation for elastic scattering. (Taken from Garreta et al., *Phys. Lett.* **135B** [1984] 266.)

§7-5 INTERMEDIATE ENERGY NUCLEON SCATTERING

In §7-3 we saw that direct reactions are good tools for investigating certain aspects of nuclear structure, as well as the interaction between free and bound nucleons. The main reason for such possibilities is that the reaction mechanism is particularly simple. This is especially true in the case of nucleon-nucleus scattering, since the internal degrees of the incident particle can usually be ignored. The condition for direct reaction is at an optimum when the incident energy is in the range of intermediate energies.

Intermediate energy nucleons are usually taken as meaning nucleons with laboratory kinetic energy in the range of 100 to 1000 MeV. At energies much lower than these values, the transit time of a nucleon through a nucleus is long enough that multiple scattering may happen frequently enough to complicate the reaction. At energies much higher than these values, good energy resolution is difficult to achieve. At the same time, the increased production rates of pions and other particles make the condition unfavorable for studying nucleon-nucleus interaction.

We shall again restrict ourselves to reactions involving two-body final states to simplify the analysis. Our main emphasis will be on proton inelastic scattering, commonly represented as (p, p') reactions, and charge exchange reactions induced by nucleons, namely, (p, n) and (n, p) reactions. Furthermore, we shall ignore elastic proton scattering here since some of the primary interests were covered in the previous section in the discussion of optical models. Besides scattering cross sections, observables related to changes in nucleon spin orientation can also be measured, as we have seen in §3-7; however, for simplicity, we shall not discuss them here. Very interesting data can also be obtained by scattering antiprotons from nuclear targets. This information helps us to understand the connection between internal degrees of freedom of nucleons and nucleon-nucleus reaction. An example of the differential cross section for scattering off ^{12}C is shown in Fig. 7-9. Unfortunately, a meaningful discussion of the topic of antiproton-nucleus scattering requires preparations that are outside the scope of this book.

Scattering amplitude. We have seen in (7-39) that scattering amplitudes in the Born approximation may be expressed in terms of the matrix elements of the nucleon-nucleon interaction potential between initial and final states of the nucleon-nucleus system,

$$f(\theta) = -\frac{\mu}{2\pi\hbar^2}\langle \chi_{k_f}(\boldsymbol{r}_0)\Phi_f(\boldsymbol{r}_1, \boldsymbol{r}_2, \ldots, \boldsymbol{r}_A)\big|\sum_{i=1}^{A} V(\boldsymbol{r}_{0i})\big|\chi_{k_i}(\boldsymbol{r}_0)\Phi_i(\boldsymbol{r}_1, \boldsymbol{r}_2, \ldots, \boldsymbol{r}_A)\rangle \tag{7-82}$$

where μ is the reduced mass of the scattering nucleon, and $\chi_{k_i}(\boldsymbol{r}_0)$ and $\chi_{k_f}(\boldsymbol{r}_0)$ are, respectively, the wave functions of the incident and scattered nucleons in the Born approximation. The wave functions, $\Phi_i(\boldsymbol{r}_1, \boldsymbol{r}_2, \ldots, \boldsymbol{r}_A)$ and $\Phi_f(\boldsymbol{r}_1, \boldsymbol{r}_2, \ldots, \boldsymbol{r}_A)$, describe the initial and final nuclear states.

There are three distinctive parts entering into a calculation of the scattering amplitude $f(\theta)$. First, we need an optical model potential from which we can calculate functions $\chi_{k_i}(\boldsymbol{r}_0)$ and $\chi_{k_f}(\boldsymbol{r}_0)$ so that all effects of the reaction other than those due to direct reaction mechanisms are accounted for on the average. This is the spirit of distorted wave approach mentioned in the previous section.

Second, we need a potential $V(\boldsymbol{r}_{0i})$ that supplies the interaction between a free nucleon and a nucleon imbedded in a nucleus. It is this potential that provides the direct reaction over and above the "background" produced by the optical model potential. Third, we need both the initial and final nuclear wave functions, Φ_i and Φ_f, in particular the relationship between them. All three parts are related to the fundamental nucleon-nucleon interaction.

Let us start with the purely nuclear structure problem of relating the nuclear final state to the initial state of the target nucleus. For simplicity we shall restrict ourselves to targets made of even-even nuclei where most of the studies have been carried out. For such targets, the spin and parity of the ground state of the initial nucleus are 0^+. The angular momentum transferred to the nucleus as a result of the scattering is then given by the spin and parity of the final nuclear state. Our basic assumption in (7-82) for the scattering amplitude is that the incident nucleon interacts with only one of the nucleons in the target nucleus. In the case of a (p, p') reaction, the process may be thought of as one in which the incident proton excites the target nucleus by promoting one of the nucleons to a higher single-particle state. In the case of a (p, n) reaction, the incident proton is captured and one of the neutrons in the target is ejected in the process. In either case, states that are strongly excited by the reactions are those made predominantly of one-particle-one-hole ($1p1h$) excitations built upon the ground state of the target nucleus.

The relation between initial and final nuclear wave functions may be expressed in terms of a "transition density." In the form of an operator, the $1p1h$-transition density may be written as

$$\boldsymbol{\rho}_{\rm tr}(1p1h) = \sum_{ph} a_{ph} |\phi_p\rangle\langle\phi_h| \tag{7-83}$$

where $\phi_h(\boldsymbol{r})$ is one of the occupied single-particle states in the ground state of the target nucleus and $\phi_p(\boldsymbol{r})$ is a single-particle state that is empty before the scattering. In principle, we should also couple $|\phi_p\rangle\langle\phi_h|$ to some definite spin and isospin so that the operator $\boldsymbol{\rho}_{\rm tr}(1p1h)$ is a spherical tensor of definite ranks. However, we shall dispense with this complication in the following discussion to simplify the argument.

If a state $|(1p1h)\, J^\pi\rangle$ is made up entirely of a linear combination of $1p1h$-excitations built upon the ground state, we can impose the normalization condition,

$$\Big|\langle(1p1h)\, J^\pi|\boldsymbol{\rho}_{\rm tr}(1p1h)|\text{ground state}\rangle\Big|^2 = 1$$

on the transition density operator. In this way, the state $|(1p1h)\, J^\pi\rangle$ can be expressed in terms of the transition density operator $\boldsymbol{\rho}_{\rm tr}(1p1h)$ acting on the ground state wave function,

$$\big|(1p1h)\, J^\pi\big\rangle = \boldsymbol{\rho}_{\rm tr}(1p1h)\big|\text{ground state}\big\rangle = \sum_{ph} a_{ph}|\phi_p\rangle\big\langle\phi_h\big|\text{ground state}\big\rangle$$

As we have seen in §6-8, such a $1p1h$-state is an eigenstate of the Hamiltonian only in the limit that two-body residual interaction can be ignored.

In general, an eigenvector of the nuclear Hamiltonian contains other components as well; the wave function of the final nuclear state may be expressed in the form:

$$|\Phi_f(\boldsymbol{r}_1, \boldsymbol{r}_2, \ldots, \boldsymbol{r}_A)\rangle = \sum_{ph} a_{ph} \sum_{i=1}^{A} |\phi_p(\boldsymbol{r}_i)\rangle\langle\phi_h(\boldsymbol{r}_i)|\Phi_i(\boldsymbol{r}_1, \boldsymbol{r}_2, \ldots, \boldsymbol{r}_A)\rangle$$
$$+ \text{ other components} \tag{7-84}$$

where Φ_i and Φ_f are, respectively, the initial and final nuclear wave functions. From this, we obtain the expansion coefficients a_{ph} of the $1p1h$-transition operator in the following form:

$$a_{ph} = \sum_{i=1}^{A} \langle\Phi_f(\boldsymbol{r}_1, \boldsymbol{r}_2, \ldots, \boldsymbol{r}_A)|\phi_p(\boldsymbol{r}_i)\rangle\langle\phi_h(\boldsymbol{r}_i)|\Phi_i(\boldsymbol{r}_1, \boldsymbol{r}_2, \ldots, \boldsymbol{r}_A)\rangle \tag{7-85}$$

The transition density is quantity between two specific nuclear states and, as such, it is independent of the probe and the reaction mechanism. For this reason, the same transition density enters into all other $1p1h$-excitation processes between the same pair of nuclear states. This gives us the opportunity to check the transition density in a (p, p') reaction against, for example, electromagnetic transitions and inelastic electron scattering. For (p, n) and (n, p) reactions, the transition densities are related to β-decay rates and scattering cross sections of charge exchange reactions induced by other probes such as pions and light nuclei.

So far we have considered the scattered nucleon to be one and the same as the incident nucleon and distinguishable from the nucleons in the target. As we have seen in the previous section, this is only the direct part of the scattering amplitude which, with the help of (7-84), may be expressed in the form

$$f_D(\theta) = -\frac{\mu}{2\pi\hbar^2} \sum_{ph} a_{ph} \langle\chi_{k_f}(\boldsymbol{r}_0)\phi_p(\boldsymbol{r})|V(\boldsymbol{r}_0, \boldsymbol{r})|\chi_{k_i}(\boldsymbol{r}_0)\phi_h(\boldsymbol{r})\rangle \tag{7-86}$$

The result is obtained after integrating over the coordinates of all other nucleons in the target not involved in this particular scattering. Components other than those related to one-particle-one-hole excitation of the target ground state disappear from the expression, since they do not contribute to the direct reaction amplitude in the limit that only $1p1h$-excitations are allowed. Their importance derives mainly from their total weight in Φ_f and, consequently, the fraction of $1p1h$-components present in the state and the overall size of the scattering amplitude $f_D(\theta)$.

To ensure proper antisymmetrization, we must also include an exchange part to the scattering amplitude which, in analogy to (7-86), may be written in the form

$$f_E(\theta) = -\frac{\mu}{2\pi\hbar^2}\sum_{ph} a_{ph}\langle\chi_{k_f}(\boldsymbol{r}_0)\phi_p(\boldsymbol{r})\big|V(\boldsymbol{r}_0,\boldsymbol{r})\big|\chi_{k_i}(\boldsymbol{r})\phi_h(\boldsymbol{r}_0)\rangle \tag{7-87}$$

Both $f_D(\theta)$ and $f_E(\theta)$ are now two-body matrix elements involving either the incident nucleon or the scattered nucleon and one of the nucleons in the nucleus. The reason we can reduce the amplitude to such a simple form comes from the direct reaction assumption that only a single interaction takes place between the incident nucleon and one of the nucleons in the target; the rest of the nucleons are merely "spectators" in the reaction.

Nucleon-nucleus interaction potential. What is the appropriate interaction potential $V(\boldsymbol{r}_0,\boldsymbol{r})$ to be used in (7-86,87) for the scattering amplitude? The simplest approach is to use an impulse approximation and equate the interaction with one occurring between free nucleons. As we have seen earlier in optical model potential calculations, this turns out to be too crude an assumption because of the influence of the nuclear medium on the target nucleons. For a semi-empirical approach, we can take a phenomenological one-boson exchange potential consisting of a sum of Yukawa forms, each with a different range to simulate the mass of different mesons exchanged. The strength of each term in such a potential may be taken as an adjustable parameter to reflect the fact that we do not have a complete knowledge of the interaction between free and bound nucleons. An example of such a potential is the Michigan three-Yukawa (M3Y) potential (Bertsch et al., *Nucl. Phys.* **A284** [1977] 399). With a sum of only three Yukawa terms, fairly good descriptions have been provided for the observed different cross sections of many (p, p') reactions. A more realistic approach is to use an effective interaction based on free nucleon-nucleon scattering with corrections for the influence of the nuclear medium, for instance, as developed by Franey and Love (*Phys. Rev. C* **31** [1985] 488). Alternatively one can take the nuclear matter approach and develop a density-dependent potential as described earlier for optical model potential studies. In both cases, good description of the observed results up to very large momentum transfers have been obtained for both differential scattering cross sections and spin observables.

Let us recapture what is happening when an intermediate energy nucleon is scattered off a nucleus. Before the incident nucleon is within the range to interact with one of the nucleons in the nucleus, it is in the field of the nuclear optical potential. The wave function of the incident nucleon is, therefore, modified by the average potential before an interaction takes place directly between the projectile and a nucleon in the target. The result of the interaction promotes one of the target nucleons to a different single-particle state and the scattered nucleon leaves the target, travelling again through the field of the optical potential. The three parts of a calculation — optical potential, nucleon-nucleon interaction, and nuclear wave functions — are three distinct parts of the problem and may be treated quite independently of each other. On the other hand, all three parts are the result of interaction between nucleons and can, in principle, be calculated from the same nucleon-nucleon interaction potential. It is therefore possible to solve the problem in a self-consistent manner and obtain all three parts from a given nucleon-nucleon potential. This is an interesting development, since there are only rare occasions in many-body problems that such an approach can actually be carried out in practice. Because of this possibility, a large amount of work, both experimental and theoretical, has been done in recent years studying intermediate nucleon-nucleus scattering.

Relativistic and other effects. In addition to the interests described above, intermediate energy nucleon-nucleus scattering is also used to understand the underlying reaction mechanism. For example, we have implicitly assumed above a nonrelativistic Schrödinger approach. However, the kinetic energy of the incident nucleon here is a large fraction of its rest mass energy and, as a result, relativistic effects must be important. Besides simple kinematic effects, which require Lorentz invariance in the place of Galilean invariance, we may need also to replace the Schrödinger equation with a Dirac equation. The main difference here may be viewed in the following way. In the Schrödinger approach, the nucleon, being a spin-$\frac{1}{2}$ particle, is described by a two-component wave function to account for the fact that the intrinsic spin of a nucleon can either point up (projection along the quantization axis $+\frac{1}{2}$) or down (projection $-\frac{1}{2}$). In a relativistic, quantum mechanical treatment, a four-component wave function is required to describe a spin-$\frac{1}{2}$ particle, with the upper two components describing the two possible directions of the nucleon spin and the lower two components accounting for the two possible directions of the antinucleon spin. A fully relativistic treatment of the nucleon-nucleus scattering therefore differs from the Schrödinger approach by the presence of the two lower components.

At low energies, the influence of the lower components on the behavior of the nucleon is very small and may be replaced by spin-dependent terms in the potential, as we have done earlier. At higher energies, such a simple substitution may not be adequate and it becomes essential to solve the Dirac equation for the many-body scattering problem. There are indications that for certain observables in intermediate energy nucleon-nucleus scattering, particularly those related to changes in the polarization direction between incident and scattered nucleons, a relativistic treatment is needed (see, *e.g.*, L.S. Celenza and C.M. Shakin, *Relativistic Nuclear Physics*, World Scientific, Singapore, 1986; and M. Danos, V. Gillet, and M. Cauvin, *Methods in Relativistic Nuclear Physics*, North-Holland, Amsterdam, 1984; and references therein). Unfortunately, we do not have as much experience in handling relativistic many-body problems compared with the knowledge accumulated in nonrelativistic approaches. As a result, it may be some time before a clear picture can emerge on the significance of relativistic effects in intermediate energy nucleon-nucleus scattering.

One of the interests in charge exchange reactions is to relate strong and weak interaction processes, as mentioned earlier in §5-5. Intermediate energy (p, n) and (n, p) reactions are ideal here, since the reaction mechanism is sufficiently simple and the nuclear matrix elements involved are the same as in nuclear β-decay. Apart from kinematic factors, the only difference between nuclear β-decay and charge exchange reactions induced by intermediate energy nucleons is the difference in their "coupling" constants. If this is so, the ratio between these two processes should be independent of the target nucleus used, and this indeed is found to be the case. As a result, sum rule and giant resonance studies have been extended into charge exchange processes, as we have seen earlier in §6-2.

Alternatively, intermediate energy nucleon-nucleus scattering can be viewed as a good way to obtain information on the interaction between free and bound nucleons. This is made possible by the fact that two of three ingredients in a reaction calculation may be checked by other means. For example, we have seen that the optical model potential is the same one as that entering into an elastic scattering. From the success in describing elastic scattering, we can establish the validity of an optical potential for using in either (p, p') or charge exchange reactions. We have also seen that the nuclear structure problem involved in the scattering process is identical to that which occurs in many other processes. By comparing the transition density with, for example, intermediate energy inelastic electron scattering, we have a fairly reliable way to find out whether the correct nuclear structure information is used. The net result is that the interaction $V(\boldsymbol{r}, \boldsymbol{r}_0)$ in (7-86,87) becomes the least well known part of the three ingredients and may therefore be regarded as the primary aim of an investigation. Furthermore, different transitions are sensitive to different parts of the interaction potential. By carefully selecting the initial and final states, it is possible to emphasize a particular part of $V(\boldsymbol{r}, \boldsymbol{r}_0)$ for examination.

Finally, if we are confident of all three aspects above, we may start to ask the finer and more detailed question of whether there are exotic effects related to the internal degrees of freedom of nucleons in nucleon-nucleus scattering. The energy involved here is certainly high enough, for example, that intermediate states involving the excitation of a nucleon into a Δ-particle can take place, particularly in view of the strong P_{33}-resonance in the pion-nucleon channel. Since a Δ is a distinguishable particle from a nucleon, it does not suffer from the Pauli exclusion principle due to the presence of other nucleons. Instead of particle-hole excitations, we can imagine Δ-hole excitations as taking place. There is already some evidence that such non-nucleonic degrees of freedom may be present in the observed data. However, the sizes of such effects are still below accuracies that we can calculate contributions from conventional nuclear effects, and no firm conclusion can be drawn yet.

High-energy nuclear physics. Experimentally, proton beams are available up to extremely high energies, measured in TeV (10^{12} eV) units. One of the difficulties of making use of such high energies is the multitude of particles produced when the protons are scattered off a target. However, if the energy is sufficiently high, say in excess of 100 GeV, production of J/ψ-particles becomes significant. As we have seen earlier in §2-6, J/ψ-particles are mesons made of a charm quark and a charm antiquark. They have relatively long half-lives, represented by very narrow widths for such high energy events. This makes the identification of J/ψ particles a relatively simple task at such energies. As a result, their production rates may be used as a test for some of the physics underlying the scattering of high-energy protons.

For protons with kinetic energy in excess of 100 GeV (or more properly, with momentum in excess of 100 GeV/c), the de Broglie wave length is shorter than 10^{-2} fm, far less than the dimension of a nucleon (radius $\sim$ 1 fm). For such short wave lengths, we expect very little difference in scattering off targets made of free nucleons and nucleons bound in nuclei. This is reminiscent of scattering at lower energies where the result is independent of whether the nucleus is imbedded in one kind of chemical molecule or another. For this reason, we expect that the J/ψ production should be the same for a hydrogen target as for a target made of some heavy element such as tungsten. The observed results, however, are quite different from such an expectation, and the cross section for nuclear targets is found to be lower. Besides J/ψ, there are also differences in the production rates for other particles such as kaons and pions. In every case, there seems to be very strong evidence to suggest that scattering off bound nucleons is different from free nucleons.

Several explanations have been proposed for the different rates. One interesting scenario is that we might be seeing the same phenomenon as in EMC effects (see §4-5) observed in high-energy lepton scattering off nuclear targets. Although the EMC effect itself is not yet well understood, the observation of the same effect with a different type of probe is interesting and suggests that some new physics may be present. Another possible explanation involves the formation and absorption of J/ψ as it traverses through a medium of strongly interacting particles. Since this is directly related to QCD, it becomes a very promising line of investigation for understanding QCD itself. The number of different measurements for such high energy nuclear reactions is still relatively small, but several others are either going on at the moment or in the planning stages. It is very likely that high-energy nuclear physics may open a new window for strong interaction studies.

§7-6 MESON-NUCLEUS REACTIONS

Interaction of mesons with nucleons and nuclei is important to nuclear physics for the obvious reason that a large and important part of the nuclear force is mediated by the exchange of mesons. In a scattering experiment, real mesons are involved. Such reactions are related to the exchange of virtual mesons between nucleons in that they share the same coupling constant. Meson scattering is also interesting from the point of view that they are bosons. Since bosons can be absorbed and created in the scattering process, we expect to learn something new about scattering that cannot be achieved with baryons and leptons. In this way meson-nucleus scattering becomes an integral part of hadron-nucleus scattering studies and forms an essential part of our understanding of hadrons.

Experimentally, intense sources of pions are available from "meson" factories, such as LAMPF (Los Alamos Meson Physics Facility), SIN (Swiss Institute for Nuclear Research), and TRIUMF (Tri-University Meson Facility) (see, *e.g.,* D.E. Nagle, M.B. Johnson and D. Measday, *Phys. Today,* [April 1987] 56). There are two features that are special to pion scattering. The first is the dominance of the P_{33}-resonance that produces a Δ-particle from a pion and a nucleon. As we have seen in §2-6, the strength of this resonance at pion laboratory energy of 195 MeV is so overwhelming that pion-nucleus reactions at energies below a few hundred MeV are dominated by the formation of a Δ. The second is that pions have three charge states, π^+, π^0, and π^-. As a result, single-charge exchange as well as double-charge exchange reactions are possible. The study of pion-nucleus reactions can therefore be carried out in a variety of ways, including pion absorption, elastic and inelastic scattering, as well as charge exchange reactions.

Pion absorption. There are two different types of pion absorption studies that can be made, stopped pion and fast pion. In order to enhance the probability, the pion to be absorbed must be slowed down so that it is essentially at rest with respect to the nucleus. One way to "stop" a π^- is to capture it first in an atomic orbit to form a π-*mesic* atom. It may happen that the π^- is initially captured in one of the higher electronic orbits of the atom. If this is true, the negative pion will eventually cascade down to a low-lying orbit through atomic electromagnetic decay. Since the pion mass is far larger than that of an electron ($m_\pi \approx 300 m_e$), a low-lying π^- atomic wave function has a significant overlap with that of the nucleus, as we have seen earlier in the analogous situation of muonic atoms in §4-6. However, being a hadron, a pion behaves quite differently from a muon, particularly in the nuclear medium. Because of the strong interaction, a pion is readily absorbed by the nucleus once it is close enough for the short-range force to be effective.

When the π^- is absorbed, all its rest mass energy of $\sim$ 140 MeV is transferred to the nucleus in the form of excitation energy. Since this amount is about sixteen times the average binding energy of a nucleon in a nucleus, it is difficult for a single nucleon to take up the full amount of energy and conserve momentum at the same time. It is therefore likely that a cluster of nucleons, such as an α-particle cluster, is involved here. Alternatively, the internal degrees of freedom within a nucleon may be excited. Even though the peak of the lowest energy resonance, the Δ-channel, is still far away in energy, the small possibility remains an important consideration.

In contrast to stopped pions, the absorption of "fast" pions may be defined as a reaction involving an incident pion and no scattered pion,

$$\pi + A \to A^*$$

Since the incident pion carries both energy and momentum, it is again impossible for a single nucleon in the nucleus to absorb the pion and conserve both energy and momentum at the same time. From the relatively large cross sections observed in the reaction

$$\pi^+ + d \to p + p$$

we can conclude that two-nucleon absorption is an important process taking place in the absorption of fast pions in nuclei. This idea is also corroborated by the relatively large cross sections observed for (π^+, pp) reactions on nuclei in general. A concise review of the subject can be found, *e.g.,* in Ashery and Schiffer (*Ann. Rev Nucl. Part. Sci.* **36** [1986] 207).

The inverse of pion absorption is pion production. When a nucleus is bombarded by electrons, protons, or other particles, pions are produced if sufficient energy is available. The reaction usually results in final states with three or more particles. Furthermore, many other exit channels are also open at these energies; as a result, both the measurement and the analysis are complicated. For this reason, we shall not be concerned with such reactions here. An example of the pion production cross section in nucleon-nucleon scattering was shown earlier in Fig. 3-4.

Pion scattering. Pion scattering is an important source of information on pion-nucleus interaction. The experiments may be divided into three categories: elastic and inelastic scattering, single-charge exchange (SCX) reactions, and double-charge exchange (DCX) reactions. Because of the strong P_{33}-resonance, the measurements may also be divided into three groups depending on whether the energy is below the resonance, on the resonance, or above the resonance. We have already seen in §4-6 that the large enhancements of the $\pi^+ + p$ and $\pi^- + n$ cross sections over those for $\pi^- + p$ and $\pi^+ + n$ scattering on the resonance make it possible to use pion scattering to distinguish between neutron and proton density distributions in a nucleus. We shall be mainly interested here in the other aspects of pion-nucleus scattering.

At energies no higher than 50 MeV, far below the P_{33}-resonance, the average interaction of pions with nuclei may be represented by an optical model potential. There are several different possible ways to construct an average potential for pion-nucleus scattering. An example is one given by Stricker, Carr, and McManus (*Phys. Rev. C* **22** [1980] 2043) which makes use of the fact that, at such low energies, the scattering is dominated by s- and p-partial waves.

Let us examine first the amplitude for pion-nucleus scattering. Since pions are isospin $t = 1$, pseudoscalar ($J^\pi = 0^-$) particles, the pion-nucleon scattering amplitude may be expressed in terms of the isospin operator $\boldsymbol{t}$ for the pion and $\boldsymbol{\tau}$ for the nucleon,

$$f_{\pi N} = b_0 + b_1 \boldsymbol{t}\cdot\boldsymbol{\tau} + (c_0 + c_1 \boldsymbol{t}\cdot\boldsymbol{\tau})\boldsymbol{k}\cdot\boldsymbol{k}' \tag{7-88}$$

where $\boldsymbol{k}$ and $\boldsymbol{k}'$ are the initial and final pion wave number vectors. The coefficients b_0 and b_1 are related to s-wave scattering from a nucleon followed by absorption on a neighboring nucleon, and the coefficients c_0 and c_1 are related to the corresponding p-wave process. These coefficients are, in general, complex, since pions can be absorbed by nucleons; their values may be found by fitting calculated results to data on pionic atoms.

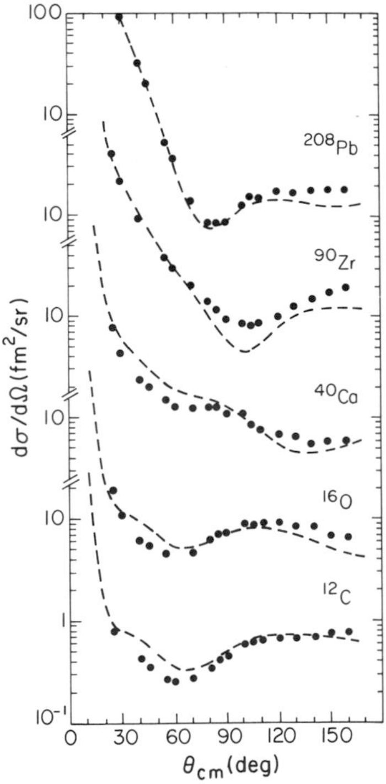

Fig. 7-10 Elastic scattering of low-energy pions from nuclei. The continuous curves are calculated results using the values of b_0, b_1, c_0, b_1 and λ given in (7-90) (adapted from K. Stricker, J.A. Carr and H. McManus, *Phys. Rev. C* **22** [1980] 2043).

The pion-nucleus optical potential that generates a scattering amplitude of the form of (7-88) may be expressed in the following operator form:

$$U_{\rm opt}(\boldsymbol{r}) = -\frac{2\pi}{\mu}\Big\{\big(b(r) + B(r)\big) - \nabla\cdot\Big[L(r)\big(c(r) + C(r)\big)\Big]\nabla + \frac{p_1 - 1}{2}\nabla^2 c(r) + \frac{p_2 - 1}{2}\nabla^2 C(r)\Big\} \tag{7-89}$$

where μ is the reduced mass of a pion. The kinematic factors,

$$p_1 = 1 + \frac{\hbar\omega}{M_N c^2} \qquad\qquad p_2 = 1 + \frac{\hbar\omega}{2M_N c^2}$$

come from transformation between frames of reference attached to the center of mass of the pion-nucleon system and the pion-nucleus system. They are functions of the total pion energy $\hbar\omega$ and nucleon mass M_N. The other factors,

$$b(r) = p_1[\bar{b}_0\rho(r) - e_\pi b_1\delta\rho(r)] \qquad\qquad B(r) = p_2 B_0\rho^2(r)$$
$$c(r) = p_1^{-1}\{c_0\rho(r) - e_\pi c_1\delta\rho(r)\} \qquad\qquad C(r) = p_2^{-1}C_0\rho^2(r)$$
$$L(r) = \left\{1 + \tfrac{4\pi}{3}\lambda[c(r) + C(r)]\right\}^{-1} \qquad\qquad \delta\rho(r) = \rho_n(r) - \rho_p(r)$$

may also depend on e_π, the charge of the pion. The densities of neutron, proton, and nucleon in the nucleus, $\rho_n(r)$, $\rho_p(r)$, and $\rho(r)$, are normalized to N, Z, and A, respectively. These factors express various first order correlations between nucleons in a nucleus. Second order correlations in s-wave are also included in (7-89) through the factor

$$\bar{b}_0 = b_0 - \frac{3k_F}{2\pi}(b_0^2 + 2b_1^2)$$

where k_F, the Fermi momentum of a nucleon in a nucleus, is taken to be 1.4 fm^{-1}. A typical set of parameters for 50 MeV incident pions has the values

$$\begin{aligned} &\lambda = 1.4 \\ &b_0 = -0.057{+}0.006i \text{ fm} && c_0 = 0.75 \ {+}0.03i \ \text{fm}^3 \\ &b_1 = -0.134{-}0.002i \text{ fm} && c_1 = 0.428{+}0.014i \ \text{fm}^3 \\ &B_0 = \ \ 0.02 \ {+}0.25i \ \text{fm}^4 && C_0 = 0.36 \ {+}1.2i \ \ \text{fm}^6 \end{aligned} \qquad (7\text{-}90)$$

As can be seen from examples shown in Fig. 7-10, such an optical potential gives a good description of the experimental data on the elastic scattering of both π^+ and π^- off a variety of nuclei at low energies.

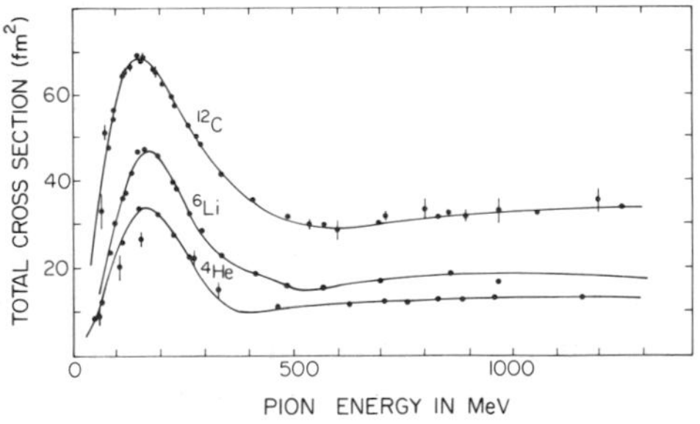

Fig. 7-11 Total scattering cross section of pions off ^{4}He, ^{6}Li and ^{12}C showing the strong reaction near the P_{33}-resonance and smooth variations at higher incident pion energies. (Adapted from C.J. Batty, G.T.A. Squier, and G.K. Turner, *Nucl. Phys.* **B67** [1973] 492.)

At pion energies far above 200 MeV, the influence of the P_{33}-resonance diminishes and the nucleus becomes much less absorptive to pions, as can be seen from the examples given shown in Fig. 7-11. In the energy range 300 to 800 MeV, pion-nucleon scattering is dominated by many overlapping resonances; we expect that pion-nucleus scattering at comparable energies may be given by a convolution of these resonances, with the nucleon Fermi motion inside the nucleus acting as a smoothing function. However, not many data are available yet for a more detailed discussion.

Measurements of pion scattering from nuclei are limited by the energy resolution that can be achieved with pions. The problem is caused partly by the fact that pions are produced by high-energy protons striking a thick target made of heavy elements. Energy selection is accomplished by passing the broad spectrum of particles produced through electromagnetic fields. Both the limited initial flux and the short lifetimes of pions put stringent limitations on what can be achieved. The same difficulties are also present in the measuring instruments since the pion energies here are still relatively low for some of the more efficient detection techniques to work well. As a result, measurements of pion-nucleus scattering are usually carried out for light nuclei where the low-lying levels are well separated in energy. Both elastic and inelastic data are available, and they have been very useful in complementing the information obtained with other probes.

Charge exchange reactions. Single-charge exchange processes, (π^+, π^0) and (π^-, π^0), are among the most extensively studied π-nucleus reactions. Examples of (π^+, π^0) scattering off ^{14}C and ^{60}Ni at different angles are shown in Fig. 7-12. Except around the P_{33}-resonance, the processes are similar to (p, n) and (n, p) reactions and their results are often compared. At energies above the resonance, (π^-, π^0)-reactions have some advantage over competing (n, p)-reactions since intense intermediate energy neutron beams with well-defined energies are difficult to obtain. On the other hand, the particle emerging from an SCX reaction is π^0, a neutral particle that is usually detected through γ-rays produced in its decay through the reaction $\pi^0 \rightarrow \gamma + \gamma$. This puts some constraint on the types of SCX measurements that can be carried out. For this reason, good-quality angular distributions of SCX are only beginning to be available.

There are two different double-charge exchange reactions that can be studied with pions, (π^+, π^-) and (π^-, π^+). They are interesting for two reasons. In the first place, these processes must involve at least two nucleons and are therefore useful for investigating nucleon correlations inside a nucleus. In the second place, the nuclear matrix element that enters into the scattering cross section calculation is related to double β-decay, a process important for understanding the nature of weak interaction itself, as described in §1-3 and §5-5. For all practical purposes, DCX reactions are unique to pions. A nucleon, being an isospin-$\frac{1}{2}$ particle, can only induce SCX reactions. The only way to induce DCX with conventional nuclear probes is to use heavy ions. Here we have the complication that the probe itself can be excited by the reaction.

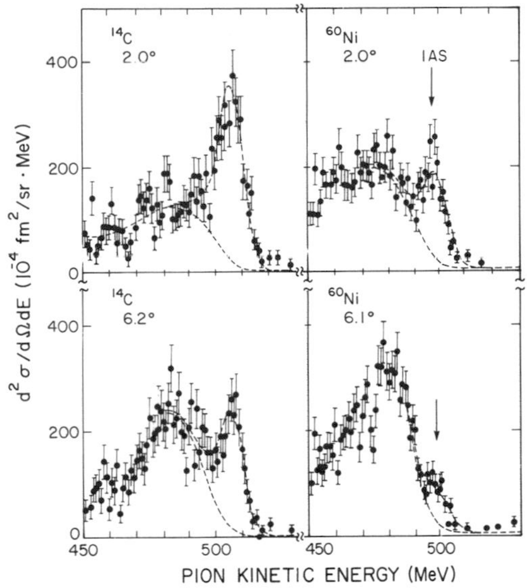

Fig. 7-12 Energies of π^0 observed in (π^+, π^0) single-charge exchange reactions induced by 500 MeV pions on ^{14}C and ^{60}Ni at scattering angles indicated. The continuous curves are polynomial fits to the background (taken from Baer et al., in *Pion-Nucleus Physics*, AIP Conf. Proc. **163**, ed. by R.J. Peterson and D.D. Strottman, Amer. Inst. Phys., New York [1988] p. 67).

Because of the small cross section, DCX studies have so far been limited mainly to strong transitions leading to isobaric analogue states in light nuclei separated by a pair of neutrons or protons. The main interest has been centered around effects involving the internal degrees of freedom of the nucleons. The results seem to indicate that an important role may be played by excitation processes involving intermediate states with two nucleons excited to become Δ-particles. In the near future, the prospects of using DCX to relate strong and weak interactions and to understand nucleon correlations are quite bright. Examples of inclusive pion double-charge exchange reaction cross sections are shown in Fig. 7-13.

Kaons and other mesons. In addition to pions, kaons have been available for scattering off nuclei. Kaons are "strange" mesons involving either an s-quark (K^- and $\overline{K}^0$, strangeness $\mathcal{S} = -1$) or an $\bar{s}$-quark (K^+ and K^0, $\mathcal{S} = +1$). Conservation of strangeness requires that when a K^--meson is absorbed by a nucleus, one of the nucleon changes into a "strange" baryon such as a Λ ($m_\Lambda c^2 = 1115.6$ MeV) or a Σ ($m_{\Sigma^0} c^2 = 1192$ MeV). The nucleus A becomes a *hypernucleus* $_{Y}A^*$ in the process. Because of the large kaon mass (~ 500 MeV/c^2), the nucleus is left in a highly excited state, much more so than the case of a pion absorption. On the other hand, since there is no light baryon with $\mathcal{S} = +1$, a K^+ meson cannot be absorbed by a nucleus.

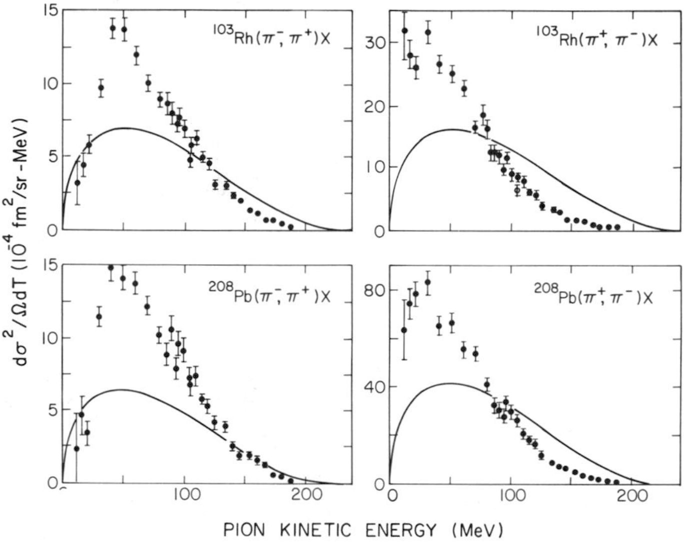

Fig. 7-13 Inclusive pion double-charge exchange reactions (π^+, π^-) and (π^-, π^+) on ^{103}Rb and ^{208}Pb at incident pion energy 240 MeV. The emerging pions were detected at $\theta_{\text{lab.}} = 130°$. The smooth curves are classical estimates made by Hufner and Thies (*Phys. Rev.* **C20** [1979] 273) based on Boltzmann equations (taken from P.A.M. Gram, in *Pion-Nucleus Physics*, AIP Conf. Proc. **163**, ed. by R.J. Peterson and D.D. Strottman, Amer. Inst. Phys., New York [1988] p. 79).

Many new and different avenues of study are opened up by a hadronic probe with nonzero strangeness. In terms of new insights into nuclear structure problems, this is similar to what studies of nuclei far away from the valley of β-stability can provide us. Such possibilities will be greatly enhanced by the new kaon factories with far more intense beams.

Besides pions, interaction between nucleons is also mediated by other mesons such as ρ and ω. For this reason, studies of reactions involving these mesons with nucleons and nuclei will be of interest. The difficulty is an experimental one; there does not seem to be any easy way to produce intense beams of mesons other than pions and kaons. Some of the information on their interactions with nucleons, and baryons in general, is obtained from their production rates through decays of heavier particles; any knowledge of their interaction with nuclei must come from indirect sources.

§7-7 HEAVY-ION REACTIONS

The term *heavy ion* is generally used to mean nuclei heavier than the helium nucleus. For nuclei with $A > 4$, the internal structure becomes sufficiently complex that, when two heavy ions scatter off each other, many new reaction channels become open, such as the transfer of clusters of nucleons and large amounts of angular momentum. With improvements in accelerator technology, intense beams of increasingly heavier nuclei are accelerated to higher and higher energies. It is now technically feasible to bring heavy nuclei, such as uranium, to energies far in excess of their rest mass energies.

For our brief introduction to the broad and expanding subject of heavy-ion reactions, we shall skip the more conventional reaction mechanisms which we have already seen for light ions. These include Coulomb excitation, elastic scattering, compound nucleus formation, and direct reactions. Instead, we shall focus our attention on those aspects that are special to heavy ions. Because of the large amount of kinetic energy, angular momentum, and nucleons involved, highly excited nuclear states and exotic nuclei very far away from the valley of stability may be made in the process. In addition, extremely strong electrostatic fields are created when two heavy ions, for example both uranium nuclei, are briefly fused together. Such strong fields offer tests of our knowledge of quantum electrodynamics under conditions never before encountered. At even higher energies, the confinement of quarks inside nucleons may become meaningless and, in the collision region between two relativistic heavy ions, a new state of matter may evolve that is of interest to quantum chromodynamics investigations.

Semi-classical treatment. The basic mechanism underlying heavy-ion reactions is not too different from that of other hadron nucleus scattering. However, because of the large masses, it is instructive to view the reaction first in the limit of classical scattering. In classical mechanics, the scattering between two particles is often discussed in terms of an *impact* parameter b, defined as the perpendicular distance between the center of force and the incident velocity vector. Physically, we may view b as the closest distance between two (point) particles when they pass by each other in the absence of any force acting between them, as shown in Fig. 7-14.

If v_0 is the magnitude of the incident velocity when the particles are far away from each other, the amount of angular momentum ℓ in the system may be expressed in terms of the impact parameter b,

$$\ell = mv_0 b = b\sqrt{2mE} \tag{7-91}$$

In quantum mechanics, ℓ is measured in units of $\hbar$ and the value of angular momentum may be written in terms of the wave number k,

$$\ell/\hbar = b\frac{\sqrt{2mE}}{\hbar} = bk \tag{7-92}$$

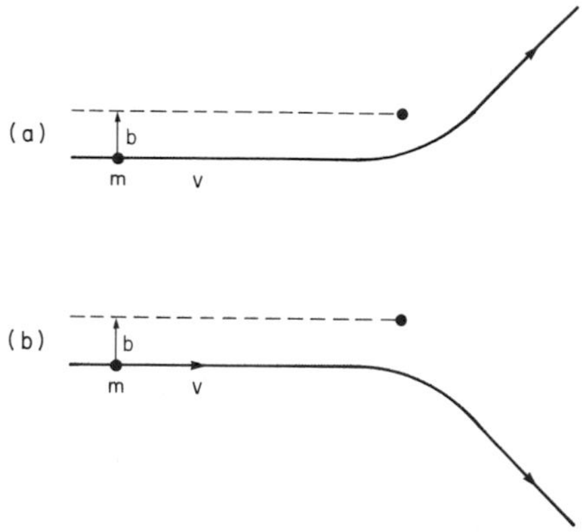

Fig. 7-14 Semi-classical treatment of heavy-ion collision in terms of the classical trajectory of a particle in a finite-range central field. Figure (*a*) is for an attractive potential and (*b*) for a repulsive potential. The impact parameter *b* is the perpendicular distance between the center of force and the incident velocity vector.

From this result we see that, for a given partial wave ℓ, the impact parameter b is inversely proportional to k, and that a higher center-of-mass energy implies a smaller impact parameter.

Because of the complexity in the interaction between two nuclei, heavy-ion collisions are often treated in semi-classical approximations. This is justified in part by the large masses and high angular momenta involved. In such cases, the classical description is often a good approximation of the underlying quantum-mechanical processes, especially in forward angles and at low energies. The other part of the justification comes from the fact that, because of the strong Coulomb repulsion, the distance of closest approach between the two nuclei is large at low bombarding energies. In such cases, nuclear effects are small and may be treated by semi-classical approximations.

In heavy-ion scattering, the short-range nuclear force does not operate between the two colliding particles if the impact parameter b is much larger than the sum of the two nuclear radii and the scattering is dominated by the repulsive Coulomb interaction. Analogous to optics, the scattering yields a Fresnel-like diffraction pattern, observed also in proton-nucleus elastic scattering at small angles (see, *e.g.,* Fig. 7-8). Under these conditions, the scattering is mainly elastic, with Coulomb excitation as the only possible inelastic scattering channel. As the impact parameter decreases with increasing incident energy, the two heavy ions come into contact with each other, and short-range nuclear force begins to influence the scattering pattern.

To describe the changes in the reaction mechanism with increasing incident energy, we may define a *grazing* or critical angular momentum,

$$\ell_{\rm gr} = kR_{\rm gr} \tag{7-93}$$

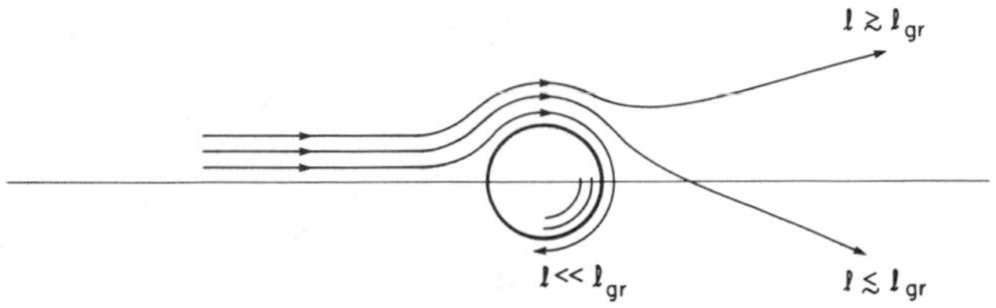

Fig. 7-15 Grazing angle in heavy-ion collision. When the orbital angular momentum is much less than the grazing angular momentum ℓ_{gr}, the two heavy ions are able to revolve around each other and their nucleons have the opportunity to interact. When $\ell > \ell_{gr}$, very little interaction can take place except when the incident energy is sufficiently high to overcome the Coulomb barrier in a head-on collision.

where R_{gr} is the grazing radius at which the colliding pair starts to feel the attractive nuclear force acting between them. This is shown schematically in Fig. 7-15. The grazing radius may be interpreted as an "interaction" distance by the way it is defined. It is usually taken to be sightly larger than the geometric "touching" distance between the two heavy ions,

$$R = r_0\left(A_1^{1/3} + A_2^{1/3}\right) \tag{7-94}$$

where A_1 and A_2 are the nucleon numbers of the two heavy ions. When b is small enough to be comparable to R_{gr}, nucleon transfer between the two heavy ions takes place through quantum mechanical tunnelling. As the impact parameter becomes smaller than R_{gr}, the two heavy ions overlap each other in space and many reaction channels open as a result of interactions between the nucleons in them. Here the process is much more complicated than in the case of nucleon-nucleus scattering we saw earlier. This comes in part because of the large amount of energy available in the composite system for exciting the nuclear degrees of freedom. In addition to the usual mechanisms of compound nucleus formation and direct reaction involving the transfer of one or two nucleons, it is possible to shift large amounts of energy, angular momentum, and nucleons between the colliding pair. For such transfers, a transport theory for nonequilibrium systems in statistical mechanics is often the more appropriate vehicle to understand the observed phenomena.

Deep inelastic scattering. Except at very low energies, the time available for two heavy ions to overlap each other and coalesce in a heavy ion scattering is very short. The interaction time t_{int} is of the order of 10^{-22} s to 10^{-20} s, usually less than the time, t_{rot} ($> 10^{-20}$ s), required for the two nuclei to go around each other once. Consequently, there is not enough time for the two systems to reach an equilibrium with each other, as in the case of compound nucleus formation. Many experimental studies have been carried out under such conditions for projectile

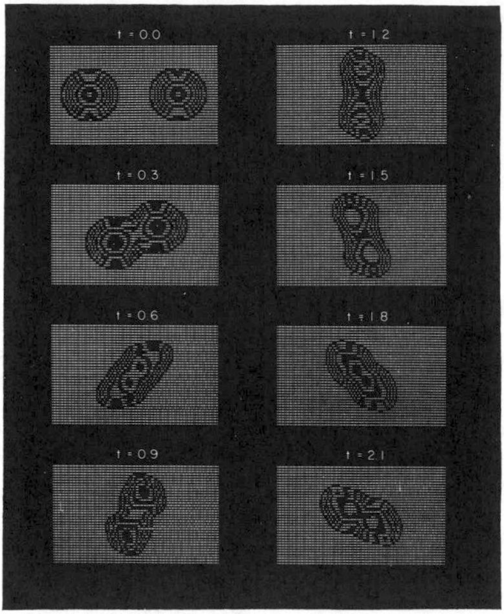

Fig. 7-16 Snapshots of time-dependent Hartree-Fock calculations of 278 MeV ^{40}Ca on ^{40}Ca for $\ell = 40$. The unit of time is 10^{-21}s. (Adapted from P. Bonche, B. Grammaticos and S. Koonin, *Phys. Rev. C* **17** [1978] 1700.)

and target nuclei with $A \gtrsim 40$. If the collision energy is slightly above the Coulomb barrier, many reaction channels are open and compete with each other. It is usually impossible to investigate each channel individually, and only the inclusive cross sections are studied. For this reason, the reaction is often referred to as ***deep inelastic*** collision, similar to the situation of high-energy electron scattering discussed earlier in §4-5. The cross section is large here, especially for heavy nuclei ($\sigma \sim 10$ to 20 fm^2); as many as 20 nucleons may be transferred from one nucleus to another, and up to 100 MeV of kinetic energy and $50\hbar$ of angular momentum shifted from relative motion between the two nuclei to the excitation energies of the final nuclei involved.

A good starting point for a macroscopic approach to deep inelastic collision is the master equation in statistical mechanics,

$$\frac{d}{dt}P_n(t) = \sum_m [W_{nm}P_m(t) - W_{mn}P_n(t)] \tag{7-95}$$

where $P_n(t)$ is the probability that, at time t, the system is in a group of closely related states n, and W_{nm} is the transition probability per unit time from the group of states m to the group n. Eq. (7-95) simply means that the probability of finding a group of states is equal to the sum of the probabilities for transfers into the group minus the probabilities for leaving the group. This is a very reasonable approach for studying all kinds of transport phenomena. However, in order for

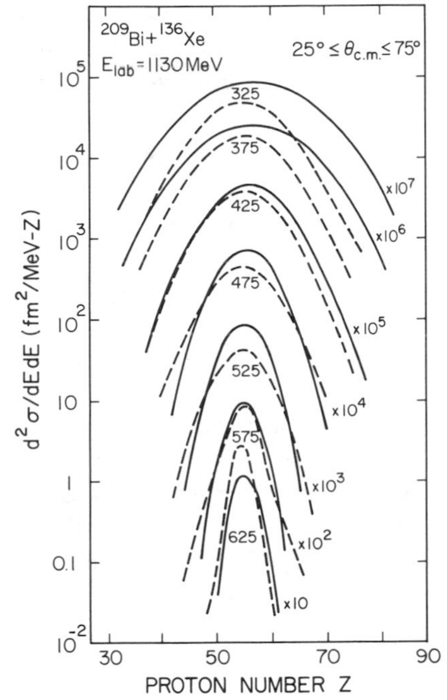

Fig. 7-17 Double differential cross section for the reaction ^{136}Xe on ^{209}Bi at E_{lab}=1130 MeV. The solid lines are the measured values of $d^2\sigma/dEdZ$ integrated over center-of-mass angles $25° \leq \theta \leq 75°$ and the dashed lines are the calculated results using a transport theory. (Taken from C.M. Ko, *Phys. Lett.* **81B** [1979] 299).

the transport description to be valid, the time scales involved in the system must obey the condition,

$$t_{\text{equ}} \ll t_{\text{coll}} \ll t_{\text{Poincaré}}$$

where t_{equ} is the time it takes for the noncollective degrees of freedom to reach internal equilibrium, t_{coll} is the time required for the collective degrees of freedom to reach equilibrium, and $t_{\text{Poincaré}}$ is the Poincaré recurrence time, the time for the system to return to its original position in phase space. The condition imposed on the relationships between different time scales is necessary here, since we are using the master equation to deal only with the collective degrees of freedom of a system of two heavy ions. This means that the noncollective degrees of freedom of the system must have reached an equilibrium already and do not participate in the transport process as a result. The extent to which these conditions are met and the success of the transport theory description of deep inelastic collision can be seen in the agreement between the calculated results and experimental observation of the collision of ^{136}Xe on ^{209}Bi at 1130 MeV laboratory energy shown in Fig. 7-17.

The entire deep inelastic collision process is, in principle, governed by the time-dependent Schrödinger equation (C-1) and described by a wave function $\Psi(\boldsymbol{r}, t)$ as the two heavy ions approach each other, collide, and evolve into the final state. A general solution to (C-1) is impossible for the complicated case of two heavy ions scattering off each other. One possibility is to use a time-dependent Hartree-Fock approach and take the time-dependent differential equation (C-1) as a difference equation that gives the change of the wave function $\Delta\Psi(\boldsymbol{r},t)$ in time interval Δt,

$$\Delta\Psi(\boldsymbol{r},t) = \frac{1}{i\hbar} H\,\Psi(\boldsymbol{r},t)\Delta t$$

At a given time t we have a system of $(A_1 + A_2)$ nucleons whose motions are given by the Hamiltonian H. The behavior of the system at time $t + \Delta t$ is given by the action of H on $\Psi(\boldsymbol{r},t)$. The difference between the wave function at time t and $t + \Delta t$,

$$\Delta\Psi(\boldsymbol{r},t) = \Psi(\boldsymbol{r},t+\Delta t) - \Psi(\boldsymbol{r},t)$$

is the result of the action of the Hamiltonian H on the system in the small time interval Δt. By solving the difference equation, we can obtain the changes in the wave function $\Delta\Psi(\boldsymbol{r},t)$ and thus the wave function $\Psi(\boldsymbol{r},t+\Delta t)$ describing the system at time $(t + \Delta t)$. In this way, the whole time development of the system may be traced out in small time steps. The Hamiltonian equation is still a very complicated one to solve in view of the large number of nucleons involved; the Hartree-Fock method is invoked as a simplification that can still give a fairly realistic description of the nuclear physics, as we have seen for the time-independent situation in §6-7. The calculated results for 278 MeV ^{40}Ca on ^{40}Ca are shown in Fig. 7-16 as an illustrative example.

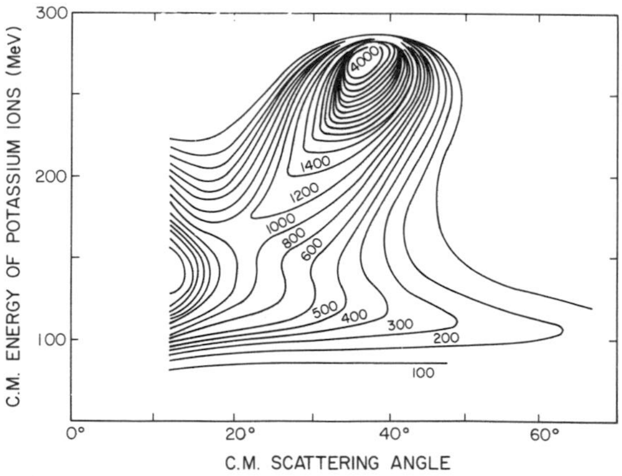

Fig. 7-18 Double differential cross section $d^2\sigma/dEd\theta$ for the reaction ^{232}Th(^{40}Ar,K) at laboratory energy 388 MeV. The contour plot, in units of 10^{-4} fm^2/MeV·rad in the plane formed by center-of-mass scattering angle θ and energy of the scattered nucleus, is a convenient way to display the reaction cross section. (Adapted from J. Wilczyński, *Phys. Lett.* **47B** [1973] 484.)

Angular momentum transfer. Heavy-ion collisions are often accompanied by large amounts of angular momentum transfer. When two heavy ions approach each other with kinetic energy in excess of the Coulomb barrier between them, there is a chance that they will fuse together and form a composite system. Because of the large amount of excitation energy in the system, the composite system lives only for a short time compared with a compound nucleus. However, since the collision involves two large masses at high velocities, the composite system possesses very high angular momentum, of the order many tens to hundreds $\hbar$. If the composite system decays by fission into two or more fragments, each with tens of nucleons, the angular momentum carried away by the relative motion between these fragments is large, and not much appears as spins of the fragments. On the other hand, if there are barriers against fission, the composite system may turn instead to nucleon evaporation and γ-ray emission to dispose of the excess energy. The angular momentum carried away by emitting a nucleon is quite small on the average. This may be estimated using an approximation that the maximum angular momentum ℓ_m carried away by a nucleon is given by $\hbar kR$, where k is the wave number of the nucleon and R is the radius of the composite system. For a neutron, the average kinetic energy carried away is around 2 MeV, since it is difficult for an individual nucleon to acquire much more energy in the collision of two heavy ions regardless of whether the composite system is fully equilibrated or not. For A around 150, the value of ℓ_m obtained in this way is around $2\hbar$ (see Problem 7-7). The actual angular momentum carried away by a nucleon is lower than this value on the average, and is more likely to be $\sim 1\hbar$.

Since fission and nucleon evaporation are relatively fast processes, a composite system cannot be considered as a single nucleus until a substantial amount of excitation energies are carried away by the fast processes and the remaining decays are dominated by the relatively slow γ-ray emissions. If fission has not taken place, there can be a substantial amount of angular momentum left in the remnant nucleus of the composite system and this will appear in the form of nuclear spin. As a result, it is likely that the residual nucleus will end up in one of its yrast levels, the lowest level in energy for a given value of J. All the subsequent decay of the nucleus will then proceed predominantly through γ-ray emission, cascading down in angular momentum following the sequence of yrast levels, as we have seen earlier in §6-3. This is shown schematically in Fig. 7-19.

The highest angular momentum a nucleus can attain depend on several different considerations. For light nuclei, the limiting factor is the highest spin to which the valence nucleons can be coupled together. For example, in the $1p$-shell, the maximum allowed J is 5 for 6 particles coupled together to $T = 0$. In the ds-shell, the maximum is $J = 14$ for 12 active nucleons. However, for heavier nuclei, both the number of active orbits and the average spin of single-particle orbits are larger. For such nuclei, the maximum spin is limited by considerations such as stability against fission.

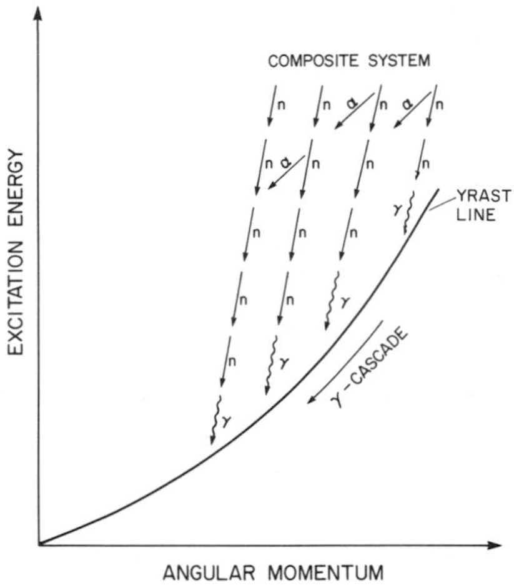

Fig. 7-19 Schematic diagram showing the population of yrast levels in heavy-ion reactions. When two heavy nuclei come into contact with each other though collision, the composite system formed as a result of the fusion contains large amounts of excess energy and angular momentum. The angular momenta carried away by neutron, α-particle, and γ-ray emission are very limited. When the fast particle emission processes stop, the nucleus is likely to be left in a high-spin state, often a member of the yrast band. Thereafter, the nucleus decays through γ-ray cascade.

High-spin nuclear states usually lie quite high in excitation energy, in part due to the energy associated with rotation. Since the density of states in such regions is high, the lifetimes of most states are short due to the large number of open decay channels. As a result, it is usually impossible to resolve individual levels. The yrast levels are, however, the exception here, since their decays are dominated by γ-ray emission to the next yrast level below, and as a result their lifetimes are long compared with other levels in the vicinity. The widths of the γ-rays are narrow as a result of the long lifetimes and they stand out as sharp lines against backgrounds formed by the multitude of decays from short-lived levels in the same region. Since a large fraction of the excitation energy is in the form of rotational energy so as to maintain the high angular momenta, the nucleus itself may still be quite "cold" in the sense that most of the nucleons are in their lowest possible single-particle orbits. In spite of the high excitation energies, the deformation itself does not need to be large, as we saw in §6-3.

The highest J-values are observed in nuclei with "super-deformed" bands. These are formed as the result of two heavy ions fused together into a highly deformed shape, with the ratio between polar and equatorial axes as large as two. Such a configuration is not usually the lowest one in energy and therefore is not found in ground state bands. However, they may be sufficiently stable and live long enough to be observed before decaying into the ground state configuration. The high J-value attained by some of these nuclear states makes them almost classical, according to the correspondence principle. Further progress may lead to even higher spins and thus allow us to trace the development of a nucleus from a purely quantum mechanical state to a classical one. This may also be of interest to the study of the transition from quantum mechanical to classical description of physical phenomena in general.

Creation of neutron-deficient nuclei. Heavy-ion fusion is a good method of creating neutron-deficient nuclei that are far away from the valley of stability. Because of Coulomb repulsion, nuclei with large numbers of protons require even larger numbers of neutrons to be stable. As the number of proton increases, an increasingly larger neutron excess is needed. For example, below $Z = 20$, the ratio $N/Z \approx 1$ in stable nuclei. For medium heavy nuclei such as zirconium ($Z = 40$), $N/Z \approx 1.3$ and for lead ($Z = 82$), $N/Z \approx 1.5$. For the heaviest nuclei created in the laboratory, the more stable isotopes have a ratio of $N/Z \approx 1.6$.

When two heavy ions fuse, the composite system retains the average N/Z ratio of the two and the neutron excess is much smaller than the value suitable for the combined system. For example, when two ^{90}Zr nuclei are joined together, the composite system is ^{180}Hg. The lightest stable mercury isotope is ^{196}Hg. This implies that the composite system is "deficient" by roughly 16 neutrons. The nucleus ^{180}Hg is an unstable one with a half-life of 2.9 s for the ground state. In order to overcome the Coulomb barrier, the composite system formed as a result of heavy-ion collision is usually in a highly excited state of the corresponding nucleus with large amounts of excess energy and angular momentum. As a result, the lifetime of the composite system is likely to be much shorter than its ground state, and the probability of decaying into other nuclei by particle emission is quite high. If our intent is to create neutron-deficient nuclei, it is desirable to have the nuclei made in one of the low-lying states. This may be achieved, for example, by selecting a reaction that can carry away some of the energy and angular momentum through the emission of one or more nucleons and γ-rays. This is the usual method by which many of the neutron-deficient nuclei are made. However, it is not possible to go on indefinitely depleting the neutron number in a nucleus; one of the ultimate limitations of neutron-deficient nuclei is that, when the neutron number is too small, proton emission may be favored, and this increases the neutron excess.

With the wide range of available heavy-ion projectiles, from lithium to uranium, a large variety of exotic nuclei can be made and studied. In addition to the 1600 naturally occurring nuclei, we have potentially five times that number that can be made in the laboratory. Since these new species may be quite different from the naturally occurring ones, there is a good chance of learning something new and fundamental about the nuclear system. In additional, many of these nuclei are important as the intermediate states in nucleosynthesis under conditions in stars that are quite different from those usually found in the laboratory. A good understanding of their formation and their lifetimes is useful for astrophysics studies as well.

Quantum electrodynamics interest. Let us return for the moment to the simple quantum mechanical problem of a hydrogen-like atom with a single electron outside a nucleus having Z protons. Nonrelativistically, the energy levels are given by the solution to a Schrödinger equation in the form:

$$E_n = -\left[\left(\frac{1}{4\pi\epsilon_0}\right)^2\right]\frac{m_e e^4}{2\hbar^2}\frac{Z^2}{n^2} = -\frac{\alpha^2 m_e c^2}{2}\frac{Z^2}{n^2} \tag{7-96}$$

where n is the principal quantum number that labels the energy levels. The lowest state is the $1s$-state ($n = 1$). For a hydrogen atom, we have $Z = 1$ and the ground state has the energy

$$E_{1s} = -\frac{\alpha^2 m_e c^2}{2} = -13.6 \text{ eV} \tag{7-97}$$

where $\alpha^2 m_e c^2/2 = R_y$ is known as the Rydberg energy, the ionization energy for a hydrogen atom in the ground state.

More generally, we can solve the Dirac equation for a hydrogen-like atom assuming a point nucleus of charge $+Ze$. The total energy of the system in the ground state is now

$$E_{1s} = m_e c^2 \sqrt{1 - (Z\alpha)^2}, \tag{7-98}$$

where $\alpha(= 1/137.036)$ is the fine structure constant. For $Z = 1$, this also yields a value of 13.6 eV for the ionization energy of a hydrogen atom, as expected. For $Z > 1$, the expression is valid up to some critical value $Z_{\text{cr}} \approx 137$. This is quite adequate for most purposes, since all nuclei found on earth have a Z value much less than this limit. Even among the man-made elements, the highest Z value known so far is around 103. However, in heavy ion collisions, the "nuclear molecules" formed by the fusion of two heavy nuclei have a possible total Z value far in excess of 137; a *supercritical* field is created as a result.

Some corrections to the result $Z_{\text{cr}} \approx 137$ are necessary due to the fact that nuclei are not point charges. The exact value of Z_{cr} depends somewhat on the charge distribution inside a nucleus. For a uniform charged sphere of radius reasonable for nuclei, a result of $Z_{\text{cr}} < 200$ is obtained. This higher value can also be exceeded in heavy ion collisions.

The physical meaning of a supercritical field may be seen from the following arguments. As the charge number of a nucleus is increased, the eigenvalues of the $1s$ and other atomic levels decrease from the value given roughly by (7-98) until the critical value is reached. When this happens, the $1s$-level becomes degenerate with the negative energy continuum filled with electrons. As a result, the charge-neutral vacuum is no longer the state of minimum energy. To lower the energy, positrons must be released and the remaining vacuum becomes a charged one. This phenomenon is referred to as the "spontaneous" decay of the neutral vacuum.

Experimentally, the signature of a supercritical field is the occurrence of narrow positron peaks when two heavy nuclei collide with each other, such as thorium ($Z = 90$) on thorium. Narrow positron peaks have actually been observed

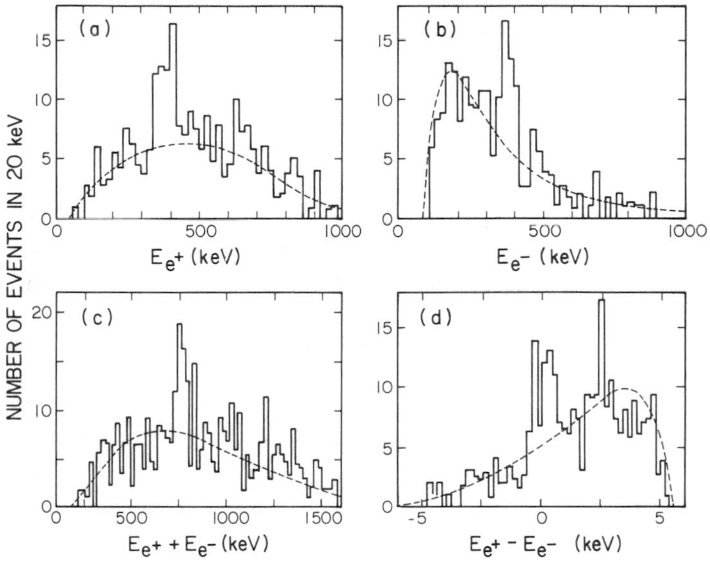

Fig. 7-20 Intensity distribution of positrons and electrons emitted in coincidence in the collision of thorium on thorium at projectile energy of $\approx$ 1300 MeV (5.7 MeV per nucleon). The different distributions are taken from different projections. Note the correlation of e^+ and e^- counts in the histogram as a function of the sum of positron and electron energies given in (*c*). (Adapted from Cowan et al., *Phys. Rev. Lett.* **56** [1986] 444).

in several experiments (see, *e.g.*, Fig. 7-20), and the positions of the positron peaks are correlated with those of electron peaks of roughly the same energy. The most obvious source of electron-positron production from internal pair conversion is ruled out from the coincidence intensity. However, the observed production does not seem to fit into the picture of a supercritical field in heavy ion collisions either. Many unknown aspects of the dynamics still remain. Regardless of the final outcome, the presence of narrow positron peaks in heavy-ion collision is an interesting topic in its own right. Several proposals have been advanced for their interpretation, including one with a new, hitherto unknown, neutral particle; however, the true explanation still eludes us at the moment.

Relativistic heavy-ion collision. One of the interests in both nuclear and particle physics these days is the collision of heavy ions at relativistic energies. In a nucleus of radius $R = 1.2A^{1/3}$, the energy density is given by

$$\rho_A \approx \frac{M_A c^2}{\frac{4\pi}{3}R^3} \approx 130 \text{ MeV/fm}^3. \tag{7-99}$$

For three quarks confined in a nucleon of radius $r \approx 1$ fm, the energy density is even higher, of the order of 250 MeV/fm^3. If heavy ions are made to collide with each other at center-of-mass kinetic energy far larger than their rest mass energies, for example, several hundred GeV per nucleon or higher, energy density far greater than the value of 250 MeV/fm^3 within a nucleon may be achieved over the dimension of a few femtometers. Again analogous to the parallel case in nuclear scattering where energies in excess of 20 MeV per nucleon can penetrate the Coulomb barrier of a nucleus, relativistic heavy-ion collisions can overcome the strong repulsive force associated with the hard core of nucleons. The usual "bags" that confine quarks inside hadrons disappear in the same way as Coulomb barriers become irrelevant between nuclei at energies in excess of the barrier height. The state of matter created under such a circumstance is expected to be quite different from nuclear matter and is generally referred to as *quark-gluon plasma*, consisting of quarks, antiquarks, and gluons. This is a very different state of matter, perhaps similar to that which existed at the beginning of the "big bang," postulated in cosmology as the event that gave birth to our universe. The study of such a state of matter will therefore be a central question for cosmology, as well as for quantum chromodynamics. It may also lead us into new areas of physics we have not yet conceived of.

The total energies involved in relativistic heavy-ion collisions attain the value of several TeV (10^{12} eV) and represent some of the highest energy available in the laboratory for subatomic physics investigations. A large number of final states are possible at such high energies. How can we recognize that the state of quark-gluon plasma is reached during such a collision? One of the possible signals is the production of particles involving quarks other than u and d found in nuclei. As we saw in the previous section on nucleon-nucleus scattering at high energies, the production of J/ψ-mesons made of c- and $\bar{c}$-quarks may be used as a guide. Such mesons are also produced in the usual collision of nucleons at high energies and cannot be taken by themselves as the signature of quark-gluon plasma. One proposal is that substantial deviation of the production rate in heavy-ion collision from that observed in nucleon-nucleon collision may indicate a different production mechanism that might be related to a new state of matter. However, this is by no means the only reason for a difference in the J/ψ production rate in heavy-ion collisions. Other methods of identifying the existence of quark-gluon plasma are also under active investigation. Certainly, the production and identification of this new state of matter will be one of the major scientific achievements of our era.

Problems

7-1. Show that in the scattering of a particle of mass M_a off a target nucleus of mass M_b, the momentum transfer q from a to b has the same form in both the laboratory and center-of-mass coordinates.

7-2. Calculate the Q-value for the reaction $^{120}\text{Sn}(d,p)^{121}\text{Sn}$ leading to the ground state of $^{121}_{50}\text{Sn}$ using binding energies obtained from a table of mass excess.

7-3. The angular distribution of an $\ell = 2$ transfer, $^{20}\text{Ne}(d,n)^{21}\text{Na}$ reaction leading to the $J^\pi = 5/2^+$ state at 2.14 MeV in ^{21}Na peaks at 36° for 6.0 MeV energy deuterons in the center of mass. Use a plane wave Born approximation to deduce the radius of ^{21}Na. Compare the result with that given by $R = 1.2A^{1/3}$ fm.

7-4. Show that

$$E_{1s} = m_e c^2 \sqrt{1 - (Z\alpha)^2}$$

gives the ionization energy of 13.6 eV for a hydrogen atom.

7-5. Show that, for direct reactions in the plane wave Born approximation, only $\ell = 0$ transfers have maxima in the differential scattering cross section at scattering angle $\theta = 0°$. For $\ell > 0$ transfers, the forward direction is a minimum and the first maximum of the differential cross section occurs at increasingly larger angles for higher ℓ transfers.

7-6. Calculate the radius of the lowest orbit of π^- in a π-mesic atomic with Z protons in the nucleus. Assuming a two-parameter Fermi form (4-21) for the distribution of nucleons in the nucleus, calculate the overlap between the wave function of the π^- and the nucleus consisting of A nucleons. Take $c = 5.0$, $z = 0.5$ fm, $Z = 50$, and $A = 120$.

7-7. If a neutron with 2 MeV of kinetic energy is evaporated from a composite system made of two heavy ions consisting of a total of 150 nucleons, find the maximum angular momentum carried away using the relation $\ell_m = \hbar k R$.

7-8. Two ^{90}Zr nuclei approach each other with kinetic energy 200 MeV in the center of mass. Calculate the total angular momentum in the system assuming that, in the absence of any interaction between them, they will pass each other at a distance of 10 fm between their centers.

7-9. Find the angular distribution of neutrons emerging from a (n,n) reaction on ^{208}Pb. If the incident energy is sufficiently low, multiple scattering may take place. Assume that each neutron suffers two elastic collisions with the nucleons in the target nucleus before leaving and that the angular distribution of each scattering is given by $|j_0(qR)|^2$, where R is the radius of the target nucleus. Take the incident neutron energy to be 10 MeV in the center of mass and ignore any energy dependence in the scattering cross section.

7-10. Show that for low-energy hard-sphere scattering, the cross section is equal to $4\pi R^2$, where R is the radius of the potential well.

Appendix A

PARITY TRANSFORMATION AND CHARGE CONJUGATION

§A-1 PARITY TRANSFORMATION

Parity or space reflection transformation is the operation whereby all three coordinate axes in the cartesian system change sign. That is, if the location of a point in space is given by coordinates (x, y, z) in a given system, the coordinates of the same point in a system related to the original one by a parity transformation P are $(-x, -y, -z)$,

$$(x, y, z) \xrightarrow[P]{} (-x, -y, -z) \tag{A-1}$$

Such a reflection of the axes changes a right-handed coordinate system to a left-handed one, as illustrated by Fig. 5-3.

In quantum mechanics, the probability of finding a particle at location $\boldsymbol{r}$ is given by the absolute square of its wave function $|\Psi(\boldsymbol{r})|^2$ at the point. Since the probability is an observable, it cannot change its value simply because we have switched from using a right-handed coordinate system to a left-handed one or vice versa. The wave function itself, however, may change under a parity transformation. The possible variations are governed by the following two conditions. The first is that $|\Psi(\boldsymbol{r})|^2$ must remain invariant. The second is that two successive parity operations must bring the system back to its original state, *i.e.*, $P^2 = 1$. As a result, the wave function $\Psi(\boldsymbol{r})$ can change at most by a sign. States whose wave functions do not change sign under a parity transformation, *i.e.*,

$$P\Psi(\boldsymbol{r}) = \Psi(-\boldsymbol{r}) = +\Psi(\boldsymbol{r}) \tag{A-2}$$

are called positive parity states and states whose wave functions change sign,

$$P\Psi(\boldsymbol{r}) = \Psi(-\boldsymbol{r}) = -\Psi(\boldsymbol{r}) \tag{A-3}$$

are called negative parity states. A wave function which does not fall into either one of these two categories does not have a definite parity.

In terms of spherical polar coordinates, the radial distance r is not affected by a parity transformation. The only changes are in the angular variables,

$$(r, \theta, \phi) \xrightarrow[P]{} (r, \pi - \theta, \pi + \phi) \tag{A-4}$$

This relation can be shown to be identical as that in (A-1), for example, by transforming both sides to cartesian coordinate systems. Because of (A-4), radial wave functions are not changed by a parity transformation. As a result, the parity of a wave function of a state is given by the angular part alone. For a state $\Psi(\boldsymbol{r})$ with definite orbital angular momentum (ℓ, m), we can decompose the wave function into radial and angular parts in the following way:

$$\Psi(\boldsymbol{r}) = R_{n\ell}(r) Y_{\ell m}(\theta, \phi)$$

The angular dependence is described by spherical harmonics $Y_{\ell m}(\theta, \phi)$, the eigenfunctions of orbital angular momentum operators given later in (B-2). The parity of a spherical harmonics of order ℓ is $(-1)^\ell$, as can be seen from the explicit form of $Y_{\ell m}(\theta, \phi)$ given by (B-7),

$$Y_{\ell m}(\theta, \phi) = \frac{(-1)^m}{2^\ell \ell!} \sqrt{\frac{(2\ell+1)}{4\pi} \frac{(\ell-m)!}{(\ell+m)!}}\, e^{im\phi} \left(1 - \eta^2\right)^{m/2} \left(\frac{d}{d\eta}\right)^{\ell+m} (\eta^2 - 1)^\ell \tag{A-5}$$

where $\eta = \cos\theta$. Since

$$\cos(\pi - \theta) = -\cos(\theta)$$

we have, under a parity transformation,

$$\eta \xrightarrow[P]{} -\eta$$

The transformation of the polar angle θ gives a phase factor $(-1)^{\ell+m}$ to $Y_{\ell m}(\theta, \phi)$. The azimuth angle ϕ enters (A-5) only in the exponential factor $e^{im\phi}$. The transformation from ϕ to $(\pi + \phi)$ produces a factor $e^{im\pi} = (-1)^m$. The combination of both transformations produces the net result,

$$Y_{\ell m}(\theta, \phi) \xrightarrow[P]{} Y_{\ell m}(\pi - \theta, \pi + \phi) = (-1)^\ell Y_{\ell m}(\theta, \phi) \tag{A-6}$$

For this reason, spherical harmonics of even order have even parity and spherical harmonics of odd order have odd parity.

In additional to parity associated with spatial wave functions, the intrinsic wave function of a particle can also have a definite parity associated with the internal structure of the particle. If the structure is known, such as the quark structure of nucleons, the intrinsic parity may be deduced from the wave function. On the other hand, in cases where the internal structure is not known, the intrinsic parity must be determined experimentally using reactions in which the parities of all other particles as well as all the relative angular momenta involved are known.

As an example, we shall see how the intrinsic parity of a pion is determined to be negative. The measurement involves the absorption of a π^- by a deuteron.

The negative pion is first captured in the s-state of the deuterium atom, forming a π-mesic atom as a result. Since the pion is a meson, it may be absorbed by the proton in the deuterium nucleus through the reaction,

$$\pi^- + d \rightarrow n + n \tag{A-7}$$

Before the reaction, the total angular momentum J of the π-mesic atom is 1 since the intrinsic spin of the pion is 0, the spin of the deuteron is 1 (see §3-1), and the orbital angular momentum of the πd system is 0 (the π^- is in the atomic s-state). Total angular momentum is conserved in the reaction (A-7), and as a result, the final state produced by the reaction must also have $J = 1$.

The two neutrons in the final state, being identical fermions, must be in an antisymmetric state to satisfy the Pauli principle. The symmetry of the wave function of the two-neutron system is determined by L, the relative orbital angular momentum, and S, the sum of the intrinsic spin of the two particles. If the spatial part of the system of two identical fermions is symmetrical ($L =$ even), the total intrinsic spin wave function must be antisymmetrical ($S = 0$). Alternatively, if the spatial part is antisymmetrical ($L =$ odd), the total intrinsic spin wave function must be symmetrical ($S = 1$).

From the fact that $\boldsymbol{J} = \boldsymbol{L} + \boldsymbol{S} = 1$, we find that the possible pairs of (L, S) values to form $J = 1$ are (0,1), (1,0), (1,1), and (2,1). The combinations (0,1) and (2,1) can be ruled out on the ground that both intrinsic spin and spatial wave functions are symmetric and are therefore in violation of the Pauli principle. Similarly, the combination (1,0) is not allowed, since both orbital and intrinsic spin parts are antisymmetric. The only possible combination remaining is $(L, S) = (1, 1)$, which is antisymmetric in the spatial part but symmetric in the intrinsic spin part of the wave function.

The parity of the right hand side of the reaction (A-7) is therefore $(-1)^L = -1$, independent of the intrinsic parity of neutrons, as there are two involved. Since parity is conserved in the reaction, the left hand side must also have negative parity. There are three components contributing to the parity of the initial state of the reaction. The parity of the ground state of deuteron is known to be even ($L = 0$ or 2, and both neutron and proton have the same intrinsic parity). The parity of the orbital wave function of the π-mesic atom is positive, as we have seen earlier. As a result, we conclude that the intrinsic parity of π^-, the third component in the initial state, must be negative in order for the parity of the total system to be negative.

For fermions, the intrinsic parity of an antiparticle is opposite to that of its corresponding particle. This can be seen from the structure of the Dirac equation where a particle and an antiparticle are described by a single four-component wave function, or through such measurements as the polarization of the two photons emitted in the decay of a positronium (e^+e^- system) in the singlet state ($J = 0$). On the other hand, for bosons the parity of both particle and antiparticle must be the same. For more details, see standard particle physics textbooks such as Perkins (*Introduction to High Energy Physics*, Addison-Wesley, Menlo Park, California, 1987); and Halzen and Martin (*Quarks and Leptons*, Wiley, New York, 1984).

§A-2 CHARGE CONJUGATION

Charge conjugation is the operation which changes the sign of the charge of a particle without affecting any of the properties unrelated to charge. In relativistic quantum mechanics, it also implies a transformation between a particle and its antiparticle. Let $|p\rangle$ and $|n\rangle$ represent the wave functions of a proton and a neutron, respectively. In terms of second quantized creation operators $a^\dagger_{tt_0}$ for a particle, we may express these wave functions in the form,

$$|p\rangle = a^\dagger_{\frac{1}{2},+\frac{1}{2}}|0\rangle \qquad |n\rangle = a^\dagger_{\frac{1}{2},-\frac{1}{2}}|0\rangle \tag{A-8}$$

where $|0\rangle$ is the wave function for the vacuum. In the expression, we have displayed only the isospin ranks and suppressed all other labels for simplicity. The wave functions of an antiproton $|\bar{p}\rangle$ and an antineutron $|\bar{n}\rangle$ may be constructed in a similar way using the creation operator $b^\dagger_{tt_0}$ for an antiparticle,

$$|\bar{n}\rangle = b^\dagger_{\frac{1}{2},+\frac{1}{2}}|0\rangle \qquad |\bar{p}\rangle = b^\dagger_{\frac{1}{2},-\frac{1}{2}}|0\rangle \tag{A-9}$$

Here we have made use of the fact that on transforming a particle to an antiparticle (and *vice versa*), the charge, and hence the projection of isospin on the quantization axis, changes sign. If particles and antiparticles are unrelated to each other, $a^\dagger_{tt_0}$ and $b^\dagger_{tt_0}$ are completely different operators defined by (A-8) and (A-9), respectively. However, particles and antiparticles can transform into each other through charge conjugation, C, and as a result, operators $a^\dagger$ and $b^\dagger$ are not independent of each other.

Since a particle and an antiparticle can annihilate each other, their wave functions must be able to be coupled to an isoscalar quantity (and a scalar in spin as well). Thus, a particle and its antiparticle must have the same isospin (and spin). Furthermore, their projections on the quantization axis must be equal to each other in magnitude but opposite in sign. Given the fact that a proton has $\boldsymbol{t} = \frac{1}{2}$ and $t_0 = +\frac{1}{2}$, an antiproton must have $\boldsymbol{t} = \frac{1}{2}$ and $t_0 = -\frac{1}{2}$. Similarly, both a neutron and an antineutron must have $\boldsymbol{t} = \frac{1}{2}$, and for a neutron, $t_0 = -\frac{1}{2}$, and for an antineutron, $t_0 = +\frac{1}{2}$.

In addition to changes in the isospin (and spin), the wave functions of a particle and an antiparticle may also differ by a phase factor. There are several ways to obtain this factor. If we take $a^\dagger_{tt_0}$ and $b^\dagger_{tt_0}$ as operators with a definite irreducible spherical tensor rank t, the phase factor is fixed by the transformation properties under a rotation in the isospin space. For second quantized operators, we have the relation

$$b^\dagger_{tt_0} = (-1)^{t-t_0} a_{t,-t_0} \tag{A-10}$$

The phase factor arises from the fact that operators $a^\dagger_{tt_0}$ and $a_{t,-t_0}$ are not Hermitian conjugate of each other without the factor $(-1)^{t-t_0}$. (For a more detailed discussion, see A. Bohr and B.R. Mottelson, *Nuclear Structure,* vol. I, Benjamin,

Reading, Massachusetts, 1969; and A. de Shalit and I. Talmi, *Nuclear Shell Theory*, Academic Press, New York, 1963.)

With the relation between the second quantized operators given by (A-10), we find that under charge conjugation,

$$\begin{aligned} |\,p\,\rangle &\xrightarrow[C]{} (-1)^{\frac{1}{2}+\frac{1}{2}}|\,\bar{p}\,\rangle = -|\,\bar{p}\,\rangle \\ |\,n\,\rangle &\xrightarrow[C]{} (-1)^{\frac{1}{2}-\frac{1}{2}}|\,\bar{n}\,\rangle = +|\,\bar{n}\,\rangle \end{aligned} \tag{A-11}$$

The same phase factor considerations apply to other particles as well. For example, since a u-quark has isospin ranks $(t, t_0) = (\frac{1}{2}, +\frac{1}{2})$ and a d-quark has ranks $(\frac{1}{2}, -\frac{1}{2})$, the transformations to their antiparticles under charge conjugation are

$$\begin{aligned} |\,u\,\rangle &\xrightarrow[C]{} (-1)^{\frac{1}{2}+\frac{1}{2}}|\,\bar{u}\,\rangle = -|\,\bar{u}\,\rangle \\ |\,d\,\rangle &\xrightarrow[C]{} (-1)^{\frac{1}{2}-\frac{1}{2}}|\,\bar{d}\,\rangle = +|\,\bar{d}\,\rangle \end{aligned} \tag{A-12}$$

These phase factors are used, for example, in writing the quark wave functions of the pions in Chapter 2.

Appendix B

SPHERICAL HARMONICS AND SPHERICAL TENSOR

§B-1 SPHERICAL HARMONICS

A Hamiltonian describing the motion of a particle in a central field $V(r)$,

$$H = -\frac{\hbar^2}{2\mu}\nabla^2 + V(r) \tag{B-1}$$

commutes with the square and the z-component of the orbital angular momentum operators for the particle, $\boldsymbol{L}^2$ and $\boldsymbol{L}_z$, respectively. In the spherical polar coordinate system, the Laplacian operator in (B-1) may be written in the form

$$\nabla^2 = \frac{1}{r^2}\frac{\partial}{\partial r}r^2\frac{\partial}{\partial r} + \frac{1}{r^2}\left\{\frac{1}{\sin\theta}\frac{\partial}{\partial\theta}\left(\sin\theta\frac{\partial}{\partial\theta}\right) + \frac{1}{\sin^2\theta}\frac{\partial^2}{\partial\phi^2}\right\} \tag{B-2}$$

The terms inside the curly brackets, which depend only on angular variables θ and ϕ, are proportional to the square of the angular momentum operator $\boldsymbol{L} = \boldsymbol{r}\times\boldsymbol{p}$,

$$\boldsymbol{L}^2 = -\hbar^2\left\{\frac{1}{\sin\theta}\frac{\partial}{\partial\theta}\left(\sin\theta\frac{\partial}{\partial\theta}\right) + \frac{1}{\sin^2\theta}\frac{\partial^2}{\partial\phi^2}\right\} \tag{B-3}$$

The components of $\boldsymbol{L}$ are

$$\begin{aligned}
\boldsymbol{L}_x &= i\hbar\left(\sin\phi\frac{\partial}{\partial\theta} + \cot\theta\cos\phi\frac{\partial}{\partial\phi}\right)\\
\boldsymbol{L}_y &= -i\hbar\left(\cos\phi\frac{\partial}{\partial\theta} - \cot\theta\sin\phi\frac{\partial}{\partial\phi}\right)\\
\boldsymbol{L}_z &= -i\hbar\frac{\partial}{\partial\phi}
\end{aligned} \tag{B-4}$$

Since $[H, \boldsymbol{L}^2] = [H, \boldsymbol{L}_z] = 0$, the eigenfunctions $\Psi(r,\theta,\phi)$ of H are also eigenfunctions of $\boldsymbol{L}^2$ and $\boldsymbol{L}_z$.

Let us write the eigenvector of (B-1) in terms of a product of radial and angular parts,

$$\Psi(r,\theta,\phi) = R(r)\,Y(\theta,\phi) \tag{B-5}$$

A central potential may be taken as a function of the radial coordinate r only. For such cases, the potential enters only in the radial equation. As a result, the angular parts of the wave function $Y(\theta, \phi)$ are eigenfunctions of both $\boldsymbol{L}^2$ and $\boldsymbol{L}_z$,

$$\boldsymbol{L}^2 Y(\theta, \phi) = \Lambda \hbar^2 Y(\theta, \phi) \qquad \boldsymbol{L}_z Y(\theta, \phi) = \mu \hbar Y(\theta, \phi) \tag{B-6}$$

where appropriate powers of $\hbar$ have been inserted in the definition of Λ and μ for later convenience.

Functions that satisfy the conditions expressed by (B-6) are proportional to the spherical harmonics $Y_{\ell m}(\theta, \phi)$. For positive m-values, $Y_{\ell m}(\theta, \phi)$ has the explicit form

$$Y_{\ell m}(\theta, \phi) = \frac{(-1)^m}{2^\ell \ell!} \sqrt{\frac{(2\ell+1)}{4\pi} \frac{(\ell-m)!}{(\ell+m)!}}\, e^{im\phi} \left(1 - \eta^2\right)^{m/2} \left(\frac{d}{d\eta}\right)^{\ell+m} (\eta^2 - 1)^\ell \tag{B-7}$$

where $\eta = \cos\theta$. Eq. (B-6) is satisfied with

$$\Lambda = \ell(\ell+1) \qquad \mu = m \tag{B-8}$$

Spherical harmonics with negative m-values are obtained from those with positive m-values given in (B-7) using the following relation:

$$Y_{\ell,-m}(\theta, \phi) = (-1)^m Y_{\ell m}^*(\theta, \phi) \tag{B-9}$$

The spherical harmonics form a complete, orthogonal, and normalized set of functions over the spherical surface satisfying the following condition for normalization and orthogonality:

$$\int_0^{2\pi} \int_0^{\pi} Y_{\ell m}^*(\theta, \phi) Y_{\ell' m'}(\theta, \phi) \sin\theta \, d\theta \, d\phi = \delta_{\ell\ell'} \delta_{mm'} \tag{B-10}$$

Derivations of these results are given in standard quantum mechanics and mathematical physics textbooks, and we shall not repeat them here.

Using (B-4), we can define the angular momentum raising operator as

$$\boldsymbol{L}_+ = \boldsymbol{L}_x + i\boldsymbol{L}_y = \hbar e^{i\phi}\left(\frac{\partial}{\partial\theta} + i\cot\theta \frac{\partial}{\partial\phi}\right) \tag{B-11}$$

and the angular momentum lowering operator as

$$\boldsymbol{L}_- = \boldsymbol{L}_x - i\boldsymbol{L}_y = -\hbar e^{-i\phi}\left(\frac{\partial}{\partial\theta} - i\cot\theta \frac{\partial}{\partial\phi}\right) \tag{B-12}$$

These operators, when acting on $Y_{\ell m}(\theta, \phi)$, produce spherical harmonics of the same ℓ except that the m-values are raised or lowered by unity,

$$\boldsymbol{L}_\pm Y_{\ell m}(\theta, \phi) = \hbar\sqrt{(\ell \mp m)(\ell \pm m + 1)}\; Y_{\ell\, m\pm1}(\theta, \phi) \tag{B-13}$$

Thus, angular momentum raising and lowering operators may be used to generate spherical harmonics and spherical tensors (see next section) of the same rank from a given m-value to another. For such purposes, it is more convenient to define a set of angular momentum operators without the factor $\hbar$ in (B-5,6) and (B-11,12), a convention adopted in the rest of the book.

As mentioned in §A-1, the symmetry of $Y_{\ell m}(\theta, \phi)$ under a parity transformation, through which $(r, \theta, \phi) \rightarrow (r, \pi - \theta, \pi + \phi)$, is given by

$$Y_{\ell m}(\pi - \theta, \pi + \phi) = (-1)^{\ell} Y_{\ell m}(\theta, \phi) \tag{B-14}$$

Another useful relation is that, for $\theta = 0$,

$$Y_{\ell m}(0, \phi) = \sqrt{\frac{2\ell + 1}{4\pi}}\, \delta_{m0} \tag{B-15}$$

Both relations may be obtained directly from the explicit forms of $Y_{\ell m}(\theta, \phi)$ given in (B-7).

If a system is independent of the azimuthal angle ϕ, only $Y_{\ell m}(\theta, \phi)$ with $m = 0$ are involved. This subset of spherical harmonics is equivalent to a set of Legendre polynomials $P_\ell(\cos\theta)$,

$$Y_{\ell 0}(\theta) = \sqrt{\frac{2\ell + 1}{4\pi}}\, P_\ell(\cos\theta) \tag{B-16}$$

where

$$P_\ell(\eta) = \frac{1}{2^\ell \ell!} \frac{d^\ell}{d\eta^\ell} (\eta^2 - 1)^\ell \tag{B-17}$$

It is a set of functions often used instead of $Y_{\ell 0}(\theta)$, for example, in the expansion of a plane wave.

The explicit forms of $Y_{\ell m}(\theta, \phi)$ for $\ell \leq 3$ are needed frequently, and are given here for convenience:

$$Y_{0,0}(\theta, \phi) = \sqrt{\frac{1}{4\pi}}$$

$$Y_{1,0}(\theta, \phi) = \sqrt{\frac{3}{4\pi}} \cos\theta$$

$$Y_{1,\pm 1}(\theta, \phi) = \mp \sqrt{\frac{3}{8\pi}} \sin\theta e^{\pm i\phi}$$

$$Y_{2,0}(\theta, \phi) = \sqrt{\frac{5}{16\pi}} (3\cos^2\theta - 1)$$

$$Y_{2,\pm 1}(\theta, \phi) = \mp \sqrt{\frac{15}{8\pi}} \cos\theta \sin\theta e^{\pm i\phi}$$

$$Y_{2,\pm 2}(\theta, \phi) = \sqrt{\frac{15}{32\pi}} \sin^2\theta e^{\pm 2i\phi}$$

$$Y_{3,0}(\theta, \phi) = \sqrt{\frac{7}{16\pi}} (5\cos^3\theta - 3\cos\theta)$$

$$
\begin{aligned}
Y_{3,\pm1}(\theta,\phi) &= \mp\sqrt{\frac{21}{64\pi}}(5\cos^2\theta-1)\sin\theta e^{\pm i\phi} \\
Y_{3,\pm2}(\theta,\phi) &= \sqrt{\frac{105}{32\pi}}\cos\theta\sin^2\theta e^{\pm 2i\phi} \\
Y_{3,\pm3}(\theta,\phi) &= \mp\sqrt{\frac{35}{64\pi}}\sin^3\theta e^{\pm 3i\phi}
\end{aligned}
\qquad \text{(B-18)}
$$

Higher order functions may be obtained using (B-18) or by expressing $Y_{\ell m}(\theta,\phi)$ first in terms of associated Legendre polynomials $P_{\ell m}$ and use recurrence relations given, for example, in Arfken (*Mathematical Methods for Physicists*, Academic Press, New York, 1970) to generate the required higher order $P_{\ell m}$.

§B-2 SPHERICAL TENSOR AND ROTATION MATRIX

A quantity $\psi_{JM}(\boldsymbol{r})$ is said to be a *spherical tensor* of angular momentum rank J if it belongs to a group consisting of $(2J+1)$ members having the same J and differing only in their M-values, projections of the tensor on the quantization axis. The possible values of M are $-J$, $-J+1$, ..., and J. Under a rotation of the coordinate axes through Euler angles (α,β,γ) shown in Fig. B-1, the $(2J+1)$ members transform among themselves according to the relation,

$$\psi_{JM'}(\boldsymbol{r}') = \sum_M \psi_{JM}(\boldsymbol{r})\,\mathcal{D}^J_{MM'}(\alpha,\beta,\gamma) \qquad \text{(B-19)}$$

where coefficients $\mathcal{D}^J_{MM'}(\alpha,\beta,\gamma)$ are the rotation matrices, or $\mathcal{D}$-functions for short. The $(2J+1)$ components of a spherical tensor ψ_{JM} for all possible values of M therefore form an irreducible group under rotation. The set of $(2\ell+1)$ spherical harmonics $Y_{\ell m}(\theta,\phi)$, with $m=-\ell, -\ell+1, \ldots, \ell$, is an example of a spherical tensor of integer rank ℓ. However, spherical tensors are more general quantities and can have half-integer ranks as well. Both wave functions and operators can be spherical tensors, since the requirements are satisfied by both types of quantities.

There are several possible ways to define the $\mathcal{D}$-function. We shall adopt the convention given by Brink and Satchler (*Angular Momentum*, 2nd edition, Clarendon Press, Oxford, 1968). A rotation of the coordinate axes in the way defined in Fig. B-1 can be achieved as three successive infinitesimal rotations represented by the operator,

$$\boldsymbol{R}(\alpha,\beta,\gamma) = e^{-i\gamma J_{z'}}e^{-i\beta J_{y_1}}e^{-i\alpha J_z} \qquad \text{(B-20)}$$

The same transformation is equivalent to a rotation first around the Z-axis by angle γ, followed by a rotation through angle β around the Y-axis, and finally a rotation through angle α around the Z-axis again, *i.e.*,

$$\boldsymbol{R}(\alpha,\beta,\gamma) = e^{-i\alpha J_z}e^{-i\beta J_y}e^{-i\gamma J_z} \qquad \text{(B-21)}$$

The $\mathcal{D}$-function in (B-19) may be written as the matrix element of $\boldsymbol{R}(\alpha,\beta,\gamma)$,

$$\mathcal{D}^J_{MM'}(\alpha,\beta,\gamma) = \langle\psi_{JM}|\boldsymbol{R}(\alpha,\beta,\gamma)|\psi_{JM'}\rangle \tag{B-22}$$

between components M and M' of a spherical tensor of rank J. The orthogonality relations among the $\mathcal{D}$-functions are given by

$$\begin{aligned}\sum_{M'}\left(\mathcal{D}^J_{M'N}(\alpha,\beta,\gamma)\right)^*\mathcal{D}^J_{M'M}(\alpha,\beta,\gamma)\\ = \sum_{M'}\mathcal{D}^J_{MM'}(\alpha,\beta,\gamma)\left(\mathcal{D}^J_{NM'}(\alpha,\beta,\gamma)\right)^* = \delta_{MN}\end{aligned} \tag{B-23}$$

and

$$\begin{aligned}\int_0^{2\pi}\int_0^{2\pi}\int_0^{\pi}\left(\mathcal{D}^{J'}_{M'M}(\alpha,\beta,\gamma)\right)^*\mathcal{D}^J_{N'N}(\alpha,\beta,\gamma)\sin\beta\,d\beta\,d\alpha\,d\gamma\\ = \frac{8\pi^2}{2J+1}\delta_{M'N'}\delta_{MN}\delta_{J'J}\end{aligned} \tag{B-24}$$

where δ_{xy} is the Kronecker delta which equals unity if $x=y$ and zero otherwise.

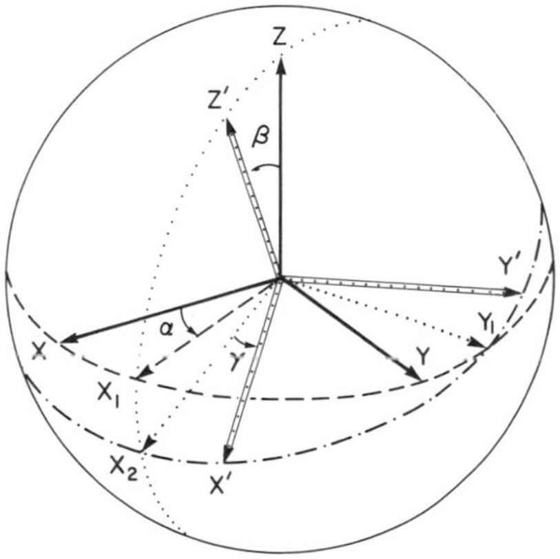

Fig. B-1 Rotation of the coordinate axes from (X,Y,Z) to (X',Y',Z') by Euler angles (α,β,γ) in three steps. First, a rotation around the Z-axis through angle α brings (X,Y,Z) to (X_1,Y_1,Z). Second, a rotation around the new Y_1-axis through angle β brings (X_1,Y_1,Z) to (X_2,Y_1,Z_2). In the last step, (X_2,Y_1,Z_2) is brought to (X',Y',Z') by a rotation around Z_2 (same as Z') through angle γ.

Since ψ_{JM} is an eigenfunction of J_z, we can make use of (B-21) to reduce the $\mathcal{D}$-function to a simpler form,

$$\begin{aligned}\mathcal{D}^J_{MN}(\alpha,\beta,\gamma) &= \langle\psi_{JM}|e^{-i\alpha J_z}e^{-i\beta J_y}e^{-i\gamma J_z}|\psi_{JN}\rangle\\ &= e^{-i(\alpha M+\gamma N)}\langle\psi_{JM}|e^{-i\beta J_y}|\psi_{JN}\rangle\end{aligned} \tag{B-25}$$

where the matrix element remaining in the final form is called the reduced rotation matrix element,

$$d^J_{MN}(\beta) \equiv \langle\psi_{JM}|e^{-i\beta J_y}|\psi_{JN}\rangle \tag{B-26}$$

and has the explicit form

$$d^J_{MN}(\beta) = \sum_Q (-1)^Q \frac{\sqrt{(J+M)!(J-M)!(J+N)!(J-N)!}}{(J+M-Q)!(J-N-Q)!Q!(Q+N-M)!} \times \left(\cos\tfrac{\beta}{2}\right)^{2J+M-N-2Q} \left(\sin\tfrac{\beta}{2}\right)^{2Q+N-M} \quad \text{(B-27)}$$

The summation is over all possible values of Q that do not lead to negative arguments in the factorials. The phase convention used here is the usual Condon and Shortley choice.

Let T_{kq} represent component q of a spherical tensor operator of rank k. The conjugate $(T_{kq})^\dagger$ of T_{kq} is defined through the relation between the Hermitian conjugate of their matrix elements,

$$\langle JM|(T_{kq})^\dagger|J'M'\rangle = \langle J'M'|T_{kq}|JM\rangle^* \quad \text{(B-28)}$$

It is easy to see that $(T_{kq})^\dagger$ is not a proper tensor. If T_{kq} transforms under a rotation according to (B-19), the transformation of its conjugate is given by

$$(T_{kq'})^\dagger = \Big(\sum_q T_{kq}\mathcal{D}^k_{qq'}(\alpha,\beta,\gamma)\Big)^\dagger = \sum_q (T_{kq})^\dagger\big(\mathcal{D}^k_{qq'}(\alpha,\beta,\gamma)\big)^* \quad \text{(B-29)}$$

In order to have the proper transformation properties, we can make use of the fact that

$$\big(\mathcal{D}^k_{pq}(\alpha,\beta,\gamma)\big)^* = (-1)^{p-q}\mathcal{D}^k_{-p\,-q}(\alpha,\beta,\gamma) \quad \text{(B-30)}$$

and define an adjoint tensor

$$\overline{T}_{kq} \equiv (-1)^{k-q}(T_{k,-q})^\dagger \quad \text{(B-31)}$$

which transforms under a rotation of the coordinate axes in the same way as given by (B-20) and is, therefore, a proper tensor. It should be pointed out that the phase factor $(-1)^q$ in (B-31) is unique, but the factor $(-1)^k$ is somewhat arbitrary. There are other conventions used in the literature: the one adopted here has the advantage that it is convenient for tensors of half-integer rank, for example, like those used in (A-11,12).

In problems where spherical symmetry is important, such as those encountered in subatomic physics, spherical tensors are useful for a variety of reasons. Some technical advantages of using spherical tensors are related to the algebra of angular momentum coupling given in the next four sections.

§B-3 ANGULAR MOMENTUM RECOUPLING COEFFICIENTS

In general the product of two spherical tensors is not a spherical tensor. For example, the product of T_{JM}, a spherical tensor of rank J, and $U_{J'M'}$, a spherical tensor of rank J', is a mixture of tensors with ranks $|J-J'|$ to $(J+J')$. We can use the product between two ordinary vectors as an illustration. A vector $\boldsymbol{r}$ is a spherical tensor of rank unity and is specified, for example, by giving its projections (r_1, r_2, r_3) on the three axes of a cartesian coordinate system. The product of $\boldsymbol{r}$ with another vector $\boldsymbol{r}'$, having projections (r_1', r_2', r_3'), contains, in general, a total of nine components. We can separate these nine products into three groups. The linear combination,

$$S \equiv r_1 r_1' + r_2 r_2' + r_3 r_3' = \boldsymbol{r} \cdot \boldsymbol{r}'$$

is a scalar since it is invariant under a rotation of the coordinate axes. Three of the quantities transform among themselves like a vector,

$$\boldsymbol{V} \equiv (r_2 r_3' - r_3 r_2', r_3 r_1' - r_1 r_3', r_1 r_2' - r_2 r_1') = \boldsymbol{r} \times \boldsymbol{r}'$$

as can be seen from the fact it has the standard form of an ordinary vector product between $\boldsymbol{r}$ and $\boldsymbol{r}'$. The remaining five components, which may be written in the form

$$T_{ij} \equiv \tfrac{1}{2}(r_i r_j' + r_j r_i') - \tfrac{1}{3}\delta_{ij} \sum_{k=1}^{3} r_k r_k'$$

form a second rank spherical tensor.

In general, a tensor of definite rank can be *projected* out of a product of two tensors using angular momentum coupling coefficients,

$$\left(T_{j_1} \times U_{j_2}\right)_{j_3 m_3} \equiv \sum_{pq} \langle j_1 m_1 j_2 m_2 | j_3 m_3 \rangle \, T_{j_1 m_1} U_{j_2 m_2} \tag{B-32}$$

where the Clebsch-Gordan coefficient $\langle j_1 m_1 j_2 m_2 | j_3 m_3 \rangle$ vanishes unless $m_3 = m_1 + m_2$ and $|j_1 - j_2| \leq j_3 \leq (j_1 + j_2)$. Several different symbols are commonly used in the literature to represent Clebsch-Gordan coefficients,

$$\langle j_1 m_1 j_2 m_2 | j_3 m_3 \rangle \equiv \langle j_1 j_2 m_1 m_2 | j_1 j_2 j_3 m_3 \rangle \equiv C^{j_1 j_2 j_3}_{m_1 m_2 m_3}$$

We shall use the first form in the above expression, since it gives the image of the overlap of $|jm\rangle$ with the product of $|j_1 m_1\rangle$ and $|j_2 m_2\rangle$.

It is often more convenient to express the coupling coefficients between two spherical tensors in terms of Wigner $3j$-symbols, related to the Clebsch-Gordan coefficients by a simple factor in the following way:

$$\begin{pmatrix} j_1 & j_2 & j_3 \\ m_1 & m_2 & m_3 \end{pmatrix} = \frac{(-1)^{j_1 - j_2 - m_3}}{\sqrt{2j_3 + 1}} \langle j_1 m_1 j_2 m_2 | j_3 \, {-m_3} \rangle \tag{B-33}$$

In terms of $3j$-symbols, the symmetry in the arguments of the coupling coefficient may be expressed in the form

$$\begin{pmatrix} j_1 & j_2 & j_3 \\ m_1 & m_2 & m_3 \end{pmatrix} = \begin{pmatrix} j_2 & j_3 & j_1 \\ m_2 & m_3 & m_1 \end{pmatrix} = \begin{pmatrix} j_3 & j_1 & j_2 \\ m_3 & m_1 & m_2 \end{pmatrix}$$

$$= (-1)^{j_1+j_2+j_3} \begin{pmatrix} j_1 & j_3 & j_2 \\ m_1 & m_3 & m_2 \end{pmatrix} = (-1)^{j_1+j_2+j_3} \begin{pmatrix} j_1 & j_2 & j_3 \\ -m_1 & -m_2 & -m_3 \end{pmatrix} \qquad \text{(B-34)}$$

In other words, $3j$-symbols are invariant under an even permutation of the three pairs of arguments and a phase factor $(-1)^{j_1+j_2+j_3}$ is needed for an odd permutation as well as for the case when all the m-values change sign.

Table B-1 Some useful Clebsch-Gordan coefficients.

$\langle \frac{1}{2} m_s \ell m_\ell | jm \rangle$

j	$m_s = +\frac{1}{2}$	$m_s = -\frac{1}{2}$
$\ell + \frac{1}{2}$	$\sqrt{\frac{\ell+\frac{1}{2}+m}{2\ell+1}}$	$\sqrt{\frac{\ell+\frac{1}{2}-m}{2\ell+1}}$
$\ell - \frac{1}{2}$	$\sqrt{\frac{\ell+\frac{1}{2}-m}{2\ell+1}}$	$-\sqrt{\frac{\ell+\frac{1}{2}+m}{2\ell+1}}$

$\langle 1 m_s \ell m_\ell | jm \rangle$

j	$m_s = +1$	$m_s = 0$	$m_s = -1$
$\ell + 1$	$\sqrt{\frac{(\ell+m)(\ell+m+1)}{2(\ell+1)(2\ell+1)}}$	$\sqrt{\frac{(\ell-m+1)(\ell+m+1)}{(\ell+1)(2\ell+1)}}$	$\sqrt{\frac{(\ell-m)(\ell-m+1)}{2(\ell+1)(2\ell+1)}}$
ℓ	$\sqrt{\frac{(\ell+m)(\ell-m+1)}{2\ell(\ell+1)}}$	$\frac{-m}{\sqrt{\ell(\ell+1)}}$	$-\sqrt{\frac{(\ell-m)(\ell+m+1)}{2\ell(\ell+1)}}$
$\ell - 1$	$\sqrt{\frac{(\ell-m)(\ell-m+1)}{2\ell(2\ell+1)}}$	$-\sqrt{\frac{(\ell-m)(\ell+m)}{\ell(2\ell+1)}}$	$\sqrt{\frac{(\ell+m+1)(\ell+m)}{2\ell(2\ell+1)}}$

$$\begin{pmatrix} j & 0 & j' \\ -m & 0 & m' \end{pmatrix} = \langle jmj' -m'|00\rangle = \frac{(-1)^{j-m}}{\sqrt{2j+1}} \delta_{jj'} \delta mm' \qquad \langle jm00|j'm'\rangle = \delta_{jj'}\,\delta_{mm'}$$

$$\begin{pmatrix} j & 1 & j \\ -m & 0 & m \end{pmatrix} = (-1)^{j-m} \frac{m}{\sqrt{j(j+1)(2j+1)}}$$

$$\begin{pmatrix} j & 2 & j \\ -m & 0 & m \end{pmatrix} = (-1)^{j-m} \frac{3m^2 - j(j+1)}{\sqrt{(2j-1)j(j+1)(2j+1)(2j+3)}}$$

$$\begin{pmatrix} j_1 & j_2 & j_3 \\ 0 & 0 & 0 \end{pmatrix} = \begin{cases} (-1)^g \sqrt{\frac{(2g-2j_1)!(2g-2j_2)!(2g-2j_3)!}{(2g+1)!}} \frac{g!}{(g-j_1)!(g-j_2)!(g-j_3)!} & \text{if } 2g = \text{even} \\ 0 & \text{if } 2g = \text{odd} \end{cases}$$

where $2g = j_1 + j_2 + j_3$.

The orthogonality relations between the coefficients are

$$\sum_{m_1 m_2} \begin{pmatrix} j_1 & j_2 & j_3 \\ m_1 & m_2 & m_3 \end{pmatrix} \begin{pmatrix} j_1 & j_2 & j_3' \\ m_1 & m_2 & m_3' \end{pmatrix} = \Delta(j_1 j_2 j_3) \frac{\delta_{j_3 j_3'} \delta_{m_3 m_3'}}{2j_3 + 1} \tag{B-35}$$

$$\sum_{m_1 m_2 m_3} \begin{pmatrix} j_1 & j_2 & j_3 \\ m_1 & m_2 & m_3 \end{pmatrix} \begin{pmatrix} j_1 & j_2 & j_3 \\ m_1 & m_2 & m_3 \end{pmatrix} = \Delta(j_1 j_2 j_3) \tag{B-36}$$

$$\sum_{j_3 m_3} (2j_3 + 1) \begin{pmatrix} j_1 & j_2 & j_3 \\ m_1 & m_2 & m_3 \end{pmatrix} \begin{pmatrix} j_1 & j_2 & j_3 \\ m_1' & m_2' & m_3 \end{pmatrix} = \Delta(j_1 j_2 j_3)\, \delta_{m_1 m_1'} \delta_{m_2 m_2'} \tag{B-37}$$

where

$$\Delta(j_1 j_2 j_3) = \begin{cases} 1 & \text{for} \quad |j_1 - j_2| \le j_3 \le j_1 + j_2 \\ 0 & \text{otherwise.} \end{cases}$$

In terms of Clebsch-Gordan coefficients, the same relations may be expressed in the forms:

$$\sum_{m_1 m_2} \langle j_1 m_1 j_2 m_2 | j_3 m_3 \rangle \langle j_1 m_1 j_2 m_2 | j_3' m_3' \rangle = \Delta(j_1 j_2 j_3)\, \delta_{j_3 j_3'} \delta_{m_3 m_3'} \tag{B-38}$$

$$\sum_{m_1 m_2 m_3} \langle j_1 m_1 j_2 m_2 | j_3 m_3 \rangle \langle j_1 m_1 j_2 m_2 | j_3 m_3 \rangle = (2j_3 + 1) \Delta(j_1 j_2 j_3) \tag{B-39}$$

$$\sum_{j_3 m_3} \langle j_1 m_1 j_2 m_2 | j_3 m_3 \rangle \langle j_1 m_1' j_2 m_2' | j_3 m_3 \rangle = \Delta(j_1 j_2 j_3) \delta_{m_1 m_1'} \delta_{m_2 m_2'} \tag{B-40}$$

The Condon and Shortley phase convention, commonly adopted nowadays, states that in coupling j_1 and j_2 to the maximum possible angular momentum $(j_1 + j_2)$ and all the projections on the z-axis are the maximum values allowed, *i.e.*, $m_1 = j_1$ and $m_2 = j_2$, the Clebsch-Gordan coefficient

$$\langle j_1 j_1 j_2 j_2 | j_1 + j_2 \; j_1 + j_2 \rangle = +1 \tag{B-41}$$

and that

$$\sum_{m_1 m_2} m_1 \langle j_1 m_1 j_2 m_2 | j_3 m_3 \rangle \langle j_1 m_1 j_2 m_2 | (j_3 - 1)\, m_3 \rangle > 0 \tag{B-42}$$

All Clebsch-Gordan coefficients are real in this convention, and explicit values of some of the more useful ones involving low angular ranks are listed in Table B-1.

§B-4 RACAH COEFFICIENT AND 9j-SYMBOL

When three spherical tensors R_{j_1}, S_{j_2}, and T_{j_3} are coupled together, the final rank J alone is not adequate to uniquely specify the product. In order to distinguish between the different possible ways to couple three quantities to a definite final rank J, an intermediate rank specifying the coupling between two of the three tensors is used. There are two equivalent ways to construct this intermediate coupling. One way is to couple the first two tensors R_{j_1} and S_{j_2} together to rank J_{12} and then couple the product to T_{j_3} to obtain the final rank J. The angular momentum structure of the product may be expressed in the form $((R_{j_1} \times S_{j_2})_{J_{12}} \times T_{j_3})_J$. Alternatively, we can couple the last two tensors S_{j_2} and T_{j_3} together first to rank J_{23} and then couple R_{j_1} to the product. This way of coupling may be represented as $(R_{j_1} \times (S_{j_2} \times T_{j_3})_{J_{23}})_J$.

The two forms are not independent of each other and the relation between them is given by

$$\left((R_{j_1} \times S_{j_2})_{J_{12}} \times T_{j_3}\right)_J = \sum_{J_{23}} \sqrt{(2J_{12}+1)(2J_{23}+1)}\, W(j_1 j_2 J j_3; J_{12} J_{23}) \times \left(R_{j_1} \times (S_{j_2} \times T_{j_3})_{J_{23}}\right)_J \tag{B-43}$$

where $W(j_1 j_2 J j_3; J_{12} J_{23})$ is the Racah coefficient. It may be expressed as the sum over products of four Clebsch-Gordan coefficients,

$$W(abcd; ef) = \frac{1}{\sqrt{(2e+1)(2f+1)}} \sum_{\alpha\beta\gamma} \langle a\alpha b\beta | e(\alpha+\beta)\rangle \langle e(\alpha+\beta)\, d(\gamma-\alpha-\beta)|c\gamma\rangle \times \langle b\beta d(\gamma-\alpha-\beta)|f(\gamma-\alpha)\rangle \langle a\alpha f(\gamma-\alpha)|c\gamma\rangle \tag{B-44}$$

However, this is not the way to evaluate a Racah coefficient numerically; it is more convenient to use explicit formulae in terms of its six arguments. These can be found, for example, in Brink and Satchler.

A more convenient form of Racah coefficients is in terms of $6j$-symbols defined by the relation:

$$\begin{Bmatrix} j_1 & j_2 & J_{12} \\ j_3 & J & J_{23} \end{Bmatrix} = (-1)^{j_1+j_2+j_3+J}\, W(j_1 j_2 J j_3; J_{12} J_{23}) \tag{B-45}$$

For example, the symmetry of Racah coefficients may be expressed in terms of $6j$-symbols in the following manner:

$$\begin{Bmatrix} j_1 & j_2 & j_3 \\ j_4 & j_5 & j_6 \end{Bmatrix} = \begin{Bmatrix} j_2 & j_3 & j_1 \\ j_5 & j_6 & j_4 \end{Bmatrix} = \begin{Bmatrix} j_3 & j_1 & j_2 \\ j_6 & j_4 & j_5 \end{Bmatrix} = \begin{Bmatrix} j_2 & j_1 & j_3 \\ j_5 & j_4 & j_6 \end{Bmatrix} = \begin{Bmatrix} j_4 & j_5 & j_3 \\ j_1 & j_2 & j_6 \end{Bmatrix} \tag{B-46}$$

The orthogonality relation between two $6j$-symbols is

$$\sum_j (2j+1) \begin{Bmatrix} j_1 & j_2 & j' \\ j_3 & j_4 & j \end{Bmatrix} \begin{Bmatrix} j_1 & j_2 & j'' \\ j_3 & j_4 & j \end{Bmatrix} = \frac{\delta_{j'j''}}{2j'+1} \tag{B-47}$$

There are also identities involving products of $3j$- and $6j$-symbols which can be found in most advanced texts on nuclear structure.

In coupling four spherical tensors together, two intermediate coupling ranks are needed to specify the product uniquely. The different ways of specifying the intermediate couplings are related to each other through $9j$-symbols,

$$\begin{aligned}\big((R_{j_1}\times S_{j_2})_{J_{12}}\times(T_{j_3}\times U_{j_4})_{J_{34}}\big)_J &= \sum_{J_{13}J_{24}}\sqrt{(2J_{12}+1)(2J_{34}+1)(2J_{13}+1)(2J_{24}+1)}\\ &\quad\times\begin{Bmatrix} j_1 & j_2 & J_{12}\\ j_3 & j_4 & J_{34}\\ J_{13} & J_{24} & J\end{Bmatrix}\big((R_{j_1}\times T_{j_3})_{J_{13}}\times(S_{j_2}\times U_{j_4})_{J_{24}}\big)_J \qquad \text{(B-48)}\end{aligned}$$

The value of a $9j$-symbol may be expressed as the sum over products of three $6j$-symbols,

$$\begin{Bmatrix} j_1 & j_2 & J_{12}\\ j_3 & j_4 & J_{34}\\ J_{13} & J_{24} & J\end{Bmatrix} = \sum_{J'}(-1)^{2J'}(2J'+1)\begin{Bmatrix} j_1 & j_3 & J_{13}\\ J_{24} & J & J'\end{Bmatrix}\begin{Bmatrix} j_2 & j_4 & J_{24}\\ j_3 & J' & J_{34}\end{Bmatrix}\begin{Bmatrix} J_{12} & J_{34} & J\\ J' & j_1 & j_2\end{Bmatrix} \qquad \text{(B-49)}$$

The symmetries of $9j$-symbols are

$$\begin{aligned}\begin{Bmatrix} j_1 & j_2 & j_3\\ j_4 & j_5 & j_6\\ j_7 & j_8 & j_9\end{Bmatrix} &= \begin{Bmatrix} j_1 & j_4 & j_7\\ j_2 & j_5 & j_8\\ j_3 & j_6 & j_9\end{Bmatrix} = \begin{Bmatrix} j_7 & j_8 & j_9\\ j_1 & j_2 & j_3\\ j_4 & j_5 & j_6\end{Bmatrix} = \begin{Bmatrix} j_4 & j_5 & j_6\\ j_7 & j_8 & j_9\\ j_1 & j_2 & j_3\end{Bmatrix}\\ &= (-1)^{j_1+j_2+j_3+j_4+j_5+j_6+j_7+j_8+j_9}\begin{Bmatrix} j_4 & j_5 & j_6\\ j_1 & j_2 & j_3\\ j_7 & j_8 & j_9\end{Bmatrix} \qquad \text{(B-50)}\end{aligned}$$

and the orthogonality relation is

$$\sum_{J_{13}J_{24}}(2J_{13}+1)(2J_{24}+1)\begin{Bmatrix} j_1 & j_2 & J_{12}\\ j_3 & j_4 & J_{34}\\ J_{13} & J_{24} & J\end{Bmatrix}\begin{Bmatrix} j_1 & j_2 & J'_{12}\\ j_3 & j_4 & J'_{34}\\ J_{13} & J_{24} & J\end{Bmatrix} = \frac{\delta_{J_{12}J'_{12}}\delta_{J_{34}J'_{34}}}{(2J_{12}+1)(2J_{34}+1)} \qquad \text{(B-51)}$$

A collection of symmetry and orthogonality relations as well as relations between $3j$-, $6j$- and $9j$-symbols can be found, *e.g.*, in the appendices of Brink and Satchler, and Wong (*Nuclear Statistical Spectroscopy*, Oxford University Press, New York, 1986).

§B-5 WIGNER-ECKART THEOREM

One of the advantages of using tensors of definite spherical ranks is offered by the Wigner-Eckart theorem. The matrix element of an operator of rank k between states of angular momenta J and J' may be separated into two parts, one part invariant under a rotation of the coordinate system used and the other part expressing the dependence of the matrix element on the coordinate system. Since only projections of tensors on the quantization axis are changed by a rotation of the axes, the invariant part of the matrix element is independent of the projections and, as a result, is a function of the nature of the operator and the states involved. The Wigner-Eckart theorem states that the dependence on the coordinate system is given by an angular momentum coupling coefficient and may be expressed in the form:

$$\langle JM|T_{kq}|J'M'\rangle = (-1)^{J-M}\begin{pmatrix} J & k & J' \\ -M & q & M' \end{pmatrix}\langle J\|T_k\|J'\rangle \tag{B-52}$$

Here the double-bar matrix element $\langle J\|T_k\|J'\rangle$ represents the invariant part and is generally known as the *reduced matrix element.* The angular dependence of the matrix element is contained in the $3j$-coefficients and is independent of the operator and the states involved other than the angular momentum ranks. All the physical content of a matrix element is contained in the reduced matrix element which may be compared with other physical quantities without being encumbered by dependence on the coordinate system used.

In terms of Clebsch-Gordan coefficients, (B-52) appears in the form

$$\langle JM|T_{kq}|J'M'\rangle = (-1)^{2k}\frac{\langle J'M'kq|JM\rangle}{\sqrt{2J+1}}\langle J\|T_k\|J'\rangle \tag{B-53}$$

There are also slightly different ways to define the reduced matrix element. In some books, the phase factor and/or the square root in the denominator are absorbed into the definition of the reduced matrix element. Note that the phase factor $(-1)^{2k}$ is essential here, since we deal with operators of half-integer ranks as well.

§B-6 LANDÉ FORMULA

Consider a vector operator $\boldsymbol{V}$. Since it is an operator of spherical tensor rank unity, its matrix element behaves, under a rotation of the coordinate system, in the same way as any other spherical tensor of the same rank, including the angular momentum operator $\boldsymbol{J}$. Using the Wigner-Eckart theorem, the matrix element of the q-th component of $\boldsymbol{V}$ may be expressed in terms of its reduced matrix element,

$$\langle JM|V_{1q}|JM'\rangle = (-1)^{J-M}\begin{pmatrix} J & 1 & J \\ -M & q & M' \end{pmatrix}\langle J\|\boldsymbol{V}\|J\rangle \tag{B-54}$$

where q has possible values ± 1 and 0. Similarly, the matrix element of $\boldsymbol{J}$ has the form

$$\langle JM|J_{1q}|JM'\rangle = (-1)^{J-M}\begin{pmatrix} J & 1 & J \\ -M & q & M' \end{pmatrix}\langle J\|\boldsymbol{J}\|J\rangle \tag{B-55}$$

Since both reduced matrix elements $\langle J\|\boldsymbol{V}\|J\rangle$ and $\langle J\|\boldsymbol{J}\|J\rangle$ are quantities independent of the coordinate system, they must be multiples of each other,

$$\langle J\|\boldsymbol{V}\|J\rangle = \mathcal{R}\langle J\|\boldsymbol{J}\|J\rangle \tag{B-56}$$

where the ratio

$$\mathcal{R} = \frac{\langle J\|\boldsymbol{V}\|J\rangle}{\langle J\|\boldsymbol{J}\|J\rangle} \tag{B-57}$$

is independent of M.

Consider, now, the matrix element of the scalar product $(\boldsymbol{J}\cdot\boldsymbol{V})$. Since it is a scalar operator, it has nonvanishing matrix elements only along the diagonal; *i.e.*, $J = J'$ and $M = M'$. In a spherical basis, the scalar product may be expressed in the form

$$\boldsymbol{J}\cdot\boldsymbol{V} = \sum_q (-1)^q \boldsymbol{J}_{1q}\boldsymbol{V}_{1,-q} \tag{B-58}$$

We can check that this is the same as scalar products defined in terms of cartesian components of the vectors by noting that

$$\begin{aligned} \boldsymbol{J}_{\pm1} &= \mp\frac{1}{\sqrt{2}}(\boldsymbol{J}_x \pm i\boldsymbol{J}_y) \qquad & \boldsymbol{J}_0 &= \boldsymbol{J}_z \\ \boldsymbol{V}_{\pm1} &= \mp\frac{1}{\sqrt{2}}(\boldsymbol{V}_x \pm i\boldsymbol{V}_y) \qquad & \boldsymbol{V}_0 &= \boldsymbol{V}_z \end{aligned} \tag{B-59}$$

This is slightly different from the definition of angular momentum raising and lowering operators $L_\pm$ given in (B-12,13), since the usual convention for $L_\pm$ does not attempt to make them components of a spherical tensor operator.

We can now make an intermediate state expansion of the diagonal matrix element of $(\boldsymbol{J}\cdot\boldsymbol{V})$,

$$\langle JM|(\boldsymbol{J}\cdot\boldsymbol{V})|JM\rangle = \sum_{M'}\sum_q \langle JM|\boldsymbol{J}_q|JM'\rangle\langle JM'|\boldsymbol{V}_{-q}|JM\rangle$$

Since the operator $\boldsymbol{J}$ can change at most the M-value but not the J-value of a function on which it acts, a sum over intermediate states of different J-values is not needed. Using the ratio $\mathcal{R}$ defined in (B-57) and the relations given by (B-54, 55), we have the relation

$$\begin{aligned} \langle JM'|\boldsymbol{V}_{-q}|JM\rangle &= (-1)^{J-M'}\begin{pmatrix} J & 1 & J \\ -M' & -q & M \end{pmatrix}\langle J\|\boldsymbol{V}\|J\rangle \\ &= (-1)^{J-M'}\begin{pmatrix} J & 1 & J \\ -M' & -q & M \end{pmatrix}\mathcal{R}\langle J\|\boldsymbol{J}\|J\rangle \\ &= \mathcal{R}\langle JM'|\boldsymbol{J}_{-q}|JM\rangle \end{aligned}$$

Using this, we obtain the result

$$\begin{aligned} \langle JM|(\boldsymbol{J}\cdot\boldsymbol{V})|JM\rangle &= \mathcal{R}\sum_{M'}\sum_q \langle JM|\boldsymbol{J}_q|JM'\rangle\langle JM'|\boldsymbol{J}_{-q}|JM\rangle \\ &= \mathcal{R}\langle JM|\boldsymbol{J}^2|JM\rangle \\ &= \mathcal{R}J(J+1) \end{aligned} \tag{B-60}$$

In other words, the ratio $\mathcal{R}$ has the value

$$\mathcal{R} = \frac{1}{J(J+1)}\langle JM|(\boldsymbol{J}\cdot\boldsymbol{V})|JM\rangle \tag{B-61}$$

As a result,

$$\langle JM|V_q|JM'\rangle = \frac{1}{J(J+1)}\langle JM|(\boldsymbol{J}\cdot\boldsymbol{V})|JM\rangle\langle JM|\boldsymbol{J}_q|JM'\rangle \tag{B-62}$$

generally known as Landé formula.

Appendix C

SCATTERING BY A CENTRAL POTENTIAL

§C-1 SCATTERING AMPLITUDE AND CROSS SECTION

The scattering of one particle off another in the center of mass of the two particles at nonrelativistic energies is described by a time-dependent Schrödinger equation,

$$i\hbar\frac{\partial}{\partial t}\Psi(\boldsymbol{r},t) = H\Psi(\boldsymbol{r},t) \tag{C-1}$$

under appropriate boundary conditions. The Hamiltonian has the form,

$$H = -\frac{\hbar^2}{2\mu}\nabla^2 + V \tag{C-2}$$

where μ is the reduced mass and V is the potential representing the interaction between the two particles. If H is independent of time t, the time dependence in the wave function may be separated from the rest of the wave function by the following decomposition:

$$\Psi(\boldsymbol{r},t) = \psi(\boldsymbol{r})e^{-iEt/\hbar}$$

For simplicity we shall consider $\psi(\boldsymbol{r})$ to be a function of spatial coordinates only and ignore any dependence of the wave function on other variables such as spin and isospin.

The function $\psi(\boldsymbol{r})$ is the eigenfunction of the time-independent Schrödinger equation,

$$-\frac{\hbar^2}{2\mu}\nabla^2\psi(\boldsymbol{r}) + (V - E)\psi(\boldsymbol{r}) = 0 \tag{C-3}$$

If the interaction between the two particles is given by a central potential $V(r)$ that depends only on the relative distance r between them, angular momentum is a constant of motion and the wave function $\psi(\boldsymbol{r})$ may be decomposed into a product of radial and angular parts, as done in §B-1.

Incident flux. The usual scattering arrangement involves a collimated beam of projectile particles travelling along the positive z-direction and incident on a target placed at the origin. Except for the Coulomb force, interactions between nuclei have a short range. For this reason, we shall consider first potentials with finite range and return to the Coulomb interaction in §C-5. Outside the range of the interaction, we can take $V(r) = 0$; the particles are free and their wave functions represented by plane waves e^{ikz}, where $k = \sqrt{2\mu E}/\hbar$ is the wave number. (For a Coulomb interaction, Coulomb wave functions must be used instead of plane waves.)

The relation between the wave function and the intensity of the incident beam is given by the quantum mechanical probability current density,

$$\boldsymbol{S}(\boldsymbol{r},t) = \text{Re}\left\{\psi^* \frac{\hbar}{i\mu}\nabla\psi\right\} = \frac{\hbar}{\mu}\,\text{Im}\left\{\psi^*\nabla\psi\right\}$$

For an incident plane wave travelling along the positive z-direction, the number of particles passing through a unit area perpendicular to the z-axis is then

$$S_i = \frac{\hbar}{\mu}\text{Im}\left\{e^{-ikz}\frac{d}{dz}e^{ikz}\right\} = \frac{\hbar k}{\mu} = v \tag{C-4}$$

where v is the velocity of the projectile far away from the interaction region. The value of incident flux S_i depends on the way the plane wave is normalized. Here we have taken it in such a way that $S_i = v$.

Scattered wave. The scattered particle outside the interaction region is described by a spherical wave e^{ikr}/r, radiating outward from the center of the interaction region. The particle density in the incident beam is usually sufficient low that we may ignore any interference between incident and scattering particles. As a result, the wave function at large r is a linear combination of a plane wave made of the incident beam and particles not scattered by the potential, and a spherical wave, made of scattered particles. The result may be expressed in the form

$$\psi(\boldsymbol{r}) \xrightarrow[r\to\infty]{} e^{ikz} + f(\theta,\phi)\frac{e^{ikr}}{r} \tag{C-5}$$

Here, $f(\theta,\phi)$ is the *scattering amplitude* which measures the fraction of incident wave scattered in the direction with polar angle θ and azimuthal angle ϕ. In general, both $\psi(\boldsymbol{r})$ and $f(\theta,\phi)$ are functions of the incident wave vector $\boldsymbol{k}$ and scattered wave vector $\boldsymbol{k}'$ as well. However, to simplify the notation, we shall suppress them unless required in the discussion. Furthermore, the probability for scattering is sufficiently small that the normalization of the incident wave is not affected by particles removed from the incident beam due to scattering.

If all the particles involved in the scattering have spin $J = 0$ or if the spins of neither the incident particles nor the target are polarized in any given direction and the orientations of the spin of the particles in the final state are not detected, we have only one direction defined by the problem itself; the direction along which the two particles approach each other. We have already chosen this direction to be the z-axis and take as the $+z$ the direction of motion of the projectile. Scatterings are usually observed in relation to this direction. If the origin of the coordinate system is taken to be the center of the system where the two particles come into contact with each other, we have also fixed the xy-plane, since it must be perpendicular to the z-axis. However, we do not have a natural way to define the orientation of the x-axis (or the y-axis) in the plane. This is a result of the invariance of our system under a rotation around the z-axis. Such a symmetry is removed if one or more of the particles have a nonzero spin and are polarized in directions other than that along the z-axis. A transverse spin polarization may be used as a reference direction to fix either the x- or the y-axis (and the other one follows by orthogonality requirement). In the absence of such a direction, the azimuthal angle ϕ cannot be determined uniquely and the wave function of the system must be independent of ϕ. In such a system, the scattering amplitude becomes a function of the polar angle θ only.

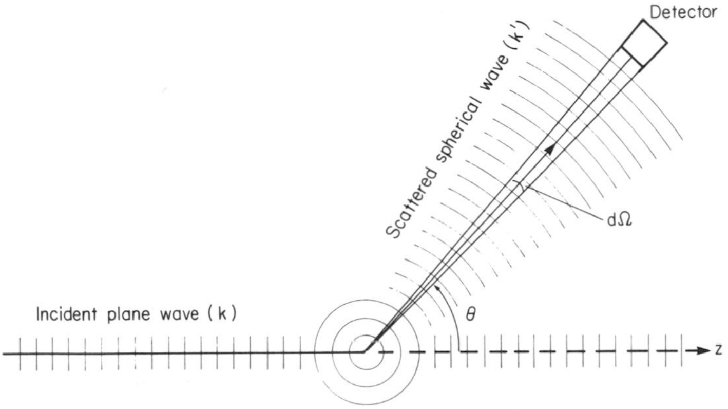

Fig. C-1 The scattering plane in which both the wave vector of the projectile $\boldsymbol{k}$ and that of the scattered particle $\boldsymbol{k}'$ lie. The scattering angle θ is the angle between $\boldsymbol{k}$ and $\boldsymbol{k}'$. The scattering is independent of the azimuthal angle ϕ unless the polarization direction of the spin of at least one of the particles, either in the initial state or in the final state, is known.

The scattering angle θ is the angle between the incident wave vector $\boldsymbol{k}$ and the scattered wave vector $\boldsymbol{k}'$, as shown in Fig. C-1. If $\theta \neq 0$, $\boldsymbol{k}$ and $\boldsymbol{k}'$ forms a plane, the scattering plane. We may define a unit vector $\boldsymbol{n}$ perpendicular to the scattering plane in the following way:

$$\hat{\boldsymbol{n}} = \frac{\boldsymbol{k} \times \boldsymbol{k}'}{|\boldsymbol{k}|\ |\boldsymbol{k}'|} \tag{C-6}$$

The orientation of $\boldsymbol{n}$ depends on the vector $\boldsymbol{k}'$ which, in turn, depends on where the detector is placed. Unless polarization is involved, the choice of the direction of $\boldsymbol{n}$ is therefore arbitrary, determined by the convenience of the experimental arrangement. However, if one or both particles involved in the initial state are polarized, or if the polarizations of one or both of the particles in the final state are detected, the spin-dependence of the interaction between the two particles may cause a difference in the scattering results depending on the direction of $\boldsymbol{n}$ relative to that of polarization. Under such conditions, the scattering amplitude is a function of θ as well as ϕ.

Differential cross section. Next we shall express the differential scattering cross section in terms of the scattering amplitude $f(\theta)$. The probability current density for the scattered spherical wave is given by the expression

$$S_r = \frac{\hbar}{\mu}\mathrm{Im}\left\{\left(f(\theta)\frac{e^{ikr}}{r}\right)^* \frac{d}{dr}\left(f(\theta)\frac{e^{ikr}}{r}\right)\right\} = \frac{v}{r^2}\left|f(\theta)\right|^2 + O(r^{-3}) \qquad \text{(C-7)}$$

If the scattered particle is observed by a detector of area da placed at distance r from the scattering center, the solid angle subtended by the detector at the origin is

$$d\Omega = \frac{da}{r^2}$$

and the number of particles detected per unit time is

$$N_r = S_r da = S_r r^2 d\Omega$$

The differential scattering cross section, $d\sigma/d\Omega$, sometimes represented also as $\sigma(\theta)$, is defined as the number of particles scattered into a solid angle $d\Omega$ at angle θ divided by the incident flux,

$$\frac{d\sigma}{d\Omega} = \frac{S_r r^2}{S_i} = \left|f(\theta)\right|^2 \qquad \text{(C-8)}$$

It has the dimension of an area divided by solid angle and gives a measure of the probability of scattering into a particular direction.

The scattering cross section is the integral of the differential cross section over all solid angles:

$$\sigma = \int \frac{d\sigma}{d\Omega} d\Omega = \int \left|f(\theta)\right|^2 2\pi \sin\theta\, d\theta \qquad \text{(C-9)}$$

It has the dimension of an area as it conveys an idea how much the incident beam is intercepted by each particle in the target. Since the typical unit of length for nuclei is the femtometer (fm), a convenient unit for scattering cross section is fm^2 ($= 10^{-30}$ m^2) and that for $d\sigma/d\Omega$ is fm^2/sr. An alternate unit, a barn (1 barn$= 10^{-28}$ m^2), is still the more commonly used unit in quoting measured values. Hadronic processes are usually of the order of a millibarn (1 mb$= 10^{-31}$ m^2 or 0.1 fm^2), whereas electromagnetic processes are of the order of 1 nanobarn (1 nb$= 10^{-37}$ m^2), and weak interaction processes are of the order of femtobarn (1 fb $= 10^{-43}$ m^2), as mentioned in Chapter I.

§C-2 PARTIAL WAVES AND PHASE SHIFTS

Expansion in partial waves. The information obtained in a scattering experiment is often expressed in terms of phase shifts. For a spherical potential, we can write the wave function as a sum over components with definite orbital angular momentum ℓ, or *partial waves*,

$$\psi(r,\theta) = \sum_{\ell=0}^{\infty} a_\ell R_\ell(r) Y_{\ell 0}(\theta) \tag{C-10}$$

where the coefficients a_ℓ are the amplitudes of each partial wave. Only $Y_{\ell m}(\theta,\phi)$ with $m=0$ are involved here, as the system we are considering is independent of the azimuthal angle ϕ (see B-16).

Since $Y_{\ell 0}(\theta)$ is an eigenfunction of the angular part of (C-3) with eigenvalue $\ell(\ell+1)$ as shown earlier in §B-1, the radial wave function for partial wave ℓ satisfies the equation,

$$-\frac{\hbar^2}{2\mu}\left\{\frac{1}{r^2}\frac{d}{dr}r^2\frac{d}{dr} - \frac{\ell(\ell+1)}{r^2}\right\}R_\ell(r) + V(r)R_\ell(r) = ER_\ell(r)$$

In terms of the modified radial wave function $u_\ell(r) \equiv rR_\ell(r)$, the equation may be simplified to the form,

$$\frac{d^2u_\ell(r)}{dr^2} - \left\{\frac{\ell(\ell+1)}{r^2} + \frac{2\mu}{\hbar^2}V(r) - k^2\right\}u_\ell(r) = 0 \tag{C-11}$$

For the short-range potentials we are concerned with here, $V(r)$ goes to zero as $r\to\infty$. The same is also true for the $\ell(\ell+1)/r^2$ term. In the asymptotic regions, we are then left with a simple second order differential equation of the form,

$$\frac{d^2u_\ell(r)}{dr^2} + k^2u_\ell(r) = 0$$

The solution for this equation is the familiar linear combination of $\sin(kr)$ and $\cos(kr)$. That is, at large r, the function $u_\ell(r)$ must take on the form

$$\begin{aligned} u_\ell(r) \xrightarrow[r\to\infty]{} & A_\ell \sin(kr - \tfrac{1}{2}\ell\pi) + B_\ell\cos(kr - \tfrac{1}{2}\ell\pi) \\ =& C_\ell \sin(kr - \tfrac{1}{2}\ell\pi + \delta_\ell) \\ =& C'_\ell\{e^{-i(kr-\frac{1}{2}\ell\pi)} - e^{2i\delta_\ell}e^{i(kr-\frac{1}{2}\ell\pi)}\} \end{aligned} \tag{C-12}$$

where A_ℓ and B_ℓ, or C_ℓ (C'_ℓ) and δ_ℓ are two constants that must be determined from boundary conditions. The reason for the additional phase factor $\frac{1}{2}\ell\pi$ in the arguments will become clear in the next paragraph.

Phase shift. The angle δ_ℓ is known as the *phase shift*. Its physical meaning can be seen by comparing (C-12) with the partial wave expansion of a plane wave,

$$e^{ikz} = \sum_{\ell=0}^{\infty} \sqrt{4\pi(2\ell+1)}\, i^\ell j_\ell(kr) Y_{\ell 0}(\theta) \tag{C-13}$$

Asymptotically, the spherical Bessel function $j_\ell(kr)$ has the form

$$j_\ell(kr) \xrightarrow[r\to\infty]{} \frac{\sin(kr - \frac{1}{2}\ell\pi)}{kr} \tag{C-14}$$

It is for the convenience of comparison with this expression that we have included the factor $\frac{1}{2}\ell\pi$ in (C-12).

For a plane wave in the asymptotic region,

$$\begin{aligned} e^{ikz} &\xrightarrow[r\to\infty]{} \sum_{\ell=0}^{\infty} \sqrt{4\pi(2\ell+1)}\, \frac{i^\ell}{kr} \sin(kr - \tfrac{1}{2}\ell\pi)\, Y_{\ell 0}(\theta) \\ &= \sum_{\ell=0}^{\infty} \sqrt{4\pi(2\ell+1)}\, \frac{i^\ell}{2ikr} \left\{ e^{i(kr-\frac{1}{2}\ell\pi)} - e^{-i(kr-\frac{1}{2}\ell\pi)} \right\} Y_{\ell 0}(\theta) \\ &= \sum_{\ell=0}^{\infty} \sqrt{4\pi(2\ell+1)} \left\{ \frac{e^{ikr}}{2ikr} - \frac{i^\ell e^{-i(kr-\frac{1}{2}\ell\pi)}}{2ikr} \right\} Y_{\ell 0}(\theta) \end{aligned} \tag{C-15}$$

where we have used the relation $e^{i\ell\pi/2} = i^\ell$ to put the expression in a form convenient for later needs. The difference between (C-12) and (C-15) is the phase shift, for example, in the argument of the sine function. Because of scattering due to potential $V(r)$, the phase of partial wave ℓ in (C-12) is shifted by a factor δ_ℓ with respect to that of a free particle represented by the plane wave of (C-15). This is a result we could have anticipated from the beginning. For a real potential, which we have implicitly assumed here, only elastic scattering can take place. Furthermore, if the potential is also a central one, orbital angular momentum ℓ is a good quantum number and the probability current density in each ℓ-partial wave "channel" is conserved. The only thing in the wave function that can change as a result of the action of $V(r)$ is the phase angle, and this is represented by the phase shift δ_ℓ. We shall return at the end of this section with a concrete example using a square well as illustration.

In general, elastic as well as inelastic scattering can take place. Such a situation is represented by a complex scattering potential with the imaginary part representing the loss of probability from the incident channel due to such inelastic events as excitation of the target nucleus and projectile particle, absorption of the incident particle by the target, and creation of new particles. In such cases, the phase shifts obtained by solving the scattering equation (C-3) are complex in general. We shall return to the case of scattering by a complex potential in §C-4.

Elastic scattering cross section. Using the result of (C-12), the scattering wave function of (C-10) in the asymptotic region may be written in the form

$$\psi(r,\theta) \xrightarrow[r\to\infty]{} \sum_{\ell=0}^{\infty} a'_\ell Y_{\ell 0}(\theta)\frac{1}{r}\sin(kr - \tfrac{1}{2}\ell\pi + \delta_\ell) \tag{C-16}$$

where the unknown coefficient a_ℓ in (C-10) and C_ℓ in (C-15) are combined into a single quantity a'_ℓ. Since this is just another asymptotic form of the same wave function as given earlier in (C-5), we arrive at the equality

$$\begin{aligned} e^{ikz} + f(\theta)\frac{e^{ikr}}{r} &= \sum_{\ell=0}^{\infty} a'_\ell Y_{\ell 0}(\theta)\frac{1}{r}\sin(kr - \tfrac{1}{2}\ell\pi + \delta_\ell) \\ &= \sum_{\ell=0}^{\infty} a'_\ell Y_{\ell 0}(\theta)\left\{(-i)^\ell e^{i\delta_\ell}\frac{e^{ikr}}{2ikr} - e^{-i\delta_\ell}\frac{e^{-i(kr-\frac{1}{2}\ell\pi)}}{2ikr}\right\} \end{aligned} \tag{C-17}$$

Using the results of (C-15) and (C-16), we can rewrite (C-17) in the following form:

$$\begin{bmatrix} \left\{\sum_{\ell=0}^{\infty}\sqrt{4\pi(2\ell+1)}\frac{1}{2ik}Y_{\ell 0}(\theta) + f(\theta)\right\}\frac{e^{ikr}}{r} \\ -\sum_{\ell=0}^{\infty}\left\{\sqrt{4\pi(2\ell+1)}i^\ell Y_{\ell 0}(\theta)\right\}\frac{e^{-i(kr-\frac{1}{2}\ell\pi)}}{2ikr} \end{bmatrix} = \begin{bmatrix} \left\{\sum_{\ell=0}^{\infty} a'_\ell\frac{1}{2ik}Y_{\ell 0}(\theta)(-i)^\ell e^{i\delta_\ell}\right\}\frac{e^{ikr}}{r} \\ -\sum_{\ell=0}^{\infty}\left\{a'_\ell Y_{\ell 0}(\theta)e^{-i\delta_\ell}\right\}\frac{e^{-i(kr-\frac{1}{2}\ell\pi)}}{2ikr} \end{bmatrix} \tag{C-18}$$

This equation has been put into a form such that terms related to e^{ikr} are on the first line and terms related to e^{-ikr} are on the second line of both sides of the equation.

Since the functions e^{ikr} and e^{-ikr} are linearly independent of each other, their coefficients on the two sides of (C-18) must equal to each other. From the coefficients for $e^{-i(kr-\ell\pi/2)}$, we obtain the result,

$$a'_\ell = \sqrt{4\pi(2\ell+1)}\, i^\ell e^{i\delta_\ell}$$

Substituting this relation back into (C-18), we obtain an expression of the scattering amplitude in terms of phase shifts,

$$\begin{aligned} f(\theta) &= \frac{\sqrt{4\pi}}{2ik}\sum_{\ell=0}^{\infty}\sqrt{2\ell+1}(e^{2i\delta_\ell}-1)Y_{\ell 0}(\theta) \\ &= \frac{\sqrt{4\pi}}{k}\sum_{\ell=0}^{\infty}\sqrt{2\ell+1}\,e^{i\delta_\ell}\sin\delta_\ell Y_{\ell 0}(\theta) \end{aligned} \tag{C-19}$$

from the equality of the coefficients for e^{ikr} on both sides of the equation. In terms of the phase shifts, the differential scattering cross section may be written in the form

$$\frac{d\sigma}{d\Omega} = \frac{4\pi}{k^2}\left|\sum_{\ell=0}^{\infty}\sqrt{2\ell+1}\, e^{i\delta_\ell}\sin\delta_\ell Y_{\ell 0}(\theta)\right|^2 \tag{C-20}$$

by substituting the results of (C-19) into (C-8).

Using the orthogonal condition of the spherical harmonics given in (B-10), the scattering cross section reduces to a particularly simple form

$$\begin{aligned}\sigma^{\text{el}} &= \frac{4\pi}{k^2}\sum_{\ell\ell'}\sqrt{(2\ell+1)(2\ell'+1)}e^{i(\delta_\ell-\delta_{\ell'})}\sin\delta_\ell\sin\delta_{\ell'}\int_0^\pi Y_{\ell 0}(\theta)Y_{\ell' 0}(\theta)2\pi\sin\theta\, d\theta \\ &= \frac{4\pi}{k^2}\sum_{\ell=0}^{\infty}(2\ell+1)\sin^2\delta_\ell \\ &= \frac{\pi}{k^2}\sum_{\ell}(2\ell+1)|1-e^{2i\delta_\ell}|^2\end{aligned} \tag{C-21}$$

Since we have taken the scattering potential $V(r)$ to be real in this section, only elastic scattering can take place. Later on, when we discuss the more general case of a complex scattering potential, inelastic scattering can also take place. The superscript is to remind us that the cross section calculated here is for elastic scattering only.

Relation to scattering potential. A more direct connection between phase shift for partial ℓ wave and scattering potential is provided by the following analysis. Eq. (C-11) may be further simplified into the form,

$$\frac{d^2u_\ell(\rho)}{d\rho^2} - \left\{\frac{V(\rho)}{E} + \frac{\ell(\ell+1)}{\rho^2} - 1\right\}u_\ell(\rho) = 0 \tag{C-22}$$

by making the substitution $\rho = kr$. For a free particle, we have $V = 0$ and the corresponding modified radial wave function $f_\ell(\rho)$ for partial wave ℓ satisfies the equation

$$\frac{d^2 f_\ell(\rho)}{d\rho^2} - \left\{\frac{\ell(\ell+1)}{\rho^2} - 1\right\}f_\ell(\rho) = 0 \tag{C-23}$$

where $f_\ell(\rho) = \rho\, j_\ell(\rho)$ and $j_\ell(\rho)$ is the spherical Bessel function of order ℓ.

The ℓ-dependent term as well as the constant term in (C-22) and (C-23) may be eliminated by multiplying (C-22) with $f_\ell(\rho)$ and subtracting from it (C-23) multiplied by $u_\ell(\rho)$. The result takes on the form

$$\frac{d}{d\rho}\left\{\frac{df_\ell}{d\rho}u_\ell - f_\ell\frac{du_\ell}{d\rho}\right\} + \frac{V(\rho)}{E}f_\ell(\rho)u_\ell(\rho) = 0 \tag{C-24}$$

As $r \to \infty$, the spherical Bessel function $j_\ell(\rho) \to \rho^{-1}\sin(\rho - \frac{1}{2}\ell\pi)$ as we have seen earlier, and we obtain the results:

$$f_\ell(\rho) \to \sin(\rho - \tfrac{1}{2}\ell\pi) \qquad \frac{df_\ell}{d\rho} \to \cos(\rho - \tfrac{1}{2}\ell\pi)$$

and

$$u_\ell(\rho) \to \sin(\rho - \tfrac{1}{2}\ell\pi + \delta_\ell) \qquad \frac{du_\ell}{d\rho} \to \cos(\rho - \tfrac{1}{2}\ell\pi + \delta_\ell)$$

The quantity within the curly brackets in (C-24) becomes

$$\begin{aligned}\frac{df_\ell}{d\rho}u_\ell - f_\ell\frac{du_\ell}{d\rho} &\longrightarrow \cos(\rho - \tfrac{1}{2}\ell\pi)\sin(\rho - \tfrac{1}{2}\ell\pi + \delta_\ell) - \sin(\rho - \tfrac{1}{2}\ell\pi)\cos(\rho - \tfrac{1}{2}\ell\pi + \delta_\ell)\\ &= \sin\delta_\ell\end{aligned}$$

where the last equality is obtain using standard trigonometry identity. Eq. (C-24) now reduces to the form,

$$\frac{d}{d\rho}\sin\delta_\ell = -\frac{V(\rho)}{E}f_\ell(\rho)u_\ell(\rho)$$

or

$$\sin\delta_\ell = -\int_0^\infty \frac{V(\rho)}{E}f_\ell(\rho)u_\ell(\rho)\,d\rho \tag{C-25}$$

This relation determines the phase shift δ_ℓ up to a multiple of 2π. The general convention to fix this uncertainty is to take $\delta_\ell = 0$ as $E \to 0$. Although (C-25) expresses δ_ℓ in terms of $V(r)$, the relation is not as direct as it appears on the surface, since $u_\ell(\rho)$ in the integrand also depends on the potential, as can be seen from (C-22).

Partial wave and scattering energy. One useful result of partial wave analysis is the fact that for low bombarding energies, only phase shifts for $\ell \approx 0$ are substantially different from zero. This can be seen from the following argument. The classical turning radius r_1 is defined as the point where the (repulsive) potential is equal to the incident energy. For partial wave channel ℓ, the effective potential is

$$\tilde{V}(r) = V(r) + \frac{\hbar^2}{2\mu}\frac{\ell(\ell+1)}{r^2} \tag{C-26}$$

as can be seen from (C-11). As a result, we may use the relation

$$E = V(r_1) + \frac{\hbar^2}{2\mu}\frac{\ell(\ell+1)}{r_1^2} \tag{C-27}$$

to determine the classical turning point r_1.

For a short-range potential, the effective potential $\tilde{V}(r)$ of (C-26) for sufficiently large ℓ is dominated by the repulsive centrifugal barrier term $\ell(\ell+1)/r^2$ at large r. (At very small r, the centrifugal term also dominates by virtue of its inverse r^2 dependence; consequently only in the intermediate range is the nuclear potential important in the effective potential.) As a result, (C-22) and (C-23) become the same for large ℓ values and we obtain the result,

$$\lim_{\ell \to \infty} u_\ell(r) = f_\ell(r)$$

Consequently,

$$\delta_\ell \xrightarrow[\ell \to \infty]{} 0$$

We shall now establish a criterion by which ℓ may be considered to be large enough such that phase shifts may be ignored for partial waves of order greater than this value.

Let the range of the potential $V(r)$ be represented by r_0. At low energies, $r_0 < r_1$, and we may ignore the contribution of $V(r_1)$ in the definition of the turning radius. Eq. (C-27) may now be approximated by the expression,

$$E \approx \frac{\hbar^2}{2\mu} \frac{\ell(\ell+1)}{r_1^2}$$

or

$$(kr_1)^2 \approx \ell(\ell+1) \tag{C-28}$$

This gives us a definition of the turning radius that is independent of $V(r)$. It also implies that the scattering takes place mainly in channels with $\ell \lesssim kr_1$. Hence for $\ell \gg kr_1$, the phase shifts $\delta_\ell \to 0$.

On the other hand, r_1 is a quantity that depends both on E and ℓ. It is therefore more convenient to use r_0, the range of the potential, instead of r_1 as the condition for determining the highest partial wave that can contribute to the scattering. Since these two quantities are of the same order of magnitude, we obtain the condition,

$$\delta_\ell \to 0 \qquad \text{for} \qquad \ell \gg kr_0 \tag{C-29}$$

Classically, no scattering occurs in the collision of a point particle with a hard sphere if the impact parameter b is greater than r_0, the radius of the hard sphere. Since $\ell = |\boldsymbol{r} \times \boldsymbol{p}| = \hbar kr$, we arrive at the conclusion that all partial waves with $\ell/\hbar > kr_0$ are not scattered. Eq. (C-29) is essentially a quantum mechanical statement of the same criterium.

The range of nuclear potentials is of the order of a femtometer. For nucleon-nucleon collisions at $E = 1$ MeV in the center of mass, $kr_0 \sim 0.2$. Hence only the $\ell = 0$ or s-wave phase shift can be significantly different from zero. This is observed to be the case as can be seen, for example, in the values of phase shifts extracted from experimental nucleon-nucleon scattering data by Arndt et al., shown in Fig. 3-3. From the figure, we find that only the s-wave phase shifts are different from zero at low energies and the sizes of the phase shifts for the higher partial waves, for example the p-wave phase shifts, do not become significant until $E > 10$ MeV. As a result, nucleon-nucleon collision is often approximated by s-wave scattering for $E < 10$ MeV.

Example of a square-well potential. It is instructive to see the relation between phase shifts and scattering potential for a simple case. We shall limit ourselves to s-wave scattering and calculate δ_0 for a square well of radius r_0 and energy $E = 1$ MeV. For an attractive potential of depth V_0,

$$V(r) = \begin{cases} -V_0 & \text{for } r < r_0 \\ 0 & \text{for } r \geq r_0 \end{cases}$$

the radial equation obtained by solving (C-11) inside the well is

$$u_0(r) = \mathcal{A} \sin \kappa r \qquad \text{for} \qquad r < r_0$$

where

$$\kappa = \frac{1}{\hbar}\sqrt{2\mu(E + V_0)}$$

The amplitude $\mathcal{A}$ will be determined later. For a repulsive well, we can replace V_0 by a negative quantity. In this case κ may become purely imaginary if $E < |V_0|$ and, instead of a sine function, the radial wave function inside the well becomes a hyperbolic sine function.

Outside the well, $V(r) = 0$, and the radial wave function is sinusoidal for either an attractive or a repulsive well,

$$u_0(r) = \sin(kr + \delta_0) \qquad \text{for} \qquad r > r_0$$

For convenience, we shall normalize the wave function to have an amplitude of unity outside the well. The requirement that the logarithmic derivative of the wave function be continuous across the boundary at $r = r_0$ gives the condition,

$$\frac{\sin \kappa r_0}{\kappa \cos \kappa r_0} = \frac{\sin(kr_0 + \delta_0)}{k \cos(kr_0 + \delta_0)}$$

From this result, the s-wave phase shift is found to be

$$\delta_0 = n\pi - kr_0 + \tan^{-1}\left(\frac{k}{\kappa} \tan \kappa r_0\right)$$

where n is to be determined by the condition that $\delta_0 = 0$ at $E = 0$, as we have done for (C-25). The amplitude of the wave function inside the well is determined by the requirement that $u_0(r)$ itself is continuous across the boundary,

$$\mathcal{A} = \frac{\sin(kr_0 + \delta_0)}{\sin(\kappa r_0)}$$

The results are plotted in Fig. C-2.

For an infinite repulsive potential, the radial wave function cannot penetrate inside the well, as shown in Fig. C-2(a), and $u(r) = 0$ for $r \leq r_0$ as a result. Instead of starting at $r = 0$, the wave function is now shifted outward by a distance r_0.

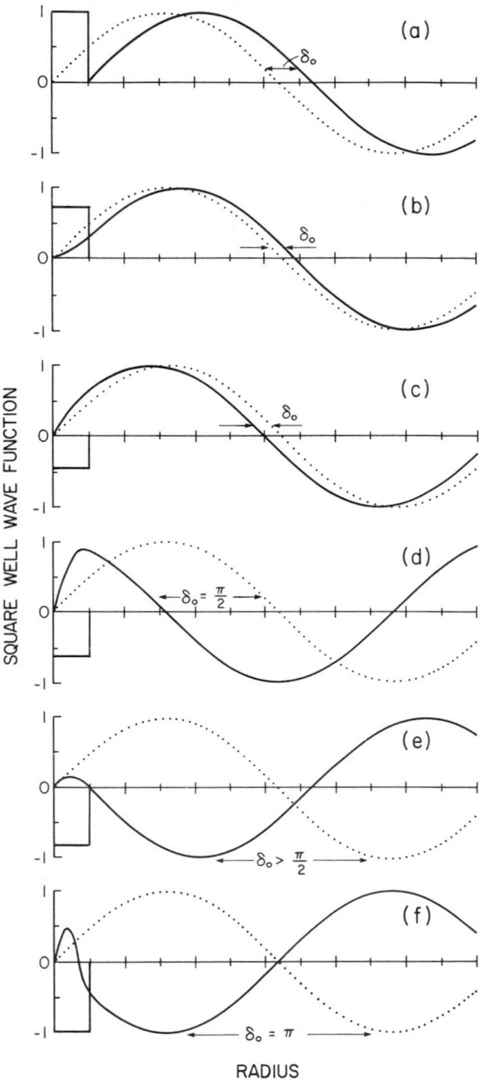

Fig. C-2 Radial wave functions for low-energy *s*-wave scattering by a square well. For comparison, the wave function for a free particle is shown by a dotted curve in each case. The result of an infinite repulsive well is shown in (*a*). For a finite repulsive well, the wave function, shown in (*b*), does not vanish inside the well but the amplitude is diminished compared with that of a free particle ($V_0 = 0$). The wave functions for attractive wells of different depth are shown in (*c*) to (*f*). In contrast with a repulsive well, the wave function inside an attractive well grows faster near the origin than that of a free particle. The difference from the wave function for a free particle is the phase shift and is positive for an attractive potential.

The phase shift is then $\delta_0 = -kr_0$. The scattering cross section from (C-21) is then

$$\sigma = \frac{4\pi}{k^2}\sin^2\delta_0 = \frac{4\pi}{k^2}\sin^2 kr_0 \approx 4\pi r_0^2 \tag{C-30}$$

a result we expect from comparisons with the scattering of two hard spheres of radius r_0 each. For a finite repulsive well, the radial wave function does not vanish completely inside the well. The amplitude rises exponentially with r at small r instead of sinusoidally for a free particle as shown in Fig. C-2(b). The phase shift is still negative, but the magnitude of δ_0 is less than that for an infinite repulsive well.

For an attractive well, the phase shift is positive. If $|V_0|$ is small, the wave function inside the well rises faster near the origin than that of a free particle. As a result, the nodes of the wave function outside the well are shifted closer to the origin as shown in Fig. C-2(c). As the attractive well becomes deeper, the phase shift grows in magnitude. At well depth corresponding to $\delta_0 = \pi/2$, shown in Fig. C-2(d), the scattering cross section becomes $4\pi/k^2$, reaching very large values at low energies. For $E = 0$, we have the result

$$\sigma = \frac{4\pi}{k^2} \longrightarrow \infty$$

The meaning of an infinite scattering cross section at zero energy is that the incident particle never emerges from the potential well; *i.e.*, a bound state is formed at $E = 0$. In fact, a bound state appears whenever the phase shift is an odd integer multiple of $\frac{1}{2}\pi$. On the other hand, when δ_0 is a multiple of π, the cross section drops to zero and nodes in the wave function appear also inside the well, as can be seen in Fig. C-2(e). In realistic situations, the potential has a more complicated form than a square well; however, the qualitative features discussed above remain true.

§C-3 EFFECTIVE RANGE ANALYSIS

Scattering length. For low bombarding energies, it is customary to parametrize experimental scattering data in terms of scattering length a and effective range r_e. Since the scattering cross section must be finite at $E = 0$, we can define a length parameter a in the form

$$\lim_{k \to 0} \sigma = 4\pi a^2 \tag{C-31}$$

Except for a sign, the *scattering length* is given in terms of the s-wave phase shift by comparing (C-31) with (C-21),

$$a = \lim_{k \to 0} \text{Re}\left\{-\frac{1}{k} e^{i\delta_0} \sin\delta_0\right\} \tag{C-32}$$

The sign convention adopted here is such that for an attractive potential the scattering length is positive if there is a bound state, for example, as in the case of isoscalar ($T = 0$) nucleon-nucleon interaction, and $a < 0$ if there is no bound state, for example, as in the case of isovector ($T = 1$) nucleon-nucleon interaction.

Effective range. The energy dependence of scattering at low energies is given by the *effective range* r_e. The origin of this parameter comes from the following rationale. For $\ell = 0$, (C-11) may be written in the form

$$\frac{d^2u_0(k,r)}{dr^2} - \left\{\frac{2\mu}{\hbar^2}V(r) - k^2\right\}u_0(k,r) = 0 \tag{C-33}$$

where we have included the wave number k explicitly in the arguments of the modified radial wave function u_0 in order to emphasize the energy dependence of the solution to (C-33). For two different energies, $E_1 = 2\hbar^2k_1^2/2\mu$ and $E_2 = \hbar^2k_2^2/2\mu$, we have two different solutions of (C-33), $u_0(k_1,r)$ and $u(k_2,r)$, respectively. These two functions satisfy the following equations:

$$u_0''(k_1,r) - \left\{\frac{2\mu}{\hbar^2}V(r) - k_1^2\right\}u_0(k_1,r) = 0$$

$$u_0''(k_2,r) - \left\{\frac{2\mu}{\hbar^2}V(r) - k_2^2\right\}u_0(k_2,r) = 0 \tag{C-34}$$

By multiplying the first equation of (C-34) with $u_0(k_2,r)$ and the second equation with $u_0(k_1,r)$, and integrating the difference over variable r, we obtain the equation,

$$\int_0^\infty \{u_0(k_2,r)\,u_0''(k_1,r) - u_0(k_1,r)\,u_0''(k_2,r)\}\,dr + (k_1^2 - k_2^2)\int_0^\infty u_0(k_1,r)u_0(k_2,r)\,dr = 0$$

The first integral may be carried out by parts and we obtain the result,

$$\{u_0(k_2,r)u_0'(k_1,r) - u_0(k_1,r)u_0'(k_2,r)\}\Big|_0^\infty = (k_2^2 - k_1^2)\int_0^\infty u_0(k_1,r)u_0(k_2,r)\,dr \tag{C-35}$$

This is true for an arbitrary potential including $V(r) = 0$.

Consider another function $v_0(k,r)$ satisfying the same equation as (C-33) except with $V(r) = 0$,

$$\frac{d^2v_0(k,r)}{dr^2} + k^2v_0(k,r) = 0 \tag{C-36}$$

Analogous to (C-35), we have

$$\{v_0(k_2,r)v_0'(k_1,r) - v_0(k_1,r)v_0'(k_2,r)\}\Big|_0^\infty = (k_2^2 - k_1^2)\int_0^\infty v_0(k_1,r)v_0(k_2,r)\,dr \tag{C-37}$$

If the range of the potential in (C-33) is short, (C-33) and (C-36) are identical to each other in the asymptotic region. As a result, we may require that their solutions have the same form at $r = \infty$,

$$v_0(k,r) \underset{r\to\infty}{=} u_0(k,r) \underset{r\to\infty}{=} \mathcal{A}\sin(kr+\delta_0) \tag{C-38}$$

where the amplitude $\mathcal{A}$ will be determined later. Since

$$u_0(k,r) \xrightarrow[r\to 0]{} 0 \qquad\qquad u_0(k,r) \xrightarrow[r\to\infty]{} v_0(k,r)$$

The left hand side of (C-35) may be expressed in terms of $v_0(k,r)$,

$$\begin{aligned}\left\{u_0(k_2,r)u_0'(k_1,r) - u_0(k_1,r)u_0'(k_2,r)\right\}\Big|_0^\infty \\ = \lim_{r\to\infty}\left\{v_0(k_2,r)v_0'(k_1,r) - v_0(k_1,r)v_0'(k_2,r)\right\}\end{aligned}$$

Using this, we can subtract (C-35) from (C-37). The contributions from $r = \infty$ on the left hand side of the two equations cancel each other out and we are left with the result:

$$\begin{aligned}v_0(k_10)\,v_0'(k_2,0) - v_0(k_2,0)v_0'(k_1,0) \\ = (k_2^2 - k_1^2)\int_0^\infty \left\{v_0(k_1,r)v_0(k_2,r) - u_0(k_1,r)u_0(k_2,r)\right\}dr\end{aligned} \tag{C-39}$$

Because of (C-38), we have $v_0(k,r) \neq 0$ at $r = 0$. This result may be used to fix the amplitude $\mathcal{A}$ such that $v_0(k,0) = 1$. As a result,

$$v_0(k,r) = \frac{\sin(kr+\delta_0)}{\sin\delta_0} \tag{C-40}$$

and (C-39) simplifies to the form

$$v_0'(k_2,0) - v_0'(k_1,0) = (k_2^2 - k_1^2)\int_0^\infty \left\{v_0(k_1,r)v_0(k_2,r) - u_0(k_1,r)u_0(k_2,r)\right\}dr$$

Alternatively, we obtain

$$\frac{k_2\cot\delta_0(k_2) - k_1\cot\delta_0(k_1)}{k_2^2 - k_1^2} = \int_0^\infty \left\{v_0(k_1,r)v_0(k_2,r) - u_0(k_1,r)u_0(k_2,r)\right\}dr$$

using (C-40).

If both E_1 and E_2 are close to some energy $E = 2\mu k^2/\hbar^2$, the above expression may be written in the form

$$\frac{d}{d(k^2)}k\cot\delta_0 = \int_0^\infty \left\{v_0^2(k,r) - u_0^2(k,r)\right\}dr$$

The effective range is defined as twice the integral in this expression at $k = 0$,

$$r_e = 2\int_0^\infty \left\{v_0^2(k,r) - u_0^2(k,r)\right\}_{k=0} dr \tag{C-41}$$

The energy dependence of the s-wave phase shift can now be expressed in the form

$$k \cot \delta_0(k) = (k \cot \delta_0)_{k=0} + \tfrac{1}{2} r_e k^2 + \cdots \tag{C-42}$$

Using the definition of scattering length a in (C-32), the first term on the right hand side of (C-42) can be shown to be equal to $-1/a$. Up to order k^2, we find

$$k \cot \delta_0(k) = -\frac{1}{a} + \frac{1}{2} r_e k^2 \tag{C-43}$$

The s-wave scattering cross section is then

$$\sigma = \frac{4\pi}{k^2} \sin^2 \delta_0(k) = \frac{4\pi}{k^2 + \left\{\frac{1}{2} r_e k^2 - \frac{1}{a}\right\}^2} \tag{C-44}$$

which reduces to (C-31) when $k \to 0$.

§C-4 SCATTERING FROM A COMPLEX POTENTIAL

When a particle is scattered from a target, part of the kinetic energy may be transferred to exciting the projectile or the target nucleus. At the same time, some of the nucleons from one particle may be transferred to the other. If enough energy is available in the collision, new particles may also be created. All such processes are inelastic in the sense that the exit channel of the reaction is different from the entrance channel. A reaction with both elastic and inelastic scattering present is caused by a complex scattering potential. The solution of the Schrödinger equation in such a case may still be represented by (C-10); however, the phase shifts are now complex quantities in general.

In order to treat a broader class of scattering problems, in the place of (C-12), we shall write the asymptotic form of the modified radial equation $u_\ell(r)$ for partial wave ℓ in terms of an incoming wave $\mathcal{I}_\ell(r)$ and an outgoing wave $\mathcal{O}_\ell(r)$,

$$u_\ell(r) \xrightarrow[r\to\infty]{} \mathcal{I}_\ell(r) - \eta_\ell \mathcal{O}_\ell(r) \tag{C-45}$$

where η_ℓ, the inelasticity parameter, is a way to measure of the contribution of inelastic scattering, as we shall see later. (The definition of $\eta_e ll$ here is a more general one than that in (3-80), where η_ℓ is a real number representing the absolute value of η_ℓ here.) Each of the factors in (C-45) has a counterpart in (C-12),

$$\eta_\ell \sim e^{2i\delta_\ell} \qquad \mathcal{I}_\ell(r) \sim e^{-i(kr - \frac{1}{2}\ell\pi)} \qquad \mathcal{O}_\ell(r) \sim e^{i(kr - \frac{1}{2}\ell\pi)} \tag{C-46}$$

The elastic scattering cross section given in (C-21) may now be expressed in the form

$$\sigma^{\text{el}} = \frac{\pi}{k^2} \sum_\ell (2\ell + 1) |1 - \eta_\ell|^2 \tag{C-47}$$

In addition to this, there are new terms contributing to the scattering process that were not present in scattering by a real potential.

One way to see the difference between scattering by a real and a complex potential is to examine the difference in the intensities of the incoming and outgoing waves for partial wave ℓ. Using the last form of (C-12), we obtain the difference as

$$1 - |\eta_\ell|^2 = 1 - |e^{2i\delta_\ell}|^2$$

If the phase shift δ_ℓ is real, the difference vanishes and only elastic scattering is nonzero. For a complex phase shift, some of the incident flux is transferred to channels other than the incident one. This part of the scattering is represented by the "reaction" cross section

$$\sigma^{\text{re}} = \frac{\pi}{k^2}\sum_\ell (2\ell+1)(1-|\eta_\ell|^2) \tag{C-48}$$

The total cross section is then the sum of elastic and reaction cross sections,

$$\begin{aligned}\sigma^{\text{tot}} = \sigma^{\text{el}} + \sigma^{\text{re}} &= \frac{\pi}{k^2}\sum_\ell (2\ell+1)\left(|1-\eta_\ell|^2 + 1 - |\eta_\ell|^2\right)\\ &= \frac{2\pi}{k^2}\sum_\ell (2\ell+1)(1-\text{Re}\,\eta_\ell)\end{aligned} \tag{C-49}$$

We may compare this result with the scattering amplitude $f(\theta)$ at $\theta = 0$. From (C-19), we have

$$f(\theta=0) = \frac{1}{2ik}\sum_{\ell=0}^{\infty}(2\ell+1)(e^{2i\delta_\ell}-1) = \frac{1}{2ik}\sum_{\ell=0}^{\infty}(2\ell+1)(\eta_\ell - 1)$$

where we have made use of the explicit values of $Y_{\ell 0}(\theta)$ for $\theta = 0$ given by (B-15). Comparing this result with the final form of (C-49), we obtain the relation,

$$\sigma^{\text{tot}} = \frac{4\pi}{k}\text{Im}\, f(0) \tag{C-50}$$

known as the *optical theorem.*

In order to discuss inelastic scattering involving nuclear particles in more detail, we need to define the concept of a *reaction channel.* It is used to describe a particular quantum mechanical state of the scattering system either before or after the scattering event. We shall examine here only two-body scattering although the formalism can be generalized to include reactions involving three or more particles in the final state. The labels required to specify a reaction channel consist of three distinctive parts: those describing the internal degrees of freedom of the projectile or the scattered particle, those describing the corresponding quantities for the target or the residual nucleus, and those describing the relative motion between these two particles. For simplicity we shall use a single letter c, the channel quantum number, to represent the complete set of labels,

$$c \equiv \{j_p\alpha_p, j_t\alpha_t; \gamma\mu : \ell m\}$$

where ℓ is the relative angular momentum and m is its projection on the quantization axis. The wave function of the projectile (or scattered particle) is represented by $\phi_{j_p\alpha_p}$, where j_p is the spin and α_p represents all the other quantum numbers required to specify the state the projectile (or the scattered particle) is in. The wave function of the target (or the residual) nucleus is given by $\psi_{j_t\alpha_t}$ where j_t is the spin and α_t represents all the other labels.

Since there are three different angular momenta involved here, it is useful to couple two of them together first. For this purpose, we shall define a function,

$$\Phi_{\gamma\mu} = \left(\phi_{j_p\alpha_p} \times \psi_{j_t\alpha_t}\right)_{\gamma\mu} \tag{C-51}$$

the product of the "intrinsic" wave functions of the projectile (or the scattered particle) and the target (or the residual) nucleus with their angular momenta coupled together to (γ, μ). It is convenient to treat the relative orbital angular momentum ℓ separately from the spins of the particles since it is not usually observed directly in a measurement. The identification of one of the two particles involved in the scattering as the projectile and the other one as the target nucleus before the scattering, and one of the particle as the scattered particle and the other one as the residual nucleus, is an artificial one without much significance in the center-of-mass system we are using. Furthermore, we have omitted references to isospin to simplify the notation.

Instead of (C-46), we shall define properly normalized incoming and outgoing waves in the following way:

$$\begin{aligned}\mathcal{I}_c(\boldsymbol{r}) &= \frac{1}{r\sqrt{v_c}}\, i^\ell Y_{\ell m}(\theta, \phi) e^{-i(kr-\frac{1}{2}\ell\pi)}\Phi_{\gamma\mu} \\ \mathcal{O}_c(\boldsymbol{r}) &= \frac{1}{r\sqrt{v_c}}\, i^\ell Y_{\ell m}(\theta, \phi) e^{+i(kr-\frac{1}{2}\ell\pi)}\Phi_{\gamma\mu}\end{aligned} \tag{C-52}$$

where v_c is the center-of-mass velocity in channel c and its presence is required to normalize the wave function. Consider first the simple case of a definite incoming channel c. The scattering wave function with this incident channel and all possible outgoing channels may be written in the form,

$$\Psi_c(\boldsymbol{r}) = \mathcal{I}_c(\boldsymbol{r}) - \sum_{c'} S_{c'c}\mathcal{O}_{c'}(\boldsymbol{r}) \tag{C-53}$$

where $S_{c'c}$ is the matrix element relating the scattering amplitude from incident channel c to exit channel c'.

In general, the scattering process is described by the s-matrix (also known as the reaction matrix and the collision matrix). The matrix element,

$$S_{c'c} = \langle \Psi_{c'}^{\text{out}}(\boldsymbol{r}) | \boldsymbol{S} | \Psi_{c}^{\text{in}}(\boldsymbol{r}) \rangle \tag{C-54}$$

is taken between wave functions in the incident channel c and outgoing channel c'. We shall return to the topic of s-matrix in the final section of this Appendix.

The general solution of the Schrödinger equation (C-3) in a region at large distances outside the range of scattering potential V is a linear combination of the wave function given in (C-53),

$$\Psi(\boldsymbol{r}) = \sum_{c} C_c \Big\{ \mathcal{I}_c(\boldsymbol{r}) - \sum_{c'} S_{c'c} \mathcal{O}_{c'}(\boldsymbol{r}) \Big\} \tag{C-55}$$

where the coefficients C_c depend on the initial conditions, *i.e.*, arrangement of the incident beam and the target.

The asymptotic form of the incident wave function, with a projectile described by the wave function $\phi_{j_p \alpha_p}$, a target nucleus described by the wave function $\psi_{j_t \alpha_t}$, and the two particles approaching each other along the z-axis with relative wave function described by a plane wave (or a Coulomb wave if both particles carry charge), is given by the form

$$\begin{aligned}
\Psi_{\text{inc}}(\boldsymbol{r}) &= \frac{1}{\sqrt{v}} e^{ikz} \Phi_{\gamma\mu} \\
&\xrightarrow[r\to\infty]{} \sqrt{\frac{4\pi}{v}} \sum_{\ell} \sqrt{(2\ell+1)} \frac{i^{\ell}}{2ikr} \Big\{ e^{-i(kr-\frac{1}{2}\ell\pi)} - e^{i(kr-\frac{1}{2}\ell\pi)} \Big\} Y_{\ell 0}(\theta) \Phi_{\gamma\mu} \\
&= \frac{i\sqrt{\pi}}{k} \sum_{\ell} \sqrt{(2\ell+1)} \Big\{ \mathcal{I}_{c(\ell,m=0)} - \mathcal{O}_{c(\ell,m=0)} \Big\}
\end{aligned} \tag{C-56}$$

in analogy with (C-15). For clarity, in addition to channel quantum number c, we have also given those labels that must be specified explicitly in brackets. The complete scattering wave function has the form given by (C-55) and must contain a term describing an incident beam identical to that given in (C-56). Hence (C-55) may be written in the form

$$\begin{aligned}
\Psi(\boldsymbol{r}) &\xrightarrow[r\to\infty]{} \frac{i\sqrt{\pi}}{k} \sum_{\ell} \sqrt{(2\ell+1)} \Big\{ \mathcal{I}_{c(\ell,m=0)} - \sum_{c'} S_{c'c(\ell,m=0)} \mathcal{O}_{c'} \Big\} \\
&= \frac{i\sqrt{\pi}}{k} \sum_{\ell} \sqrt{(2\ell+1)} \Big\{ \mathcal{I}_{c(\ell,m=0)} - \mathcal{O}_{c(\ell,m=0)} \\
&\qquad + \mathcal{O}_{c(\ell,m=0)} - \sum_{c'} S_{c'c(\ell,m=0)} \mathcal{O}_{c'} \Big\} \\
&= \Psi_{\text{inc}}(\boldsymbol{r}) + \frac{i\sqrt{\pi}}{k} \sum_{\ell} \sqrt{(2\ell+1)} \Big\{ \mathcal{O}_{c(\ell,m=0)} - \sum_{c'} S_{c'c(\ell,m=0)} \mathcal{O}_{c'} \Big\}
\end{aligned} \tag{C-57}$$

We shall now work out the differential scattering cross section from this expression.

Since the incident probability current density is normalized to unity because of (C-52), the differential scattering cross section is given by the expression,

$$\left(\frac{d\sigma}{d\Omega}\right)_{\gamma\mu\alpha;\gamma'\mu'\beta} = \frac{\pi}{k^2}\Big|\sum_{\ell\ell'}\sqrt{(2\ell+1)}\,S_{c'(\ell'\gamma'\mu'\beta)\,c(\ell,m=0,\gamma\mu\alpha)}Y_{\ell 0}(\theta)\Big|^2 \qquad \text{(C-58)}$$

where we have integrated over all the internal variables in the initial state, described by the product wave function $\Phi_{\gamma\mu}(j_p\alpha_p; j_t\alpha_t)$, and in the final state, described by the product wave function $\Phi_{\gamma'\mu'}(j_s\beta_s; j_r\beta_r)$. The expression is basically the same equation as (C-20) except that elements of the s-matrix between incident and final scattering states are used to replace the phase shifts. The summation over ℓ', the orbital angular momentum in the outgoing channel, is required since in a scattering experiment only the states of the scattered particle and the residual nucleus are observed; their relative angular momentum ℓ' is not usually identified. On integrating over the angles, we obtain the scattering cross section as

$$\sigma_{\gamma\mu\alpha;\gamma'\mu'\beta} = \frac{\pi}{k^2}\sum_{\ell}(2\ell+1)\big|S_{c'(\ell\gamma'\mu'\beta)\,c(\ell,m=0,\gamma\mu\alpha)}\big|^2 \qquad \text{(C-59)}$$

in the same way as was done to arrive at (C-21). The reaction cross section is represented in this expression by terms with exit channels with $\beta \neq \alpha$.

For elastic scattering, the scattering amplitude is given by the expression,

$$T_{c'(\ell'm'\gamma'\mu'\beta)\,c(\ell,m=0,\gamma\mu\alpha)} = \delta_{\ell\ell'}\delta_{m'0}\delta_{\gamma\gamma'}\delta_{\mu\mu'}\delta_{\alpha\beta} - S_{c(\ell'm'\gamma'\mu'\beta)\,c(\ell,m=0,\gamma\mu\alpha)}$$

which, in its more general form, is known as the t-matrix. The elastic scattering cross section is then

$$\begin{aligned}\sigma^{\text{el}}_{\gamma\mu\alpha;\gamma\mu\alpha} &= \frac{\pi}{k^2}\sum_{\ell}(2\ell+1)\big|1 - S_{c(\ell\gamma\mu\alpha)\,c(\ell,m=0,\gamma\mu\alpha)}\big|^2\\ &= \frac{\pi}{k^2}\sum_{\ell}(2\ell+1)\Big\{1 - 2\text{Re}\,S_{\gamma\mu\alpha;\gamma\mu\alpha(m=0)}\\ &\qquad + \sum_{\ell'}\big|S_{c(\ell\gamma\mu\alpha)\,c(\ell,m=0,\gamma\mu\alpha)}\big|^2\Big\}\end{aligned} \qquad \text{(C-60)}$$

We can recover from this the relation given by (C-49) for total scattering cross section by adding to (C-60) the contribution from the reaction cross section contained in (C-59) and summing over all possible exit channels,

$$\begin{aligned}\sigma^{\text{tot}}_{\gamma\mu\alpha;\gamma\mu\alpha} = \frac{\pi}{k^2}\sum_{\ell}(2\ell+1)\Big\{1 - 2\text{Re}\,S_{c(\ell\gamma\mu\alpha)\,c(\ell,m=0,\gamma\mu\alpha)}\\ + \sum_{\ell'\gamma'\mu'\beta}\big|S_{c'(\ell'\gamma'\mu'\beta)\,c(\ell,m=0,\gamma\mu\alpha)}\big|^2\Big\}\end{aligned} \qquad \text{(C-61)}$$

Because of the unitary property of the s-matrix,

$$\sum_{\ell'\gamma'\mu'\beta} \left|S_{c'(\ell'\gamma'\mu'\beta)\,c(\ell\gamma\mu\alpha)}\right|^2 = 1$$

where the summation is taken all the possible channels, we have the result,

$$\sigma^{\text{tot}}_{\gamma\mu\alpha;\gamma\mu\alpha} = \frac{2\pi}{k^2}\sum_{\ell}(2\ell+1)\Big\{1 - \operatorname{Re} S_{c(\ell\gamma\mu\alpha)\,c(\ell\gamma\mu\alpha)}\Big\} \tag{C-62}$$

From this we recover the optical theorem in the same way as was done when deriving (C-50) from (C-49).

§C-5 COULOMB SCATTERING

The discussions in §C-2 and §C-3 apply only to short-range potentials. For nuclear scattering this is quite adequate except for Coulomb interaction between charged particles. The Coulomb potential between two nuclei with charges Z_1e and Z_2e has the form

$$V_c(r) = \left[\frac{1}{4\pi\epsilon_0}\right]\frac{Z_1Z_2e^2}{r} = \frac{Z_1Z_2\alpha\hbar c}{r}. \tag{C-63}$$

where the factor inside the square brackets converts the expression from cgs to SI units. Since the range of this potential is infinite, the techniques used in §C-2 no longer apply. This is, however, not a problem since an exact solution of Coulomb scattering is available (see, *e.g.*, A. Messiah, *Quantum Mechanics*, North-Holland, Amsterdam, 1966; P.M. Morse and H. Feshbach, *Methods of Theoretical Physics*, McGraw-Hill, New York, 1953; J.M. Blatt and V.F. Weisskopf, *Theoretical Nuclear Physics*, Wiley, New York, 1952). A short summary of the results is given here for completeness.

For scattering involving only the Coulomb potential, the Schrödinger equation can be written in the form

$$\left\{\nabla^2 + k^2 - \frac{2\gamma k}{r}\right\}\psi_c(\boldsymbol{r}) = 0 \tag{C-64}$$

where

$$k^2 = \frac{2\mu E}{\hbar^2} \qquad\qquad \gamma = \frac{Z_1Z_2\alpha\mu c}{\hbar k}.$$

The regular solution of (C-64) has the form

$$\psi(\boldsymbol{r}) = e^{ikz}f(r-z) \tag{C-65}$$

where $kz = kr\cos\theta = \boldsymbol{k}\cdot\boldsymbol{r}$. The function $f(\zeta)$ satisfies the differential equation,

$$\left\{\zeta\frac{d^2}{d\zeta^2} + (1-\zeta)\frac{d}{d\zeta} + i\gamma\right\}f(\zeta) = 0 \tag{C-66}$$

where $\zeta = ik(r-z)$. It is a type of Laplace equation

$$\left\{u\frac{d^2}{du^2} + (\beta - u)\frac{d}{du} - \alpha\right\} f(u) = 0 \tag{C-67}$$

with solution involving the confluent hypergeometric series

$$F(\alpha|\beta|u) = 1 + \frac{\alpha}{\beta}\frac{u}{1!} + \frac{\alpha(\alpha+1)}{\beta(\beta+1)}\frac{u^2}{2!} + \cdots \tag{C-68}$$

The normalized Coulomb wave function takes on the form,

$$\psi_c(\boldsymbol{r}) = e^{-\frac{1}{2}\pi\gamma}\Gamma(1+i\gamma)e^{ikz}F\big(-i\gamma|1|ik(r-z)\big) \tag{C-69}$$

The definition of the Gamma function $\Gamma(1+i\gamma)$ and its properties may be found in the *Handbook of Mathematical Functions* (edited by M. Abramowitz and I. A. Segun, Dover, New York, 1965).

At the origin, $F(\alpha|\beta|u) = 1$ and only the normalization factor remains,

$$\psi_c(0) = e^{-\frac{1}{2}\pi\gamma}\Gamma(1+i\gamma).$$

Using the identity that

$$\left|\Gamma(1+i\gamma)\right|^2 = \frac{\pi\gamma}{\sinh \pi\gamma}$$

we obtain the result

$$\left|\psi_c(0)\right|^2 = \frac{2\pi\gamma}{e^{2\pi\gamma}-1} \tag{C-70}$$

which gives the form of the Fermi function $F(Z, E_e)$ given in (5-97) for nuclear β-decay if the charge distribution in the daughter nucleus can be considered to be concentrated at the origin in the form of a point charge.

For scattering, we are more concerned with the asymptotic behavior of the wave functions. As in (C-5), we wish to express the functions at large distances away from the origin as a sum of incident wave $\psi_i(\boldsymbol{r})$ and scattered wave $\psi_s(\boldsymbol{r})$,

$$\psi_c(\boldsymbol{r}) = \psi_i(\boldsymbol{r}) + \psi_s(\boldsymbol{r}) \tag{C-71}$$

For $|r-z| \to \infty$, we have the result

$$\psi_i(\boldsymbol{r}) \longrightarrow e^{i\{kz+\gamma\ln k(r-z)\}}\left\{1 + \frac{\gamma^2}{ik(r-z)} + \cdots\right\}$$

$$\psi_s(\boldsymbol{r}) \longrightarrow \frac{1}{r}e^{i\{kr-\gamma\ln 2kr\}}f^c(\theta) + O(r^{-2}) \tag{C-72}$$

The Coulomb scattering amplitude $f^c(\theta)$ is given by

$$f^c(\theta) = -\frac{\gamma}{2k\sin^2\frac{1}{2}\theta}e^{i\{\gamma\ln(\sin^2\frac{1}{2}\theta)+2\delta_0^c\}} \tag{C-73}$$

where $\delta_0^c = \arg\Gamma(1+i\gamma)$ is the Coulomb phase shift for $\ell = 0$. Using this result, we obtain the Rutherford scattering formula,

$$\left(\frac{d\sigma}{d\Omega}\right)_{\text{Ruth.}} = \left\{\frac{Z_1 Z_2 \alpha\hbar c}{4E\sin^2\frac{\theta}{2}}\right\}^2 \tag{C-74}$$

This is the same expression as (4-1) except here the kinetic energy is represented by the symbol E to conform with the practice used in nonrelativistic scattering rather than T in (4-1), where we need to make a distinction from the total relativistic energy.

We can also make a partial wave expansion of the solution to (C-64) in a form similar to that given in (C-10),

$$\psi_c(\boldsymbol{r}) = \sum_\ell \sqrt{4\pi(2\ell+1)}\,\frac{i^\ell}{kr}\,u_\ell^c(r)Y_{\ell 0}(\theta) \tag{C-75}$$

The modified Coulomb radial wave function $u_\ell^c(r)$ satisfies the radial equation,

$$\left\{\frac{d^2}{d\rho^2} + 1 - \frac{2\gamma}{\rho} - \frac{\ell(\ell+1)}{\rho^2}\right\}u_\ell^c(r) = 0 \tag{C-76}$$

where $\rho = kr$. The solution of this equation may also be expressed as a sum of $F_\ell(\gamma,\rho)$ and $G_\ell(\gamma,\rho)$, the regular and irregular Coulomb wave functions (see, *e.g.*, Abramowitz and Segun),

$$u_\ell^c(\rho) = C_1 F_\ell(\gamma,\rho) + C_2 G_\ell(\gamma,\rho)$$

However, for scattering problems, it is more convenient to use the form

$$u_\ell^c(r) = e^{i\delta_\ell^c} F_\ell(\gamma,\rho) \tag{C-77}$$

where

$$\delta_\ell^c = \arg\Gamma(\ell+1+i\gamma) \tag{C-78}$$

is the Coulomb phase shift for partial wave ℓ.

Asymptotically, the Coulomb wave function has the form

$$F_\ell(\gamma,\rho) \xrightarrow[r\to\infty]{} \sin\xi_\ell \qquad\qquad G_\ell(\gamma,\rho) \xrightarrow[r\to\infty]{} \cos\xi_\ell \tag{C-79}$$

where

$$\xi_\ell = \rho - \gamma\ln 2\rho - \tfrac{1}{2}\ell\pi + \delta_\ell^c$$

Applying this result to the right hand side of (C-77), we can now write the asymptotic form of the modified radial wave function in a manner similar to the final form of (C-12),

$$u_\ell^c(r) \xrightarrow[r\to\infty]{} \frac{i^{\ell+1}}{2kr}\left\{e^{-i(kr-\gamma\ln 2kr))} - e^{2i\delta_\ell^c}e^{i(kr-\gamma\ln 2kr-\ell\pi))}\right\} \tag{C-80}$$

From this we obtain the Coulomb scattering amplitude in terms of the phase shifts

$$f^c(\theta) = \frac{1}{2ik}\sum_\ell \sqrt{4\pi(2\ell+1)}\,(e^{2i\delta_\ell^c} - 1)Y_{\ell 0}(\theta) \tag{C-81}$$

in a form similar to that given in (C-19).

§C-6 FORMAL SOLUTION TO THE SCATTERING EQUATION

There are two main purposes for initiating a short discussion here on the formal solution to the scattering equation. The first is to define some of the terminology commonly used in scattering and related problems. The second is to make a connection with methods used in standard references in nuclear scattering.

We shall write the time-independent Hamiltonian in the form

$$H = H_0 + V \tag{C-82}$$

Normally H_0 consists of the kinetic energy operator only as in (C-2)

$$H_0 = -\frac{\hbar^2}{2\mu}\nabla^2 \tag{C-83}$$

However, we may also choose to include in H_0 a part of the interaction potential such as the Coulomb potential and the optical potential (see §7-4). The potential V in (C-82) will then represent the *residual interaction*, the remainder of V that is not included in H_0. For our purpose here, we shall further assume that any long-range part of the potential is included in H_0.

The eigenfunction of the scattering equation is the solution of the equation

$$(H_0 - E)\psi_{\boldsymbol{k}}^{\pm}(\boldsymbol{r}) = -V\psi_{\boldsymbol{k}}^{\pm}(\boldsymbol{r}) \tag{C-84}$$

where the superscript + on $\psi_{\boldsymbol{k}}(\boldsymbol{r})$ indicates that the solution satisfies the *outgoing* boundary condition and the superscript − for the *incoming* boundary condition. Our concern will be mainly with the former. The subscript $\boldsymbol{k}$, with magnitude $k = \sqrt{2\mu E}/\hbar$, displays the explicit dependence of the solution on energy.

The solution of the homogeneous equation,

$$(H_0 - E)\phi_{\boldsymbol{k}}(\boldsymbol{r}) = 0 \tag{C-85}$$

forms a complete set satisfying the orthogonality condition,

$$\int \phi_{\boldsymbol{k}'}^*(\boldsymbol{r})\phi_{\boldsymbol{k}}(\boldsymbol{r})\,d\boldsymbol{r} = \delta(\boldsymbol{k} - \boldsymbol{k}')$$

and having the closure property

$$\int \phi_{\boldsymbol{k}}^*(\boldsymbol{r}')\phi_{\boldsymbol{k}}(\boldsymbol{r})\,d\boldsymbol{k} = \delta(\boldsymbol{r} - \boldsymbol{r}')$$

For the simple case of (C-83), we have the plane wave, $\phi_{\boldsymbol{k}}(\boldsymbol{r}) \sim \exp(i\boldsymbol{k}\cdot\boldsymbol{r})$ as the solution of (C-85). On the other hand if, for example, the Coulomb potential is included as a part of H_0, we have the Coulomb wave function as the solution instead.

Using the method of Green's function, the solution of the scattering equation is given by the integral equation

$$\psi_{\boldsymbol{k}}^{+}(\boldsymbol{r}) = \phi_{\boldsymbol{k}}(\boldsymbol{r}) + \frac{2\mu}{\hbar^2}\int G^{+}(\boldsymbol{r},\boldsymbol{r}')V(\boldsymbol{r}')\psi_{\boldsymbol{k}}^{+}(\boldsymbol{r}')\,d\boldsymbol{r}' \tag{C-86}$$

The first term is the solution to the homogeneous equation of (C-85). The Green's function $G^{+}(\boldsymbol{r},\boldsymbol{r}')$ in the second term satisfies the equation,

$$(H_0 - E)G^{+}(\boldsymbol{r},\boldsymbol{r}') = -\frac{\hbar^2}{2\mu}\delta(\boldsymbol{r}-\boldsymbol{r}') \tag{C-87}$$

with outgoing boundary conditions. In the simple case that H_0 involves only the kinetic energy as given in (C-83),

$$G^{+}(\boldsymbol{r},\boldsymbol{r}') = -\frac{1}{4\pi}\frac{e^{ik|\boldsymbol{r}-\boldsymbol{r}'|}}{|\boldsymbol{r}-\boldsymbol{r}'|} \tag{C-88}$$

We shall use this form of the Green's function exclusively for the examples below.

It is easy to check that $\psi_{\boldsymbol{k}}^{+}(\boldsymbol{r})$ given in (C-86) is a solution to (C-84). On applying $(H_0 - E)$ to both sides of (C-86), we obtain

$$(H_0 - E)\psi_{\boldsymbol{k}}^{+}(\boldsymbol{r}) = (H_0 - E)\phi_{\boldsymbol{k}}(\boldsymbol{r}) + \frac{2\mu}{\hbar^2}(H_0 - E)\int G^{+}(\boldsymbol{r},\boldsymbol{r}')V(\boldsymbol{r}')\psi_{\boldsymbol{k}}^{+}(\boldsymbol{r}')\,d\boldsymbol{r}'$$

The first term on the right hand side vanishes because of (C-85). In the second term, since $(H_0 - E)$ operates only on the variable $\boldsymbol{r}$ and not on $\boldsymbol{r}'$, we may bring the operator inside the integral without affecting the result. Furthermore, since $\boldsymbol{r}$ appears only in $G(\boldsymbol{r},\boldsymbol{r}')$, we obtain, by using (C-87), the result

$$(H_0 - E)\psi_{\boldsymbol{k}}^{+}(\boldsymbol{r}) = -\int \delta(\boldsymbol{r}-\boldsymbol{r}')V(\boldsymbol{r}')\psi_{\boldsymbol{k}}^{+}(\boldsymbol{r}')\,d\boldsymbol{r}' = -V(\boldsymbol{r})\psi_{\boldsymbol{k}}^{+}(\boldsymbol{r}) \tag{C-89}$$

the same equality given in (C-84).

It is easy to see how the scattering amplitude may be obtained from (C-86) using the explicit form of the Green's function given in (C-88). Let $\hat{\boldsymbol{r}} = \boldsymbol{r}/|r|$ be a unit vector in the direction $\boldsymbol{r}$. In the asymptotic region,

$$|\boldsymbol{r}-\boldsymbol{r}'| \approx r - \hat{\boldsymbol{r}}\cdot\boldsymbol{r}'$$

since the integral over $\boldsymbol{r}'$ is effective only in the region of small r' where the short-range potential $V(\boldsymbol{r}')$ is nonvanishing. As a result, we may approximate the Green's function of (C-88) by the form

$$G^{+}(\boldsymbol{r},\boldsymbol{r}') \xrightarrow[r\to\infty]{} -\frac{1}{4\pi}\frac{e^{ikr}}{r}e^{-ik\hat{\boldsymbol{r}}\cdot\boldsymbol{r}'} = -\frac{1}{4\pi}\frac{e^{ikr}}{r}\phi_{\boldsymbol{k}'}^{*}(\boldsymbol{r}') \tag{C-90}$$

where we have taken $\boldsymbol{k}'$ to be along the direction of $\hat{\boldsymbol{r}}$. Eq. (C-86) is now reduced to the form

$$\psi_{\boldsymbol{k}}^{+}(\boldsymbol{r}) = \phi_{\boldsymbol{k}}(\boldsymbol{r}) - \frac{e^{ikr}}{r}\frac{\mu}{2\pi\hbar^2}\int \phi_{\boldsymbol{k}'}^{*}(\boldsymbol{r}')V(\boldsymbol{r}')\psi_{\boldsymbol{k}}^{+}(\boldsymbol{r}')\,d\boldsymbol{r}' \tag{C-91}$$

Comparing this result with the form of (C-5), the scattering amplitude is identified as

$$f(\theta) = -\frac{\mu}{2\pi\hbar^2}\int \phi^*_{\boldsymbol{k}'}(\boldsymbol{r}')V(\boldsymbol{r}')\psi^+_{\boldsymbol{k}}(\boldsymbol{r}')\,d\boldsymbol{r}' = -\frac{\mu}{2\pi\hbar^2}\langle\phi_{\boldsymbol{k}'}|V|\psi^+_{\boldsymbol{k}}\rangle \qquad \text{(C-92)}$$

The result here is an exact one in the asymptotic region and is different from that of first Born approximation given in (7-40), since $\psi^+_{\boldsymbol{k}}$, the solution of the scattering equation (C-84), appears in $f(\theta)$ in the place of $\phi_{\boldsymbol{k}}$. The differential scattering cross section is then

$$\frac{d\sigma}{d\Omega} = \left|f(\theta)\right|^2 = \frac{\mu^2}{4\pi^2\hbar^4}\left|\langle\phi_{\boldsymbol{k}'}|V|\psi^+_{\boldsymbol{k}}\rangle\right|^2 \qquad \text{(C-93)}$$

The usefulness of this result is limited by the fact that it requires a knowledge of $\psi^+_{\boldsymbol{k}}(\boldsymbol{r}')$, the complete solution to the scattering problem.

Eq. (C-91) is an integral equation or a "formal" solution of the scattering equation since $\psi^+_{\boldsymbol{k}}$ itself appears on the right hand side as well. Its usefulness lies mainly in analytical works such as a Born series expansion of the scattering wave function and scattering amplitude. In order to simplify the notation, we shall write (C-86) in the following form,

$$\psi^+_{\boldsymbol{k}} = \phi_{\boldsymbol{k}} + G^+V\psi^+_{\boldsymbol{k}} \qquad \text{(C-94)}$$

where, instead of the Green's function $G^+(\boldsymbol{r},\boldsymbol{r}')$, we have used G^+, an operator form of it,

$$G^+(\boldsymbol{r},\boldsymbol{r}') = \langle\boldsymbol{r}|G^+|\boldsymbol{r}'\rangle$$

In terms of H_0 and E, the Green's function operator G^+ may be expressed in the form

$$G^+ = \lim_{\epsilon\to 0}\frac{1}{E - H_0 + i\epsilon} \qquad \text{(C-95)}$$

where the factor $+i\epsilon$, with ϵ as some small quantity, is required to ensure that the operator corresponds to the outgoing boundary condition. The derivation of (C-95) may be found in standard quantum mechanics texts such as Merzbacher (*Quantum Mechanics*, 2nd ed., Wiley, New York, 1961), Messiah (*Quantum Mechanics*, North-Holland, Amsterdam, 1966), and Schiff (*Quantum Mechanics*, 3rd ed., McGraw-Hill, New York, 1968).

It is easy to see that (C-95) has the correct form by substituting it into (C-94). The result

$$\psi^+_{\boldsymbol{k}} = \phi_{\boldsymbol{k}} + \frac{1}{E - H_0 + i\epsilon}V\psi^+_{\boldsymbol{k}} \qquad \text{(C-96)}$$

is one way to write the Lippmann-Schwinger equation. The equation may be reduced into a more familiar form by operating from the left with $(E - H_0 + i\epsilon)$ and taking the limit $\epsilon \to 0$,

$$(E - H_0)\psi^+_{\boldsymbol{k}} = (E - H_0)\phi_{\boldsymbol{k}} + V\psi^+_{\boldsymbol{k}}$$

The first term on the right hand side vanishes because of (C-85) and the rest of the equation is identical as (C-84).

If we replace $\psi_{\boldsymbol{k}}^{+}$ on the right hand side of (C-94) by its value in (C-94) itself and repeat the process, we obtain an infinite series expansion of $\psi_{\boldsymbol{k}}^{+}$ in terms of $\phi_{\boldsymbol{k}}$,

$$\begin{aligned}\psi_{\boldsymbol{k}}^{+} &= \phi_{\boldsymbol{k}} + G^{+}V\left(\phi_{\boldsymbol{k}} + G^{+}V\psi_{\boldsymbol{k}}^{+}\right)\\ &= \phi_{\boldsymbol{k}} + G^{+}V\phi_{\boldsymbol{k}} + G^{+}VG^{+}V\left(\phi_{\boldsymbol{k}} + G^{+}V\psi_{\boldsymbol{k}}^{+}\right)\\ &= \left(1+\sum_{n=1}^{\infty}\left(G^{+}V\right)^{n}\right)\phi_{\boldsymbol{k}} \end{aligned} \tag{C-97}$$

Using this result we can obtain a Born series expansion of the scattering amplitude by substituting the expansion of $\psi_{\boldsymbol{k}}^{+}$ in terms of $\phi_{\boldsymbol{k}}$ into (C-92).

We have seen earlier that the scattering amplitude, $(-\mu/2\pi\hbar^{2})\langle\phi_{\boldsymbol{k}'}|V|\psi_{\boldsymbol{k}}^{+}\rangle$, given in (C-92) is not useful directly for calculating scattering cross sections because of its dependence on $\psi_{\boldsymbol{k}}^{+}$. For many purposes it is more convenient to define a transition matrix or t-matrix such that the following relation is satisfied:

$$\langle\phi_{\boldsymbol{k}'}|t|\phi_{\boldsymbol{k}}\rangle = \langle\phi_{\boldsymbol{k}'}|V|\psi_{\boldsymbol{k}}^{+}\rangle \tag{C-98}$$

In terms of the t-matrix, the scattering amplitude is a function of matrix elements involving only $\phi_{\boldsymbol{k}}$, the solution of the homogeneous equation given in (C-85). For the simple case of H_0 consisting of the kinetic energy operator only, the elements of the t-matrix involve only plane wave states. Again, this is useful for formal work, since the t-matrix itself cannot be written down unless we have solved the scattering problem first.

Using the series expansion of $\psi_{\boldsymbol{k}}^{+}$ given in (C-97), we can write the elements of t-matrix in the form

$$\langle\phi_{\boldsymbol{k}'}|t|\phi_{\boldsymbol{k}}\rangle = \langle\phi_{\boldsymbol{k}'}|V\left(1+\sum_{n=1}^{\infty}\left(G^{+}V\right)^{n}\right)|\phi_{\boldsymbol{k}}\rangle \tag{C-99}$$

Since the equality holds for arbitrary $\phi_{\boldsymbol{k}}$ and $\phi_{\boldsymbol{k}'}$, we obtain a relation between the operators involved

$$t = V\left(1+\sum_{n=1}^{\infty}\left(G^{+}V\right)^{n}\right)$$

This can be written in a more compact form. Since the summation is taken up to infinity, we can take one product of G^{+} with V out of the summation and rewrite the equation in the form

$$t = V + VG^{+}V + VG^{+}V\sum_{n=1}^{\infty}\left(G^{+}V\right)^{n} = V + VG^{+}\left\{V + V\sum_{n=1}^{\infty}\left(G^{+}V\right)^{n}\right\}$$

The quantity inside the curly brackets is nothing but the transition operator t itself, and we obtain the result,

$$t = V + VG^+t \tag{C-100}$$

a form that is convenient for the starting point of many other derivations.

The s-matrix may be expressed in terms of the t-matrix by the relation,

$$\langle\phi_p|\boldsymbol{S}|\phi_q\rangle = \delta_{pq} - 2\pi i\delta(E_p - E_q)\,\langle\phi_p|\boldsymbol{t}|\phi_q\rangle \tag{C-101}$$

The definition of the s-matrix is usually introduced through the time-development operator $U(t, t_0)$ in the interaction representation of quantum mechanics (see, *e.g.,* J.J. Sakurai, *Advanced Quantum Mechanics*, Addison-Wesley, Reading, Massachusetts, 1967; L.I. Schiff, *Quantum Mechanics*, 3rd ed., McGraw-Hill, New York, 1968).

For most elementary applications, the time dependence of a quantum mechanical state is expressed in the Schrödinger representation. Here, the operators are time-independent; all the time-dependence resides with the wave functions $\Psi_s(t)$. Using (C-1), we have the result,

$$i\hbar\frac{\partial}{\partial t}\Psi_s(t) = H\Psi_s(t), \tag{C-102}$$

where the subscript s emphasizes that the wave function is in the Schrödinger representation. In order to simplify the notation, we have suppressed all arguments other than time. Alternatively, one can work in the Heisenberg representation where, in contrast, the wave function is time-independent and all time-dependence is built into the operators.

In the *interaction representation*, the time-dependence of a system is partly in the operator and partly in the wave function. The Hamiltonian is divided into two parts

$$H = H_0 + H_I$$

The wave functions and operators in this representation are related to those in the Schrödinger representation through the transformations:

$$\Psi(t) = e^{iH_0t/\hbar}\Psi_s(t) \tag{C-103}$$

$$\hat{O}(t) = e^{iH_0t/\hbar}\hat{O}_s e^{-iH_0t/\hbar} \tag{C-104}$$

As a result, the time development of a state in the interaction representation is given by the equation,

$$i\hbar\frac{\partial}{\partial t}\Psi(t) = H_I(t)\Psi(t) \tag{C-105}$$

as can be seen by substituting the inverse of (C-103) into (C-102). For many purposes, this can be simpler than working in the Schrödinger representation, especially if H_I is only a small part of the complete Hamiltonian.

We can now define the time-development operator $\boldsymbol{U}(t_0, t)$ that takes a state from time t_0 to time t in the interaction representation,

$$\Psi(t) = \boldsymbol{U}(t, t_0)\Psi(t_0). \tag{C-106}$$

On substituting this definition in to (C-103,4), we obtain the equation for $\boldsymbol{U}(t_0, t)$,

$$i\hbar\frac{\partial}{\partial t}\boldsymbol{U}(t, t_0) = H_I(t)\boldsymbol{U}(t, t_0) \tag{C-107}$$

The solution of this equation may be given in the form of an integral equation,

$$\boldsymbol{U}(t, t_0) = 1 - i\hbar\int_{t_0}^{t} H_I(t)\boldsymbol{U}(t, t_0)\, dt \tag{C-108}$$

The s-matrix operator is defined by the following relation:

$$\boldsymbol{S} = \lim_{\substack{t\to+\infty \\ t'\to-\infty}} \boldsymbol{U}(t, t') \tag{C-109}$$

It is easy to see that the matrix elements of the operator $\boldsymbol{S}$ between a specific initial state and a specific final state are proportional to the scattering amplitude, since both quantities are related to the probability of finding a system in the final state at $t = +\infty$ if it started out from the initial state at $t = -\infty$.

In terms of phase shifts, the element of the s-matrix for partial wave ℓ is given by

$$\langle\ell|\boldsymbol{S}|\ell\rangle \sim e^{2i\delta_\ell} \tag{C-110}$$

The analogous relation for the t-matrix element is

$$\langle\ell|\boldsymbol{t}|\ell\rangle \sim e^{i\delta_\ell}\sin\delta_\ell \tag{C-111}$$

The advantage of using the s-matrix for scattering problems is its unitarity and other symmetry properties that are convenient in more advanced treatments.

Appendix D

TRANSFORMATION BETWEEN CENTER OF MASS AND LABORATORY SYSTEMS

For most of the discussion on scattering we have used a coordinate system that is fixed with respect to the center of mass of the particles involved. However, measurements of reaction cross sections are usually carried out in a reference system fixed with respect to the laboratory. The transformation between quantities measured in the center of mass and in the laboratory systems is a straightforward problem in mechanics. We shall carry out here a derivation of the relations using a fully relativistic treatment. This produces a set of more general results from which nonrelativistic results may be obtained by taking the limit $v \ll c$.

In order to simplify the notation, let us consider a reaction denoted by $A(a,b)B$, where the projectile is represented by letter a and the target by letter A. We shall also restrict ourselves to cases in which the target is initially at rest in the laboratory and consider only final states involving two particles. Where necessary, we shall take particle b to be the lighter one and particle B to be the heavier one of the two particles in the final state.

Lorentz transformation. Let the four-momentum of a particle be given by

$$\varrho = \left(\frac{iE}{c}, p_x, p_y, p_z\right)$$

where E is the total energy, the sum of rest mass and kinetic energies, and $\boldsymbol{p} \equiv (p_x, p_y, p_z)$ is the three-momentum. The kinetic energy of a particle is then

$$T = E - mc^2 \tag{D-1}$$

where m is the rest mass of the particle.

The Lorentz transformation between two frames of reference moving with relative velocity $v = \beta c$ along the z-direction is

$$\begin{pmatrix} \frac{iE}{c} \\ p_x \\ p_y \\ p_z \end{pmatrix} = \begin{pmatrix} \gamma & 0 & 0 & i\gamma\beta \\ 0 & 1 & 0 & 0 \\ 0 & 0 & 1 & 0 \\ -i\gamma\beta & 0 & 0 & \gamma \end{pmatrix} \begin{pmatrix} \frac{iE'}{c} \\ p'_x \\ p'_y \\ p'_z \end{pmatrix} \tag{D-2}$$

where

$$\gamma \equiv \frac{1}{\sqrt{1-\beta^2}} \xrightarrow[v \ll c]{} 1 + \frac{1}{2}\beta^2 + \frac{3}{8}\beta^4 + \cdots$$

In the laboratory frame of reference, the total energy of a moving particle is

$$E = \gamma mc^2 \xrightarrow[v \ll c]{} mc^2 + \frac{1}{2}mv^2 + \frac{3}{8}mv^2\beta^2 + \cdots \tag{D-3}$$

and the momentum along the direction of motion is

$$p_z = \gamma\beta mc \xrightarrow[v \ll c]{} mv + \frac{1}{2}mv\beta^2 + \cdots \tag{D-4}$$

where v is the velocity of the particle. From (D-3), we obtain

$$E^2 = (\gamma mc^2)^2 = \frac{(mc^2)^2}{1-\beta^2} = (mc^2)^2 + (\gamma\beta mc^2)^2 = (mc^2)^2 + (pc)^2 \tag{D-5}$$

the relation between the energy, rest mass, and momentum of a particle.

Initial state. For scattering of particle a off particle A, we can define the direction of the incident particle a to be the z-axis. Since the target A initially is at rest in the laboratory, the four-momenta of the two particles in the laboratory frame of reference are

$$\varrho'_a = \left(\frac{iE_a}{c}, 0, 0, p_a\right) \qquad \varrho'_A = (im_A c, 0, 0, 0) \tag{D-6}$$

where $E_a = m_a c^2 + T$ according to (D-1) and p_a is the magnitude of the three-momentum of particle a in the laboratory.

In the center of mass, the three-momenta of the two particles, by definition, are equal to each other in magnitude but opposite in direction. As a result, their four-momenta are simply

$$\varrho_a = \left(\frac{i\epsilon_a}{c}, k_x, k_y, k_z\right) \qquad \varrho_A = \left(\frac{i\epsilon_A}{c}, -k_x, -k_y, -k_z\right) \tag{D-7}$$

where ϵ_a and ϵ_A are, respectively, the center-of-mass energies of particles a and A, and $\boldsymbol{k} \equiv (k_x, k_y, k_z)$ is the three-momentum of a. The total center-of-mass energy of the system is then

$$\mathcal{E} = \epsilon_a + \epsilon_A \xrightarrow[v \ll c]{} m_a c^2 + \frac{1}{2}\frac{k^2}{m_a} + \cdots + m_A c^2 + \frac{1}{2}\frac{k^2}{m_A} + \cdots \tag{D-8}$$

Using the fact that the squares of the three-momentum of the two particles must be equal to each other, we obtain from (D-5) the relation,

$$\epsilon_a^2 - (m_a c^2)^2 = \epsilon_A^2 - (m_A c^2)^2 \tag{D-9}$$

Combining (D-8) with (D-9), we obtain the result

$$\begin{aligned}\mathcal{E}^2 + (m_a c^2)^2 - (m_A c^2)^2 &= \epsilon_a^2 + \epsilon_A^2 + 2\epsilon_a\epsilon_A + (m_a c^2)^2 - (m_A c^2)^2 \\ &= \epsilon_a^2 + (m_a c^2)^2 + \{\epsilon_a^2 - (m_a c^2)^2\} + 2\epsilon_a\epsilon_A \\ &= 2\epsilon_a^2 + 2\epsilon_a\epsilon_A \\ &= 2\epsilon_a\mathcal{E}\end{aligned}$$

or

$$\epsilon_a = \frac{\mathcal{E}^2 + (m_a c^2)^2 - (m_A c^2)^2}{2\mathcal{E}} \tag{D-10}$$

Similarly, we obtain the corresponding relation

$$\epsilon_A = \frac{\mathcal{E}^2 - (m_a c^2)^2 + (m_A c^2)^2}{2\mathcal{E}} \tag{D-11}$$

by interchanging the roles of a and A.

In the center of mass, the four-momentum of the entire system is $(\frac{i\mathcal{E}}{c}, 0, 0, 0)$ and in the laboratory, $(\frac{i(E_a + m_A c^2)}{c}, 0, 0, p_a)$. The transformation between them is given by (D-2), and this gives us the results

$$E_a + m_A c^2 = \gamma\mathcal{E} \qquad\qquad p_a = \frac{\gamma\beta}{c}\mathcal{E} \tag{D-12}$$

From these relations, we obtain the values of γ and β for the motion of the center of mass as observed in the laboratory

$$\gamma = \frac{E_a + m_A c^2}{\mathcal{E}} \tag{D-13}$$

$$\beta = \frac{p_a c}{E_a + m_A c^2} \xrightarrow[v \ll c]{} \frac{1}{c}\frac{m_a}{m_a + m_A}\boldsymbol{v}'_a \tag{D-14}$$

where $\boldsymbol{v}'_a$ is the velocity of the incident particle in the laboratory. In the nonrelativistic limit, the relations between the velocities are given by

$$\begin{aligned} \boldsymbol{v}'_{cm} &= \frac{m_a}{m_a + m_A}\boldsymbol{v}'_a & \boldsymbol{v}_a &= \boldsymbol{v}'_a - \boldsymbol{v}_{cm} = \frac{m_A}{m_a + m_A}\boldsymbol{v}'_a \\ \boldsymbol{v}_A &= -\boldsymbol{v}_{cm} = -\frac{m_A}{m_a + m_A}\boldsymbol{v}'_a & \boldsymbol{k} &= \frac{m_a m_A}{m_a + m_A}\boldsymbol{v}'_a \end{aligned} \tag{D-15}$$

where the primed quantities are measured in the laboratory and the unprimed quantities in the center of mass.

Since target A is initially at rest in the laboratory, as shown in Fig. D-1, we may use the value of γ in (D-13) to obtain the relation

$$\epsilon_A = \gamma m_A c^2 = \frac{E_a + m_A c^2}{\mathcal{E}}\, m_A c^2 \tag{D-16}$$

Combining this result with that of (D-11), we obtain the equality,

$$\frac{E_a + m_A c^2}{\mathcal{E}} m_A c^2 = \frac{\mathcal{E}^2 - (m_a c^2)^2 + (m_A c^2)^2}{2\mathcal{E}}$$

or

$$\mathcal{E}^2 = 2E_a m_A c^2 + (m_a c^2)^2 + (m_A c^2)^2 \tag{D-17}$$

This relates the total energy in the center of mass with the energy of the incident particle.

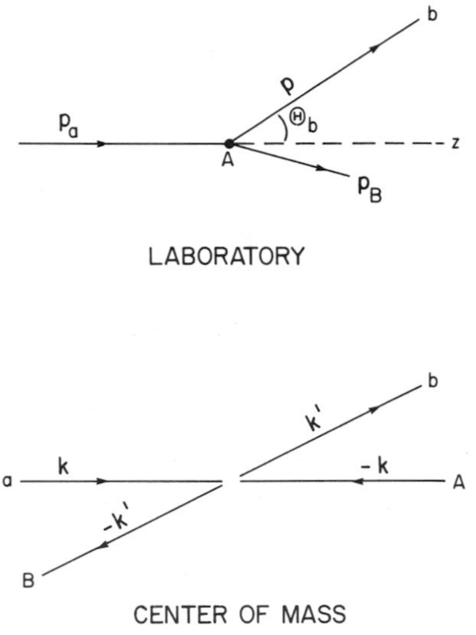

Fig. D-1 Three-momenta before and after the collision in the laboratory and the center-of-mass systems.

Using (D-15), the sum of the kinetic energies of particles a and A in the center of mass becomes

$$\mathcal{E} - m_a c^2 - m_A c^2 \xrightarrow[v \ll c]{} \left(\frac{1}{2m_a} + \frac{1}{2m_A}\right)\left(\frac{m_a m_A}{m_a + m_A}\right)^2 v_a'^2$$

$$= \frac{1}{2}\frac{m_a m_A}{m_a + m_A} v_a'^2 \tag{D-18}$$

This result is less than the kinetic energy of particle a in the laboratory by the amount

$$\frac{1}{2}(m_a + m_A) v_{cm}^2 = \frac{1}{2}\frac{m_a^2}{m_a + m_A} v_a'^2 \tag{D-19}$$

corresponding to the kinetic energy associated with the motion of the center of mass.

Final state. After the reaction, particles b and B emerge. For an elastic scattering, we have $m_b = m_a$ and $m_B = m_A$. On the other hand, in an inelastic scattering, part of the kinetic energy is transferred to exciting the target nucleus or scattered particle. As a result, $m_Bc^2 = m_Ac^2 + Q$, where Q is the amount of internal excitation energy absorbed by the target. (Alternatively, the excitation can be in particle b or both particles b and B.)

Let the center-of-mass four-momenta of the two particles in the final state be

$$\varrho_b = \left(\frac{i\epsilon_b}{c}, k'_x, k'_y, k'_z\right) \qquad \varrho_B = \left(\frac{i\epsilon_B}{c}, -k'_x, -k'_y, -k'_z\right) \tag{D-20}$$

where ϵ_b and ϵ_B are, respectively, the energies of particles b and B, and $\boldsymbol{k}' \equiv (k'_x, k'_y, k'_z)$ is the three-momentum of particle b in the center of mass. Since the total center-of-mass energy $\mathcal{E}$ is conserved in a reaction,

$$\mathcal{E} = \epsilon_b + \epsilon_B \tag{D-21}$$

Analogous to (D-10) and (D-11), we have the relations

$$\epsilon_b = \frac{\mathcal{E}^2 + (m_bc^2)^2 - (m_Bc^2)^2}{2\mathcal{E}} \tag{D-22}$$

$$\epsilon_B = \frac{\mathcal{E}^2 - (m_bc^2)^2 + (m_Bc^2)^2}{2\mathcal{E}} \tag{D-23}$$

From (D-5), we obtain the magnitude of the momentum of particle b (or B)

$$k' = \sqrt{\frac{\epsilon_b^2}{c^2} - m_b^2c^2} \tag{D-24}$$

In the center of mass, the velocity of the scattered particle b in units of c is β_b. It may be expressed in terms of its energy ϵ_b by the relation

$$\epsilon_b = \gamma_b m_b c^2 \tag{D-25}$$

where

$$\gamma_b = \frac{1}{\sqrt{1-\beta_b^2}}$$

Upon taking squares of both sides of (D-25) and rearranging the terms, we obtain the result,

$$(\epsilon_b\beta_b)^2 = \epsilon_b^2 - (m_bc^2)^2 = (k'c)^2$$

This gives us the relation

$$\epsilon_b = \frac{k'c}{\beta_b} \tag{D-26}$$

which we will need later in deducing the relation between laboratory and center-of-mass angles. In the nonrelativistic limit, the total kinetic energy remains the same before and after the scattering and, as a result, we have $k' = k$.

Scattering angle. Let us define the x-axis such that the momentum vector of the scattered particle $\boldsymbol{k}'$ is in the xz-plane in the center of mass. Because of this choice of the orientation of the spatial coordinates, the y-component of the three-momentum of particle b is zero and the four-momentum of particle b becomes $(\frac{i\epsilon_b}{c}, k'_x, 0, k'_z)$. In the laboratory, the four-momentum may be written as $(\frac{iE_b}{c}, p_x, p_y, p_z)$ where $\boldsymbol{p}$ is the three-momentum of b, the scattered particle. Using the Lorentz transformation given in (D-1), we obtain the relation between these two sets of quantities in the following form:

$$p_x = k'_x \qquad p_y = 0 \qquad p_z = \gamma\beta\frac{\epsilon_b}{c} + \gamma k'_z$$

The scattering angle Θ_b along which particle b moves in the laboratory is given by the angle between $\boldsymbol{p}$ and the z-axis defined by the direction of the incident particle a. In terms of the components of $\boldsymbol{p}$, we have

$$\tan\Theta_b = \frac{p_x}{p_z} = \frac{k'_x}{\gamma(k'_z + \beta\frac{\epsilon_b}{c})} \tag{D-27}$$

If the scattering angle in the center of mass is denoted by θ_b, the various quantities appearing the final form of (D-27) may be related to k', the magnitude of $\boldsymbol{k}'$, in the following manner:

$$k'_x = k'\sin\theta_b \qquad k'_z = k'\cos\theta_b \qquad \beta\frac{\epsilon_b}{c} = k'\frac{\beta}{\beta_b}$$

where the last equation is obtained from (D-26). Substituting these into (D-27), we obtain the result

$$\tan\Theta_b = \frac{\sin\theta_b}{\gamma\left(\cos\theta_b + \frac{\beta}{\beta_b}\right)} \tag{D-28}$$

which gives the relation between the scattering angle in the center of mass and in the laboratory systems.

When the scattered particle b is more massive than the residual particle B, we have the situation that $\beta > \beta_b$ must be true in order to satisfy four-momentum conservation. The scattered particle is then restricted to $\Theta_b < 90°$. The maximum laboratory scattering angle $\Theta_b^{\max}$ is given by the relation,

$$\tan\Theta_b^{\max} = \beta_b\sqrt{\frac{1-\beta^2}{\beta^2-\beta_b^2}} \tag{D-29}$$

obtained from the condition that $\cos\theta_b \geq -\beta_b/\beta$ in the center of mass.

For $v \ll c$, we have $\gamma \to 1$, and the ratio of β to β_b may be expressed in terms of magnitude of the velocity of the center of mass and the velocities of the particles in the center of mass,

$$\frac{\beta}{\beta_b} = \frac{v'_{cm}}{v_b} = \frac{v'_{cm}}{v_a} = \frac{m_a}{m_A} \tag{D-30}$$

using (D-15). As a result, the relation between laboratory and center-of-mass scattering angles reduces to

$$\tan\Theta_b \xrightarrow[v \ll c]{} \frac{\sin\theta_b}{\cos\theta_b + \frac{m_a}{m_A}} \tag{D-31}$$

in the nonrelativistic limit.

Solid angle relation. The solid angle $d\Omega'$, subtended by a detector in laboratory, and $d\Omega$, the equivalent quantity in the center of mass, are different since the two representations are moving at a given velocity with respect to each other. The ratio between their magnitudes is given by the following relation:

$$\frac{d\Omega'}{d\Omega} = \frac{d(\cos\Theta_b)}{d(\cos\theta_b)}$$

as can be seen from Fig. D-2. The relation between the two cosines may obtained using (D-28),

$$\cos\Theta_b = \frac{\gamma(\cos\theta_b + \frac{\beta}{\beta_b})}{\sqrt{\sin^2\theta_b + \gamma^2(\cos\theta_b + \frac{\beta}{\beta_b})^2}} \tag{D-32}$$

On differentiating we obtain the result,

$$\frac{d\Omega'}{d\Omega} = \frac{\gamma(1 + \frac{\beta}{\beta_b}\cos\theta_b)}{\left\{\sin^2\theta_b + \gamma^2\left(\cos\theta_b + \frac{\beta}{\beta_b}\right)^2\right\}^{3/2}} \tag{D-33}$$

From this factor, we obtain

$$\left(\frac{d\sigma}{d\Omega}\right)_{\rm cm} = \left(\frac{d\sigma}{d\Omega'}\right)_{\rm lab}\frac{d\Omega'}{d\Omega} = \left(\frac{d\sigma}{d\Omega'}\right)_{\rm lab}\frac{\gamma(1 + \frac{\beta}{\beta_b}\cos\theta_b)}{\left\{\sin^2\theta_b + \gamma^2\left(\cos\theta_b + \frac{\beta}{\beta_b}\right)^2\right\}^{3/2}} \tag{D-34}$$

as the relation between the differential scattering cross sections in the center of mass and in the laboratory.

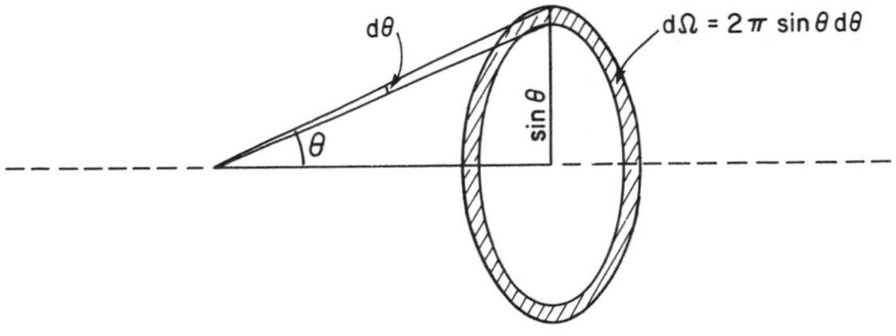

Fig. D-2 Relation between solid angle $d\Omega$ and scattering angle θ.

Mandelstam variables. Three Lorentz scalars, s, t, and u, may be defined for the reaction $A(a,b)B$ in terms of the various four-momenta involved,

$$\begin{aligned} s &= -(\varrho_a + \varrho_A)^2 = -(\varrho_b + \varrho_B)^2 \\ t &= -(\varrho_a - \varrho_b)^2 = -(\varrho_A - \varrho_B)^2 \\ u &= -(\varrho_a - \varrho_B)^2 = -(\varrho_A - \varrho_b)^2 \end{aligned} \tag{D-35}$$

These three quantities are also known as the Mandelstam variables.

In the center of mass, $\varrho_a + \varrho_A = \varrho_b + \varrho_B = i(\epsilon_a + \epsilon_A)/c = i\mathcal{E}/c$, as can be seen from the definitions of ϱ_a and ϱ_A given in (D-7), ϱ_b and ϱ_B in (D-20), and $\mathcal{E}$ in (D-8,21). This gives us the result,

$$s = \frac{(\epsilon_a + \epsilon_A)^2}{c^2} = \frac{\mathcal{E}^2}{c^2} \tag{D-36}$$

The square of the four-momentum transfer, t, is the square of the momentum transferred from particle a to c (or from A to B). Substituting the values of ϱ_a and ϱ_b from (D-7) and (D-20), we obtain the result

$$\begin{aligned} t &= \frac{(\epsilon_a - \epsilon_b)^2}{c^2} - (\boldsymbol{k} - \boldsymbol{k}')^2 \\ &= \frac{\epsilon_a^2}{c^2} - k^2 + \frac{\epsilon_b^2}{c^2} - k'^2 - 2\frac{\epsilon_a \epsilon_b}{c^2} + 2kk' \cos\theta_b \\ &= m_a^2 c^2 + m_b^2 c^2 - 2\frac{\epsilon_a \epsilon_b}{c^2} + 2k\,k' \cos\theta_b \\ &\xrightarrow[m_a = m_b]{} -2kk' + 2k\,k' \cos\theta_b = -4kk' \sin^2 \tfrac{1}{2}\theta_b \end{aligned} \tag{D-37}$$

where k and k' are, respectively, the magnitudes of $\boldsymbol{k}$ and $\boldsymbol{k}'$. Similarly u is the square of the four-momentum transfer from particle a to B (or A to c) and may be written in the form

$$u = \frac{(\epsilon_a - \epsilon_B)^2}{c^2} - (\boldsymbol{k} - \boldsymbol{k}')^2 = m_a^2 c^2 + m_B c^2 - \frac{2\epsilon_a \epsilon_B}{c^2} + 2kk' \cos\theta_B \tag{D-38}$$

These three scalars are not independent of each other. Since $\theta_b + \theta_B = \pi$ in the center of mass, $\cos\theta_b + \cos\theta_B = 0$. As a result,

$$t + u = 2m_a^2 c^2 + m_b^2 c^2 + m_B^2 c^2 - \frac{2\epsilon_a \epsilon_b}{c^2} - \frac{2\epsilon_a \epsilon_B}{c^2} = 2m_a^2 c^2 + m_b^2 c^2 + m_B^2 c^2 - \frac{2\epsilon_a \mathcal{E}}{c^2}$$

Furthermore, from (D-10), we have the relation

$$(m_a c^2)^2 - (m_A c^2)^2 + \mathcal{E}^2 = 2\epsilon_a \mathcal{E}$$

Using this, we obtain the result,

$$s + t + u = (m_a^2 + m_A^2 + m_b^2 + m_B^2)c^2 \tag{D-39}$$

a constant regardless of the frame of reference. This means that only two of the three Mandelstam variables s, t and u are independent variables.

For elastic scattering, $m_a = m_b$, $k = k'$, and $\epsilon_a = \epsilon_b$. Eq. (D-37) may be rewritten in the form

$$t = -2k^2(1 - \cos\theta_b) = -4k^2 \sin^2 \tfrac{1}{2}\theta_b \tag{D-40}$$

The momentum transfer q is reduced to

$$q = \sqrt{-t} = 2k \sin \tfrac{1}{2}\theta_b \tag{D-41}$$

where k is the magnitude of the center-of-mass momentum of the incident particle.

Appendix E

TRANSITION PROBABILITY IN TIME-DEPENDENT PERTURBATION THEORY

Consider a time-dependent Hamiltonian $H(t)$ which may be written as a sum of two parts:

$$H(t) = H_0 + H'(t) \tag{E-1}$$

where H_0 is independent of time and all the time dependence is contained in $H'(t)$. We are interested here in the case in which the strength of $H'(t)$ is sufficiently weak that it may be considered as a perturbation to H_0.

Let $\phi_n(\boldsymbol{r})$ represents the eigenfunction of H_0,

$$H_0\phi_n(\boldsymbol{r}) = E_n\phi_n(\boldsymbol{r}) \tag{E-2}$$

We shall assume that all $\phi_n(\boldsymbol{r})$ together form a complete set of orthonormal functions. Again, we have suppressed any indications of possible dependence of $\phi_n(\boldsymbol{r})$ on spin, isospin, and other variables so as to simplify the notation. The eigenfunctions $\psi(\boldsymbol{r}, t)$ of the time-dependent Schrödinger equation for H_0 alone,

$$i\hbar\frac{\partial\psi(\boldsymbol{r}, t)}{\partial t} = H_0\psi(\boldsymbol{r}, t) \tag{E-3}$$

may be expressed in terms of $\phi_n(\boldsymbol{r})$,

$$\psi(\boldsymbol{r}, t) = \sum_n c_n\phi(\boldsymbol{r})e^{-iE_nt/\hbar} \tag{E-4}$$

Here, the expansion coefficients,

$$c_n = \int \phi_n^*(\boldsymbol{r})e^{iE_nt/\hbar}\psi(\boldsymbol{r}, t)\, dV \tag{E-5}$$

are independent of time, since we have not yet included $H'(t)$ as a part of the Hamiltonian.

For the complete Hamiltonian, the eigenfunctions $\Psi(\boldsymbol{r},t)$ may still be expanded in terms of $\phi_n(\boldsymbol{r})$ except that the expansion coefficients are now time-dependent,

$$\Psi(\boldsymbol{r},t) = \sum_n c_n(t)\phi_n(\boldsymbol{r})e^{-iE_n t/\hbar} \tag{E-6}$$

The coefficient $c_n(t)$ may be interpreted as the probability amplitude for finding the system in the unperturbed state n at time t. On substituting the results given in (E-6) into the time-dependent Schrödinger equation for $H(t)$,

$$i\hbar\frac{\partial\Psi(\boldsymbol{r},t)}{\partial t} = \big(H_0 + H'(t)\big)\Psi(\boldsymbol{r},t) \tag{E-7}$$

we obtain the equation governing coefficients $c_n(t)$,

$$i\hbar\sum_n\left\{\frac{dc_n(t)}{dt} - c_n(t)i\frac{E_n}{\hbar}\right\}\phi_n(\boldsymbol{r})\,e^{-iE_n t/\hbar} = \big(H_0 + H'(t)\big)\sum_n c_n(t)\phi(\boldsymbol{r})e^{-iE_n t/\hbar}$$

By taking the product with $\phi_k^*(\boldsymbol{r})\exp\{iE_k t/\hbar\}$ on both sides of the equation and integrating over all independent variables except t, we obtain the result

$$\begin{aligned} i\hbar\sum_n&\left\{\frac{dc_n(t)}{dt} - ic_n(t)\frac{E_n}{\hbar}\right\}e^{i(E_k-E_n)t/\hbar}\,\langle\phi_k(\boldsymbol{r})|\phi_n(\boldsymbol{r})\rangle \\ &= \sum_n c_n(t)\Big\{\langle\phi_k(\boldsymbol{r})|H_0|\phi_n(\boldsymbol{r})\rangle + \langle\phi_k(\boldsymbol{r})|H'(t)|\phi_n(\boldsymbol{r})\rangle\Big\}e^{i(E_k-E_n)t/\hbar} \end{aligned} \tag{E-8}$$

Since $\phi_n(\boldsymbol{r})$ is a part of an orthonormal set of eigenfunctions for H_0, we have the conditions,

$$\langle\phi_k(\boldsymbol{r})|\phi_n(\boldsymbol{r})\rangle = \delta_{kn} \qquad\qquad \langle\phi_k(\boldsymbol{r})|H_0|\phi_n(\boldsymbol{r})\rangle = E_n\delta_{kn}$$

Putting these results into (E-8), we obtain a differential equation for $c_k(t)$,

$$i\hbar\frac{dc_k(t)}{dt} = \sum_n\langle\phi_k|H'(t)|\phi_n(t)\rangle\,e^{i\omega_{kn}t} \tag{E-9}$$

where $\omega_{kn} = (E_k - E_n)/\hbar$.

As initial conditions, let us assume that at $t = 0$ the system is in state $\phi_0(\boldsymbol{r})$; that is,

$$c_n(0) = \begin{cases} 1 & \text{for } n = 0 \\ 0 & \text{for } n \neq 0 \end{cases}$$

If the perturbation is sufficiently weak, we expect that

$$c_k(t) \approx \begin{cases} 1 & \text{for } k = 0 \\ 0 & \text{for } k \neq 0 \end{cases}$$

for all time t of interest. As a result we can approximate (E-7) by retaining only the $n = 0$ term in the sum on the right hand side. This gives us the result,

$$i\hbar\frac{dc_k(t)}{dt} = \langle\phi_k|H'(t)|\phi_0(t)\rangle\, e^{i\omega_{k0}t} \tag{E-10}$$

Furthermore, if the time variation of $H'(t)$ is slow compared with $\exp(i\omega_{k0}t)$, we may take H' to be a constant. In this approximation, (E-10) may be solved explicitly and the result may be expressed in the form:

$$c_k(t) = \frac{\langle\phi_k|H'|\phi_0(t)\rangle}{E_k - E_0}(1 - e^{i\omega_{k0}t}) \tag{E-11}$$

From this we obtain

$$\left|c_k(t)\right|^2 = 2\left|\langle\phi_k(\boldsymbol{r})|H'|\phi_0(\boldsymbol{r})\rangle\right|^2 \frac{1-\cos\omega_{k0}t}{(E_k - E_0)^2} \tag{E-12}$$

as the probability for finding the system in state n at time t if it started from state 0 at time $t = 0$.

The total probability to some interval labelled by f is given by a summation over the probabilities to all the final states k in the interval,

$$\begin{aligned}\sum_{k\in f}\left|c_k(t)\right|^2 &= 2\sum_{k\in f}\left|\langle\phi_k(\boldsymbol{r})|H'|\phi_0(\boldsymbol{r})\rangle\right|^2 \frac{1-\cos\omega_{k0}t}{(E_k - E_0)^2} \\ &= \frac{2}{\hbar^2}\int\left|\langle\phi_k(\boldsymbol{r})|H'|\phi_0(\boldsymbol{r})\rangle\right|^2 \frac{1-\cos\omega_{k0}t}{\omega_{k0}^2}\rho(E_k)\,dE_k \end{aligned} \tag{E-13}$$

The summation over all possible final states is changed in the last step into an integration over energy multiplied by the density of final states $\rho(E_k)$ for reasons that will soon become clear.

The decay constant, or transition probability per unit time, $\mathcal{W}$, corresponds to the rate of finding the system in the group of final states labelled by f and may be expressed in the following form:

$$\mathcal{W} = \frac{d}{dt}\sum_{k\in f}\left|c_k(t)\right|^2 = \frac{2}{\hbar^2}\int\left|\langle\phi_k(\boldsymbol{r})|H'|\phi_0(\boldsymbol{r})\rangle\right|^2 \frac{\sin\omega_{k0}t}{\omega_{k0}}\rho(E_k)\,dE_k \tag{E-14}$$

Since the function $\sin\omega_{k0}t/\omega_{k0}$ oscillates very quickly except where $\omega_{k0} \approx 0$, only a small region around $E_k = E_0$ can contribute to the integral. In this small energy interval we may regard the matrix element $\langle\phi_k(\boldsymbol{r})|H'|\phi_0(\boldsymbol{r})\rangle$ and the state density $\rho(E_k)$ ($= \rho(E_f)$ to be constant, and may be taken outside the integration. Furthermore, the limits of the integration over E_k may be replaced by $\pm\infty$ under these conditions without losing too much accuracy. The final form of the transition probability per unit time takes on the form,

$$\mathcal{W} = \frac{2\pi}{\hbar}\left|\langle\phi_f(\boldsymbol{r})|H'|\phi_0(\boldsymbol{r})\rangle\right|^2\rho(E_f) \tag{E-15}$$

where we have made use of the fact that

$$\int_{-\infty}^{+\infty} \frac{\sin \omega_{k0} t}{\omega_{k0}} \, d\omega_{k0} = \pi$$

This formula is the starting point for calculations of transition probabilities used in Chapter 5. Since Fermi called it the "Golden Rule of time-dependent perturbation theory," it is often referred to as *Fermi's Golden Rule.*

Appendix F

SIMPLE MANIPULATIONS WITH SECOND QUANTIZED OPERATORS

It is sometimes convenient to use second quantized notation for many problems in quantum mechanics. A brief introduction to the technique is given here on aspects that are used implicitly in the main text for both boson and fermion operators.

§F-1 BOSON OPERATORS

One of the more elementary applications of second quantization techniques for bosons is in the case of one-dimension harmonic oscillator where operators $\boldsymbol{b}^\dagger$ and $\boldsymbol{b}$ are used, respectively, to create and to annihilate a quantum of energy $\hbar\omega$. The Hamiltonian may be written in the form

$$H = \frac{p^2}{2m} + \tfrac{1}{2}m\omega^2 x^2 \tag{F-1}$$

where m is the mass and ω is the angular frequency of the oscillator. The commutation relation between the coordinate operator x and the momentum operator p is

$$\{x, p\} = xp - px = i\hbar \tag{F-2}$$

Instead of x and p, we can define two dimensionless operators

$$\boldsymbol{b} = \frac{1}{\sqrt{2}}\left(\sqrt{\frac{m\omega}{\hbar}}x + i\sqrt{\frac{1}{m\hbar\omega}}p\right)$$

$$\boldsymbol{b}^\dagger = \frac{1}{\sqrt{2}}\left(\sqrt{\frac{m\omega}{\hbar}}x - i\sqrt{\frac{1}{m\hbar\omega}}p\right) \tag{F-3}$$

We shall try to identify that these two operators are, respectively, the annihilation and creation operators of one harmonic oscillator quantum $\hbar\omega$.

From (F-3), we find that

$$\boldsymbol{b}\boldsymbol{b}^\dagger = \frac{1}{2}\Big(\frac{m\omega}{\hbar}x^2 + \frac{1}{m\hbar\omega}p^2 + \frac{i}{\hbar}(px - xp)\Big) = \frac{1}{2}\Big(\frac{m\omega}{\hbar}x^2 + \frac{1}{m\hbar\omega}p^2 + 1\Big) \quad \text{(F-4)}$$

where we have made use of the commutation relation (F-2) to arrive at the final result. Similarly

$$\boldsymbol{b}^\dagger\boldsymbol{b} = \frac{1}{2}\Big(\frac{m\omega}{\hbar}x^2 + \frac{1}{m\hbar\omega}p^2 - 1\Big) \quad \text{(F-5)}$$

By taking the difference between (F-4) and (F-5), we obtain the commutation relation

$$\{\boldsymbol{b}, \boldsymbol{b}^\dagger\} \equiv \boldsymbol{b}\boldsymbol{b}^\dagger - \boldsymbol{b}^\dagger\boldsymbol{b} = 1 \quad \text{(F-6)}$$

By comparing the sum of (F-4) and (F-5) with (F-1), we may rewrite the harmonic oscillator Hamiltonian in terms of $\boldsymbol{b}$ and $\boldsymbol{b}^\dagger$,

$$H = \tfrac{1}{2}(\boldsymbol{b}^\dagger\boldsymbol{b} + \boldsymbol{b}\boldsymbol{b}^\dagger)\hbar\omega \quad \text{(F-7)}$$

The Schrödinger equation for a one-dimensional harmonic oscillator may now be expressed in the following form:

$$\tfrac{1}{2}(\boldsymbol{b}^\dagger\boldsymbol{b} + \boldsymbol{b}\boldsymbol{b}^\dagger)\psi = \epsilon\,\psi \quad \text{(F-8)}$$

where ϵ is the energy in units of $\hbar\omega$.

Let $|\,n\,\rangle$ be an eigenvector of the Hamiltonian with η_n harmonic oscillator quanta, *i.e.*,

$$H|\,n\,\rangle = \eta_n|\,n\,\rangle \quad \text{(F-9)}$$

From (F-6), we obtain the commutation relation of $\boldsymbol{b}$ with H,

$$\{\boldsymbol{b}, H\} = \boldsymbol{b}$$

On applying this operator relation to the state $|\,n\,\rangle$, we find that

$$(\boldsymbol{b}H - H\boldsymbol{b})|\,n\,\rangle = \eta_n\boldsymbol{b}|\,n\,\rangle - H\boldsymbol{b}|\,n\,\rangle = \boldsymbol{b}|\,n\,\rangle$$

or

$$H\,\boldsymbol{b}|\,n\,\rangle = (\eta_n - 1)\,\boldsymbol{b}|\,n\,\rangle \quad \text{(F-10)}$$

From this result, we may interpret that $\boldsymbol{b}|\,n\,\rangle$ is an eigenvector of the Hamiltonian having $(\eta_n - 1)$ quanta, one unit less than the state $|\,n\,\rangle$. The relation between states having η_n and η_{n-1} quanta is, then,

$$\boldsymbol{b}|\,n\,\rangle = \alpha_n|\,n-1\,\rangle \quad \text{(F-11)}$$

where the normalization factor α_n will be determined later. From (F-11), we see that $\boldsymbol{b}$ is an operator which acts on a state of η_n quanta and produces a state of $(\eta_n - 1)$ quanta. It may therefore be identified as an *annihilation* operator for harmonic oscillator quantum. Similarly, by using the commutation relation

$$\{\boldsymbol{b}^\dagger, H\} = \boldsymbol{b}^\dagger \quad \text{(F-12)}$$

we obtain the analogous result to (F-10),

$$H\,\boldsymbol{b}^\dagger|\,n\,\rangle = \big(\eta_n + 1\big)\,\boldsymbol{b}^\dagger|\,n\,\rangle \tag{F-13}$$

and this identifies $\boldsymbol{b}^\dagger$ as a *creation* operator for an harmonic oscillator quantum.

Each time we apply the annihilation operator we obtain a state with one less quantum. However, since the harmonic oscillator spectrum has a lower bound, we cannot keep on applying the annihilation operator indefinitely. At some stage we will reach the lowest state $|\,0\,\rangle$ such that

$$\boldsymbol{b}|\,0\,\rangle = 0 \tag{F-14}$$

However, the ket $|\,0\rangle$ itself is not zero. This may be seen from the following argument. On substituting (F-6) into (F-7) we obtain the relation

$$\boldsymbol{b}^\dagger\boldsymbol{b} = H - \tfrac{1}{2} \tag{F-15}$$

where H is now in units of $\hbar\omega$. Because of (F-14), the application of the left hand side of (F-15) to $|\,0\,\rangle$ gives zero,

$$\boldsymbol{b}^\dagger\boldsymbol{b}|\,0\,\rangle = 0 \tag{F-16}$$

On the other hand, the right hand side of (F-15) gives the result,

$$\big(H - \tfrac{1}{2}\big)|\,0\,\rangle = \big(\eta_0 - \tfrac{1}{2}\big)|\,0\,\rangle \tag{F-17}$$

where η_0 is the eigenvalue of H for the state $|\,0\,\rangle$. Putting together (F-16) and (F-17), we obtain

$$\eta_0 = \tfrac{1}{2} \tag{F-18}$$

as the energy associated with the zero-point motion of the ground state of a one-dimensional harmonic oscillator.

The creation operator $\boldsymbol{b}^\dagger$ may be applied repetitively on $|\,0\,\rangle$ and, with each application, we obtain a state with one more harmonic oscillator quantum. After n times, we obtain a state $|\,n\,\rangle$ having $(n + \frac{1}{2})$ quanta,

$$H|\,n\,\rangle = \big(n + \tfrac{1}{2}\big)|\,n\,\rangle \tag{F-19}$$

In other words,

$$\eta_n = n + \tfrac{1}{2}$$

We can now work out the normalization constant α_n in (F-9). Using (F-9), the matrix element of $\boldsymbol{b}$ is given by

$$\langle n'|\boldsymbol{b}|n\rangle = \alpha_n\delta_{n',n-1} = \langle n|\boldsymbol{b}^\dagger|n'\rangle^* \tag{F-20}$$

From (F-15) and (F-19), we obtain the result,

$$\langle n|\boldsymbol{b}^\dagger\boldsymbol{b}|n\rangle = \langle n|H - \tfrac{1}{2}|n\rangle = n \tag{F-21}$$

On the other hand, by inserting a complete set of intermediate states between the creation and annihilation operators, we obtain

$$\langle n|\boldsymbol{b}^\dagger \boldsymbol{b}|n\rangle = \sum_{n'} \langle n|\boldsymbol{b}^\dagger|n'\rangle\langle n'|\boldsymbol{b}|n\rangle = |\alpha_n|^2 \tag{F-22}$$

Comparing this with (F-21), we obtain the result

$$\alpha_n = \sqrt{n}$$

From this, we deduce that

$$\langle n'|\boldsymbol{b}|n\rangle = \sqrt{n}\,\delta_{n',n-1} \tag{F-23}$$

and

$$\langle n'|\boldsymbol{b}^\dagger|n\rangle = \sqrt{n+1}\,\delta_{n',n+1} \tag{F-24}$$

Furthermore, by applying $\boldsymbol{b}^\dagger$ operator n times, we obtain

$$|\,n\,\rangle = \frac{1}{\sqrt{n!}}(\boldsymbol{b}^\dagger)^n|\,0\,\rangle \tag{F-25}$$

as the normalized state of n quanta.

§F-2 FERMION OPERATORS

For fermions, we can define $\boldsymbol{a}_r$ to be an operator that annihilates a particle in single-particle state r and $\boldsymbol{a}_r^\dagger$, its adjoint, to be an operator that creates a particle in single-particle state r. Let $|\,\Phi\,\rangle$ be an antisymmetrized A-particle state constructed from a product of single-particle states,

$$\big|\Phi\big\rangle = \big|\phi_1\phi_2\cdots\phi_A\big\rangle \tag{F-26}$$

where we have used the shorthand notation defined in (6-77). By definition,

$$\boldsymbol{a}_r\big|\Phi\big\rangle = 0$$

if single-particle state ϕ_r is vacant. Alternatively, a state of $(A-1)$ particles is produced if ϕ_r is occupied. Similarly, since each single-particle state can accommodate at most one fermion,

$$\boldsymbol{a}_r^\dagger\big|\Phi\big\rangle = 0$$

if the single-particle state ϕ_r is already occupied and a state of $(A+1)$ particles is produced if ϕ_r is vacant.

The commutation relations between second-quantized single-particle fermion operators $\boldsymbol{a}_r$ and $\boldsymbol{a}_r^\dagger$ are given by

$$\big[\boldsymbol{a}_r^\dagger, \boldsymbol{a}_s\big] \equiv \boldsymbol{a}_r^\dagger\boldsymbol{a}_s + \boldsymbol{a}_s\boldsymbol{a}_r^\dagger = \delta_{rs} \tag{F-27}$$

$$\big[\boldsymbol{a}_r^\dagger, \boldsymbol{a}_s^\dagger\big] = \big[\boldsymbol{a}_r, \boldsymbol{a}_s\big] = 0 \tag{F-28}$$

These are different from (F-6) by the fact that the Pauli exclusion principle acting between identical fermions is built into the commutation relation here. This may be demonstrated by the following example. Consider a two-particle state $|\,\Phi(r,s)\,\rangle$ made of antisymmetrized products of single-particle states ϕ_r and ϕ_s. In terms of single-particle creation operators,

$$\big|\Phi(r,s)\big\rangle = \boldsymbol{a}_r^\dagger \boldsymbol{a}_s^\dagger |\,0\,\rangle$$

where $|\,0\,\rangle$ denotes the vacuum, a state without any particle. If we interchange the order of the two single-particle creation operators, we obtain the state

$$\big|\Phi(s,r)\big\rangle = \boldsymbol{a}_s^\dagger \boldsymbol{a}_r^\dagger |\,0\,\rangle$$

By (F-27),

$$\boldsymbol{a}_r^\dagger \boldsymbol{a}_s^\dagger = -\boldsymbol{a}_s^\dagger \boldsymbol{a}_r^\dagger$$

Hence

$$\big|\Phi(r,s)\big\rangle = \begin{cases} 0 & \text{if } r = s \\ -|\,\Phi(s,r)\,\rangle & \text{otherwise} \end{cases}$$

We can see that these are the properties of an antisymmetrized state of two particles by writing out the two-particle state explicitly in terms of a linear combination of products of single-particle wave functions,

$$\big|\Phi(r,s)\big\rangle \equiv \sqrt{\tfrac{1}{2}}\Big(\big|\phi_r(1)\big\rangle\big|\phi_s(2)\big\rangle - \big|\phi_s(1)\big\rangle\big|\phi_r(2)\big\rangle\Big)$$

where arguments 1 and 2 label the two particles.

In principle, both $\boldsymbol{a}_r$ and $\boldsymbol{a}_r^\dagger$ are operators with definite spherical tensor ranks. For simplicity we shall ignore considerations relating to angular momentum coupling here. A discussion of second quantized spherical tensor operators can be found, for example, in Wong (*Nuclear Statistical Spectroscopy*, Oxford University Press, New York, 1986). Since $\boldsymbol{a}_r^\dagger$ connects a state of A particles to a state of $(A+1)$ particles, it is related to the operator for the one-nucleon transfer stripping reaction discussed in §7-3. The spectroscopic factor for transferring a nucleon into the single-particle state ϕ_r is proportional to the square of the matrix element $\langle\text{daughter state}|\boldsymbol{a}_r^\dagger|\text{target state}\rangle$. Similarly, the spectroscopic factor for picking up a nucleon from the state ϕ_s is proportional to the square of the matrix element $\langle\text{daughter state}|\boldsymbol{a}_s|\text{target state}\rangle$. Two-nucleon transfer reactions are connected with operators of the form $\boldsymbol{a}_r^\dagger \boldsymbol{a}_s^\dagger$ and $\boldsymbol{a}_r \boldsymbol{a}_s$.

A one-body operator for nucleons and other fermions is one that gives zero if it operates on a zero-particle state and, in general, is non-zero if the state has one or more particles. In second-quantized notation, a one-body operator has the form $\boldsymbol{a}_r^\dagger \boldsymbol{a}_s$. If $r = s$ (and if $\boldsymbol{a}_r^\dagger$ and $\boldsymbol{a}_s$ are coupled to angular momentum zero), it gives unity if the single-particle state r is occupied and zero otherwise. Hence the number operator, $\boldsymbol{n}$, which counts the number of particles present, may be written in the form

$$\boldsymbol{n} = \sum_r \boldsymbol{a}_r^\dagger \boldsymbol{a}_r \tag{F-29}$$

Electromagnetic transition operators are also one-body operators. Their structure in terms of single-particle creation and annihilation operators is slightly more complicated than the number operator $\boldsymbol{n}$ since a particle may be moved from a state s to a state r by the transition. It can be expressed as a sum over the product $\boldsymbol{a}_r^\dagger \boldsymbol{a}_s$ for all possible single-particle states r and s. Furthermore, the product must be coupled to angular momentum and isospin ranks appropriate for the multipolarity of the transition.

A two-body operator involves a product of two single-particle creation operators and two single-particle annihilation operators written in the form $\boldsymbol{a}_r^\dagger \boldsymbol{a}_s^\dagger \boldsymbol{a}_t \boldsymbol{a}_u$. Note the particular order of writing all the creation operators on the left of all the annihilation operators, known as *normal* order. Because of the commutation relation given by (F-27), the operator $\boldsymbol{a}_r^\dagger \boldsymbol{a}_t \boldsymbol{a}_s^\dagger \boldsymbol{a}_u$ is a mixture of one-and two-body operators. This is evident from the fact that, using (F-27), we obtain the result

$$\boldsymbol{a}_r^\dagger \boldsymbol{a}_t \boldsymbol{a}_s^\dagger \boldsymbol{a}_u = \boldsymbol{a}_r^\dagger (\delta_{st} - \boldsymbol{a}_s^\dagger \boldsymbol{a}_t) \boldsymbol{a}_u = \boldsymbol{a}_r^\dagger \boldsymbol{a}_u \delta_{s,t} - \boldsymbol{a}_r^\dagger \boldsymbol{a}_s^\dagger \boldsymbol{a}_t \boldsymbol{a}_u \tag{F-30}$$

Hence $\boldsymbol{a}_r^\dagger \boldsymbol{a}_t \boldsymbol{a}_s^\dagger \boldsymbol{a}_u$ contains a one-body part $\boldsymbol{a}_r^\dagger \boldsymbol{a}_u$ for $s = t$.

Second quantization is a powerful technique for solving many-body problems. In calculating the many-body matrix element of an operator, we can express the operator in terms of single-particle creation and annihilation operators as illustrated by earlier examples. Furthermore, we can also express the wave functions in terms of a product of creation operators acting on the vacuum, as we have done above for the simple case of a two-particle state. Thus, the entire many-body matrix element may be expressed as the vacuum expectation value of a product of single-particle creation and annihilation operators. In this way, the matrix elements can be evaluated by manipulating the single-particle operators using commutation relations given by (F-27,28). Angular momentum recoupling is generally not a problem except to note that $\boldsymbol{a}_r^\dagger$ is not a proper spherical tensor, as mentioned in §A-2 in connection with particle-antiparticle transformation. This problem may be avoided by defining adjoint operators that are proper spherical tensors. However, we shall not go into the technical matter of second-quantized spherical tensor operators here.

References

Books on Nuclear Physics

Blatt, J.M., and Weisskopf, V.F. *Theoretical Nuclear Physics.* Wiley, New York, 1952.

Blin-Stoyle, R.J. *Fundamental Interactions and the Nucleus.* North-Holland, Amsterdam, 1973.

Bohr, A., and Mottelson, B.R. *Nuclear Structure,* vol. I. Benjamin, Reading, Massachusetts, 1969.

Bohr, A., and Mottelson, B.R. *Nuclear Structure,* vol. II. Benjamin, Reading, Massachusetts, 1975.

Celenza, L.S., and Shakin, C.M. *Relativistic Nuclear Physics.* World Scientific, Singapore, 1986.

Danos, M., Gillet, V., and Cauvin, M. *Methods in Relativistic Nuclear Physics.* North-Holland, Amsterdam, 1984.

de Shalit, A., and Feshbach, H. *Theoretical Nuclear Physics.* Wiley, New York, 1974.

de Shalit, A., and Talmi, I. *Nuclear Shell Theory.* Academic Press, New York, 1963.

Eisenberg, J.M., and Greiner, W. *Excitation Mechanisms of the Nucleus.* North-Holland, Amsterdam, 1970.

Satchler, G.R. *Direct Nuclear Reactions.* Oxford University Press, Oxford, 1983.

Tobocman, W. *Theory of Direct Nuclear Reaction.* Oxford University Press, London, 1961.

Wildermuth, K., and Tang, Y.C. *A Unified Theory of the Nucleus.* Vieweg, Braunschweig, Germany, 1977.

Wong, S.S.M. *Nuclear Statistical Spectroscopy.* Oxford University Press, New York, 1986.

Books on Quantum Mechanics

Cohen-Tannoudji, C., Diu B., and Laloë, F. *Quantum Mechanics,* English translation by S.R. Hemley, N. Ostrowsky and D. Ostrowsky. Wiley, New York, 1977.

Fetter, A.L., and Walecka, J.D. *Quantum Theory of Many-Particle Systems.* McGraw-Hill, New York, 1971.

Merzbacher, E. *Quantum Mechanics,* 2nd ed. Wiley, New York, 1961.

Messiah, A. *Quantum Mechanics.* North-Holland, Amsterdam, 1966.

Sakurai, J.J. *Advanced Quantum Mechanics.* Addison-Wesley, Reading, Massachusetts, 1967.

Schiff, L.I. *Quantum Mechanics,* 3rd ed. McGraw-Hill, New York, 1968.

Books on Astrophysics, Mathematical Physics, and other Topics

Arfken, G. *Mathematical Methods for Physicists.* Academic Press, New York, 1970.

Audouze, J., and Vauclair, S. *An Introduction to Nuclear Astrophysics.* Reidel Publishing Co., Dordrecht, Holland, 1980.

Brink, D.M., and Satchler, G.R. *Angular Momentum*, 2nd ed. Clarendon Press, Oxford, 1968.

Clayton, D.D. *Principles of Stellar Evolution and Nucleosynthesis.* McGraw-Hill, New York, 1968.

Huang, K. *Statistical Mechanics.* Wiley, New York, 1963.

Lee, T.D. *Particle Physics and Introduction to Field Theory.* Harwood, Chur, Switzerland, 1981.

Morita, M. *Beta Decay and Muon Capture.* Benjamin, Reading, Massachusetts, 1973.

Morse, P.M., and Feshbach, H. *Methods of Theoretical Physics.* McGraw-Hill, New York, 1953.

Pathria, R.K. *Statistical Mechanics.* Pergamon, Oxford, 1972.

Perkins, D.H. *Introduction to High Energy Physics*, 3rd ed. Addison Wesley, Menlo Park, California, 1987.

Handbooks and Tables

Abramowitz, M., and Segun, I.A., editors. *Handbook of Mathematical Functions.* Dover, New York, 1965.

de Vries, H., de Jager, C.W., and de Vries, C. *Atomic Data and Nucl. Data Tables* **36** (1987) 495.

Fano, U. *Tables for the Analysis of Beta Spectra.* Applied Mathematics Series 13, National Bureau of Standards, Washington (1951) 266.

Gleit, G.E., et al. *Nucl. Data Sheets* **5** (1963) set 5.

Lederer, C.M., and Shirley, V.S., editors. *Table of Isotopes*, 7th ed. Wiley, New York, 1978.

Particle Data Group. *Rev. of Particle Properties* in *Phys. Lett.* **204B** (1988) 1.

Wapstra, A.H., Audi, G., and Hoekstra, R. *Atomic Data and Nucl. Data Tables* **39** (1988) 281.

Journal articles and conference proceedings

Akulinichev, S.V., et al. *Phys. Rev. Lett.* **55** (1985) 2239.

Alder, K., et al. *Rev. Mod. Phys.* **28** (1956) 432; **30** (1958) 353.

Alvarz, L.W. *Physics Today* **40** (July, 1987) 24.

Aprahamian, A., et al. *Phys. Rev. Lett.* **59** (1987) 535.

Arima, A., and Iachello, F. *Adv. Nucl. Phys.* **13** (1984) 139.

Arndt, R.A., Hyslop, J.S. III, and Roper, L.D. *Phys. Rev. D* **35** (1987) 128.

Ashery, D., and Schiffer, J.P. *Ann. Rev Nucl. Part. Sci.* **36** (1986) 207.

Aubert, J.J., et al. *Phys. Lett.* **123B** 788 (1983) 275.

Azgui, F., et al. *Nucl. Phys.* **A439** (1985) 537.

Baer, H.W., et al., in *Pion-Nucleus Physics*, AIP Conf. Proc. **163**, ed. by R.J. Peterson and D.D. Strottman, Amer. Inst. Phys., New York, 1988.

Batty, C.J., Squier, G.T.A., and Turner, G.K. *Nucl. Phys.* **B67** (1973) 492.

Bauhoff, W., von Geramb, H.V., and Palla, G. *Phys. Rev. C* **27** (1983) 2466.

Bertsch, G., et al. *Nucl. Phys.* **A284** (1977) 399.

Bonche, P., Grammaticos, B., and Koonin, S. *Phys. Rev. C* **17** (1978) 1700.

Bertozzi, W., et al. *Phys. Lett.* **41B** (1972) 408.

Bertrand, F.E. *Nucl. Phys.* **A354** (1981) 129c.

Bethe, H.A. *Rev. Mod. Phys.* **9** (1937) 69.

Bethe, H.A. *Scientific American* **189** (September 1953) 58.

Brekke, L., and Rosner, J.L. *Comm. Nucl. Part. Phys.* **18** (1988) 83.

Bromley, D.A. *Scientific American* **76** (December 1978) 58.

Bystricky, J., Lehar, F., and Winternitz, P. *J. de Phys.* **39** (1978) 1.

Cohen, S., and Kurath, D. *Nucl. Phys.* **73** (1965) 1.

Cowan, T., et al. *Phys. Rev. Lett.* **56** (1986) 444.

Davis, R. *Phys. Rev.* **97** (1955) 766.

Devons, S., and Duerdoth, I. *Adv. Nucl. Phys.* **2** (1969) 295.

Elliot, S.R., Hahn, H.F., and Moe, R.R. *Phys. Rev. Lett.* **59** (1987) 1649.

Franey, M.A., and Love, W.G. *Phys. Rev. C* **31** (1985) 488.

Frois, B., et al. *Phys. Rev. Lett.* **38** (1977) 152.

Frosch, R.F., et al. *Phys. Rev.* **174** (1968) 1380.

Gabioud, B., et al. *Phys. Lett.* **103B** (1981) 9.

Galonsky, A., in *The (p,n) Reaction and the Nucleon-Nucleon Force*, ed. C.D. Goodman et al. Plenum Press, New York, 1980.

Garreta, D., et al. *Phys. Lett.* **135B** (1984) 266.

Glasgow, D.W., et al., in *Nuclear Data for Basic and Applied Science,* ed. by P.G. Young et al. Gordon and Breach, New York, 1985.

Gogny, D., in *Nuclear Self-Consistent Fields*, ed. by G. Ripka and M. Porneuf. North-Holland, Amsterdam, 1975.

Goldhaber, M., Grodzins, L., and Sunyan, A.W. *Phys. Rev.* **109** (1958) 1015.

Gram, P.A.M., in *Pion-Nucleus Physics*, AIP Conf. Proc. **163**, ed. by R.J. Peterson and D.D. Strottman. Amer. Inst. Phys., New York, 1988.

Greene, G.L., et al. *Phys. Rev. Lett.* **56** (1986) 819.

Hauser, W., and Feshbach, H. *Phys. Rev.* **87** (1952) 366.

Hofstadter, R., and Herman, R. *Phys. Rev. Lett.* **6** (1961) 293.

Hufner, J., and Thies, M. *Phys. Rev.* **C20** (1979) 273.

Ingram, C.H.Q., in *Meson-Nuclear Physics – 1979*, AIP Conf. Proc. **54**, ed. by E.V. Hungerford III. Amer. Inst. Phys. New York, 1979.

Jänecke, J., and Masson, P.J. *Atomic Data and Nucl. Data Tables* **39** (1988) 265.

Jones, G., in *Pion Production and Absorption in Nuclei – 1981*, AIP Conf. Proc. **79**, ed. by R.D. Bent. Amer. Inst. Phys., New York, 1982.

Ko, C.M. *Phys. Lett.* **81B** (1979) 299.

Littauer, R.M., Schopper, H.F., and Wilson, R.R. *Phys. Rev. Lett.* **7** (1961) 144.

Lubimov, V.A., et al. *Phys. Lett.* **94B** (1980) 266.

Machleidt, R., in *Rel. Dynamics and Quark-Nuclear Phys.,* ed. by M.B. Johnson and A. Picklesimer. Wiley, New York, 1985.

Machleidt, R., Holinde, K., and Elster, Ch. *Phys. Rep.* **149** (1987) 1.

Myers, W.D., and Swiatecki, W.J. *Ann. Rev. Nucl. Part. Sci.* **32** (1982) 309.

Negele, J.W. *Phys. Rev. C* **1** (1970) 1260.

Nagle, D.E., Johnson M.B., and Measday, D. *Phys. Today* (April 1987) 56.

Noyes, H.P. *Ann. Rev. Nucl. Sci.* **22** (1972) 465.

Okubo, S., and Marshak, R.E. *Ann. Phys.* **4** (1958) 166.

Peterson, P.C., et al. *Phys. Rev. Lett.* **57** (1986) 949.

Schwandt, P., et al. *Phys. Rev. C* **26** (1982) 55.

Sirlin, A. *Phys. Rev. D* **35** (1987) 3423.

Sloan, T., Smadja, G., and Voss, R. *Phys. Rep.* **162** (1988) 45.

Stapp, H.P., Ypsilantis, T.J., and Metropolis, N. *Phys. Rev.* **105** (1957) 302.

Stricker, K., Carr, J.A., and McManus, H. *Phys. Rev. C* **22** (1980) 2043.

Strutinsky, V.M. *Soviet J. Nucl. Phys.* **3** (1966) 449.

Vinh Mau, R., in *Mesons in Nuclei*, ed. by M. Rho and D.H. Wilkinson. North-Holland, Amsterdam, 1979.

von Geramb, H.V., Brieva, F.A., and Rook, J.R., in *Microscopic Optical Potentials,* Lecture Notes in Phys. **89**, ed. by H.V. von Geramb. Springer-Verlag, Berlin, 1979.

Wilczyński, J. *Phys. Lett.* **47B** (1973) 484.

Wildenthal, B.H., and Chung, W., in *Mesons in Nuclei,* edited by M. Rho and D.H. Wilkinson. North-Holland, Amsterdam, 1979.

Wilkinson, D. *J. Phys. Soc. Jpn.* **55** (1986) Suppl. 347.

Wu, C.S., et al. *Phys. Rev.* **105** (1957) 1413.

Table of Nuclear Mass Excess in keV†

A	*Z*	Mass exc.	*A*	*Z*	Mass exc.	*A*	*Z*	Mass exc.	*A*	*Z*	Mass exc.
1	0 n	8071.38	1	1 H	7289.02	2	1 H	13135.82	3	1 H	14949.91
3	2 He	14931.31	4	1 H	25840.00	4	2 He	2424.91	4	3 Li	25120.00
5	2 He	11390.00	5	3 Li	11680.00	6	2 He	17592.60	6	3 Li	14085.70
6	4 Be	18374.00	7	2 He	26110.00	7	3 Li	14907.00	7	4 Be	15768.90
7	5 B	27870.00	8	2 He	31598.00	8	3 Li	20945.60	8	4 Be	4941.71
8	5 B	22920.40	8	6 C	35094.00	9	2 He	40810.00	9	3 Li	24954.10
9	4 Be	11347.70	9	5 B	12415.90	9	6 C	28913.90	10	3 Li	33840.00
10	4 Be	12607.10	10	5 B	12050.99	10	6 C	15699.10	11	3 Li	40900.00
11	4 Be	20174.00	11	5 B	8668.20	11	6 C	10650.40	11	7 N	24890.00
12	4 Be	25077.00	12	5 B	13369.50	12	6 C	0.00	12	7 N	17338.10
12	8 O	32060.00	13	4 Be	35000.00	13	5 B	16562.50	13	6 C	3125.03
13	7 N	5345.52	13	8 O	23113.00	14	4 Be	40100.00	14	5 B	23664.00
14	6 C	3019.90	14	7 N	2863.43	14	8 O	8006.54	15	5 B	28970.00
15	6 C	9873.20	15	7 N	101.49	15	8 O	2855.40	15	9 F	16770.00
16	6 C	13694.00	16	7 N	5682.10	16	8 O	-4737.03	16	9 F	10680.00
16	10 Ne	23989.00	17	6 C	21035.00	17	7 N	7871.00	17	8 O	-809.08
17	9 F	1951.78	17	10 Ne	16480.00	18	6 C	24920.00	18	7 N	13117.00
18	8 O	-782.20	18	9 F	873.40	18	10 Ne	5319.00	19	7 N	15871.00
19	8 O	3332.10	19	9 F	-1487.43	19	10 Ne	1751.00	19	11 Na	12928.00
20	8 O	3796.90	20	9 F	-17.35	20	10 Ne	-7047.80	20	11 Na	6839.00
20	12 Mg	17570.00	21	8 O	8066.00	21	9 F	-47.50	21	10 Ne	-5737.40
21	11 Na	-2189.80	21	12 Mg	10913.00	22	8 O	9440.00	22	9 F	2830.00
22	10 Ne	-8027.20	22	11 Na	-5185.20	22	12 Mg	-397.00	22	13 Al	18090.00
23	9 F	3350.00	23	10 Ne	-5156.00	23	11 Na	-9532.30	23	12 Mg	-5473.70
23	13 Al	6767.00	24	10 Ne	-5950.00	24	11 Na	-8420.40	24	12 Mg	-13933.50
24	13 Al	-55.00	24	14 Si	10755.00	25	10 Ne	-2060.00	25	11 Na	-9360.30
25	12 Mg	-13192.80	25	13 Al	-8915.80	25	14 Si	3827.00	26	10 Ne	440.00
26	11 Na	-6904.00	26	12 Mg	-16214.10	26	13 Al	-12210.40	26	14 Si	-7145.00
27	11 Na	-5600.00	27	12 Mg	-14586.30	27	13 Al	-17197.20	27	14 Si	-12385.40
27	15 P	-750.00	28	11 Na	-1140.00	28	12 Mg	-15019.20	28	13 Al	-16851.00
28	14 Si	-21492.90	28	15 P	-7161.00	28	16 S	4130.00	29	11 Na	2650.00
29	12 Mg	-10661.00	29	13 Al	-18215.80	29	14 Si	-21895.40	29	15 P	-16951.90
29	16 S	-3160.00	30	11 Na	8210.00	30	12 Mg	-9100.00	30	13 Al	-15890.00
30	14 Si	-24433.60	30	15 P	-20200.90	30	16 S	-14063.00	31	11 Na	11830.00
31	13 Al	-15050.00	31	14 Si	-22950.60	31	15 P	-24441.20	31	16 S	-19045.40
31	17 Cl	-7060.00	32	11 Na	16550.00	32	12 Mg	-1750.00	32	14 Si	-24081.10
32	15 P	-24305.50	32	16 S	-26016.37	32	17 Cl	-13330.00	32	18 Ar	-2180.00
33	11 Na	21470.00	33	14 Si	-20492.00	33	15 P	-26338.10	33	16 S	-26586.63
33	17 Cl	-21003.80	33	18 Ar	-9380.00	34	11 Na	26650.00	34	14 Si	-19958.00
34	15 P	-24557.70	34	16 S	-29932.37	34	17 Cl	-24440.07	34	18 Ar	-18379.00
35	14 Si	-14320.00	35	15 P	-24857.90	35	16 S	-28846.91	35	17 Cl	-29013.74
35	18 Ar	-23048.90	35	19 K	-11167.00	35	20 Ca	4450.00	36	15 P	-20251.00
36	16 S	-30664.25	36	17 Cl	-29522.10	36	18 Ar	-30230.71	36	19 K	-17425.00
36	20 Ca	-6440.00	37	16 S	-26896.48	37	17 Cl	-31761.78	37	18 Ar	-30948.70
37	19 K	-24798.70	37	20 Ca	-13159.00	38	16 S	-26861.00	38	17 Cl	-29798.26
38	18 Ar	-34715.10	38	19 K	-28802.00	38	20 Ca	-22059.00	39	17 Cl	-29802.80
39	18 Ar	-33241.00	39	19 K	-33806.30	39	20 Ca	-27275.60	39	21 Sc	-14300.00
40	16 S	-22520.00	40	17 Cl	-27530.00	40	18 Ar	-35039.10	40	19 K	-33534.50
40	20 Ca	-34846.10	40	21 Sc	-20526.00	40	22 Ti	-9063.00	41	17 Cl	-27400.00
41	18 Ar	-33066.50	41	19 K	-35558.40	41	20 Ca	-35137.50	41	21 Sc	-28643.00

†Based on A.H. Wapstra, G. Audi, and R. Hoekstra, *Atomic Data and Nucl. Data Tables* **39** (1988) 281.

A	*Z*	Mass exc.	*A*	*Z*	Mass exc.	*A*	*Z*	Mass exc.	*A*	*Z*	Mass exc.
41	22 Ti	-15690.00	42	18 Ar	-34420.00	42	19 K	-35020.90	42	20 Ca	-38547.00
42	21 Sc	-32121.90	42	22 Ti	-25121.00	43	17 Cl	-23130.00	43	18 Ar	-31980.00
43	19 K	-36593.00	43	20 Ca	-38408.60	43	21 Sc	-36187.80	43	22 Ti	-29320.00
44	18 Ar	-32260.00	44	19 K	-35810.00	44	20 Ca	-41469.20	44	21 Sc	-37815.90
44	22 Ti	-37548.20	44	24 Cr	-13450.00	45	18 Ar	-29720.00	45	19 K	-36614.00
45	20 Ca	-40812.80	45	21 Sc	-41069.50	45	22 Ti	-39006.40	45	23 V	-31875.00
45	24 Cr	-19410.00	46	18 Ar	-29720.00	46	19 K	-35418.00	46	20 Ca	-43140.70
46	21 Sc	-41758.80	46	22 Ti	-44125.40	46	23 V	-37075.00	46	24 Cr	-29472.00
47	18 Ar	-25910.00	47	19 K	-35696.00	47	20 Ca	-42345.50	47	21 Sc	-44330.40
47	22 Ti	-44931.80	47	23 V	-42004.00	47	24 Cr	-34553.00	48	19 K	-32122.00
48	20 Ca	-44214.00	48	21 Sc	-44492.00	48	22 Ti	-48487.10	48	23 V	-44474.70
48	24 Cr	-42818.00	48	25 Mn	-29211.00	48	26 Fe	-18130.00	49	19 K	-30770.00
49	20 Ca	-41289.00	49	21 Sc	-46558.00	49	22 Ti	-48558.10	49	23 V	-47956.20
49	24 Cr	-45328.30	49	25 Mn	-37611.00	49	26 Fe	-24580.00	50	20 Ca	-39570.00
50	21 Sc	-44537.00	50	22 Ti	-51426.00	50	23 V	-49219.00	50	24 Cr	-50257.30
50	25 Mn	-42625.40	50	26 Fe	-34470.00	51	20 Ca	-35010.00	51	21 Sc	-43218.00
51	22 Ti	-49726.90	51	23 V	-52199.00	51	24 Cr	-51447.70	51	25 Mn	-48238.90
51	26 Fe	-40217.00	52	22 Ti	-49464.00	52	23 V	-51438.80	52	24 Cr	-55414.40
52	25 Mn	-50702.90	52	26 Fe	-48331.00	52	27 Co	-34287.00	52	28 Ni	-22640.00
53	22 Ti	-46830.00	53	23 V	-51846.00	53	24 Cr	-55282.50	53	25 Mn	-54686.80
53	26 Fc	-50943.10	53	27 Co	-42639.00	53	28 Ni	-29380.00	54	23 V	-49889.00
54	24 Cr	-56930.10	54	25 Mn	-55553.10	54	26 Fe	-56250.30	54	27 Co	-48007.90
54	28 Ni	-39210.00	55	23 V	-49150.00	55	24 Cr	-55105.10	55	25 Mn	-57708.30
55	26 Fe	-57476.90	55	27 Co	-54025.60	55	28 Ni	-45330.00	56	24 Cr	-55290.00
56	25 Mn	-56907.50	56	26 Fe	-60603.70	56	27 Co	-56037.70	56	28 Ni	-53901.00
56	29 Cu	-38584.00	56	30 Zn	-26130.00	57	25 Mn	-57487.00	57	26 Fe	-60178.50
57	27 Co	-59342.60	57	28 Ni	-56077.40	57	29 Cu	-47350.00	57	30 Zn	-32700.00
58	25 Mn	-55830.00	58	26 Fe	-62151.80	58	27 Co	-59844.10	58	28 Ni	-60225.00
58	29 Cu	-51662.30	58	30 Zn	-42210.00	59	25 Mn	-55476.00	59	26 Fe	-60661.40
59	27 Co	-62226.20	59	28 Ni	-61153.60	59	29 Cu	-56353.50	59	30 Zn	-47260.00
60	25 Mn	-52900.00	60	26 Fe	-61406.00	60	27 Co	-61646.80	60	28 Ni	-64470.80
60	29 Cu	-58344.00	60	30 Zn	-54185.00	61	26 Fe	-58919.00	61	27 Co	-62897.10
61	28 Ni	-64219.60	61	29 Cu	-61982.00	61	30 Zn	-56343.00	62	26 Fe	-58896.00
62	27 Co	-61423.00	62	28 Ni	-66745.50	62	29 Cu	-62797.00	62	30 Zn	-61170.00
62	31 Ga	-51999.00	63	26 Fe	-55190.00	63	27 Co	-61839.00	63	28 Ni	-65512.80
63	29 Cu	-65578.80	63	30 Zn	-62211.90	63	31 Ga	-56690.00	64	27 Co	-59791.00
64	28 Ni	-67098.00	64	29 Cu	-65423.60	64	30 Zn	-66002.20	64	31 Ga	-58837.00
64	32 Ge	-54430.00	65	27 Co	-59160.00	65	28 Ni	-65124.80	65	29 Cu	-67262.10
65	30 Zn	-65910.40	65	31 Ga	-62654.90	65	32 Ge	-56410.00	66	28 Ni	-66029.00
66	29 Cu	-66256.70	66	30 Zn	-68898.80	66	31 Ga	-63724.00	66	32 Ge	-61620.00
66	33 As	-52070.00	67	28 Ni	-63743.00	67	29 Cu	-67303.00	67	30 Zn	-67879.60
67	31 Ga	-66878.60	67	32 Ge	-62656.00	67	33 As	-56650.00	68	28 Ni	-63483.00
68	29 Cu	-65540.00	68	30 Zn	-70006.50	68	31 Ga	-67085.40	68	32 Ge	-66978.00
68	33 As	-58880.00	69	28 Ni	-60460.00	69	29 Cu	-65741.00	69	30 Zn	-68417.30
69	31 Ga	-69322.60	69	32 Ge	-67097.00	69	33 As	-63080.00	69	34 Se	-56300.00
70	29 Cu	-63390.00	70	30 Zn	-69561.00	70	31 Ga	-68905.90	70	32 Ge	-70561.80
70	33 As	-64340.00	71	30 Zn	-67323.00	71	31 Ga	-70139.40	71	32 Ge	-69906.20
71	33 As	-67894.00	72	30 Zn	-68131.00	72	31 Ga	-68589.30	72	32 Ge	-72583.60
72	33 As	-68228.00	72	34 Se	-67897.00	73	30 Zn	-65410.00	73	31 Ga	-69705.00
73	32 Ge	-71295.20	73	33 As	-70955.00	73	34 Se	-68215.00	73	35 Br	-63600.00
73	36 Kr	-56890.00	74	30 Zn	-65708.00	74	31 Ga	-68060.00	74	32 Ge	-73423.60
74	33 As	-70861.40	74	34 Se	-72215.30	74	35 Br	-65301.00	74	36 Kr	-62130.00
74	37 Rb	-51670.00	75	30 Zn	-62530.00	75	31 Ga	-68466.00	75	32 Ge	-71858.10

A	Z		Mass exc.	A	Z		Mass exc.	A	Z		Mass exc.	A	Z		Mass exc.
75	33	As	-73035.40	75	34	Se	-72171.50	75	35	Br	-69142.00	75	36	Kr	-64214.00
75	37	Rb	-57210.00	76	30	Zn	-62290.00	76	31	Ga	-66440.00	76	32	Ge	-73214.80
76	33	As	-72290.50	76	34	Se	-75254.40	76	35	Br	-70291.00	76	36	Kr	-68965.00
76	37	Rb	-60530.00	77	32	Ge	-71216.00	77	33	As	-73918.80	77	34	Se	-74601.90
77	35	Br	-73237.00	77	36	Kr	-70194.00	77	37	Rb	-64917.00	77	38	Sr	-57880.00
78	32	Ge	-71863.00	78	33	As	-72819.00	78	34	Se	-77028.40	78	35	Br	-73455.00
78	36	Kr	-74147.00	78	37	Rb	-66980.00	79	31	Ga	-62720.00	79	32	Ge	-69490.00
79	33	As	-73639.00	79	34	Se	-75920.00	79	35	Br	-76070.60	79	36	Kr	-74445.00
79	37	Rb	-70839.00	80	30	Zn	-51890.00	80	31	Ga	-59380.00	80	32	Ge	-69380.00
80	33	As	-72165.00	80	34	Se	-77762.50	80	35	Br	-75891.60	80	36	Kr	-77894.00
80	37	Rb	-72176.00	80	38	Sr	-70190.00	81	31	Ga	-57990.00	81	32	Ge	-66310.00
81	33	As	-72536.00	81	34	Se	-76392.20	81	35	Br	-77978.00	81	36	Kr	-77697.00
81	37	Rb	-75459.00	81	38	Sr	-71470.00	81	39	Y	-65950.00	81	40	Zr	-58790.00
82	32	Ge	-65380.00	82	33	As	-70078.00	82	34	Se	-77596.50	82	35	Br	-77499.00
82	36	Kr	-80592.00	82	37	Rb	-76203.00	82	38	Sr	-75998.00	82	39	Y	-68180.00
82	40	Zr	-64180.00	83	33	As	-69880.00	83	34	Se	-75343.00	83	35	Br	-79010.00
83	36	Kr	-79982.00	83	37	Rb	-79049.00	83	38	Sr	-76781.00	83	39	Y	-72370.00
83	40	Zr	-66350.00	84	34	Se	-75952.00	84	35	Br	-77776.00	84	36	Kr	-82430.00
84	37	Rb	-79748.00	84	38	Sr	-80641.00	84	39	Y	-74230.00	85	34	Se	-72420.00
85	35	Br	-78607.00	85	36	Kr	-81477.00	85	37	Rb	-82164.40	85	38	Sr	-81099.00
85	39	Y	-77845.00	85	40	Zr	-73150.00	86	34	Se	-70540.00	86	35	Br	-75640.00
86	36	Kr	-83262.00	86	37	Rb	-82744.20	86	38	Sr	-84518.80	86	39	Y	-79279.00
87	35	Br	-73856.00	87	36	Kr	-80706.00	87	37	Rb	-84593.10	87	38	Sr	-84875.50
87	39	Y	-83014.20	87	40	Zr	-79348.00	87	41	Nb	-74180.00	87	42	Mo	-67440.00
88	35	Br	-70720.00	88	36	Kr	-79688.00	88	37	Rb	-82601.00	88	38	Sr	-87916.80
88	39	Y	-84294.20	88	40	Zr	-83626.00	89	36	Kr	-76720.00	89	37	Rb	-81709.00
89	38	Sr	-86211.00	89	39	Y	-87703.00	89	40	Zr	-84871.00	89	41	Nb	-80580.00
89	42	Mo	-75005.00	90	35	Br	-64650.00	90	36	Kr	-74947.00	90	37	Rb	-79350.00
90	38	Sr	-85942.80	90	39	Y	-86488.90	90	40	Zr	-88770.50	90	41	Nb	-82659.00
90	42	Mo	-80170.00	91	36	Kr	-71370.00	91	37	Rb	-77786.00	91	38	Sr	-83652.00
91	39	Y	-86349.00	91	40	Zr	-87893.70	91	41	Nb	-86640.00	91	42	Mo	-82208.00
91	43	Tc	-75990.00	92	36	Kr	-68650.00	92	37	Rb	-74811.00	92	38	Sr	-82923.00
92	39	Y	-84833.00	92	40	Zr	-88457.40	92	41	Nb	-86451.60	92	42	Mo	-86809.00
92	43	Tc	-78939.00	93	36	Kr	-64160.00	93	37	Rb	-72688.00	93	38	Sr	-80160.00
93	39	Y	-84245.00	93	40	Zr	-87120.20	93	41	Nb	-87210.70	93	42	Mo	-86805.00
93	43	Tc	-83607.00	93	44	Ru	-77270.00	94	37	Rb	-68518.00	94	38	Sr	-78836.00
94	39	Y	-82348.00	94	40	Zr	-87268.30	94	41	Nb	-86368.50	94	42	Mo	-88413.30
94	43	Tc	-84158.00	94	44	Ru	-82569.00	94	45	Rh	-72940.00	95	37	Rb	-65813.00
95	38	Sr	-75050.00	95	39	Y	-81214.00	95	40	Zr	-85659.60	95	41	Nb	-86783.90
95	42	Mo	-87709.50	95	43	Tc	-86018.00	95	44	Ru	-83451.00	95	45	Rh	-78340.00
96	37	Rb	-61150.00	96	38	Sr	-72880.00	96	39	Y	-78300.00	96	40	Zr	-85442.00
96	41	Nb	-85606.00	96	42	Mo	-88792.40	96	43	Tc	-85819.00	96	44	Ru	-86073.00
96	45	Rh	-79626.00	96	46	Pd	-76180.00	97	37	Rb	-58290.00	97	38	Sr	-68810.00
97	39	Y	-76270.00	97	40	Zr	-82950.00	97	41	Nb	-85608.20	97	42	Mo	-87542.20
97	43	Tc	-87222.00	97	44	Ru	-86113.00	97	45	Rh	-82590.00	97	46	Pd	-77800.00
98	37	Rb	-54090.00	98	38	Sr	-66380.00	98	39	Y	-72520.00	98	40	Zr	-81283.00
98	41	Nb	-83528.00	98	42	Mo	-88113.40	98	43	Tc	-86429.00	98	44	Ru	-88225.00
98	45	Rh	-83168.00	98	46	Pd	-81301.00	99	37	Rb	-50860.00	99	38	Sr	-62150.00
99	39	Y	-70170.00	99	40	Zr	-77790.00	99	41	Nb	-82328.00	99	42	Mo	-85967.20
99	43	Tc	-87324.30	99	44	Ru	-87617.70	99	45	Rh	-85519.00	99	46	Pd	-82193.00
99	47	Ag	-76760.00	100	38	Sr	-60200.00	100	39	Y	-67290.00	100	40	Zr	-76590.00
100	41	Nb	-79929.00	100	42	Mo	-86186.00	100	43	Tc	-86017.30	100	44	Ru	-89219.70

A Z	Mass exc.	A Z	Mass exc.	A Z	Mass exc.	A Z	Mass exc.
100 45 Rh	-85590.00	100 46 Pd	-85221.00	100 47 Ag	-78170.00	101 40 Zr	-73380.00
101 41 Nb	-78950.00	101 42 Mo	-83513.00	101 43 Tc	-86337.00	101 44 Ru	-87950.40
101 45 Rh	-87410.00	101 46 Pd	-85430.00	101 47 Ag	-81190.00	101 48 Cd	-75660.00
102 40 Zr	-71770.00	102 41 Nb	-76350.00	102 42 Mo	-83559.00	102 43 Tc	-84569.00
102 44 Ru	-89099.40	102 45 Rh	-86821.00	102 46 Pd	-87918.00	102 47 Ag	-82080.00
103 40 Zr	-68290.00	103 41 Nb	-75240.00	103 42 Mo	-80760.00	103 43 Tc	-84601.00
103 44 Ru	-87260.40	103 45 Rh	-88024.30	103 46 Pd	-87471.00	103 47 Ag	-84787.00
103 48 Cd	-80650.00	103 49 In	-74607.00	104 41 Nb	-72260.00	104 42 Mo	-80370.00
104 43 Tc	-82490.00	104 44 Ru	-88093.00	104 45 Rh	-86952.00	104 46 Pd	-89393.00
104 47 Ag	-85114.00	104 48 Cd	-83977.00	105 41 Nb	-70940.00	105 42 Mo	-77360.00
105 43 Tc	-82350.00	105 44 Ru	-85932.00	105 45 Rh	-87849.00	105 46 Pd	-88416.00
105 47 Ag	-87078.00	105 48 Cd	-84339.00	105 49 In	-79493.00	105 50 Sn	-73240.00
106 42 Mo	-76270.00	106 43 Tc	-79790.00	106 44 Ru	-86326.00	106 45 Rh	-86365.00
106 46 Pd	-89907.00	106 47 Ag	-86941.00	106 48 Cd	-87135.00	106 49 In	-80617.00
106 50 Sn	-77450.00	107 44 Ru	-83710.00	107 45 Rh	-86862.00	107 46 Pd	-88374.00
107 47 Ag	-88407.00	107 48 Cd	-86990.00	107 49 In	-83568.00	108 44 Ru	-83760.00
108 45 Rh	-85080.00	108 46 Pd	-89523.00	108 47 Ag	-87605.00	108 48 Cd	-89253.00
108 49 In	-84112.00	108 50 Sn	-82050.00	109 45 Rh	-85021.00	109 46 Pd	-87605.00
109 47 Ag	-88721.00	109 48 Cd	-88507.00	109 49 In	-86487.00	109 50 Sn	-82633.00
109 51 Sb	-76253.00	109 52 Te	-67620.00	110 45 Rh	-82940.00	110 46 Pd	-88345.00
110 47 Ag	-87459.00	110 48 Cd	-90351.00	110 49 In	-86410.00	110 50 Sn	-85834.00
110 52 Te	-72300.00	111 46 Pd	-86030.00	111 47 Ag	-88217.00	111 48 Cd	-89254.00
111 49 In	-88391.00	111 50 Sn	-85943.00	111 52 Te	-73470.00	112 46 Pd	-86333.00
112 47 Ag	-86624.00	112 48 Cd	-90581.20	112 49 In	-87995.00	112 50 Sn	-88658.00
112 51 Sb	-81603.00	112 52 Te	-77270.00	113 46 Pd	-83680.00	113 47 Ag	-87040.00
113 48 Cd	-89050.00	113 49 In	-89368.00	113 50 Sn	-88330.00	113 51 Sb	-84424.00
113 53 I	-71120.00	113 54 Xe	-62090.00	114 46 Pd	-83460.00	114 47 Ag	-84960.00
114 48 Cd	-90021.80	114 49 In	-88571.00	114 50 Sn	-90560.00	114 51 Sb	-84680.00
115 47 Ag	-84950.00	115 48 Cd	-88091.30	115 49 In	-89539.00	115 50 Sn	-90034.10
115 51 Sb	-87004.00	115 52 Te	-82360.00	116 46 Pd	-80140.00	116 47 Ag	-82760.00
116 48 Cd	-88720.00	116 49 In	-88252.00	116 50 Sn	-91526.10	116 51 Sb	-86819.00
116 52 Te	-85290.00	116 53 I	-77550.00	116 55 Cs	-62290.00	117 47 Ag	-82250.00
117 48 Cd	-86416.00	117 49 In	-88945.00	117 50 Sn	-90399.30	117 51 Sb	-88644.00
117 52 Te	-85110.00	117 55 Cs	-66260.00	118 47 Ag	-79580.00	118 48 Cd	-86709.00
118 49 In	-87232.00	118 50 Sn	-91654.20	118 51 Sb	-87998.00	118 52 Te	-87653.00
118 55 Cs	-68270.00	119 47 Ag	-78590.00	119 48 Cd	-83940.00	119 49 In	-87733.00
119 50 Sn	-90068.20	119 51 Sb	-89475.00	119 52 Te	-87182.00	119 53 I	-83780.00
119 54 Xe	-78750.00	119 55 Cs	-72240.00	120 47 Ag	-75770.00	120 48 Cd	-83973.00
120 49 In	-85800.00	120 50 Sn	-91103.80	120 51 Sb	-88423.00	120 52 Te	-89386.00
120 53 I	-83771.00	120 54 Xe	-81810.00	120 55 Cs	-73820.00	121 47 Ag	-74550.00
121 48 Cd	-80950.00	121 49 In	-85841.00	121 50 Sn	-89203.90	121 51 Sb	-89591.60
121 52 Te	-88551.00	121 53 I	-86270.00	121 54 Xe	-82510.00	121 55 Cs	-77110.00
122 49 In	-83580.00	122 50 Sn	-89946.00	122 51 Sb	-88327.00	122 52 Te	-90307.20
122 53 I	-86073.00	122 54 Xe	-85050.00	122 55 Cs	-78140.00	123 49 In	-83420.00
123 50 Sn	-87820.50	123 51 Sb	-89223.80	123 52 Te	-89171.30	123 53 I	-87937.00
123 54 Xe	-85258.00	123 55 Cs	-81070.00	124 49 In	-81060.00	124 50 Sn	-88237.30
124 51 Sb	-87619.90	124 52 Te	-90525.10	124 53 I	-87368.00	124 54 Xe	-87659.60
124 55 Cs	-81740.00	125 49 In	-80420.00	125 50 Sn	-85898.40	125 51 Sb	-88258.00
125 52 Te	-89024.80	125 53 I	-88846.10	125 54 Xe	-87191.50	125 55 Cs	-84113.00
125 56 Ba	-79550.00	126 49 In	-77810.00	126 50 Sn	-86021.00	126 51 Sb	-86400.00
126 52 Te	-90067.10	126 53 I	-87916.00	126 54 Xe	-89174.00	126 55 Cs	-84347.00
127 49 In	-77010.00	127 50 Sn	-83504.00	127 51 Sb	-86705.00	127 52 Te	-88286.00

A	*Z*		Mass exc.	*A*	*Z*		Mass exc.	*A*	*Z*		Mass exc.	*A*	*Z*		Mass exc.
127	53	I	-88982.00	127	54	Xe	-88319.00	127	55	Cs	-86243.00	127	56	Ba	-82790.00
128	49	In	-74020.00	128	50	Sn	-83330.00	128	51	Sb	-84610.00	128	52	Te	-88992.00
128	53	I	-87736.00	128	54	Xe	-89860.80	128	55	Cs	-85928.00	128	56	Ba	-85470.00
128	57	La	-78820.00	129	49	In	-73020.00	129	50	Sn	-80620.00	129	51	Sb	-84624.00
129	52	Te	-87006.00	129	53	I	-88507.00	129	54	Xe	-88698.10	129	55	Cs	-87506.00
129	56	Ba	-85080.00	129	57	La	-81360.00	130	49	In	-70010.00	130	50	Sn	-80130.00
130	51	Sb	-82330.00	130	52	Te	-87348.00	130	53	I	-86897.00	130	54	Xe	-89881.50
130	55	Cs	-86853.00	130	56	Ba	-87291.00	131	49	In	-68490.00	131	50	Sn	-77380.00
131	51	Sb	-82020.00	131	52	Te	-85206.00	131	53	I	-87457.00	131	54	Xe	-88428.00
131	55	Cs	-88076.00	131	56	Ba	-86714.00	131	57	La	-83750.00	131	58	Ce	-79730.00
132	50	Sn	-76610.00	132	51	Sb	-79730.00	132	52	Te	-85222.00	132	53	I	-85715.00
132	54	Xe	-89292.00	132	55	Cs	-87171.00	132	56	Ba	-88447.00	132	57	La	-83740.00
133	50	Sn	-71190.00	133	51	Sb	-79020.00	133	52	Te	-82970.00	133	53	I	-85888.00
133	54	Xe	-87659.00	133	55	Cs	-88086.00	133	56	Ba	-87570.00	134	51	Sb	-74020.00
134	52	Te	-82430.00	134	53	I	-83990.00	134	54	Xe	-88125.00	134	55	Cs	-86906.00
134	56	Ba	-88965.00	134	57	La	-85252.00	134	58	Ce	-84750.00	135	52	Te	-77870.00
135	53	I	-83821.00	135	54	Xe	-86506.00	135	55	Cs	-87662.00	135	56	Ba	-87867.00
135	57	La	-86667.00	135	58	Ce	-84641.00	135	59	Pr	-80920.00	136	52	Te	-74460.00
136	53	I	-79550.00	136	54	Xe	-86429.00	136	55	Cs	-86354.00	136	56	Ba	-88903.00
136	57	La	-86030.00	136	58	Ce	-86500.00	136	59	Pr	-81370.00	136	60	Nd	-79160.00
137	52	Te	-69480.00	137	53	I	-76507.00	137	54	Xe	-82383.00	137	55	Cs	-86556.00
137	56	Ba	-87732.00	137	57	La	-87130.00	137	58	Ce	-85910.00	137	59	Pr	-83200.00
137	60	Nd	-79700.00	137	61	Pm	-74020.00	138	53	I	-72290.00	138	54	Xe	-80110.00
138	55	Cs	-82896.00	138	56	Ba	-88272.00	138	57	La	-86531.00	138	58	Ce	-87574.00
138	59	Pr	-83137.00	139	53	I	-68880.00	139	54	Xe	-75690.00	139	55	Cs	-80710.00
139	56	Ba	-84924.00	139	57	La	-87238.00	139	58	Ce	-86973.00	139	59	Pr	-84844.00
139	60	Nd	-82060.00	139	61	Pm	-77540.00	139	62	Sm	-72080.00	140	54	Xe	-72990.00
140	55	Cs	-77053.00	140	56	Ba	-83273.00	140	57	La	-84327.00	140	58	Ce	-88088.00
140	59	Pr	-84700.00	140	60	Nd	-84471.00	140	61	Pm	-78380.00	141	54	Xe	-68320.00
141	55	Cs	-74472.00	141	56	Ba	-79732.00	141	57	La	-82983.00	141	58	Ce	-85445.00
141	59	Pr	-86026.00	141	60	Nd	-84203.00	141	61	Pm	-80472.00	141	62	Sm	-75943.00
141	63	Eu	-69980.00	142	54	Xe	-65500.00	142	55	Cs	-70538.00	142	56	Ba	-77847.00
142	57	La	-80027.00	142	58	Ce	-84542.00	142	59	Pr	-83798.00	142	60	Nd	-85960.00
142	61	Pm	-81090.00	142	62	Sm	-78986.00	142	63	Eu	-71590.00	143	55	Cs	-67745.00
143	56	Ba	-73979.00	143	57	La	-78200.00	143	58	Ce	-81616.00	143	59	Pr	-83078.00
143	60	Nd	-84012.00	143	61	Pm	-82970.00	143	62	Sm	-79526.00	143	63	Eu	-74380.00
144	55	Cs	-63370.00	144	56	Ba	-71840.00	144	57	La	-74940.00	144	58	Ce	-80441.00
144	59	Pr	-80760.00	144	60	Nd	-83758.00	144	61	Pm	-81425.00	144	62	Sm	-81975.00
144	63	Eu	-75646.00	145	55	Cs	-60210.00	145	56	Ba	-68120.00	145	57	La	-73020.00
145	58	Ce	-77110.00	145	59	Pr	-79636.00	145	60	Nd	-81442.00	145	61	Pm	-81278.00
145	62	Sm	-80660.00	145	63	Eu	-78000.00	145	64	Gd	-72950.00	146	55	Cs	-55700.00
146	56	Ba	-65060.00	146	57	La	-69200.00	146	58	Ce	-75730.00	146	59	Pr	-76760.00
146	60	Nd	-80935.00	146	61	Pm	-79458.00	146	62	Sm	-81000.00	146	63	Eu	-77125.00
146	64	Gd	-76099.00	146	65	Tb	-67860.00	147	55	Cs	-52300.00	147	56	Ba	-61500.00
147	57	La	-67250.00	147	58	Ce	-72190.00	147	59	Pr	-75470.00	147	60	Nd	-78156.00
147	61	Pm	-79052.00	147	62	Sm	-79276.00	147	63	Eu	-77555.00	147	64	Gd	-75367.00
147	65	Tb	-70880.00	147	66	Dy	-64330.00	148	55	Cs	-47580.00	148	57	La	-63810.00
148	58	Ce	-70430.00	148	59	Pr	-72490.00	148	60	Nd	-77418.00	148	61	Pm	-76874.00
148	62	Sm	-79346.00	148	63	Eu	-76239.00	148	64	Gd	-76278.00	148	65	Tb	-70680.00
148	66	Dy	-68000.00	149	58	Ce	-66800.00	149	59	Pr	-70988.00	149	60	Nd	-74385.00
149	61	Pm	-76073.00	149	62	Sm	-77146.00	149	63	Eu	-76455.00	149	64	Gd	-75135.00
149	65	Tb	-71499.00	149	68	Er	-54950.00	150	58	Ce	-64990.00	150	59	Pr	-68000.00

A	*Z*	Mass exc.	*A*	*Z*	Mass exc.	*A*	*Z*	Mass exc.	*A*	*Z*	Mass exc.
150	60 Nd	-73693.0	150	61 Pm	-73606.0	150	62 Sm	-77060.0	150	63 Eu	-74800.0
150	64 Gd	-75771.0	150	65 Tb	-71113.0	150	66 Dy	-69324.0	150	67 Ho	-62210.0
151	60 Nd	-70956.0	151	61 Pm	-73398.0	151	62 Sm	-74587.0	151	63 Eu	-74663.0
151	64 Gd	-74199.0	151	65 Tb	-71633.0	151	66 Dy	-68764.0	151	67 Ho	-63720.0
152	60 Nd	-70160.0	152	61 Pm	-71270.0	152	62 Sm	-74773.0	152	63 Eu	-72899.0
152	64 Gd	-74718.0	152	65 Tb	-70770.0	152	66 Dy	-70127.0	152	67 Ho	-63750.0
152	68 Er	-60640.0	153	61 Pm	-70669.0	153	62 Sm	-72569.0	153	63 Eu	-73378.0
153	64 Gd	-72893.0	153	65 Tb	-71322.0	153	66 Dy	-69152.0	153	67 Ho	-65023.0
154	61 Pm	-68410.0	154	62 Sm	-72465.0	154	63 Eu	-71748.0	154	64 Gd	-73717.0
154	65 Tb	-70150.0	154	66 Dy	-70399.0	154	67 Ho	-64647.0	154	68 Er	-62622.0
154	69 Tm	-54700.0	155	62 Sm	-70201.0	155	63 Eu	-71829.0	155	64 Gd	-72081.0
155	65 Tb	-71261.0	155	66 Dy	-69166.0	155	67 Ho	-66064.0	155	68 Er	-62220.0
155	69 Tm	-56730.0	156	62 Sm	-69374.0	156	63 Eu	-70096.0	156	64 Gd	-72546.0
156	65 Tb	-70102.0	156	66 Dy	-70536.0	156	69 Tm	-56980.0	156	70 Yb	-53410.0
157	62 Sm	-66870.0	157	63 Eu	-69472.0	157	64 Gd	-70834.0	157	65 Tb	-70772.0
157	66 Dy	-69434.0	157	67 Ho	-66890.0	157	68 Er	-63420.0	158	63 Eu	-67220.0
158	64 Gd	-70701.0	158	65 Tb	-69480.0	158	66 Dy	-70418.0	158	67 Ho	-66200.0
158	70 Yb	-56022.0	158	71 Lu	-47490.0	159	63 Eu	-66058.0	159	64 Gd	-68572.0
159	65 Tb	-69542.0	159	66 Dy	-69176.0	159	67 Ho	-67338.0	159	68 Er	-64570.0
159	71 Lu	-49770.0	160	64 Gd	-67953.0	160	65 Tb	-67846.0	160	66 Dy	-69682.0
160	67 Ho	-66391.0	160	68 Er	-66063.0	160	69 Tm	-60460.0	160	72 Hf	-46080.0
161	64 Gd	-65517.0	161	65 Tb	-67471.0	161	66 Dy	-68064.0	161	67 Ho	-67207.0
161	68 Er	-65203.0	161	69 Tm	-62100.0	162	64 Gd	-64240.0	162	65 Tb	-65680.0
162	66 Dy	-68189.0	162	67 Ho	-66050.0	162	68 Er	-66346.0	162	69 Tm	-61550.0
162	72 Hf	-49178.0	162	73 Ta	-40060.0	163	65 Tb	-64700.0	163	66 Dy	-66389.0
163	67 Ho	-66386.0	163	68 Er	-65177.0	163	69 Tm	-62738.0	163	70 Yb	-59370.0
163	71 Lu	-54770.0	163	73 Ta	-42600.0	164	65 Tb	-62090.0	164	66 Dy	-65976.0
164	67 Ho	-64990.0	164	68 Er	-65952.0	164	69 Tm	-61990.0	164	74 W	-38380.0
165	66 Dy	-63621.0	165	67 Ho	-64907.0	165	68 Er	-64530.0	165	69 Tm	-62938.0
165	70 Yb	-60175.0	165	71 Lu	-56260.0	166	66 Dy	-62593.0	166	67 Ho	-63079.0
166	68 Er	-64933.0	166	69 Tm	-61894.0	166	70 Yb	-61589.0	166	71 Lu	-56110.0
166	74 W	-41898.0	166	75 Re	-32130.0	167	66 Dy	-59940.0	167	67 Ho	-62291.0
167	68 Er	-63298.0	167	69 Tm	-62550.0	167	70 Yb	-60596.0	167	71 Lu	-57470.0
167	75 Re	-34910.0	168	67 Ho	-60260.0	168	68 Er	-62998.0	168	69 Tm	-61319.0
168	70 Yb	-61575.0	168	71 Lu	-57090.0	168	76 Os	-30130.0	169	67 Ho	-58805.0
169	68 Er	-60930.0	169	69 Tm	-61280.0	169	70 Yb	-60371.0	169	71 Lu	-58078.0
169	72 Hf	-54810.0	170	67 Ho	-56250.0	170	68 Er	-60117.0	170	69 Tm	-59802.0
170	70 Yb	-60770.0	170	71 Lu	-57311.0	170	76 Os	-33933.0	170	77 Ir	-23530.0
171	68 Er	-57727.0	171	69 Tm	-59217.0	171	70 Yb	-59314.0	171	71 Lu	-57834.0
171	77 Ir	-26420.0	172	68 Er	-56491.0	172	69 Tm	-57382.0	172	70 Yb	-59262.0
172	71 Lu	-56741.0	172	72 Hf	-56390.0	172	73 Ta	-51470.0	172	78 Pt	-21240.0
173	69 Tm	-56265.0	173	70 Yb	-57558.0	173	71 Lu	-56886.0	174	69 Tm	-53870.0
174	70 Yb	-56951.0	174	71 Lu	-55575.0	174	72 Hf	-55851.0	174	78 Pt	-25324.0
174	79 Au	-14330.0	175	69 Tm	-52300.0	175	70 Yb	-54702.0	175	71 Lu	-55171.9
175	72 Hf	-54488.0	175	79 Au	-17210.0	176	70 Yb	-53501.0	176	71 Lu	-53394.1
176	72 Hf	-54582.6	176	73 Ta	-51470.0	176	80 Hg	-11890.0	177	70 Yb	-50996.0
177	71 Lu	-52394.2	177	72 Hf	-52892.2	177	73 Ta	-51726.0	178	70 Yb	-49705.0
178	71 Lu	-50338.0	178	72 Hf	-52446.5	178	73 Ta	-50530.0	178	74 W	-50440.0
178	75 Re	-45780.0	178	80 Hg	-16321.0	179	71 Lu	-49110.0	179	72 Hf	-50475.1
179	73 Ta	-50365.0	179	74 W	-49306.0	179	75 Re	-46620.0	180	71 Lu	-46690.0
180	72 Hf	-49791.9	180	73 Ta	-48939.0	180	74 W	-49647.0	180	75 Re	-45840.0
181	72 Hf	-47416.2	181	73 Ta	-48444.0	181	74 W	-48256.0	182	72 Hf	-46062.0

A	Z		Mass exc.	A	Z		Mass exc.	A	Z		Mass exc.	A	Z		Mass exc.
182	73	Ta	-46436.0	182	74	W	-48250.0	182	75	Re	-45450.0	182	76	Os	-44542.0
182	82	Pb	-6874.0	183	72	Hf	-43290.0	183	73	Ta	-45299.0	183	74	W	-46369.1
183	75	Re	-45813.0	184	72	Hf	-41500.0	184	73	Ta	-42844.0	184	74	W	-45709.5
184	75	Re	-44220.0	184	76	Os	-44259.0	184	77	Ir	-39540.0	185	73	Ta	-41402.0
185	74	W	-43393.0	185	75	Re	-43826.0	185	76	Os	-42813.0	186	73	Ta	-38620.0
186	74	W	-42515.0	186	75	Re	-41933.0	186	76	Os	-43003.0	186	77	Ir	-39172.0
186	78	Pt	-37790.0	187	74	W	-39910.0	187	75	Re	-41222.0	187	76	Os	-41224.0
188	74	W	-38673.0	188	75	Re	-39022.0	188	76	Os	-41142.0	188	77	Ir	-38333.0
188	78	Pt	-37827.0	189	74	W	-35480.0	189	75	Re	-37985.0	189	76	Os	-38993.0
189	77	Ir	-38462.0	189	78	Pt	-36491.0	190	74	W	-34310.0	190	75	Re	-35580.0
190	76	Os	-38714.0	190	77	Ir	-36710.0	190	78	Pt	-37331.0	190	79	Au	-32889.0
191	75	Re	-34360.0	191	76	Os	-36401.0	191	77	Ir	-36715.0	191	78	Pt	-35701.0
191	79	Au	-33870.0	191	80	Hg	-30690.0	192	76	Os	-35892.0	192	77	Ir	-34843.0
192	78	Pt	-36303.0	192	79	Au	-32787.0	193	76	Os	-33405.0	193	77	Ir	-34544.0
193	78	Pt	-34487.0	193	81	Tl	-27450.0	194	76	Os	-32442.0	194	77	Ir	-32539.0
194	78	Pt	-34787.0	194	79	Au	-32295.0	194	80	Hg	-32255.0	195	76	Os	-29700.0
195	77	Ir	-31700.0	195	78	Pt	-32821.0	195	79	Au	-32594.0	195	80	Hg	-31070.0
195	81	Tl	-28270.0	196	76	Os	-28300.0	196	77	Ir	-29460.0	196	78	Pt	-32671.0
196	79	Au	-31166.0	196	80	Hg	-31852.0	196	83	Bi	-17970.0	197	77	Ir	-28292.0
197	78	Pt	-30446.0	197	79	Au	-31165.0	197	80	Hg	-30566.0	197	81	Tl	-28400.0
197	83	Bi	-19640.0	198	78	Pt	-29932.0	198	79	Au	-29606.0	198	80	Hg	-30979.0
198	81	Tl	-27520.0	198	83	Bi	-19540.0	199	78	Pt	-27432.0	199	79	Au	-29119.0
199	80	Hg	-29572.0	199	81	Tl	-28140.0	199	82	Pb	-25270.0	199	83	Bi	-20920.0
200	78	Pt	-26627.0	200	79	Au	-27280.0	200	80	Hg	-29529.0	200	81	Tl	-27073.0
200	83	Bi	-20400.0	200	85	At	-8940.0	201	78	Pt	-23750.0	201	79	Au	-26413.0
201	80	Hg	-27688.0	201	81	Tl	-27205.0	201	82	Pb	-25300.0	201	83	Bi	-21470.0
201	85	At	-10740.0	202	79	Au	-24420.0	202	80	Hg	-27370.0	202	81	Tl	-26006.0
202	82	Pb	-25957.0	202	83	Bi	-20800.0	202	85	At	-10770.0	203	79	Au	-23153.0
203	80	Hg	-25292.0	203	81	Tl	-25784.0	203	82	Pb	-24810.0	203	83	Bi	-21580.0
203	84	Po	-17350.0	203	85	At	-12290.0	204	80	Hg	-24716.0	204	81	Tl	-24369.0
204	82	Pb	-25132.0	204	83	Bi	-20730.0	204	85	At	-11900.0	204	87	Fr	650.0
205	80	Hg	-22312.0	205	81	Tl	-23846.0	205	82	Pb	-23793.0	205	83	Bi	-21084.0
205	84	Po	-17555.0	205	85	At	-13030.0	205	87	Fr	-1270.0	206	80	Hg	-20969.0
206	81	Tl	-22278.0	206	82	Pb	-23809.0	206	83	Bi	-20052.0	206	84	Po	-18205.0
206	85	At	-12490.0	206	87	Fr	-1420.0	207	80	Hg	-16270.0	207	81	Tl	-21049.0
207	82	Pb	-22476.0	207	83	Bi	-20079.0	207	84	Po	-17169.0	207	85	At	-13290.0
207	86	Rn	-8670.0	207	87	Fr	-2960.0	208	81	Tl	-16774.0	208	82	Pb	-21772.0
208	83	Bi	-18894.0	208	84	Po	-17492.0	208	85	At	-12560.0	208	87	Fr	-2710.0
209	81	Tl	-13652.0	209	82	Pb	-17638.0	209	83	Bi	-18282.0	209	84	Po	-16390.0
209	85	At	-12902.0	209	86	Rn	-8973.0	209	87	Fr	-3830.0	209	89	Ac	8890.0
210	81	Tl	-9262.0	210	82	Pb	-14752.0	210	83	Bi	-14815.0	210	84	Po	-15977.0
210	85	At	-11995.0	210	86	Rn	-9623.0	210	87	Fr	-3400.0	210	89	Ac	8620.0
211	82	Pb	-10494.0	211	83	Bi	-11873.0	211	84	Po	-12457.0	211	85	At	-11674.0
211	86	Rn	-8780.0	211	87	Fr	-4200.0	211	88	Ra	800.0	211	89	Ac	7080.0
212	82	Pb	-7571.0	212	83	Bi	-8142.0	212	84	Po	-10394.0	212	85	At	-8640.0
212	86	Rn	-8682.0	212	87	Fr	-3600.0	212	89	Ac	7240.0	213	83	Bi	-5244.0
213	84	Po	-6676.0	213	85	At	-6603.0	213	86	Rn	-5722.0	213	87	Fr	-3572.0
213	88	Ra	311.0	213	89	Ac	6100.0	214	82	Pb	-188.1	214	83	Bi	-1218.0
214	84	Po	-4493.0	214	85	At	-3403.0	214	86	Rn	-4343.0	214	87	Fr	-983.0
214	88	Ra	75.0	214	89	Ac	6380.0	215	83	Bi	1710.0	215	84	Po	-542.9
215	85	At	-1269.0	215	86	Rn	-1193.0	215	87	Fr	292.0	215	88	Ra	2509.0
215	89	Ac	5970.0	215	90	Th	10890.0	215	91	Pa	17680.0	216	84	Po	1760.0

A	*Z*	Mass exc.	*A*	*Z*	Mass exc.	*A*	*Z*	Mass exc.	*A*	*Z*	Mass exc.
216	85 At	2231.0	216	86 Rn	231.0	216	87 Fr	2960.0	216	88 Ra	3269.0
216	89 Ac	8060.0	216	91 Pa	17680.0	217	85 At	4383.0	217	86 Rn	3634.0
217	87 Fr	4293.0	217	88 Ra	5864.0	217	89 Ac	8685.0	217	90 Th	12160.0
217	91 Pa	17020.0	218	84 Po	8351.7	218	85 At	8090.0	218	86 Rn	5199.0
218	87 Fr	7036.0	218	88 Ra	6627.0	218	89 Ac	10820.0	218	90 Th	12348.0
218	91 Pa	18600.0	219	85 At	10520.0	219	86 Rn	8828.3	219	87 Fr	8609.0
219	88 Ra	9363.0	219	89 Ac	11540.0	219	90 Th	14450.0	220	86 Rn	10590.0
220	87 Fr	11456.0	220	88 Ra	10250.0	220	89 Ac	13730.0	220	90 Th	14647.0
221	87 Fr	13266.0	221	88 Ra	12938.0	221	89 Ac	14500.0	221	90 Th	16917.0
222	86 Rn	16367.0	222	87 Fr	16380.0	222	88 Ra	14303.0	222	89 Ac	16603.0
222	90 Th	17182.0	222	91 Pa	21940.0	223	87 Fr	18381.0	223	88 Ra	17232.4
223	89 Ac	17817.0	223	90 Th	19357.0	223	91 Pa	22310.0	224	87 Fr	21620.0
224	88 Ra	18804.0	224	89 Ac	20204.0	224	90 Th	19980.0	224	91 Pa	23780.0
225	87 Fr	23840.0	225	88 Ra	21988.0	225	89 Ac	21626.0	225	90 Th	22283.0
225	91 Pa	24310.0	226	87 Fr	27210.0	226	88 Ra	23662.7	226	89 Ac	24303.0
226	90 Th	23183.0	226	91 Pa	26015.0	226	92 U	27170.0	227	87 Fr	29590.0
227	88 Ra	27172.6	227	89 Ac	25848.7	227	90 Th	25803.9	227	91 Pa	26824.0
228	88 Ra	28936.0	228	89 Ac	28890.0	228	90 Th	26749.0	228	91 Pa	28856.0
228	92 U	29209.0	229	88 Ra	32660.0	229	89 Ac	30900.0	229	90 Th	29581.0
229	91 Pa	29887.0	229	92 U	31181.0	229	93 Np	33740.0	230	90 Th	30858.7
230	91 Pa	32168.0	230	92 U	31600.0	230	93 Np	35220.0	231	89 Ac	35910.0
231	90 Th	33812.1	231	91 Pa	33422.4	231	92 U	33780.0	231	93 Np	35620.0
232	90 Th	35444.4	232	91 Pa	35924.0	232	92 U	34587.0	232	94 Pu	38349.0
233	90 Th	38729.4	233	91 Pa	37485.8	233	92 U	36915.0	233	94 Pu	40020.0
234	90 Th	40607.0	234	91 Pa	40334.0	234	92 U	38141.9	234	93 Np	39952.0
234	94 Pu	40335.0	235	90 Th	44250.0	235	91 Pa	42330.0	235	92 U	40915.5
235	93 Np	41039.2	235	94 Pu	42160.0	236	91 Pa	45340.0	236	92 U	42441.7
236	93 Np	43370.0	236	94 Pu	42879.0	237	91 Pa	47640.0	237	92 U	45387.3
237	93 Np	44868.3	237	94 Pu	45090.0	238	91 Pa	50910.0	238	92 U	47305.9
238	93 Np	47451.6	238	94 Pu	46160.1	238	95 Am	48420.0	238	96 Cm	49380.0
239	92 U	50570.8	239	93 Np	49306.6	239	94 Pu	48584.8	239	95 Am	49385.0
240	92 U	52711.0	240	93 Np	52321.0	240	94 Pu	50122.5	240	95 Am	51498.0
240	96 Cm	51702.0	241	93 Np	54260.0	241	94 Pu	52952.0	241	95 Am	52931.1
241	96 Cm	53700.0	242	93 Np	57410.0	242	94 Pu	54713.8	242	95 Am	55463.4
242	96 Cm	54800.7	242	98 Cf	59320.0	243	93 Np	59922.0	243	94 Pu	57751.0
243	95 Am	57169.5	243	96 Cm	57177.2	243	97 Bk	58683.0	244	94 Pu	59802.0
244	95 Am	59877.2	244	96 Cm	58449.2	244	97 Bk	60700.0	244	98 Cf	61460.0
245	94 Pu	63175.0	245	95 Am	61891.9	245	96 Cm	60998.0	245	97 Bk	61809.6
245	98 Cf	63380.0	246	94 Pu	65391.0	246	95 Am	64990.0	246	96 Cm	62614.0
246	98 Cf	64087.4	246	100 Fm	70120.0	247	96 Cm	65528.0	247	97 Bk	65484.0
247	98 Cf	66130.0	247	99 Es	68550.0	248	96 Cm	67388.0	248	97 Bk	68107.0
248	98 Cf	67237.0	248	99 Es	70290.0	248	100 Fm	71888.0	249	96 Cm	70746.0
249	97 Bk	69842.8	249	98 Cf	69717.9	249	99 Es	71110.0	250	96 Cm	72985.0
250	97 Bk	72951.0	250	98 Cf	71167.0	250	100 Fm	74060.0	251	96 Cm	76642.0
251	97 Bk	75222.0	251	98 Cf	74129.0	251	99 Es	74506.0	251	100 Fm	75978.0
252	98 Cf	76030.0	252	99 Es	77290.0	252	100 Fm	76814.0	252	102 No	82857.0
253	98 Cf	79296.0	253	99 Es	79007.0	253	100 Fm	79339.0	254	98 Cf	81338.0
254	99 Es	81994.0	254	100 Fm	80900.0	254	102 No	84711.0	255	99 Es	84083.0
255	100 Fm	83788.0	255	101 Md	84835.0	255	102 No	86848.0	256	100 Fm	85482.0
256	101 Md	87550.0	256	102 No	87793.0	256	104 Rf	94234.0	257	100 Fm	88585.0
257	102 No	90220.0	259	102 No	94018.0	259	103 Lr	95840.0	259	104 Rf	98280.0
260	103 Lr	98130.0	260	106 Nh	106580.0	263	106 Nh	110090.0			

SYMBOLS USED

A: nucleon number
$\boldsymbol{A}$: vector potential
α: fine structure constant
Angular momentum coupling coefficients
  $\langle rpsq|tm\rangle$: Clebsch-Gordan coefficient
  $\begin{pmatrix} j_1 & j_2 & j_3 \\ m_1 & m_2 & m_3 \end{pmatrix}$: $3j$-symbol

b: barn (1 b $= 10^{-28}$ m^2)
  μb: mincrobarn (1μb $= 10^{-6}$ b)
$B(E\lambda)$: reduced electric transition probability
$B(M\lambda)$: reduced magnetic transition probability

c: speed of light
$\chi_{1/2}$: intrinsic spin wave function

D: average level spacing
δ: mixing ratio
$\delta(r)$: Dirac delta function
δ_ℓ: phase shift
$\delta_{\rm osc}$: deformation parameter
$\mathcal{D}^j_{nm}(\alpha\beta\gamma)$: rotation matrix
$^3D_1$: triplet D-state
$\Delta(Z,N)$: mass excess

e: unit of charge
e^-: electron
e^+: positron
E: energy
  E_B: binding energy
  $\epsilon_{\rm F}$: Fermi energy
  ϵ_r: single-particle energy of state r
  eV: electron volt
  GeV: billion electron volt (10^9 eV)
  MeV: million electron volt (10^6 eV)
ϵ_0: permittivity of free space

$f(Z,E_e)$: Fermi integral
$f(\theta)$: scattering amplitude
fm: fentometer (1 fm $= 10^{-15}$ m)
$F(q^2)$: form factor
$\langle F\rangle$: Fermi matrix element

g: gyromagnetic ratio
G: coupling constant
$G(\boldsymbol{r},\boldsymbol{r}')$: Green's function
γ: photon (gamma-ray)
Γ: width
$\langle GT\rangle$: Gamow-Teller matrix element

H, h: Hamiltonian
  W^{JT}_{rstu}: two-body matrix element
  V: two-body interaction potential
  $V_{\rm eff}$: effective potential
h, $\hbar$: Planck's constant

$\mathcal{I}$: moment of inertial

$j_\ell(kr)$: spherical Bessel function
$\boldsymbol{J}$: spin (total angular momentum)
  $\boldsymbol{J}^\pm$: angular momentum raising and lowering operator
k: wave number
k_F: Fermi momentum

$\boldsymbol{L}$, $\boldsymbol{\ell}$: angular momentum
  M, m: projection of $\boldsymbol{L}$, $\boldsymbol{\ell}$ on the quantization axis

M, m: mass
m_e: electron mass
m_ν: neutrino mass
M_p: proton mass
M_n: neutron mass

$\boldsymbol{\mu}$: magnetic dipole moment operator
μ: magnetic dipole moment
μ_N: nuclear magneton
μ_0: permeability of free space
$\boldsymbol{n}$: number operator
N: neutron number
n: neutron
ν: neutrino

$\boldsymbol{O}$: operator
$\boldsymbol{O}_{\lambda\mu}(E\lambda)$: electric multipole operator
$\boldsymbol{O}_{\lambda\mu}(M\lambda)$: magnetic multipole operator
$\boldsymbol{O}(\beta)$: β-decay operator

p: proton
$\overline{p}$: antiproton
P, Q: projection operators
P_{ij}: permutation operator
$\boldsymbol{p}$: linear momentum
π: parity
$\pi^{\pm}, \pi^0$: pion
P_{12}: permutation operator

q: quark
$\overline{q}$: antiquark
q: momentum transfer
Q: charge number
Q-value: energy released in a reaction
$\boldsymbol{Q}_0$: electric quadrupole operator
Q: electric quadrupole moment

r_0: range of interaction
$r_0 A^{1/3}$: average nuclear radius
r_e: effective range

$\rho(\boldsymbol{r})$: density distribution
$\rho(E)$: density of states
$R_{n\ell}(r)$: radial wave function

$\boldsymbol{s}$: intrinsic spin
$\boldsymbol{S}$: sum of intrinsic spin
S: strangeness
S_{12}: tenor force operator
${}^3S_1$: triplet-S state
$\boldsymbol{\sigma}$: Pauli spin matrix
σ: cross section
$\frac{d\sigma}{d\Omega}$: differential cross section
SU_N: special unitary group of N-dimension

t: time
$\overline{T}$: lifetime
$T_{1/2}$: half-life
T: kinetic energy
$\boldsymbol{T}$, $\boldsymbol{\tau}$: isospin
$T_{\pm}$, $\tau_{\pm}$: isospin raising/lowering operator
T_0, τ_0: third component of $\boldsymbol{T}$, τ
$T_{\lambda\mu}$: spherical tensors of rank λ, μ
$\mathcal{T}$: current density

$u_\ell(r)$: modified radial wave function
ν: neutrino
$V(\boldsymbol{r})$: potential

$\mathcal{W}$: transition probability
$Y_{\ell m}(\theta, \phi)$: spherical harmonics
Z: proton number

$\langle \boldsymbol{O} \rangle$: expectation value of operator $\boldsymbol{O}$
$\langle J_f M_f \xi | \boldsymbol{O}_{\lambda\mu} | J_i M_i \zeta \rangle$: matrix element of operator $\boldsymbol{O}_{\lambda\mu}$

$\langle J_f \xi || \boldsymbol{O}_\lambda || J_i \zeta \rangle$: reduced matrix elememt
$A(a, b)B$: nuclear reaction $a+A \to b+B$

Index

D

E

F

M

N

Q

R

S